K. Ebert, H. Ederer,
T. L. Isenhour

Computer Applications in Chemistry

© VCH Verlagsgesellschaft mbH, D-6940 Weinheim (Federal Republic of Germany), 1989

Distribution

VCH Verlagsgesellschaft, P. O. Box 10 1161, D-6940 Weinheim (Federal Republic of Germany)

Switzerland: VCH Verlags-AG, P. O. Box, CH-4020 Basel (Switzerland)

United Kingdom and Ireland: VCH Publishers (UK) Ltd., 8 Wellington Court, Wellington Street, Cambridge CB1 1HW (England)

USA and Canada: VCH Publishers, Suite 909, 220 East 23rd Street, New York, NY 10010-4606 (USA)

ISBN 3-527-27807-9 (VCH Verlagsgesellschaft)   ISBN 0-89573-864-3 (VCH Publishers)

K. Ebert, H. Ederer,
T.L. Isenhour

Computer Applications in Chemistry

An Introduction for PC Users,
with Two Diskettes
in BASIC and PASCAL

Prof. Dr. Klaus Ebert
Dr. Hanns Ederer
Institut für Heiße Chemie
Kernforschungszentrum
Postfach 3640
D-7500 Karlsruhe 1
and
Physikalisch-Chemisches Institut und
SFB 123 der Universität
Im Neuenheimer Feld 253
D-6900 Heidelberg
Federal Republic of Germany

Prof. Dr. Thomas L. Isenhour
Dean of Arts and Sciences
Eisenhower Hall
Kansas State University
Manhattan, Kansas 66506
USA

Published jointly by
VCH Verlagsgesellschaft, Weinheim (Federal Republic of Germany)
VCH Publishers, New York, NY (USA)

Editorial Director: Dr. Hans F. Ebel
Production Manager: Claudia Grössl

Library of Congress Card No.: 89-14770

British Library Cataloguing-in-Publication Data:
Ebert, Klaus
Computer applications in chemistry.
1. Chemistry. Applications of microcomputer systems.
I. Title, II. Ederer, Hanns, III. Isenhour, Thomas L.
542'.8
ISBN 3-527-27807-9 W. Germany
ISBN 0-89573-864-3 United States

Deutsche Bibliothek Cataloguing-in-Publication Data:
Ebert, Klaus:
Computer applications in chemistry : an introduction for PC
users / K. Ebert ; H. Ederer ; T. L. Isenhour. – Weinheim ;
Basel (Switzerland) ; Cambridge ; New York, NY : VCH, 1989
ISBN 3-527-27807-9
NE: Ederer, Hanns:; Isenhour, Thomas L.:

Printing: SDV, Saarbrücker Druckerei und Verlag GmbH, D-6600 Saarbrücken
Bookbinding: Großbuchbinderei J. Schäffer, D-6718 Grünstadt
Printed in the Federal Republic of Germany
Printed on acid-free paper

Preface

Computers belong to the integral tools chemists need today in doing their jobs, in industry as well as in research or elsewhere. From the variety of computers available, without doubt PC's are the most qualified to meet the needs of chemists, which are indeed manifold. Most important, however, is the fact that PC's offer a more easy access to computer work than any other member of the computer family. Every chemist wishes to have an understanding of how to use computers to solve his problems and direct access. In this way a PC is comparable with the glass equipment which is needed to synthesize new substances, or with the apparatus required to carry out measurements of a certain property of a chemical substance.

The present book, which has already appeared in two German and one Russian edition, is intended to assist the chemist in making full use of the PC for his requirements. It tries to be something similar to a book of recipes used by the chemist to synthesize new substances.

The book can be used in many different ways. On the one hand, it is a *textbook* originating from a course of lectures, and thus the chapters are organized in logical sequence. It is intended to be a *practical manual*, also, as it contains a considerable number of chemical and physical problems with mathematical solutions and appropriate example programs. It is certainly a *reference book*, providing programs for specific problems in chemistry and related subjects. Last, but not least, it can be used as a *study guide*, providing chemists with the knowledge they need to modify computer programs or develop new programs for their own needs in the laboratory or at the desk.

The discussion about the most suitable computer language is nearly as old as computer science itself. BASIC is the computer language — when used as an interpreter — which requires the least additional knowledge of the operating system and the computer hardware and is therefore qualified to be used by chemists, including those with little knowledge in computer science. The present book uses a BASIC-dialect called BASICA from IBM, which is understood by IBM PC's (XT and AT) and by compatibles using GWBASIC of Microsoft. In Turbo-Basic the statements used in BASICA are a subset of the complete version and therefore all programs also run with the Turbo-Basic compiler. The same is true for the QuickBASIC dialect of Microsoft. It has to be admitted that BASIC is too unsophisticated for experts. In addition, it does not offer the possibility for complex data types to be included, and it provides no local variables within independent subroutines. Last but not least, it is difficult to write well-structured programs in BASIC.

On the way to becoming a computer expert — using FORTRAN or ASSEMBLER — there will probably be a halt at the PASCAL level, which indeed has a number of significant advantages. Moreover, it is likely that in the not too distant future PASCAL will displace BASIC as the language employed by the novice. Therefore all programs in the book have been translated into Turbo-Pascal, and these versions are listed in Chapter 15.

Software is no longer transferred by typing. Electronic transmission is the modern means of communication. Therefore, all the programs in this book have been recorded as ASCII files on two floppy discs and are attached to the book: Disc I contains the programs in the Microsoft GWBASIC (or IBM BASICA) version and Disc II contains the translated datasets in Turbo-Pascal. We would not advise the beginner to work with PASCAL straight away, but we would also like to point out that once a certain level of expertise has been reached and enough experience has been accumulated, PASCAL should be considered as the option for faster, more professional programming.

This is certainly not a book on mathematics or numerical analysis; therefore the mathematical relations and derivations are given without proof. However, it should not be difficult to find those in appropriate textbooks on numerical mathematics. We have introduced computer graphics in a later chapter of the book. Computer graphics is especially useful and important for presenting results of chemical problems. Also, the use of graphics makes the work of programming more pleasing and displaying the results of a solved problem more satisfying. Therefore, one should try to use computer graphics at an early stage and as frequently as possible. Their value will soon become apparent.

PC's have proved their efficiency and usefulness also in modelling and simulation studies, which are becoming more and more important in experimental disciplines. As an introduction to this field some numerical methods are given, which can also be used to solve certain general problems. Other fields of computer applications in experimental sciences are process optimization and data processing. Some simple applications to these fields are included in the book, and these may help to make best use of the PC in such types of work.

The potential of modern PC's is nearly unlimited, but the decision of how and where to implement them rests with us. Hopefully, this book will help us to perform this task successfully.

Karin Sattel-Schwind deserves special recognition and thanks for the layout and the typesetting of the manuscript with TeX. Johannes Kinkel did the tough work of translating the BASIC-programs into PASCAL and Peter Schmich proofread the manuscript and helped to prepare the index.

<table>
<tr><td>Heidelberg</td><td></td><td>Klaus Ebert</td></tr>
<tr><td>Karlsruhe</td><td>Summer 1989</td><td>Hanns Ederer</td></tr>
<tr><td>Kansas</td><td></td><td>Thomas L. Isenhour</td></tr>
</table>

Contents

BASIC and PASCAL diskettes attached to inside back cover

1 Introduction

1.1 General Remarks

This introduction should be as short as possible; the reader will start working with the computer immediately. The programs, commands and examples, which were developed on an IBM PC-AT computer, have been kept general so that they can be used on other computers, as long as they implement BASIC — possibly with slight modifications.

Most personal computers use BASIC, which stands for Beginners All Purpose Symbolic Instruction Code. The introduction to BASIC in this book is elementary and some of the programs are somewhat tedious and slow. It is our experience that it is important for beginners to start with simple programs that lead quickly to useful results. More effective programming, which makes use of all the features of a program language, follows automatically as the beginner becomes familiar with the computer. Once a problem has been solved and a suitable program developed, it is often useful to take a second look at the program and try to make it more efficient and to make the input and output more convenient. Various other publications on computers and BASIC may be helpful for advanced programmers.

1.2 Problem Analysis

Frequently those who have not worked with computers believe that one simply has to tell the computer about the problem and it will readily solve it. This idea is not wrong in principle, but 'telling' the computer has to be done in simple, small, trivial, and precise steps. The conversion of the problem into those steps, which the computer understands, is called analysis.

The computer does not actually solve our problem but rather processes our analysis of the problem. Every process step is simple and could be executed by hand. The real advantage is:

> The computer is fast, patient and very reliable.

The program is the set of instructions for carrying out operations that provide a solution to the problem. The program must be written in a form which is 'understood' by the computer and it has to be put into the memory of the computer before it can be run.

We will begin by solving a problem from mechanics: How long does it take an object to fall from a height, given the acceleration and initial velocity?

Physics answers this question with a formula, which includes five quantities: height h, acceleration g, initial velocity v_0, time t and initial height h_0.

$$h = -\frac{g}{2}t^2 + v_0 t + h_0$$

This formula, however, does not answer our problem. As we are interested in knowing the time t, we have to rearrange the formula and solve it as follows

$$t^2 - \frac{2v_0 t}{g} + \frac{2 \cdot (h - h_0)}{g} = 0$$

If the following abbreviations are used

$$a = -\frac{2v_0}{g} \qquad \text{and} \qquad b = \frac{2 \cdot (h - h_0)}{g}$$

the equation above is rewritten as

$$t^2 + at + b = 0$$

which has the two solutions

$$t_{1,2} = -\frac{a}{2} \pm \sqrt{\frac{a^2}{4} - b}$$

This is the 'mathematical answer' to the problem which has to be translated into the form of a program.

The flow diagram of Fig. 1.1 and the attached program for solving the above problem can be understood almost without comment. The asterisk (*) designates the symbol for a multiplication. Contrary to algebra, this symbol must always be written. In treating complicated problems, it is useful to set up a flow diagram which defines the strategy for solving the problem. The actual computer program is then developed following the flow diagram. The normal order of operations is from the top downwards; any deviations from this order are designated by arrows. The common form of a flow diagram is illustrated in more detail with the next example. The sum of the first hundred squared integer numbers are to be calculated

$$S = 1^2 + 2^2 + 3^2 + 4^2 + \ldots + 99^2 + 100^2 = \sum_{i=1}^{100} i^2$$

An appropriate flow diagram is shown in Fig. 1.2.

1.3   The Most Important Commands

This section assumes that we are in the BASIC-interpreter environment and do not consider the Disk Operating System (DOS). Consult your manual if you need help with the operating system. The BASIC-environment we used for this book was entered by typing the command BASICA from DOS 3.00.

At this point those BASIC-interpreter commands shall be introduced which are most important. Additional commands will be presented at the appropriate places in the text. It should be noted that the commands and assignments are given in the syntax of BASICA for the IBM PC. Many other PC's use the same commands. The DOS and the BASIC language manuals for your computer should be consulted for possible differences.

A few comments about writing in capital and lower case characters, error messages, and the cursor are given. In this book all programs and commands are written in capitals. This is required by some computers. If you don't know whether your computer requires capitals or commonly uses lower case characters, just find it out by trying or consult your handbook. Errors during the input of commands and program errors (misprints) are indicated by the computer by error messages. Frequently, abbreviated messages are

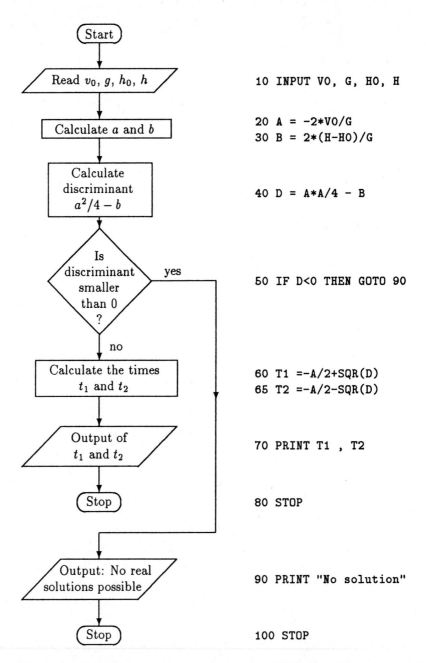

Fig. 1.1. Flow diagram inclusive BASIC-statements to calculate the falling time t of an object with height h, initial height h_0, initial velocity v_0 and acceleration g

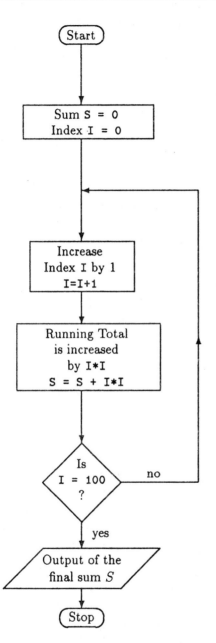

The oval is the symbol for start, end or stop.

A rectangle is used to designate a calculation, an assignment or other operations.

The two variables S and I are set to zero at the beginning of the program.

Index I will run from 1 to 100. The symbol "=" designates an assignment i.e. I is incremented by 1 each time through.

The interim sum S is increased at each run through by a row element I*I.

The diamond is the symbol for a decision or a condition which works as a branching point within the program. There are two choices each time through.

A parallelogram means that input data are introduced to the program (reading) or output results are produced (writing or printing).

Fig. 1.2. Flow diagram to add the first 100 squared integers with explanation of the symbols used.

Note: the corresponding BASIC-program is shown in section 4.1.

used which are listed in the computer manual. Error messages are very useful in finding and correcting errors.

Clear the Screen

> CLS

Syntax: CLS

To clear the screen, the word CLS has to be typed and the <ENTER>-key (or <RETURN>-key) hit. Nothing is removed from the computer memory.

Clear a Program From Computer Memory

> NEW

Syntax: NEW

The command NEW, followed by hitting the <ENTER>-key, clears the memory of the current program, however, it does not effect what is written on the screen.

List a Program on the Screen

> LIST

Syntax: LIST

To see a program from the computer memory on the screen, the command LIST has to be typed and the <ENTER>-key hit. The LIST-command has options and can either call up a certain part of the program or let the program scroll on the screen. The BASIC-manual gives detailed information about these options.

Stop the Program Listing

CNTR-NUM-LOCK

Syntax: <CNTR>-<NUM-LOCK>

To stop the listing of a program the <CNTR>-key has to be held down and the <NUM-LOCK>-key pushed. The listing may be resumed by hitting any key.

Start a Program

RUN

Syntax: RUN

To start a program, RUN has to be typed and the <ENTER>-key pushed.

Stop a Program

CNTR-BREAK

Syntax: <CNTR>-<BREAK>

To stop a running program from the keyboard the <CNTR>-key has to be held down and the <BREAK>-key pushed.

If a floppy disk or a hard disk is used, the following three commands are necessary. (The hard disk is much faster in transferring information than the floppy disk and can contain more information.)

List the Program Names in a Catalog

FILES

Syntax: FILES

Typing the command FILES (and hitting the <ENTER>-key) causes all names of the programs stored on the disk or on other magnetic media to be shown on the screen.

Load a Program From Disk

```
LOAD "NAME"
```

Syntax: LOAD "NAME"
NAME : valid filename

To load a given program from the disk to the memory of the computer, LOAD and the name of the program (in quotation marks) have to be written and the <ENTER>-key has to be pushed.

Save a Program to the Disk

```
SAVE "NAME"
```

Syntax: SAVE "NAME"
NAME : valid filename

To save a program from the computer memory on disk and simultaneously add the name to the disk catalog, the command SAVE followed by the name of the data set is used. If the particular name has already been used for a previous program on the disk or if there is not enough space left on the disk for saving the new program, some computers print an appropriate message on the screen. This is not true with BASICA or GWBASIC for the IBM PC and compatibles. They allow to use a program name twice, by copying the new program over the old one. If the command is written as SAVE "NAME",A then the program is stored in ASCII-code and not in the usual compressed BASIC-code.

With the commands mentioned above a number of important operations can be carried out. Nevertheless, a look into the manual might be useful to find additional commands i.e. those that allow erasing parts of programs, copying programs or changing names of programs.

2 Formulas

2.1 Introduction to Computer Languages

To solve problems in natural science, mathematical calculations are often carried out. If the same formula has to be used frequently with varying values, setting up a standard computer program may be useful. For simple formulas a programmable pocket calculator may be used, however, doing the same job on a PC is usually much simpler, and has the advantage that the program can be saved for later use or transferred elsewhere.

As an example, let us try to program the following simple mathematical formula

$$y = \frac{e^{-b \cdot t}}{1 + t}$$

The program for the pocket calculator SR56 from Texas Instruments looks like this

	;	Input of b as a number
*	;	to be multiplied with
R/S	;	Stop, input of t
STO	;	Store t in
1	;	memory number 1
=	;	Calculate $b \cdot t$
+/-	;	Calculate $-b \cdot t$
EXP(X)	;	Calculate $e^{-b \cdot t}$
STO	;	Store this result in
0	;	memory number 0
RCL	;	Recall t from
1	;	memory number 1
+	;	Add

1	;	1 in order to
=	;	Calculate $t + 1$
2nd	;	
1/X	;	Calculate $1/(1 + t)$
*	;	Multiply this with
RCL	;	$e^{-b \cdot t}$ from
0	;	memory number 0
=	;	Final result
R/S	;	Stop

With some experience it is not too difficult to write this program, but it is somewhat complicated and it is easy to make an error.

PC's use the so-called higher programming languages, which combine frequently used sequences of elementary commands in a computer statement. This shortens the program and improves its clarity. The longer these sequences are, the more commands are needed, and the higher is the level of the language. There are a number of common computer languages. Most of them were developed for certain applications. In natural science ADA, FORTRAN, PL1, PASCAL, C or BASIC are used mostly. Most big computers (main frames) handle these languages quite well. The programs for the above formula in three different languages read as follows:

FORTRAN

```
        REAL*4 B , T , Y
        READ(5,100) B , T
100     FORMAT(E12.5,E12.5)
        Y = EXP(-B*T)/(1+T)
        WRITE(6,200) Y
200     FORMAT(E12.5)
        STOP
        END
```

PL1

```
    FORMULA1: PROCEDURE OPTIONS(MAIN) ;
    DECLARE (B,T,Y) DECIMAL FLOAT ;
    GET LIST(B,T) ;
    Y = EXP(-B*T)/(1+T) ;
    PUT LIST(Y) ;
    END FORMULA1 ;
```

BASIC
```
10   INPUT B , T
20   Y = EXP(-B*T)/(1+T)
30   PRINT B , T , Y
```

A closer look at each of these programs shows that the actual mathematical equation is contained in one line only. All the other lines contain definitions and input and output commands.

These examples show that one first needs to know how to program a formula. One also needs to know something about the operating system to get the programs to run and to obtain the results in an appropriate form.

Undoubtedly, BASIC is the most 'beginner friendly' language; the commands are short forms of English words, the structure of the program is clear and there are also advantages in writing (editing) the program.

BASIC is suitable for treating simple, but not very complex problems in natural science. For very complicated problems and large programs with numerous subroutines, FORTRAN, PASCAL and PL1 have special advantages. For this reason, BASIC is used for PC's as the first language and FORTRAN, PASCAL or the other higher languages are used by the more advanced programmer.

Once one has learned BASIC, a switch over into another higher programming language is relatively simple. The principles of programming are similar and one has to learn only a few more words and some additional grammar.

As soon as the BASIC-program given above has been written on the screen, and the <ENTER>-key has been pressed, it can — using the commands RUN, SAVE, and LOAD explained in the last chapter — be started, saved or reloaded into the computer.

Data can also be defined in the program instead of using the INPUT-command. The previous BASIC-program can then be changed to read:

```
5    B = -7
10   T = 11.1
20   Y = EXP(-B*T)/(1+T)
30   PRINT B , T , Y
```

For B and T specific numbers were defined which make the program less general than that with the INPUT-command, as this program would have to be changed to vary values of B and T.

2.2   Line Numbers in BASIC

There is an important feature in programming in standard BASIC:

> Each program line starts with a line number.

The program is always executed in the sequence of these numbers, except for program jumps. However, it is not necessary to write the program in that order, the BASICA editor puts it into the correct order anyway.

In writing a program one should always make steps in the successive line numbers in units of 10 or 100 to be able to insert additional lines at a later time. As the maximum line number allowed is quite large, there is usually no difficulty. Most of the BASIC-environments are able to renumber a program by the command RENUM, usually in steps of 10; this is convenient when more space is needed for inserting additional lines. The maximum line number is given in your manual; for BASICA it is 65529.

Renumber the Line Numbers of a Program

RENUM

Syntax: RENUM z1, z2, a

z :   line number
a :   step (integer number)

The command RENUM renumbers the line numbers in steps of 10 beginning with line number 10. If these default values should be avoided, the parameters new line number, old line number and step for the distance between the new line numbers have to be used.

Example: RENUM 1000,55,100

2.3   Correcting Programs

Correcting and modifying programs without completely rewriting them is essential. First the screen should be cleared and the program listed. It will appear in the correct order of the line numbers.

To eliminate a line, the line number is typed and the <ENTER>-key pressed.

To rewrite a line, simply type the same line number and the new statement. As soon as the <ENTER>-key is hit, the old line is replaced by the new line.

Single characters can be corrected in a line which is listed on the screen, by simply placing the cursor to the proper position using the arrow keys and typing in the correct letter. As soon as the <ENTER>-key is pressed the correction is carried out. For inserting or deleting letters the cursor must be placed at the position in question and the <INSERT>- or the <DELETE>-key has to be hit. Consult the BASIC-manual for details of line editing.

2.4   Description of Variables

In the previous BASIC-programs, we used the variables B, T and Y to represent numbers. Practically all BASIC-dialects allow one or two letter/number variables which must start with a letter, e.g.

$$A, DE, A1, A2, \ldots, B0, \ldots, Z9$$

In most of the BASIC-dialects variables with more letters can be used — up to 40 for the BASICA of the IBM PC. Keywords in BASIC such as IF, TO, ON, etc. cannot be used as variable names. Look up in your manual for 'reserved names' or the list of commands and functions.

A convenient feature of BASIC is the 'SPACE' symbol

> Additional space symbols (blanks) are
> (usually) of no significance in BASIC.

Blanks allow improvements in the structure and the clarity of the programs, and therefore should be used freely. BASIC-dialects, which allow variable names of more than two letters demand the space as a separator between commands and variables. Why?

2.5   First BASIC Elements to Program Formulas

Three BASIC-statements, which have been used already in the introductory examples, will be explained now in more detail. Both the use of the correct syntax and the understanding of the semantic of the statements is a prerequisite for correct programming.

Data Input — Reading Data From the Keyboard

INPUT

Syntax: z `INPUT V1, V2, V3, ..., VN`

z : line number
V : variables

The `INPUT`-statement is used to enter data into the computer during program execution. The syntax of the statement requires a statement number z, the keyword `INPUT` and a list of variables. Formulas are not permitted.

Examples: correct   incorrect

`20 INPUT B , C`	`15 INPUT B*B`
`30 INPUT E,E0,E1,EFU`	`25 INPUT 3`
`50 INPUT Z`	`55 INPUT`

If several variables shall be written into one `INPUT` line they have to be separated by commas. When during the execution of a program a line with the `INPUT`-command is reached, a question mark appears on the screen, and the program stops. The computer now waits for the appropriate number of values which must be entered in the correct sequence and separated by commas; e.g. `INPUT A, B, C, D` is answered by typing `3, 99.5, 4, 75.7` and hitting the <`ENTER`>-key. Then the assignments `A=3, B=99.5, C=4` and `D=75.7` are performed.

**Data Output — Printing Data on
the Screen**

$$\boxed{\text{PRINT}}$$

Syntax: z `PRINT` A1, A2, A3, ..., AN

z   :   line number
A   :   variables, constants and numerical expressions (formulas)

The list of output items after the keyword `PRINT` may consist of variables,
formulas or numbers. The separators in the list are commas or semicolons.

Examples: correct   incorrect
```
20 PRINT I                15 PRINT I=3
30 PRINT A,2*A,A*A        25 PRINT SQR(-2)
50 PRINT 3.14,A*3.14      55 PRINT 1/0
70 PRINT J;J+1;J+2
105 PRINT EXP(A);
312 PRINT SIN(4*A/K+2)
313 PRINT
4300 PRINT X,Y;X-Y,X*Y
```
The `PRINT`-statement transfers the values from the variable/expression list
to the screen. If the variables are separated by a comma a formatted output
will appear on the screen which improves the clarity. If the variables are
separated by a semicolon the values are written on the screen one after the
other. A `PRINT`-command without a list causes a blank line to be printed
on the screen. If the `PRINT`-statements don't end with a separator then each
`PRINT` begins a new line.

The Assignment

$$\boxed{=}$$

Syntax: z `V = A`

z   :   line number
V   :   variable
A   :   numerical expression

At the left hand side of the assignment symbol (=) must be a variable. At the right hand side an expression must be written which can be a variable or a constant.

Examples: correct   incorrect

          20 I=9   15 9=I

          30 I=I+1   25 I+1 = I+2

          50 V9=V*V   55 1.23 = 0.123E1

          70 X = (A+X)/(B+SIN(X))   75 X*X*X = 19.2

          90 J = 0   95 0.5 = 1/2

In all incorrect examples, the left hand side of the assignment statement contains something other than a plain variable.

The assignment looks like an algebraic equation, but there are a lot of legitimate assignment statements that would make no sense if viewed as algebraic equation (e.g. I=I+1). V=A means that the numerical value of A is assigned to the variable V. Only one variable is allowed on the left side of the equal sign, but on the right side any possible expression (i.e. formula) can appear. If the same variable appears on both sides of the equal sign, the numerical value of the variable on the right hand side is used to evaluate the formula. The result is the new value of the variable at the left hand side. To emphasize the difference between an algebraic equation and an assignment statement, symbols other than '=' are used in some languages — for example ':=' in PASCAL, or '←' in APL.

2.6   The Notation of Numbers

Numbers or constants are written in decimal notation, for decimal exponents write E followed by the exponent. The exponential notation is similar to scientific notation, except that the base 10 is replaced by the letter E.

Table 2.1. Comparison of scientific notation with computer notation of numbers

Scientific Notation	Computer Notation
$6.022 \cdot 10^{23}$	6.022E23
$2.990 \cdot 10^{10}$	2.99E10
$6.626 \cdot 10^{-34}$	6.626E-34
10^{-10}	1E-10

The following points should be carefully observed to avoid errors:

- The decimal point may appear anywhere in a number.
- Commas must not appear anywhere in a constant.
- A number can be preceded by a + or − sign.
- The exponent can be either positive or negative but cannot have a decimal point.
- Most versions of BASIC allow a number to have as many as 8 or 9 significant figures.
- BASICA or GWBASIC is accepting floating-point constants as large as 1.7E38 and as small as 2.9E-39.
- Integer constants may range between −32768 and +32767.

The number 0.08 can be read by a computer in any of the following notations:

0.08	.08	.08E0	0.00008E3
0.8E-1	.008E1	8E-2	800.0E-4

but not as:

8/100	8*10^(-2)	8*E-2

because the latter expressions contain mathematical operations and not just plain numbers.

2.7   Arithmetic Expressions

In writing arithmetic expressions two important points should be noted:

- The multiplication symbol (*) must always be written.
- Use parentheses as necessary to guarantee the correct order of calculations. Arithmetic operations have the same order in BASIC as in algebra: exponentiation is first, multiplication or division second, and addition or subtraction third.

The BASIC-symbols for the arithmetic operations are shown in Table 2.2.

Table 2.2. BASIC-symbols used for elementary arithmetic operations

Algebraic Operation	BASIC Symbol
Addition	+
Subtraction	–
Multiplication	*
Division	/
Exponentiation	^

In Table 2.3 some examples of arithmetic expressions and the appropriate notations in BASIC are given.

Table 2.3. Examples of BASIC-notations for some algebraic expressions

Arithmetic Expression	BASIC Expression
$a - b$	A-B
ab	A*B
$2b$	2*B
$2c + 3b^2$	2*C+3*B^2 or 2*C+3*B*B
$7 \cdot \frac{a-b}{c-d}$	(A-B)/(C-D)*7 or 7*(A-B)/(C-D) or 7/(C-D)*(A-B)
$\frac{a-b}{(c-d)\cdot 7}$	(A-B)/(C-D)/7 or (A-B)/((C-D)*7) or 1/7/(C-D)*(A-B)
x^2	X^2 or X*X
$\sqrt{x}$	SQR(X) or X^(1/2) or X^(0.5)
$\left[(a+b)^2 + c\right]^2$	((A+B)^2+C)^2
$\frac{E}{RT}$	E/(R*T) or E/R/T or E/T/R or 1/(T*R)*E

2.8   Built-in Standard Functions

In treating natural science problems a number of mathematical functions are needed frequently. Some of them are available without programming in most computer languages. These built-in library functions are accessed simply by stating their name followed by whatever information — usually arithmetic expressions which again may include standard functions — must be supplied to the function, enclosed in parentheses. The most common standard functions are summarized in Table 2.4.

Table 2.4. Table of standard functions in BASIC

Function	Description
SIN(X)	Sine of x ; $\sin(x)$, x in radians
COS(X)	Cosine of x ; $\cos(x)$, x in radians
TAN(X)	Tangent of x ; $\tan(x)$, x in radians
ATN(X)	Arctangent of x ; $\arctan(x)$ Inverse function of $\tan(x)$
SQR(X)	Square root of x ; $\sqrt{x}$; $x > 0$
EXP(X)	Natural exponential of x ; e^x
LOG(X)	Natural logarithm of x ; $\log_e(x)$
ABS(X)	Absolute value of x ; $|x|$
INT(X)	Largest integer that algebraically does not exceed x
SGN(X)	Sign of $x = \begin{cases} 1 & \text{if } x > 0 \\ 0 & \text{if } x = 0 \\ -1 & \text{if } x < 0 \end{cases}$
RND(X)	Generates pseudo random numbers between 0 and 1

There are many other standard functions that are not built-in as library functions in the BASIC-interpreter. However, most of them can be programmed easily using the built-in functions. Table 2.5 contains some examples.

Table 2.5. Table of standard functions not built-in in BASIC and their equivalent as arithmetic expression

Function	Expression in BASIC
Logarithm to base b: $\log_b(x)$	`LOG(X)/LOG(B)`
Secant: $\sec(x)$	`1/COS(X)`
Cosecant: $\csc(x)$	`1/SIN(X)`
Cotangent: $\cot(x)$	`1/TAN(X)`
Inverse Sine: $\arcsin(x)$	`ATN(X/SQR(1-X^2))`
Inverse Cosine: $\arccos(x)$	`1.570796-ATN(X/SQR(1-X^2))`
Inverse Secant: $\text{arcsec}(x)$	`ATN(SQR(X^2-1))+(X<0)*3.141593`
Inverse Cosecant: $\text{arccsc}(x)$	`ATN(1/SQR(X^2-1))+(X<0)*3.141593`
Inverse Cotangent: $\text{arccot}(x)$	`1.57096-ATN(X)`
Hyperbolic Sine: $\sinh(x)$	`(EXP(X)-EXP(-X))/2`
Hyperbolic Cosine: $\cosh(x)$	`(EXP(X)+EXP(-X))/2`
Hyperbolic Tangent: $\tanh(x)$	`(EXP(X)-EXP(-X))/(EXP(X)+EXP(-X))`
Hyperbolic Cotangent: $\coth(x)$	`(EXP(X)+EXP(-X))/(EXP(X)-EXP(-X))`
Hyperbolic Secant: $\text{sech}(x)$	`2/(EXP(X)+EXP(-X))`
Hyperbolic Cosecant: $\text{csch}(x)$	`2/(EXP(X)-EXP(-X))`
Inverse Hyperbolic Sine: $\text{arsinh}(x)$	`LOG(X+SQR(X^2+1))`
Inverse Hyperbolic Cosine: $\text{arcosh}(x)$	`LOG(X+SQR(X^2-1))`
Inverse Hyperbolic Tangent: $\text{artanh}(x)$	`LOG((1+X)/(1-X))/2`
Inverse Hyperbolic Cotangent: $\text{arcoth}(x)$	`LOG((X+1)/(X-1))/2`
Inverse Hyperbolic Secant: $\text{arsech}(x)$	`LOG((SQR(1-X^2)+1)/X)`
Inverse Hyperbolic Cosecant: $\text{arcsch}(x)$	`LOG((SGN(X)*SQR(1+X^2)/X)`
Archimedes' or Rudolf's number π	`3.141592654 or 4*ATN(1)`

2.9   Unconditional Program Jumps

Our first programming example in section 2.1 had the disadvantage, as users will recognize, that the program must be started new for each value of B and T. This can be avoided by the GOTO-command at the end of the program which causes the program to jump back to the start. With the GOTO-command the program can be directed to jump to any line by simply referring to the line number.

Unconditional Branching

GOTO

Syntax: z1 GOTO z2

z : line numbers

The GOTO-statement is called an 'unconditional jump' or 'unconditional branching' because when the program arrives at this point, it jumps to line z2 and continues operation — that is control has been transferred from line z1 to z2. Jumps in either direction are possible. In case line number z2 is missing in the program, an error statement appears on the screen.

Examples: correct   incorrect

      20 GOTO 3000   15 GOTO A
      90 GOTO 10   100 GOTO 20 , 200
                               200 GOTO 30*40

Problem 1

What happens if the following statement is contained in a program?

      100 GOTO 100

If you have difficulty figuring out what will happen — just try it out!!!
The following hint may be helpful: Programs can be stopped by hitting the <CNTR>- and <BREAK>-key simultaneously.

The program of the problem in section 2.1 which is enhanced by a GOTO-command reads:

      10   INPUT B , T
      20   Y = EXP(-B*T)/(1+T)
      30   PRINT B , T , Y
      40   GOTO 10

After each run (that is after printing the result) the program jumps back to the INPUT-statement in line 10 and waits, after printing a question mark on the screen, for the next INPUT of values for B and T. If you want to terminate the program, just hit the <CNTR>-<BREAK> keys.

Problem 2

Write a program for the calculation of the reaction rate constant k using the Arrhenius equation. The kinetic parameters E_a (activation energy) und k_0 (preexponential factor) should have a fixed value within the program; the temperature T should be entered with an INPUT-statement.

$$k = k_0 \cdot e^{-\frac{E_a}{RT}}$$

Problem 2a

Write programs for some of the not built-in standard functions of Table 2.5.

Problem 3

Test the following programs: Try to figure out beforehand what will happen and compare with the results given by the computer.

```
10  I=I+1                    10  INPUT X
20  PRINT I,I*I              20  X=SQR(ABS(SIN(X)))
30  GOTO 10                  30  PRINT X,
                             40  GOTO 20
```

The correct interpretation of the second program is difficult; we will come back to this problem later. A problem easier to understand results when line 20 is substituted by:

```
20  X=SQR(X)
```

Try to make errors on purpose! For example, divide by 0, apply functions to illegal arguments, i.e., try to take the square root of a negative number. Try to find out — with the help of a program or in the direct mode (Chapter 2.10) — the biggest and the smallest number which your computer can handle.

2.10   The Direct Mode

Practically all BASIC-interpreters can calculate in the direct mode, which allows you to execute a single computer statement. If you write a command without a line number it will be executed after hitting the <ENTER>-key. This command will not be retained, however, the computer answers directly, as can be shown with the following examples:

```
PRINT LOG(5.1)*SIN(3)    <ENTER>
   .2299184
```

```
I = 217                    <ENTER>
PRINT I*I                  <ENTER>
  47089
```

2.11 Area of a Triangle From the Length of the Three Sides

This calculation can be carried out without using trigonometric functions. First S, which is the half of the value of the perimeter, is obtained from the lengths of the three sides a, b and c by

$$S = \frac{a+b+c}{2}$$

For the triangle area F, the following formula applies

$$F = \sqrt{S(S-a)(S-b)(S-c)}$$

A suitable program for this problem is:

```
0 REM "TRIANGLE"         EBERT/EDERER           880504
1 REM ************************************************
2 REM *** Calculation of the area of a triangle;   ***
3 REM *** the three sides a, b and c are given     ***
9 REM ************************************************
100    INPUT "Side  a in cm";A
200    INPUT "Side  b in cm";B
300    INPUT "Side  c in cm";C
400    S = .5*(A+B+C)
500    F = SQR(S*(S-A)*(S-B)*(S-C))
1000   PRINT
1100   PRINT "Area of the triangle =";F;"cm^2"
2000   END

RUN
Side  a in cm? 3
Side  b in cm? 4
Side  c in cm? 5

Area of the triangle = 6 cm^2
```

```
RUN
Side  a in cm? 10
Side  b in cm? 4
Side  c in cm? 5
Illegal function call in 500
```

This program looks different when compared to the previous ones. It is rather extended, the following four lines would have been sufficient:

```
100    INPUT A,B,C
400    S = .5*(A+B+C)
500    F = SQR(S*(S-A)*(S-B)*(S-C))
1100   PRINT F
```

The simple program is similar to the extended one. In lines 400 and 500 the formula is calculated. In lines 100–300 the values for the sides are input and in line 1100 the result — the value of the area of the triangle — is printed on the screen. The difference between both programs is understood as soon as the performance of the programs is compared. The extended program is much more 'friendly' as it tells the user what to do and what the results mean. It is also much easier to use. The short version can normally be used by the programmer himself only because he alone knows what sort of information the computer needs and what the result really means. However, programmers also forget what they have done, and therefore have to analyze their programs before using again some time later.

What is new with the friendly program?

Comment or Remark

REM

Syntax: z REM blabla

z   : line number
blabla   : any text

The remark statement causes no action by the computer, which proceeds directly to the next line. Therefore, REM can be used to write comments into the program. This is very useful because it offers the programmer convenient means to document a program and a good comment may save analyzing the program text later. It might be laborious and annoying to write such comments, but it is a common experience to find out later that comments are needed. Therefore we give the following advice:

> Too many comments are better than too few!

The comments at the head of the program TRIANGLE — and of all following programs — give the name of the program, the date written (this may be important for comparing changed or updated programs), the names of the authors and a short general description of the contents of the program.

A further important difference between both programs are the INPUT- and PRINT-commands with added text which has to be written in quotation marks. Whereas REM is the most common way to include documentation within the program — these explanations can be read when the program is listed — the text attached to the INPUT- and PRINT-statements is shown for information during the execution of the program.

Data Output — Extended by Text

> PRINT

Syntax: z PRINT "text" A1, A2, A3, ..., AN

z   :   line number
A   :   variables, expressions, constants or strings in quotation marks;
        separators: commas or semicolons

Instead of 'text' the expressions 'character chain' or 'string' are often used. The 'text' has to be set between quotation marks. The successive output items must be separated by either commas or semicolons. A comma advances the output to the next column of the formatted screen, a semicolon adds the next item immediately without space.

Examples: PRINT "The result is ";R*R*R;" liters"
               PRINT "Pi = "; 3.141593

Data Input — Extended by Text

$$\boxed{\text{INPUT}}$$

Syntax: z `INPUT` "text"; V1, V2, V3, ..., VN

z   : line number
V   : variables
text : any text in quotation marks

The position of 'text' follows immediately after the keyword `INPUT`. If the semicolon after the character string is substituted by a comma, the question mark otherwise appearing after the execution of the program statement is suppressed.

It should be noted that these additions are not possible in all BASIC-dialects. Consult your BASIC-manual for the correct syntax for extended `INPUT`-commands.

Examples: correct   incorrect

```
20 INPUT "Length"; L      15 INPUT "Length";
30 INPUT "x , y ";x,y      25 INPUT "x , y;" x,y
```

If the data items typed don't correspond in number to the variables listed in the `INPUT`-statement the BASIC-interpreter answers with:

```
Redo from start
```

Another general remark should be added for user-friendly programming. In natural and engineering sciences, dimensional quantities are frequently important. The computer, however, processes numbers only, therefore the dimensions must be included in the program. The program must tell the user which dimensions the values should have. For convenience, one should not require the wavelength of a monochromatic light beam to be in nautical miles (even if one does calculate the number of waves per nautical mile) but in common units like Ångströms or nanometers. The necessary conversions should be done within the program.

Line 2000 of the program `TRIANGLE` contains the command `END`. It indicates the end of the program. During the execution of a program, the end statement terminates the execution and the BASIC-interpreter is in the editing and direct mode. If there are more program lines after an `END`-statement, the execution may be resumed after this statement by the command `CONT`.

The End of a Program

$$\boxed{\text{END}}$$

Syntax: z END

z : line number

END-statements are not really necessary, but it is a common practice to terminate all programs with an END. But, if you wish to compile a BASIC-program, an END-statement is required for most compilers to indicate the physical end of a program.

Problem 4

Write a program to calculate the volume V and the surface S of a sphere and the ratio of these quantities. Given the radius of this sphere, r.

Recall: $S = 4\pi r^2$ $V = \dfrac{4\pi r^3}{3}$

Problem 5

Program some formulas that are of interest to you — make these programs 'user friendly'.

2.12 Calculation of the Mean Free Path of Gas Molecules

In textbooks on physical chemistry the equation for the mean free path l of gas molecules can be found.

$$l = \frac{1}{\sqrt{2}\pi\sigma^2 N}$$

l : Mean free path length
σ : Molecular diameter
N : Number of particles per unit volume

The molecules are considered as rigid spherical particles with diameter σ. The number of particles per unit volume N can be calculated from the ideal gas law.

$$N = \frac{N_A \cdot p}{RT}$$

The program should calculate the mean free path l which is dependent of pressure, temperature and molecular diameter. This quantity l is required for designing vacuum equipment or estimating the catalytic effect of a reactor wall to the rate of homogeneous gas reactions.

```
0 REM "FREEPATH"          EBERT/EDERER          880504
1 REM ***********************************************
2 REM *** Calculation of the mean free path      ***
3 REM *** of molecules in a gas. The temperature, ***
4 REM *** the pressure and the diameter of the    ***
5 REM *** molecules are given.                    ***
9 REM ***********************************************
10    PI = 4*ATN(1)
100   INPUT "Pressure in atm ";P
200   INPUT "Temperature in K ";T
400   INPUT "Diameter of the molecule in m ";S
1000 REM ***********************************************
1010 REM *** Calculation of the number of molecules **
1020 REM *** in 1 cubic meter.                      **
1030 REM *** R = 8.2056E-5 m^3*atm/(mol*K)          **
1040 REM ***********************************************
1100  N = 6.022E+23*P/(T*8.2056E-05)
2000 REM ***********************************************
2010 REM *** Calculation of the mean free path      ***
2020 REM *** of a gas molecule.                     ***
2040 REM ***********************************************
2500  L = 1/(PI*SQR(2)*S*N*S)
2600  PRINT
3000  PRINT "Mean free path length = ";
3100  PRINT L;" m"
63999 END

RUN
Pressure in atm ? 1
Temperature in K ? 293
Diameter of the molecule in m ? 3.64E-10

Mean free path length =  6.782184E-08  m
```

In this program the calculation of two formulas is performed. Line 1100 calculates, using the ideal gas law, the number of molecules per cubic meter and line 2500 calculates the mean free path. In setting up the equations and writing the program, one has to take care to use correct dimensions. A correct alternative for line 3100 would be:

```
3100 PRINT L * 100; " cm"
```

Most lines in this program contain comments. There are three input and three output lines, the 'naked' PRINT-command in line 2600 results in a blank line in the output print. The example shown above uses the molecular diameter of argon, which is 3.64 Å.

Problem 6

Change the program FREEPATH so, that the input should be in millibars, degrees centigrade and Ångströms and the output in Ångströms.

Problem 7

Use the program FREEPATH (or a modified version) to calculate the pressure in a vacuum apparatus for which the mean free path of nitrogen ($\sigma = 3.77\,\text{Å}$) equals 20 cm.

2.13 Maxwell's Distribution Law of Velocities

The equation which is known as Maxwell's distribution law of velocities is one of the most fundamental relations of chemistry and physics. The distribution of molecular speeds plays an important role in determining the rates of chemical reactions and the thermodynamic properties of matter.

$$\frac{dN}{du} = N_A \cdot 4\pi u^2 \cdot \left(\frac{m}{2\pi kT}\right)^{\frac{3}{2}} \cdot e^{-\frac{mu^2}{2kT}}$$

with the normalization

$$\int_0^\infty \frac{dN}{du}\,du = N_A = 6.022045 \cdot 10^{23}\,\text{mol}^{-1}$$

k :   Boltzmann constant $= 1.380662 \cdot 10^{-23}$ J/K
T :   Temperature in Kelvin
m :   Mass of a molecule
N_A :   Avogadro's constant
$\frac{dN}{du}$:   Particle density
u :   Molecular speed
$\frac{dN}{du}\,du$:   The number of particles in the
               velocity range from u to $u + du$

Before solving the problem and programming the formula, one has to decide which units for each value will be input or output. The units for the input are defined in lines 200–500, for the output in lines 4600–5500. The inputs and outputs use the 'usual' dimensions for these parameters — for the calculation, the quantities are converted into the cgs-system. The velocity u is input by line 500 in m/sec (variable U) and is converted to cm/s (variable UH) by the first part of line 3000. In the second part of line 3000 the gas constant is converted from J/(K·mol) (variable R) to erg/(K·mol) (variable RH). Furthermore the equation above is divided by the Avogadro's constant N_A, converting molecules to moles. The expression m/k was multiplied above and below the line by N_A and became M/R. Hence, the quotient remains unchanged, but the 'mass of a molecule' is converted to 'molecular weight' and Boltzmann's constant k becomes the gas constant R.

```
0 REM "MAXWELL "          EBERT/EDERER          880504
1 REM ********************************************
2 REM *** Calculation of the MAXWELL distribution ***
3 REM *** of velocities of gas molecules.        ***
9 REM ********************************************
100    PI = 4*ATN(1)
200    INPUT "Temperature in K ";T
400    INPUT "Molecular weight in amu ";M
500    INPUT "Molecular velocity in m/sec ";U
2000 REM ********************************************
2010 REM ***   R      =    8.31441 J/(K*mol)        ***
2020 REM *** 1 J      =    1E7 erg                  ***
2030 REM *** 1 erg    =    1 cm^2*g/(s^2)           ***
2040 REM *** L        =    6.022e23 particles/mole  ***
2500 REM ********************************************
2600    R = 8.31441
3000    UH = U*100  :  RH = R*1E+07
```

```
4000    DN = 4*PI*UH*UH*(M/(2*PI*RH*T))^(3/2)
4100    DN = DN*EXP(-M*UH/RH*UH/T/2)
4200    DN = DN*100 : REM because output is in m
4600    PRINT "Temperature        = ";T;"K"
4700    PRINT "Molecular weight   = ";M;"amu"
4800    PRINT "Molecular velocity = ";U;"m/sec"
5000    PRINT
5200    PRINT "Particle density   = ";DN;"mol/(m/sec)"
5500    PRINT:PRINT
6000  GOTO 500
63999 END


RUN
Temperature in K ? 300
Molecular weight in amu ? 28
Molecular velocity in m/sec ? 200
Temperature         =  300 K
Molecular weight    =  28 amu
Molecular velocity =  200 m/sec

Particle density    =  9.589698E-04 mol/(m/sec)


Molecular velocity in m/sec ? 400
Temperature         =  300 K
Molecular weight    =  28 amu
Molecular velocity =  400 m/sec

Particle density    =  1.955937E-03 mol/(m/sec)


Molecular velocity in m/sec ?
Break in 500
```

Listing Programs in Parts

LIST

Syntax: LIST z1 – z2

z : line numbers

The program MAXWELL is too long and will not fit on the screen. However one can list the program in parts by using the command LIST z1–z2, where z1 and z2 are the first and the last line numbers of the program segment of interest. Alternatively the program listing can be stopped or scrolled slowly over the screen by some interpreters and an appropriate keyboard.

Examples: LIST 9 - 500
             LIST - 900
             LIST 500
             LIST 2500 -

A few technical improvements within the program MAXWELL will be given. Line 3000 reads:

    3000 UH = U * 100 : RH = R * 1E7

This is identical with the two statements

    3000 UH = U * 1000
    3010 RH = R * 1E7

In BASIC several statements can be written consecutively in one line, if they are separated by colons. One has to remember, however, that in BASIC the GOTO-command jumps always to the beginning of a line.

The long equation is solved in steps. In line 4000 the first part is calculated and the interim sum is assigned to the variable DN. In line 4100 the second part of the formula is calculated, multiplied by DN and saved as a 'new' DN. In line 4200 the result is finally multiplied by 100 to convert it into meters.

For clarification we recommend:

> Long formulas should be programmed in parts and then linked together again.

Problem 8

Use the program MAXWELL to calculate particle densities for Argon at 1000 °C for several different velocities and draw a graph of the distribution of the velocities.

Problem 9

Use the program `MAXWELL` to find out the velocity (by trial, not by programming — this follows later) at a given temperature and molecular weight for which the distribution of the velocities is greatest. How is the maximum displaced as temperature changes? These problems can be treated with mathematical analytical methods as well, but you should recognize that these are solvable with simple numerical methods. Sometimes these numerical methods are faster and less likely to involve errors. In some cases the numerical method provides the only possible solution.

2.14 Planck's Law for Black Body Radiation

Planck's law for the volume density of the energy of a black body is the most fundamental of the radiation laws. The distribution of energy E_ν with frequency ν and of energy E_λ with wavelength λ is given by

$$\frac{dE_\nu}{d\nu} = \frac{8\pi h}{c^3} \cdot \frac{\nu^3}{e^{\frac{h\nu}{kT}} - 1}$$

$$\frac{dE_\lambda}{d\lambda} = \frac{8\pi hc}{\lambda^5} \cdot \frac{1}{e^{\frac{h\nu}{kT}} - 1}$$

h	:	Planck's constant = $6.626176 \cdot 10^{-34}$ Js
k	:	Boltzmann constant = $1.380662 \cdot 10^{-23}$ J/K
c	:	Speed of light = $2.99792458 \cdot 10^8$ m/s
T	:	Temperature in Kelvin
ν	:	Frequency
λ	:	Wavelength
$\frac{dE_\nu}{d\nu} \, d\nu$	:	Energy per unit volume in a frequency interval $d\nu$ around frequency ν.
$\frac{dE_\lambda}{d\lambda} \, d\lambda$	:	The same with wavelength λ

The equation which is used for this calculation is written in line 2000 of the program `PLANCK`. A dimension analysis is necessary to define the dimensions of the input and output quantities.

There is nothing new with respect to programming the code; the main problem is the dimension analysis.

```
0 REM "PLANCK  "          EBERT/EDERER          880504
1 REM ***************************************************
2 REM *** Calculation of the density of the radia- ***
3 REM *** tion energy per unit volume as a func-   ***
4 REM *** tion of the wave frequency according to  ***
5 REM *** PLANCK's law for black body radiation.   ***
9 REM ***************************************************
100    INPUT "Frequency in 1/sec ";V
200    INPUT "Temperature in K ";T
300    H = 6.6262824D-34
400    K = 1.38054E-23
500    C = 29979245812#
600    PI= 4*ATN(1)
1000 REM ***************************************************
1100 REM *** Planck's constant                    ***
1200 REM ***       h = 6.6262824 E-34   J*sec      ***
1300 REM *** Boltzmann constant                   ***
1400 REM ***       k = 1.38054 E-23   J/K          ***
1500 REM *** Light velocity                       ***
1600 REM ***       c = 2.9979245812 E10  cm/sec    ***
1700 REM ***************************************************
2000   U = 8*PI*(V/C)^3*H/(EXP(H*V/(K*T))-1)
2900   PRINT:PRINT
3000   PRINT "Volume density of the ";
3100   PRINT "energy of radiation"
3200   PRINT "at frequency";V;"1/sec"
3300   PRINT "and at temperature";T;"K "
3400   PRINT "= ";U;"J*sec/(cm*cm*cm)"
3500   END

RUN
Frequency in 1/sec ? 1E13
Temperature in K ? 1000


Volume density of the energy of radiation
at frequency 1E+13 1/sec
and at temperature 1000 K
=  1.003322E-24 J*sec/(cm*cm*cm)
Ok


RUN
Frequency in 1/sec ? 5E13
Temperature in K ? 273
```

```
Volume density of the energy of radiation
at frequency 5E+13 1/sec
and at temperature 273 K
=   1.175517E-26 J*sec/(cm*cm*cm)
Ok


RUN
Frequency in 1/sec ? 5E13
Temperature in K ? 1000


Volume density of the energy of radiation
at frequency 5E+13 1/sec
and at temperature 1000 K
=   7.70914E-24 J*sec/(cm*cm*cm)
Ok
```

Problem 10

Alter the program PLANCK to calculate the volume density of the light quanta (photons) instead of the volume density of the energy.

Problem 10a

The wavelength λ_m at which the maximum of $\frac{dE_\lambda}{d\lambda}$ occurs varies with temperature according to Wien's displacement law

$$\lambda_m T = \text{const.} = 0.2898\,[\text{cm} \cdot \text{K}]$$

Verify this rule by trial and error experiments with the program PLANCK.

2.15   Conversion of Degree Celsius into Degree Fahrenheit

The mathematical formula for the conversion which is very simple

$$[°\text{F}] = \frac{9}{5} \cdot [°\text{C}] + 32$$

is written in line 400 of the program C-F. Nevertheless, such a simple program can be very useful for the chemist in handling unusual temperature data.

```
0 REM "C-F     "         EBERT/EDERER        880504
1 REM ****************************************************
2 REM *** Conversion of degree Celsius into        ***
3 REM ***                degree Fahrenheit         ***
9 REM ****************************************************
150    PRINT
200    INPUT "Temperature in Celsius ";C
400      F = C*9/5 + 32
500      PRINT C;"C corresponds to ";F;"F"
550      PRINT
600    GOTO 200
1000   END


RUN
Temperature in Celsius ? 0
 0 C corresponds to   32 F

Temperature in Celsius ? 100
 100 C corresponds to   212 F

Temperature in Celsius ? -40
-40 C corresponds to -40 F

Temperature in Celsius ?
Break in 200
```

Problem 11

Write a program to convert Fahrenheit into Celsius or Fahrenheit into Kelvin.

Write several programs that convert English units into System International units (i.e. pounds per square inch into bars).

2.16   Diffusion Potential of Electrolyte Solutions

The next example deals with the calculation of diffusion potentials — potentials which arise at the boundary of two solutions with different concentrations of an electrolyte. This program described below calculates the potential between the aqueous solutions at different electrolyte concentrations at room temperature. It is a simple problem to extend the program to any other temperature.

$$\Delta E_{\text{diff}}[\text{Volt}] = \frac{u^- - u^+}{u^+ + u^-} \cdot \frac{0.058}{n_e} \cdot \log \frac{c_1}{c_2}$$

In this program the concentrations should be entered in mol/l and the Hittorf transport numbers also have to be entered.

As a reminder: If the ion mobility of the anions and the cations are designated with u^- and u^+ respectively, then the Hittorf's transport number is

$$H = \frac{u^-}{u^+ + u^-}$$

and is therefore dimensionless. n_e is the electrochemical valence and the number 0.058 results from the expression RT/F, where F is the Faraday constant.

```
0 REM "DIFF-POT"          EBERT/EDERER           880504
1 REM ***********************************************
2 REM *** Electrochemical diffusion potential      ***
3 REM *** between two aqueous solutions of diffe-   ***
4 REM *** rent concentrations at room temperature. ***
9 REM ***********************************************
200    INPUT "Hittorf number ";H
300    INPUT "Electrochemical valence ";N
400    INPUT "Concentration 1 ";C1
500    INPUT "Concentration 2 ";C2
600    E = (2*H-1)*.058/N*LOG(C1/C2)
800    PRINT "Diff.pot. = ";E;"Volt"
63999 END

RUN
Hittorf number ? 0.4
Electrochemical valence ? 2
Concentration 1 ? 0.1
Concentration 2 ? 0.01
Diff.pot. = -1.335499E-02 Volt
Ok

RUN
Hittorf number ? 0.3
Electrochemical valence ? 1
Concentration 1 ? 1.0
Concentration 2 ? 1E-3
Diff.pot. = -.1602599 Volt
Ok
```

2.17   Effusion Velocity of a Gas

As a final example for programming equations, the effusion velocity u of an ideal gas from a container under a given pressure p_i shall be calculated. The velocity u is dependent on the pressure difference inside (p_i) and outside (p_o) the container, the molecular weight M of the gas, the absolute temperature T inside the container and, because of adiabatic expansion, on the quotient of the molar heats $\kappa = C_p/C_v$. The formula is as follows

$$u = \sqrt{\frac{2RT}{M} \cdot \frac{\kappa}{\kappa - 1} \cdot \left(1 - \frac{p_o}{p_i}\right)^{\frac{\kappa-1}{\kappa}}}$$

First, a dimensional analysis has to be carried out to determine the quantities for input and output that are required.

Recommendation:

> The calculation in the program should always be carried out in the same system of fundamental quantities (i.e. cgs). For input and output appropriate conversions should be performed.

```
0 REM "EFFUSION"          EBERT/EDERER          880504
1 REM *************************************************
2 REM *** Calculation of the effusion velocity of  ***
3 REM *** a gas from a vessel under pressure        ***
4 REM *** according to the generalized BERNOULLI-   ***
5 REM *** equation.                                 ***
9 REM *************************************************
200    INPUT "Temperature in K "; T
400    INPUT "Molecular weight "; M
500    INPUT "Cp/Cv "          ; K
510    REM *************************************************
515    REM *** Cp/Cv is the ratio of the heat         ***
518    REM *** capacities at constant pressure and    ***
521    REM *** at constant volume.                    ***
530    REM *************************************************
```

```
600   INPUT "External pressure ad. lib. " ; P2
620   INPUT "Internal pressure ad. lib. " ; P1
2000 REM *******************************************
2010 REM ***    R  =  8.31441 J/(K*mol)        ***
2020 REM *** 1  J  =  1 E7 erg                  ***
2030 REM *** 1 erg =  1 cm^2*g/(s^2)            ***
2040 REM *** L     =  6.022 E23 particles/mole  ***
2500 REM *******************************************
2600  R  = 8.31441*1E+07
4000  K9 = (K-1)/K
4050  P9 = P2/P1
4100  R9 = R*T/M
4200  U  = SQR(2/K9*R9*(1-P9^K9))/100
5000  PRINT "Effusion velocity in m/sec"
5200  PRINT "=    ";U
63999 END


RUN
Temperature in K ? 293
Molecular weight ? 28.96
Cp/Cv ? 1.402
External pressure ad. lib. ? 1.2
Internal pressure ad. lib. ? 2.4
Effusion velocity in m/sec
=     325.2045
Ok

RUN
Temperature in K ? 1000
Molecular weight ? 28.96
Cp/Cv ? 1.402
External pressure ad. lib. ? 1.0
Internal pressure ad. lib. ? 2.0
Effusion velocity in m/sec
=     600.79
Ok
```

This program contains nothing really new. So the analysis of the program
is left to the reader.

3 Series

This chapter deals with problems in physics and chemistry that can be solved mathematically by calculating series. Series are defined as sums with an infinite number of terms. Solving such problems quickly is essentially limited to personal computers. The program must be written so that the computer calculates as many terms as necessary. Intermediate results are often printed on the screen, and the operation can be terminated at any time by the user.

If the printing of intermediate results is too fast and the figures are not readable on the screen, then one can use the so called 'break', which every computer has and which slows down or stops the motion of the figures and the text.

stop:		\<CNTL>-\<NUM-LOCK>
	or	\<CNTL>-\<S>
continue:		any key

The program can be stopped by hand as soon as the results are sufficiently accurate and, while this procedure is not mathematically exact, it is frequently effective and convenient. An alternative is a sophisticated termination criterion, which must be included into the program and may also not be mathematically perfect. Program execution is discontinued by pressing the \<CNTL>-\<BREAK> keys and resumed by the command CONT which is an abbreviation of 'continue'.

New BASIC-elements are not introduced in this section. The first example is simple and can be calculated by elementary mathematics. In contrary, the third example is a rather complicated heat transfer problem using the partial differential equation of the second Fourier law. Again, the solution is given as an infinite series.

3.1   Geometric Series

Given are the initial term a and a multiplication factor q. The geometric series is defined as the following infinite series

$$S = a + aq + aq^2 + aq^3 + aq^4 + aq^5 + \ldots = \sum_{i=0}^{\infty} aq^i$$

With $|q| < 1$ the analytic formula for the sum of series with an infinite number of terms is

$$S = \frac{a}{1-q}$$

and for n terms

$$S_n = \sum_{i=0}^{n} aq^i = \frac{a(q^n - 1)}{q - 1}$$

An appropriate program reads as follows:

```
0 REM "GEO-SER "          EBERT/EDERER          880509
1 REM ************************************************
2 REM *** Calculation of the sum and the partial   ***
3 REM *** sums of a geometric series.              ***
4 REM *** S = A + A*Q + A*Q^2 + A*Q^3 + A*Q^4 + .. ***
9 REM ************************************************
100    INPUT "Initial term of the series " ; A
200    INPUT "Multiplication factor "        ; Q
300    S=0
400    H=A
500    S=S+H
600      H=H*Q
700      PRINT I;H;S
800      I=I+1
900    GOTO 500
2000   END

RUN
Initial term of the series ? 2
Multiplication factor ? 0.35
  0  .7  2
  1  .245  2.7
  2  .08575  2.945
```

```
3   .0300125  3.03075
4   1.050437E-02  3.060763
5   3.676531E-03  3.071267
6   1.286786E-03  3.074944
7   4.50375E-04  3.07623
8   1.576313E-04  3.076681
9   5.517094E-05  3.076838
10  1.930983E-05  3.076893
11  6.75844E-06  3.076913
12  2.365454E-06  3.076919
13  8.279088E-07  3.076922
14  2.897681E-07  3.076923
15  1.014188E-07  3.076923
^C
Break in 700
```

The initial term a of the series is input for variable A and the multiplication factor q is input for variable Q. Variable S always contains the intermediate sum and variable H the intermediate term, which will be added to this sum in the next step.

$$H = aq^i$$

After reading a and q in lines 100 and 200, the sum variable S is set to 0 in line 300 and the first term to be added is set to a (line 400).

```
300    S=0
400    H=A
```

In line 500 the first sum is performed

```
500    S=S+H      this is 0+A=A on the first pass.
```

In line 600 the second term is calculated

```
600    H=H*Q      this is A*Q on the first pass of the program.
```

In line 700 the running variable I, the next term and the intermediate sum S are printed. In line 800 the running variable I is increased by 1. The program jumps from line 900 back to line line 500 and the second intermediate sum is formed.

```
500    S=S+H      = A + A*Q
```

In line 600 the third term is calculated.

```
600    H=H*Q      = A*Q*Q
```

It can easily be seen that within the loop of this program the terms are calculated one after the other and added, and the intermediate sums are printed on the screen.

The section of the program between lines 500 and 900 is called an 'infinite' loop.

The example chosen converges relatively fast and therefore the program can be terminated after a few passes. If the screen is chosen for the output, the run can be terminated as soon as the result does not change any more.

The alert reader might notice that in the program GEO-SER no initial value is assigned to the variable I. I is just increased in line 800 by 1. This is careless and should generally be avoided. In BASICA however this is not very important as at the command RUN all variables are set to 0 before the program is started.

Problem 12

Alter the program GEO-SER to calculate the following series

$$S = a + aq + \frac{aq^4}{4} + \frac{aq^9}{9} + \frac{aq^{16}}{16} + \ldots + \frac{aq^{i^2}}{i^2} + \ldots = a + \sum_{i=1}^{\infty} \frac{aq^{i^2}}{i^2}$$

Try to calculate some other series, especially those which have an analytical solution — you may find them in mathematical textbooks — and compare both solutions. Examples are

$$\sum_{i=1}^{\infty} \frac{1}{i^2} = \frac{\pi^2}{6} \quad \text{or} \quad \sum_{i=1}^{\infty} \frac{1}{i^4} = \frac{\pi^4}{90} \quad \text{or} \quad \sum_{i=1}^{\infty} \frac{1}{i} \cdot (-1)^{i+1} = \log 2$$

3.2   Fourier Series

Fourier series are used in many areas of natural and engineering sciences, e.g. to solve differential equations (see the next example in Chapter 3.3) and to display periodic functions such as the saw tooth in electronics. They are infinite series which are composed of trigonometric functions whose periods become continuously smaller. A Fourier series which only contains sines, is shown in a general form.

$$f(x) = a_1 \sin(1x) + a_2 \sin(2x) + a_3 \sin(3x) + \ldots + a_i \sin(ix) + \ldots$$

$$= \sum_{n=1}^{\infty} a_n \sin(n \cdot x)$$

The coefficients a_n are frequently a function of n. Our first example for a Fourier series is the Fourier description of a symmetric zig-zag curve. The formula for such a Fourier series is

$$f(x) = 1 + \sum_{n=1}^{\infty} \left(\frac{2}{n\pi}\right)^2 \cdot \Big(\cos(n\pi) - 1\Big) \cdot \cos\left(\frac{n\pi x}{2}\right)$$

The program for this rather complex formula is short and simple:

```
0 REM "FOU-SER "            EBERT/EDERER            880509
1 REM ************************************************
2 REM *** Calculation of the sum and the partial    ***
3 REM *** sums of a FOURIER series.                  ***
4 REM *** A zigzag function is described as a sum    ***
5 REM *** of cosines.                                ***
9 REM ************************************************
10     PI=4*ATN(1)
100    INPUT "Which x shall be tried ";X
200    F=1
300    N=0
400    N=N+1
500      H=4/N/N/PI/PI*(COS(N*PI)-1)
550      H=H*COS(N*PI*X/2)
600      F=F+H
700      PRINT N;X;F;H
800    GOTO 400
2000   END

RUN
Which x shall be tried ? 1.85
 1  1.85  1.788173  .7881733
 2  1.85  1.788173  0
 3  1.85  1.856658  6.848468E-02
 4  1.85  1.856658  1.774882E-09
 5  1.85  1.869066  1.240769E-02
 6  1.85  1.869066  0
 7  1.85  1.867768  -1.297862E-03
 8  1.85  1.867768  0
```

```
 9   1.85   1.862539  -5.228624E-03
10   1.85   1.862539   0
11   1.85   1.856827  -5.71176E-03
12   1.85   1.856827  -3.190904E-10
13   1.85   1.852046  -4.781481E-03
14   1.85   1.852046   0
15   1.85   1.848717  -3.328311E-03
16   1.85   1.848717   0
17   1.85   1.846896  -1.821544E-03
^C
Break in 550
```

The infinite loop runs between the lines 400 and 600, the intermediate sum is calculated in line 600 and stored in variable F, and in lines 500 and 550 the term associated with the next n is calculated.

In testing this program one realizes that for certain x-values it converges relatively fast, but for other x-values it converges rather slowly. The convergence is the slower the nearer the x-values approach the peaks.

The program is evidence that it is possible to approximate complicated periodic functions by the sums of sine or cosine functions with different periods. This is the famous Fourier transformation which is preferently applied for the evaluation of data from analytical equipment such as NMR and IR, and which generally produces excellent results.

Problem 13

Try to program the Fourier series for the following rectangle function

$$f(x) = \frac{4}{\pi} - \sum_{n=1}^{\infty} \frac{1}{2n-1} \cdot \sin\left(\frac{(2n-1)\pi x}{L}\right)$$

and the saw tooth function

$$f(x) = \frac{2}{\pi} - \sum_{n=1}^{\infty} \frac{(-1)^{n+1}}{n} \cdot \sin\left(\frac{n\pi x}{L}\right)$$

The parameter L which appears in both formulas is the length of a half period. If the program is completely analogous to the program FOU-SER, L can be set to 1 and can therefore be 'ignored'.

3.3   Heat Transfer According to the Second Fourier Law

Transport phenomena — such as diffusion, heat transfer, momentum transfer, etc. — can be described in the stationary case by the so called 'first
laws'. The second laws which describe the more complicated local and time
variations (e.g. temperature) are partial differential equations.[1,2]

$t < 0$: $-5\,°C$ [   ] $+25\,°C$

$t \geq 0$: $+10\,°C$ [   ] $+40\,°C$

The physical problem is as follows: a homogeneous rod of 3 cm length kept
at the left end $(L = 0)$ at $-5\,°C$, and at the right end $(L = 3)$ at $+25\,°C$.
After a sufficiently long time — assuming homogeneous properties — a linear
temperature gradient will be established. For time $t = 0$ the temperature
profile can be expressed by

$$T = 10L - 5$$

At time $t = 0$ both ends are suddenly increased to the temperatures of $10\,°C$
at $L = 0$ and $40\,°C$ at $L = 3$, and kept at those values. Question: How does
the temperature T depend on position L and time t?

The mathematical evaluation follows the second Fourier law

$$\frac{\partial T(t, L)}{\partial t} = k \cdot \frac{\partial^2 T(t, L)}{\partial L^2}$$

and the boundary conditions mentioned above.

The solution for such problems is given in many standard textbooks.
It is a Fourier series which on paper gives as little an answer to the real
physical problem as the original differential equation. The time t and position L dependence of the temperature $T(t, L)$ of our rod is described by
the following Fourier series: Numerical results of this equation are difficult
if not impossible to obtain without a computer.

---

[1] R.B. Bird, W.E. Stewart, E.N. Lightfoot, *Transport Phenomena*, New York,
London (1960), Wiley & Sons Inc.
[2] H.S. Carslow, J.C. Jaeger, *Conduction of Heat in Solids*, Oxford University
Press (1959)

$$T = 10 + 10L + \sum_{n=1}^{\infty} \frac{30}{n\pi} \Big(\cos(n\pi) - 1 \Big) \cdot e^{-\frac{2n^2\pi^2 t}{3 \cdot 3 \cdot k}} \cdot \sin\left(\frac{n\pi L}{3}\right)$$

This equation can be programmed in the same scheme as the above program FOU-SER.

```
0 REM "HEATCOND"          EBERT/EDERER          880516
1 REM ****************************************************
2 REM *** Calculation of the temperature of a rod  ***
3 REM *** as a function of time and position       ***
4 REM *** using a FOURIER series.                  ***
6 REM *** Temperature at the left  end 0 cm is 10C ***
7 REM *** Temperature at the right end 3 cm is 40C ***
8 REM *** Time < 0 ; T is raised from -5C to 25C   ***
9 REM ****************************************************
10      PI=4*ATN(1)
100     PRINT "Calculation of temperature"
200     INPUT "at which position x " ; X
300     INPUT "at which time      t " ; Z
400     T=10*X+10
500     N=0
600     N=N+1
700       H=30/N/PI*(COS(N*PI)-1)
800       H=H*EXP(-2*N*N*PI*PI*Z/9/100)
900       H=H*SIN(N*PI*X/3)
1000      T=T+H
1100      PRINT N;T;H
1200    GOTO 600
2000    END

RUN
Calculation of temperature
at which position x ? 2
at which time      t ? 4
 1   14.84935 -15.15065
 2   14.84935  0
 3   14.84935 -1.883697E-06
 4   14.84935  6.055395E-08
 5   15.21837  .369018
 6   15.21837  0
 7   15.18626 -3.210003E-02
 8   15.18626  0
^C
Break in 700
```

The infinite loop of the program proceeds from line 600 to line 1200. The assigned terms of the Fourier series are computed in lines 700, 800 and 900 (because of the considerable length of the equation it has been split up and distributed over three lines). The numerical value of the heat transfer coefficient k was set to 100.

Problem 14

Use the program HEATCOND to calculate the temperature profiles at times $t = 5$ and $t = 50$ and make drawings of these profiles. Calculate the variation of the temperature with time at the positions $L = 1.5\,\text{cm}$ and $L = 0.2\,\text{cm}$ and plot these time dependences.

3.4   Partition Functions

As an example we will write a program for the rotational partition functions. In treating statistical thermodynamics in physical chemistry textbooks, rotational partition functions of a harmonic oscillator are frequently calculated as an example, and from this the corresponding thermodynamic quantities are derived. This example shows that apparently complicated problems can be solved using simple numerical methods.

As a reminder: if e_i represents the different energy states of a molecule, then the formula for the partition function q is

$$q = \sum_i e^{-\frac{e_i}{kT}}$$

where $\sum_i$ means summation over all quantum states. The different rotational energy states of a diatomic molecule are defined as

$$e_{\text{rot}}(J) = J \cdot (J+1)\frac{h^2}{8\pi^2 I}$$

J : Rotational quantum number ($J = 0, 1, 2, 3 \ldots$)
I : Moment of inertia, $I = \mu r^2$ for a diatomic molecule
h : Planck's constant

The degeneracy for the energy state J is $2J + 1$.

The formula for the rotational partition function is given by

$$q_{\text{rot}} = \sum_{J=0}^{\infty} (2J+1)\, e^{-\frac{J(J+1)h^2}{8\pi^2 I \cdot kT}}$$

k :   Boltzmann constant
I :   Moment of inertia, $I = \mu r^2$ for a diatomic molecule
r :   Distance between the atoms
μ :   Reduced mass $\left(\frac{m_1 \cdot m_2}{m_1 + m_2}\right)$
m_i :   Mass of atom i (atomic weight/Avogadro's constant)

The rotational partition function is dependent on the temperature, the masses of the atoms and the distance between the atoms. The rigid rotator neglects the increase of the atomic distance at higher quantum numbers. For this effect to be considered, a single line has to be added to the program in which the distance r is calculated using the quantum number J.

```
0 REM "ROTATION"          EBERT/EDERER          880509
1 REM ************************************************
2 REM *** Calculation of the partition function of ***
3 REM *** a rigid rotator.                         ***
4 REM *** e(rot) = J(J+1)*h*h/(8*pi^2*I)           ***
5 REM *** h      = 6.626176 E-34 J*sec             ***
6 REM *** I      = momentum of inertia = m*r^2     ***
7 REM *** For diatomic molecules the degeneration  ***
8 REM *** factor = 2*J + 1 .                       ***
9 REM ************************************************
100    PI = 4*ATN(1)
1000   H  = 6.626176E-34
1050   L  = 6.022E+23
1100   K  = 1.38066E-23 : REM in J/Kelvin
1200   INPUT "Temperature in K ";       T
1300   INPUT "Atomic weight 1 in amu ";M1
1400   INPUT "Atomic weight 2 in amu ";M2
1500   INPUT "Distance in Angstroem "; A
2000   REM Calculation of the momenta of inertia
2100   M=M1*M2/(M1+M2)
2200   I=M/1000*(A*1E-10)^2
3000   S=0
3100   J=0
3200   E=J*(J+1)*H*L/(8*PI*PI*I)*H
3300     Q=(2*J+1)*EXP(-E/K/T)
```

```
3400     S=S+Q
3500     PRINT
3510     PRINT "Number of the level    ="; J
3520     PRINT "Energy level           ="; E;"J"
3530     PRINT "Frequency of occupation ="; Q
3540     PRINT "Partition function     ="; S
3800     J=J+1
3900  GOTO 3200
20000 END
RUN
Temperature in K ? 1000
Atomic weight 1 in amu ? 14
Atomic weight 2 in amu ? 16
Distance in Angstroem ? 1.146

Number of the level    = 0
Energy level           = 0 J
Frequency of occupation = 1
Partition function     = 1

Number of the level    = 1
Energy level           = 6.829846E-23 J
Frequency of occupation = 2.985196
Partition function     = 3.985196

Number of the level    = 2
Energy level           = 2.048954E-22 J
Frequency of occupation = 4.926346
Partition function     = 8.911542

Number of the level    = 3
Energy level           = 4.097907E-22 J
Frequency of occupation = 6.795287
Partition function     = 15.70683
       ...              ...
Number of the level    = 80
Energy level           = 2.21287E-19 J
Frequency of occupation = 1.762455E-05
Partition function     = 404.6356

Number of the level    = 81
Energy level           = 2.268192E-19 J
Frequency of occupation = 1.195259E-05
Partition function     = 404.6356
^C
Break in 3530
```

An infinite loop runs in this program from line 3200 to line 3900. In line 2100 the reduced mass and in line 2200 the moment of inertia are calculated. The calculation is performed in the meter–kilogram–seconds–system to use the Joule as an energy unit. (Why is in line 2200 a division by 1000 carried out?) In line 3200 the energy states corresponding to the rotational quantum numbers are evaluated. In line 3300, finally, the terms of the partition functions, which were multiplied by the degeneracy, are calculated. The intermediate sum is found in line 3400 and the results are printed as a block in lines 3510 to 3540. Next, the quantum number is increased by 1 and the program jumps to line 3200 to start the calculation of the next energy state.

To calculate the rotational partition function with specified input parameters, the program should be running until the results cease to vary noticeably.

Problem 15

From the partition function q_{rot}, the contribution of the rotator to the internal energy U_{rot} can be calculated according to the following formula

$$U_{rot} = kT^2 \frac{\partial \log q_{rot}}{\partial T}$$

Determine the rotational portion of the internal energy of the H–D molecule (bond-length $0.7436\,\text{Å}$) as a function of the temperature and plot this dependence as a graph (U_{rot} approaches the value RT at high temperatures). Use the following approximation

$$\frac{\partial \log q_{rot}}{\partial T} \approx \frac{\log q_{rot}(T_1) - \log q_{rot}(T_2)}{T_1 - T_2}$$

Calculate the partition function for two adjacent temperatures (for example with one degree difference) using the above approximation.

To solve this problem the program ROTATION can be used. This implies no additional effort in programming, but it is laborious because the different quotients and the formula for U_{rot} must be calculated by means of a pocket calculator. Modify the computer program so that the internal energy is calculated directly.

Problem 16

Write a program for calculating the partition functions of a harmonic and an inharmonic oscillator and compare the results.

Problem 17

Calculate the partition functions and determine their contributions to the thermo-dynamic functions (as e.g. S, U, C_p; the formulas may be found in physical chemistry textbooks). If you require the derivative of the partition function with respect to temperature, calculate the partition function for two adjacent temperatures and form the difference quotient using the approximation given in Problem 15.

Problem 18

q_{rot} for a diatomic gas may be derived by an approximated integration instead of an exact summation. The following equation applies

$$q_{rot} = \frac{8\pi^2 I \cdot kT}{h^2}$$

Extend the program ROTATION to obtain as output in line 3550 the intermediate sum of the exact partition function as well as the approximate value of the above formula. Compare both results as a function of temperature and plot both partition functions in a diagram.

3.5   Monte-Carlo Method to Calculate the Archimedes Number

Another example of the application of infinite loops is the Monte-Carlo method. A new BASIC-element, the IF-statement, is needed, which will be explained in detail later (Chapter 4.6). As the name implies, the Monte-Carlo method is based on chance calculations. Let us consider the following graph.

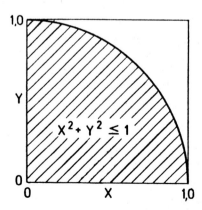

To shoot random points into the square as a target, accidental x- and y-values are needed between 0 and 1. These values are generated in line 300 and 400 by the RND-functions. The probability of hitting the shaded section of the circle is equal to the quotient w of the area of the quarter circle to the square area.

$$w = \frac{r^2\pi}{4r^2} = \frac{\pi}{4}$$

```
0 REM "PI      "          EBERT/EDERER          880511
1 REM ************************************************
2 REM *** Monte-Carlo-Calculation of the constant  ***
3 REM ***                 PI                        ***
9 REM ************************************************
50     RANDOMIZE INT(TIMER/3)
100    S=0
200    N=0
300    X=RND
400      Y=RND
600      N=N+1
700      IF X*X+Y*Y>1 THEN GOTO 300
800      S=S+1
1100     PRINT N,S,S/N*4
1200   GOTO 300
1300   END
```

An infinite loop runs from line 300 to line 1200. The variable N counts the number of the shots to the target, the variable S contains the number of the successful shots which lie inside the quarter circle.

The decision whether a point is successful or not is made in line 700. The shot is outside the quarter circle if

$$x^2 + y^2 > 1$$

and line 800, in which the successful results are counted, is omitted. According to the above formula the number of successful shots must be divided by the total number of shots and multiplied by 4 to obtain an approximate value of the quantity π.

We will use Monte-Carlo methods for other applications later. Usually, these methods are rather simple to program. The programs can often be developed directly from the physical model and no extensive mathematical analysis is needed. A disadvantage of this method can be seen from this little example. The accuracy of the results improves only very slowly. The reason for this is the fluctuations which are caused by this random method. If n random tries are made, the accuracy is proportional to the inverse of the square root of n.

$$\text{accuracy} \sim \frac{1}{\sqrt{n}}$$

4   Loops

This chapter deals with different problems from physics, chemistry and mathematics which can be solved by finite program loops.

4.1   The Calculation of Sums

Let us consider as an example the addition of squared integers, a problem which was already presented in the introductory chapter.

$$S = 1^2 + 2^2 + 3^2 + 4^2 + \ldots + (n-1)^2 + n^2 = \sum_{i=1}^{n} i^2$$

The flow diagram shown in Fig. 4.1 gives a structure for the computational operations. The flow diagram on the left corresponds to the second flow diagram in Chapter 1.2 (Fig. 1.2). An n-fold summation uses a counter I, which is increased by one during each pass and is compared to the given number n. As long as I is smaller than or equal to n, the program jumps back to start a new loop run. As soon as I becomes larger than n the program proceeds to print the results and the program ends.

In the flow diagram on the right a loop element is introduced which does essentially the same job. It consists of two boxes with the keywords FOR and NEXT. In BASIC this kind of loop is often called the FOR-NEXT-loop (in FORTRAN or PL1 this element is called a DO-loop because other keywords are used). The FOR-box tells which variable (in the example, I) will run from which initial value to which final value (in the example, from 1 to N) and in which step size (keyword STEP). According to these specifications the

With IF-statement   With FOR-NEXT-loop

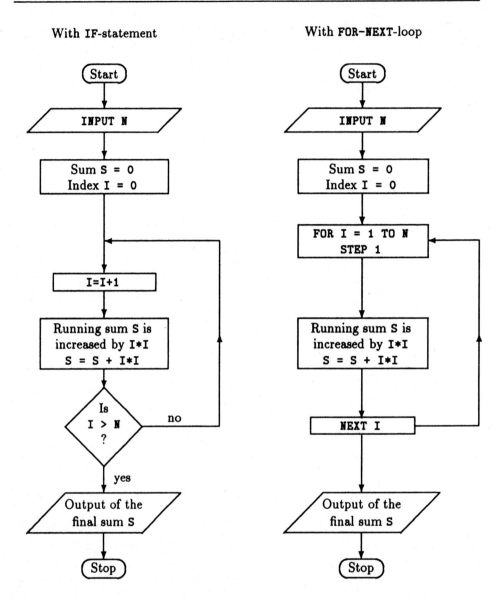

Fig. 4.1. Two flow diagrams for calculating the sum of n squared integers

number of loop passes is set. For each pass all the statements between the
FOR- and the NEXT-statement are executed. When the FOR-variable exceeds
the final value, control is passed to the first statement following the NEXT.

FOR-NEXT-loops lead to shorter and clearer programs, and therefore the
recommendation:

> Favor the use of FOR-NEXT-loops
> and avoid GOTO-statements.

The FOR-NEXT-Loops

> FOR-NEXT

Syntax: z1 FOR V=a1 TO a2 STEP a3
              any number of arbitrary statements
          z2 NEXT V

V : variable, called the loop or running variable
a : arithmetic expressions, variables or constants
a1 : initial value of variable V
a2 : final possible value of V
a3 : step size, the variable V is varied with
z : line numbers

Between lines z1 and z2 (the FOR- and NEXT-lines) any number of lines and
any kind of statement can be written. Just as a loop always begins with a
FOR statement, it always ends with a NEXT-statement. The NEXT-statement
closes a loop and it consists of a line number, followed by the keyword NEXT,
followed by a running variable name.

Examples: 10 FOR I=1 TO 15
            20   PRINT I, SQR(I)
            30 NEXT I
            40 END

```
100 FOR T5=10 TO -10 STEP -2.5
110   FOR A=0 TO 20 STEP 3
120     PRINT T5 , A
130   NEXT A
140 NEXT T5
150 END

500 X=2
520 FOR Y=X*X TO 10*SQR(X) STEP X/4
540   PRINT Y , LOG(Y)
560   X=X+SIN(Y)
580   PRINT X
600 NEXT Y
650 END
```

In the second example one loop encloses another. The maximum number of embedded or nested loops is dependent on the operating system of the computer and on the BASIC-interpreter or compiler which is used, but is generally large enough for most problems. However, examples may be constructed that reach the limits of BASIC. The reader should try to find this out.

As far as structure is concerned, loops can be compared with boxes. They are packed together or nested in two ways only. Either they can follow each other or they can be nested. There must never be an overlap. These restrictions can be summarized in the following rules:

- Each nested loop must begin with its own FOR-statement and end with its own NEXT-statement.
- An outer loop and an inner (nested) loop cannot have the same running variable.
- Each inner (nested) loop must be completely embedded within an outer loop (i.e., the loops cannot overlap).
- Control can be transferred from an inner loop to a statement in an outer loop or to a statement outside of the entire nest. However control cannot be transferred to a statement within an inner loop.

> The loop is a most effective instrument of programming. Use it generously!

The program for the above summation problem reads as follows:

```
0 REM "SUM    "          EBERT/EDERER          880514
1 REM ***************************************************
2 REM *** Calculation of the sum of the first n    ***
3 REM *** squared integer numbers.                 ***
4 REM *** The number n is arbitrary.               ***
9 REM ***************************************************
100    PRINT "Calculation of the sum of the first n"
200    PRINT "squared integer numbers"
300    INPUT "  n    ";N
400    PRINT:PRINT "This will need approx";N*.002;"sec"
1000   S=0
1100   FOR I=1 TO N
1200     S=S+I*I
1300   NEXT I
2000   PRINT "Sum = ";S
2100   END

RUN
Calculation of the sum of the first n
squared integer numbers
    n    ? 120

This will need approx .24 sec
Sum =   583220
Ok

RUN
Calculation of the sum of the first n
squared integer numbers
    n    ? 1000

This will need approx 2 sec
Sum =   3.338332E+08
Ok
```

The central loop in which the sum is calculated is written in lines 1100–1300. In line 2000 the result is printed. In line 1000, just before the start of the loop, the variable S, which is used to store the intermediate sum, is set to 0.

As the loop runs from 1 to N, the value of N has to be entered, which happens in line 300. In front of this line in the program there is a PRINT-command which serves to give information to the user during the run. The output of line 400 is a special service of the program: an estimation is carried out as to how much program running time is needed to produce the result. Program runs which take a longer time, make some users 'nervous' if nothing happens on the screen. To avoid this, the output of intermediate results, running indices of loops or of calculating times are suitable possibilities to shorten waiting times in front of the screen and to present additional useful information to the user.

Problem 19

Write a program (or rewrite the program SUM) to calculate the following sums.

$$S = \sum_{i=1}^{20} i^i \qquad\qquad S = \sum_{x=-10}^{30} \log(|\sin x^2| + 1)$$

$$S = \sum_{i=1}^{1000} \frac{1}{i} \qquad\qquad S = \sum_{x=-100}^{100} 1 + e^{-\frac{x+0.5}{100}}$$

$$S = \sum_{i=0}^{500} \frac{1}{1+i+i^3} \qquad\qquad \text{and other sums, like:}$$

$$S = 0.5^{0.5} + 1^1 + 1.5^{1.5} + 2^2 + 2.5^{2.5} + \ldots + 6^6$$

Remember: the absolute value of x is formed by the BASIC standard function ABS(X).

Problem 20

Write a program for computing finite products, i.e. the factorial function. The factorial of the positive integer n equals the product of all positive integers up to and including n.

$$n! = 1 \cdot 2 \cdot 3 \cdot \cdots \cdot (n-1) \cdot n = \prod_{j=1}^{n} j$$

The definition is supplemented by the value

$$0! = 1$$

The factorial may also be defined as recurrence relation

$$0! = 1 \qquad\qquad (n+1)! = n!(n+1)$$

4.2   The Compilation of Tables – Molecular Velocity

For many programmed equations not just one value is required, but several. For reasons of clarity and comparability the results should be arranged in the form of tables. As an example, the average velocity $\bar{u}$ of gas molecules — characterized by the molecular weight M — should be listed as a function of absolute temperature T. The corresponding formula is:

$$\bar{u} = \sqrt{\frac{3RT}{M}}$$

R : $8.314\,[\mathrm{JK^{-1}mol^{-1}}] = 8314\,[\mathrm{g\,m^2K^{-1}mol^{-1}s^{-2}}]$
T : Absolute temperature in K
M : Molecular weight in g/mol

Putting the values for T and M in the specified dimensions into the equation, then the average molecular velocity $\bar{u}$ is obtained in $\mathrm{ms^{-1}}$.

```
0 REM "MOL-VEL "          EBERT/EDERER          880509
1 REM ***************************************************
2 REM *** Setting up a table with mean velocities  ***
3 REM *** of gas molecules for different substan-  ***
4 REM *** ces and different temperatures according ***
6 REM *** to the formula:   v = (3RT/m)^.5         ***
9 REM ***************************************************
100    DATA "Hydrogen",  2
200    DATA "Benzene",  78
300    N=2
350    PRINT
400    PRINT "Mean molecular velocity in m/sec"
450    PRINT
500    PRINT "Temperature  ";
600    R=8314.34
1000   FOR I=1 TO N
1100     READ A$,X
1200     PRINT A$;" ";
1300   NEXT I
1400   PRINT
3000   FOR T=150 TO 600 STEP 50
3050     PRINT T;"         ";
3100     RESTORE
```

```
3200    FOR I=1 TO N
3300      READ A$,X
3400      V=(3*R*T/X)^.5
3500      PRINT V;
3600    NEXT I
3700    PRINT
3800  NEXT T
4000  END


RUN


Mean molecular velocity in m/sec

Temperature  Hydrogen Benzene
 150         1367.745  219.0145
 200         1579.336  252.896
 250         1765.751  282.7464
 300         1934.284  309.733
 350         2089.264  334.5502
 400         2233.518  357.6491
 450         2369.004  379.3442
 500         2497.149  399.8638
 550         2619.033  419.3806
 600         2735.49   438.029
Ok
```

First, look at the program MOL-VEL without taking notice of the new BASIC-elements. The calculation of the average molecular velocity takes place in line 3400. This line is surrounded by two loops nested in one another. The external T-loop includes lines 3000 to 3800. In this loop the value of temperature is increased by 50 degrees centigrade during each pass. The temperatures are printed together with seven blanks on the screen in line 3050. The semicolon in this line causes the next output to continue on the same line. The internal I-loop runs from line 3200 to 3600. The variable I counts the different substances. In line 3300 the name of the ith substance and its molecular weight is read in — the statements used will be explained a little later. After the calculation the resulting molecular velocity is printed in line 3500, again without a line feed. If, for a certain temperature, the velocities of the N substances have been calculated and printed, the I-loop is finished and the blank PRINT-command causes the output device to begin a new line.

This program does not contain an INPUT-command. The number of substances N is assigned by line 300. The names of the individual substances and their molecular weights were assigned to variables in a new input mode.

Input of Data Which Are Included in the Program

```
DATA
READ
RESTORE
```

Syntax: z1 DATA c1, c2, c3 ..., cn
         z2 READ V1, V2, V3 ...,VM
         z3 RESTORE

z : line numbers
c : constants — numbers or strings
V : variables

By means of DATA-statements, which can be placed in any positions in the program, data representing constants (numbers or text) can be written into the program. The computer contains an internal pointer for these data. With the command READ V a datum is read into the variable V and the internal pointer jumps to the next data position. The purpose of the DATA-statement is to assign appropriate values to the variables listed in the READ-statement. The DATA-statement is comprised of a statement number, followed by the keyword DATA, followed by a set of numbers and/or strings, separated by commas. Each number and/or string in a DATA-statement must correspond to a variable of the same type in a READ-statement.

The command RESTORE puts the internal pointer to the first data in the program. In the example below five numbers are first summed up in variable S and then the product of the same five numbers is calculated in variable P.

```
10   DATA 3, 5, 2.5, 7, 8.1
15   S=0
20   FOR I=1 to 5
30   READ X
40   S=S + X
50   NEXT I
```

```
60   P=1
70   RESTORE
80   FOR I=1 to 5
90   READ Y
100  P = P*Y
110  NEXT I
120  PRINT S,P
```

In the program MOL-VEL there are two DATA-assignments, in the lines 100 and 200, each with two constants. The first constant contains the name of the substance and the second its molecular weight. The text constant (string) within the DATA-statement must be written in quotation marks. A string is a sequence of characters including letters, numbers, blanks or certain special characters. The number constants are written without quotation marks in the DATA-assignment.

A text or string cannot be written into number variables, but there is another type of variable, the so-called text-, string- or $-variable, to handle this.

Text-, String- or $-Variables

$

Syntax: NAME$

NAME   :   valid BASIC-name

A string variable must be written as a variable name directly followed by a dollar sign (e.g. B$, R4$, K0$). A text can now be assigned to such a string variable. To input text in such a variable, the INPUT- or READ-command may be used, for instance INPUT A$ or READ A$. Using the PRINT-statement the text can be put onto the screen, e.g. PRINT A$.

In the program MOL-VEL there is another loop from line 1000 through 1300. I runs from 1 to N in increments of 1. Because N=2 in our example program, it runs only from 1 to 2. In the first pass the text HYDROGEN is put into the text variable A$ and later, in line 1200, is printed on the screen. At the same time, the number 2 — the molecular weight of hydrogen — is stored in the variable X. The internal pointer points now to the constant BENZENE. In the second pass, BENZENE and 78 are read into the variables A$ and X respectively, and in line 1200 A\$ is put on the screen. This loop, together with line 500 and the line feed in 1400, serves to print the head of the table. RESTORE in line 3100 brings back the internal pointer to the first constant (i.e. HYDROGEN). Then the real calculation starts and the results are printed out continuously in a table. If tables with other or more substances should be generated, the DATA-assignments in lines 100 and 200 and, if necessary, the number N in 300, may be changed.

Problem 21

Change the program MOL-VEL so that a table for three substances of your choice is generated, in which the corresponding values of the average molecular velocities at temperatures decreasing from 750 to 150 K in increments of 25 °C are compiled.

Problem 22

Write a program to set up a table of rate constants k, which have been calculated by the Arrhenius equation. The table should contain the rate constants at ten different temperatures and three different activation energies E_a.

$$k = k_0 \cdot e^{-\frac{E_a}{RT}}$$

Use for example $k_0 = 10^{13}\,\text{s}^{-1}$ as the value for the pre-exponential factor of a first order reaction rate constant.

4.3   Mean Values and Standard Deviation

A further example of the use of finite loops is the calculation of statistical quantities as mean values and standard deviations. The data should be put in again using the DATA- and READ-statements. This has the advantage that the input data are incorporated into the program and therefore can be called upon at any time, checked and, if necessary, modified.

For n experimental data x_i the arithmetic mean value $\bar{x}$ is given by

$$\bar{x} = \frac{x_1 + x_2 + \ldots + x_{n-1} + x_n}{n} = \frac{1}{n} \sum_{i=1}^{n} x_i$$

and the standard deviation s by

$$s = \sqrt{\frac{1}{n-1} \sum_{i=1}^{n} (x_i - \bar{x})^2}$$

The program for the calculation of these values is called **MEAN**.

```
0 REM "MEAN    "          EBERT/EDERER        880514
1 REM *************************************************
2 REM *** Calculation of the mean value  and the    ***
3 REM *** standard deviation of values. Those       ***
4 REM *** values are written in a DATA-statement    ***
5 REM *** beginning with line # 10000.              ***
6 REM *** N is the number of values and it is       ***
7 REM *** defined in line # 100.                    ***
9 REM *************************************************
100    N=9
200    M=0
300    FOR I=1 TO N
400      READ X
500      M=M+X
600    NEXT I
700    M=M/N
900    RESTORE
1000   S=0
1100   FOR I=1 TO N
1200     READ X
1300     S=S+(M-X)^2
1400   NEXT I
2000   S=SQR(S/(N-1))
2500   PRINT "Mean value         = ";M
3000   PRINT "Standard deviation = ";S
10000 DATA 5,5.5,3,7.2,8.1,5.95
10100 DATA 4.7,4,6.05
20000 END

RUN
Mean value         =  5.5
Standard deviation =  1.56305
Ok
```

The number of experimental values n is written in line 100. The quantities of the values x_i to be averaged are put into the program by the DATA-assignments in lines 10000 and 10100. During the running of the program, the loop from lines 300 to 600 forms the sum of all data by reading them into the variable X one by one using the statement READ X, and finally adding the data up in the variable M. The sum of all data is then divided by N and the result is assigned to the variable M (line 700). Therefore beginning with line 700 the arithmetic mean value is stored in the variable M.

In the loop from line 1100 through 1400 the sum of the squares of the deviations $(\bar{x} - x_i)^2$ is calculated. At the beginning of this loop the internal pointer is set to the first figure by RESTORE, and the variable S, in which the summing takes place, is set to 0 (line 1000). In line 2000, behind the loop, the sum of the squares of the deviations is divided by $n - 1$, then the root is calculated and the result assigned to the variable S. Finally the mean value and the standard deviation are printed on the screen.

Note:

If one and the same quantity has been measured frequently, and the mean value and the standard deviation have been calculated, the number n of experimental measurements has to be supplemented to the results. This is necessary to be able to judge the quality of both statistical quantities.

Problem 23

Modify the program MEAN (or write it new) so that weighted mean values and the standard deviation can be calculated. Weighted values mean, that to each data x_i, a weighting factor w_i is attached which may be considered to be proportional to the quality of the measured value x_i.

$$\bar{x}_{\text{weighted}} = \frac{\sum\limits_{i=1}^{n} w_i x_i}{\sum\limits_{i=1}^{n} w_i}$$

$$s_{\text{weighted}} = \sqrt{\frac{\sum\limits_{i=1}^{n} w_i (x_i - \bar{x})^2}{\sum\limits_{i=1}^{n} w_i}}$$

Problem 24

Calculate — using the program **MEAN** — the mean value and the standard deviation of the following series of data:

a)   17.1, 23.2, 21.1, 19.8, 19.3, 20.0, 20.3

b)   456, 444, 445, 411, 456, 438, 476, 435, 455, 444, 429

Include in the output of the program, as mentioned above, the number of experimental measurements.

Problem 25

Write a program for the calculation of mean values and standard deviations and test the measured values for possible outliers. It does not have to be a sophisticated test for outliers, it is sufficient to look for the value with the largest deviation and state the distance from the mean value. For this task the IF-command should be used which is explained in more detail in Chapter 4.6.

Problem 25a

Write a program for the calculation of the geometric mean x_g.

$$x_g = \sqrt[n]{\prod_{i=1}^{n} x_i}$$

This equation may be converted into:

$$\log x_g = \frac{1}{n} \sum_{i=1}^{n} \log x_i$$

Test the program with the set of the following data:

1 000 344 ,   999 999 ,   1 111 111 ,   989 989 ,   1 101 101 ,   1 063 636 ,
953 539 ,   1 020 411 ,   973 377 ,   1 055 505

4.4   Monte-Carlo Integration

For the solution of multiple integrals, Monte-Carlo integrations are frequently superior to the usual numerical methods such as Euler's, Simpson's, etc. which are treated in some later examples.

Let us consider first a problem in one dimension, viz. the determination of the definite integral of a function $f(x)$ in a certain interval of the x-axis. The area of the function $f(x)$ over the interval (a, e) can be expressed as the area of the rectangle with the mean height $\bar{h}$ of the function $f(x)$ in the interval (a, e) (Fig. 4.2).

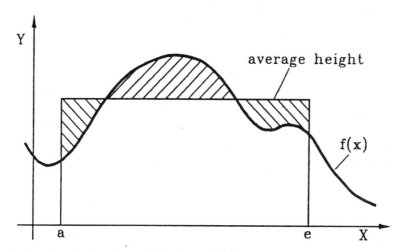

Fig. 4.2. Graph of a function $f(x)$ whose definite integral between a and e will be estimated by a Monte-Carlo procedure that produces an average of the function values

The mean height $\bar{h}$ shall be calculated by a Monte-Carlo method by chosing random values of x_{rnd} in the interval (a, e) and calculating for these points the values of the function $f(x_{\mathrm{rnd}})$ and summing them up. After division by the number of random trials, the mean height $\bar{h}$ is obtained. This calculated value approaches the true mean value the better the more random values are used.

If we want to calculate the integral of the function of two independent variables $f(x, y)$ over an area in the (x, y)-plane, we choose random points in this area, determine the function values for these (x, y)-values, form the average function value, and multiply this mean function value by the area over which the integration was performed. Such a two-dimensional example is the function $g(x, y)$ which describes the density of daisies in a meadow. If one solves the definite integral of the density function over a certain area of the meadow, the number of daisies in this area is obtained.

The program `MC-INT` contains the integration of a three-dimensional function $f(x, y, z)$ over a three-dimensional box. For illustration, we consider $f(x, y, z)$ as a function, which describes the mass density distribution in a three-dimensional space. The integration over a box (three-dimensional region) implies the determination of the mass within the three-dimensional box.

The three-dimensional function whose definite integral is to be calculated is:

$$f(x, y, z) = x^2 + y^2 + z^2$$

This function was chosen because it is possible to solve the integral analytically; this result can then be compared with the numerical result.

$$\int_{x_a}^{x_e} \int_{y_a}^{y_e} \int_{z_a}^{z_e} (x^2 + y^2 + z^2) \, dx \, dy \, dz = \frac{xyz}{3}(x^2 + y^2 + z^2) \left.\right|_{x_a}^{x_e} \left.\right|_{y_a}^{y_e} \left.\right|_{z_a}^{z_e}$$

In this function the upper and lower limits (with positive and negative signs) have to be inserted, in the same way as in the one-dimensional integration. The program `MC-INT` for a Monte-Carlo integration of this function is as follows:

```
0 REM "MC-INT  "          EBERT/EDERER           880518
1 REM ************************************************
2 REM *** Monte-Carlo integration of a multi-      ***
3 REM *** dimensional function.                    ***
4 REM *** e.g. f(x,y,z) = x*x + y*y + z*z          ***
5 REM *** The function to be integrated is pro-    ***
6 REM *** grammed in line # 1000 .                 ***
9 REM ************************************************
50    RANDOMIZE INT(TIMER/3)
100   INPUT "x-interval "; X0,X9
120   X5=X9-X0
200   INPUT "y-interval "; Y0,Y9
220   Y5=Y9-Y0
300   INPUT "z-interval "; Z0,Z9
320   Z5=Z9-Z0
350   V=X5*Y5*Z5
400   I=0
500   S=0
900   I=I+1
```

```
910      X=RND*X5+X0
920      Y=RND*Y5+Y0
930      Z=RND*Z5+Z0
1000     F=X*X+Y*Y+Z*Z
1100     S=S+F
1200     PRINT "Estimated integral = ";S/I*V,I
1300  GOTO 900
2000  END

RUN
x-interval ? 0,1
y-interval ? 0,1
z-interval ? 0,1
Estimated integral =  1.601509            1
Estimated integral =  1.395435            2
Estimated integral =  1.282691            3
Estimated integral =  1.327442            4
Estimated integral =  1.247966            5
Estimated integral =  1.12869             6
Estimated integral =  1.16279             7
Estimated integral =  1.105992            8
Estimated integral =  1.086622            9
Estimated integral =  1.047838            10
Estimated integral =  1.063518            11
Estimated integral =  1.040665            12
Estimated integral =  1.07767             13
Estimated integral =  1.063864            14
Estimated integral =  1.052711            15
Estimated integral =  1.013171            16
Estimated integral =  .9748961            17
Estimated integral =  .9978019            18
Estimated integral =  1.007599            19
Estimated integral =  1.000479            20
Estimated integral =  1.022234            21
^C
Break in 1200

RUN
x-interval ? -1,1
y-interval ? 0,1
z-interval ? -1,0
Estimated integral =  2.05289             1
Estimated integral =  2.06324             2
Estimated integral =  1.535993            3
Estimated integral =  1.778739            4
Estimated integral =  1.552155            5
```

```
Estimated integral =    1.465704              6
Estimated integral =    1.4675                7
Estimated integral =    1.368669              8
Estimated integral =    1.553707              9
Estimated integral =    1.639911             10
Estimated integral =    1.840717             11
Estimated integral =    1.883959             12
Estimated integral =    1.822875             13

   . . . . . . . . . . . . . . . . .
Estimated integral =    1.984575            173
Estimated integral =    1.992805            174
Estimated integral =    2.000212            175
Estimated integral =    2.002354            176
Estimated integral =    2.001573            177
Estimated integral =    2.00422             178
Estimated integral =    2.002703            179
Estimated integral =    1.998658            180
Estimated integral =    2.001824            181
Estimated integral =    2.004777            182
Estimated integral =    2.016829            183
^C
Break in 1200
```

In lines 100, 200 and 300 the upper and lower limits of the independent variables x, y and z are written into the BASIC-variables X0, X9, Y0, Y9, Z0 and Z9. In lines 120, 220 and 320 the length of the x-, y- and z-intervals are calculated. In line 350 the three intervals were multiplied by each other. With this volume V (in the three-dimensional case) the mean value has to be multiplied to obtain an approximation of this integral.

Variable I contains the number of function calculations and variable S contains the sum of all function values. The mean value is equal to the quotient S/I; the estimated integral is then S/I*V. After I and S were set to 0, a random point within the specified integral limits was calculated in lines 910, 920 and 930. The command RND*X5 generates random numbers between 0 and X5 which is the length of the x-interval. If one adds X0, the lower limit of the x-interval, then random numbers between X0 and X9 are obtained, which is exactly within the desired interval. After the generation of random numbers in all three intervals, the function $f(x_{rnd}, y_{rnd}, z_{rnd})$ is calculated for this triple random number. The function value F is summed up in line 1200 and the approximated value of the integral is printed. The infinite loop runs from line 900 through 1300.

A test run of the program MC-INT will show that it takes a rather long time to converge to the correct values of 1.00 and 2.00, respectively. But, because this method works even with very complicated integrals which are not solvable analytically, one should consider it for certain applications. (A PC can often do the job over night.)

Problem 26

Use program MC-INT to integrate other three-dimensional functions such as $\sin(x + y + z)$ or $\exp(-|xyz|)$.

Problem 27

Rewrite program MC-INT so that functions with two or four independent variables can be integrated.

Problem 28

Write a Monte-Carlo program so that not only multidimensional intervals can be used as integration regions but also multidimensional spheres. The two-dimensional sphere is a circle, the three-dimensional sphere the sphere itself. For higher dimensions the random numbers have to be chosen so that the following equation is satisfied.

$$x_1^2 + x_2^2 + x_3^2 + x_4^2 + \ldots + x_n^2 < r^2$$

A chemical example of application of this rather abstract integration is the determination of the mean reaction rate on a catalyst pellet. The pellet should have the form of a sphere with a diameter of 1 mm. Because of the release of heat from an exothermic first order reaction, the catalyst pellet has an inhomogeneous temperature gradient:

$$T = 700 - 200 \cdot r^2 \qquad (0 \leq r \leq 1)$$

T is the temperature in K and r is the radial distance in mm from the center of the pellet. This means that the temperature T in the center of the pellet is 700 K, and at the surface 500 K, with a parabolic gradient in between. The Arrhenius equation for the first order rate constant should be:

$$k = 10^{13} \cdot e^{-\frac{20000}{T}} \quad \left[\sec^{-1}\right]$$

The quantity 20000 represents the activation energy divided by the gas constant. The concentration of the reactants is assumed to be 0.01 mol/l. This problem is a direct application of the integration over a sphere.

4.5   Molecular Formulas of a Chemical Compound From the Elemental Analysis

This example is restricted to the chemical elements carbon C, hydrogen H, nitrogen N, and oxygen O. An extension to other elements is possible without much difficulty. The following problem is a further example for the application of loops and the usefulness of printing out results in the form of tables.

```
0 REM "C-H-N-O "           EBERT/EDERER          880518
1 REM ************************************************
2 REM *** C-H-N-O analysis of a chemical compound  ***
3 REM *** containing C, H, N and O. Possible       ***
4 REM *** compositions as chemical molecular       ***
5 REM *** formulas shall be calculated.            ***
9 REM ************************************************
200    INPUT "Weight% C ";C
300    INPUT "Weight% H ";H
400    INPUT "Weight% N ";N
500    O=100-C-H-N
600    PRINT "Weight% O ";O
1200   C1=C/12.011
1300   H1=H/1.008
1400   N1=N/14.007
1500   O1=O/15.999
1700   PRINT
1800   PRINT " C        H          N          O"
2000   FOR I=1 TO 15
2100     X=I/C1
2200     C2=C1*X
2300     H2=INT(H1*X*100)/100
2400     N2=INT(N1*X*100)/100
2500     O2=INT(O1*X*100)/100
3000     PRINT C2;TAB(5);H2;
3100     PRINT TAB(15);N2;TAB(25);O2
4000   NEXT I
5000   END

RUN
Weight% C ? 77.7
Weight% H ? 7.46
Weight% N ? 0
Weight% O  14.84
```

```
C      H          N      O
1    1.14        0      .14
2    2.28        0      .28
3    3.43        0      .43
4    4.57        0      .57
5    5.72        0      .71
6    6.86        0      .86
7    8           0      1
8    9.149999    0      1.14
9    10.29       0      1.29
10   11.44       0      1.43
11   12.58       0      1.57
12   13.72       0      1.72
13   14.87       0      1.86
14   16.01       0      2
15   17.16       0      2.15
Ok


RUN
Weight% C ? 36.4
Weight% H ? 6.1
Weight% N ? 21.2
Weight% O   36.3

C      H          N      O
1    1.99        .49    .74
2    3.99        .99    1.49
3    5.99        1.49   2.24
4    7.98        1.99   2.99
5    9.979999    2.49   3.74
6    11.98       2.99   4.49
7    13.97       3.49   5.24
8    15.97       3.99   5.98
9    17.97       4.49   6.73
10   19.96       4.99   7.48
11   21.96       5.49   8.23
12   23.96       5.99   8.979999
13   25.95       6.49   9.729999
14   27.95       6.99   10.48
15   29.95       7.49   11.23
Ok
```

The weight percents of C, H, N, and O are required. The input of the weight percents for the first three elements happens in lines 200–400, in line 500 the weight percent for oxygen is calculated from the difference to 100.

If the weight percents are divided by the atomic weight, the molar fraction of the elements (in the unit moles/100 g substance) are calculated. What is left over is the normalization of these molar portions to integer figures for carbon, which represents moles or atoms, just as one wishes. This normalization to 1, 2, ..., 15 atoms occurs in the loop of lines 2000 through 4000. In 2100 this normalization factor is calculated and stored in variable X. In lines 2300 through 2500 the portions of H, N, and O are multiplied by this normalization factor.

The rounding to two significant figures after the comma is done by multiplication by 100, eliminating the digits after the decimal points (this is performed by the built-in function INT) and by dividing by 100. The printed formula is therefore clearer. The print occurs in lines 3000 and 3100. The function TAB is new; it works the same way as in any ordinary typewriter. TAB(14) means that the cursor proceeds to the column position 14 before it starts to print. The semicolon serves to prevent a line feed after the PRINT-command. The missing semicolon at the end of line 3100 leads to a line feed.

If the table is written on the screen, the combination of the numbers can be inspected. Those consisting of integer — or nearly integer — combinations represent the correct molecular formula. (It is of course possible that they have to be multiplied by an integer number.)

Problem 29

Extend the program C-H-N-O, to include the elements S and Cl. It is advantageous to introduce a variable upper limit for the number of C atoms instead of the fixed number 15 which is used in the program. Try to incorporate this into the modified program as well.

4.6   Conditional Branching

At this point another BASIC-statement shall be introduced, the IF-THEN-command. It was used in earlier examples.

Conditional Jump

IF-THEN

Syntax: z1 IF a1 ◇ a2 THEN z2
z1 IF a1 ◇ a2 THEN GOTO z2
z1 IF a1 ◇ a2 THEN any statement

z : line numbers
a : expressions, variables or constants
◇ : comparison operator

The two expressions a1 and a2 are compared; a1 and a2 can also be constants or simple variables. If the comparison is answered with 'yes' (if the comparison is true) a jump to line z2 will take place (GOTO z2 is executed), otherwise the statement following IF-THEN will be executed next. The IF-THEN-statement can be used in most of the BASIC-dialects in various ways. A line number, a GOTO-command or any other BASIC-statement may follow THEN.

Table 4.1. Comparison operators used by the IF-THEN-statement

◇	Relation
=	equal
<>	not equal
>	greater than
<	less than
>=	greater than or equal
<=	less than or equal

Examples: 200 INPUT Z
220 IF Z=1 THEN 200 : REM function is not defined
240 IF Z<1 THEN 300 : REM imaginary solution
260 E=SQR(1/(Z-1))
280 PRINT E
290 STOP
300 E=SQR(-1/(Z-1))
320 PRINT E; "*I"
340 STOP

```
10   INPUT I
20   IF I<=0 THEN GOTO 100
30   PRINT LOG(I)
40   STOP
100  PRINT "false input"
120  STOP
```

```
10 INPUT A,B,C
20 IF A*A<4*B THEN 50
30 PRINT C+SQR(A*A-4*B)
40 GOTO 10
50 PRINT "new input"
60 GOTO 10
```

```
5  DATA 4
10 INPUT A
20 IF A-1=0 THEN PRINT "new input" : GOTO 10
30 IF A=2 THEN INPUT B
40 IF A=3 THEN READ X
60 PRINT X
80 IF A=B THEN STOP
90 STOP
```

In modern BASIC-implementations more advanced branching constructions with keywords IF, THEN, ELSE, ENDIF, ENDELSE and others are possible. But these branching statements may be special for the BASIC-dialect used. For this reason we use only the simple IF-THEN-statement which is understood in each BASIC-dialect. It should be mentioned here that there is a multiple branching statement using the keyword ON.

In spite of the various possibilities which the IF-THEN-statements offer, it is recommended to use the IF's as sparsely as possible, otherwise the program becomes unnecessarily hard to understand.

4.7   Molecular Formulas From High Resolution Mass Spectrometry

We introduce in this chapter the determination of molecular formulas of chemical substances from the molecular peak of the high resolution mass spectrometer. The idea which underlies a treatment of this problem is simple. If the mass of the molecular peak is measured in a mass spectrometer, only a more or less big interval for the molar mass (because of the small error) can be specified. Then the molar mass of all — possible and impossible — compounds are calculated and checked as to whether they lie in the specified interval. If this is true, the compound and its molar mass is printed out on the screen. The chemist must now decide whether the specified combination is significant and which combination is the correct or the most probable one.

```
0 REM "MS      "           EBERT/EDERER          880514
1 REM ************************************************
2 REM *** Interpretation program of high resolu-  ***
3 REM *** tion mass spectrometry.                 ***
4 REM *** All possible chemical compositions      ***
5 REM *** are calculated which have a molecular   ***
6 REM *** weight within a given mass interval.    ***
7 REM *** In this example the elements H, C, N, O ***
8 REM *** and S are considered.                   ***
9 REM ************************************************
100    M1= 1.0078: REM Atomic mass of H
200    M2=31.9721: REM Atomic mass of S
300    M3=15.9949: REM Atomic mass of O
400    M4=14.0031: REM Atomic mass of N
500    M5=12.0000: REM Atomic mass of C
1000   INPUT "Mass interval ";A,B
2000   M=B
3000   FOR I2=0  TO M STEP M2
3100     FOR I3=I2 TO M STEP M3
3200       FOR I4=I3 TO M STEP M4
3300         FOR I5=I4 TO M STEP M5
4000           J1=INT((A-I5)/M1+1)
4100           J2=INT((B-I5)/M1)
4200           IF J1>J2 THEN 6100
5000           FOR I1=J1 TO J2
5100             IF I1<0 THEN 6000
5200             IF I1>M/4+2 THEN 6100
5300             S=I5+I1*M1
5310             P2=INT(I2/M2+.1)
5320             P3=INT((I3-I2)/M3+.1)
5330             P4=INT((I4-I3)/M4+.1)
5340             P5=INT((I5-I4)/M5+.1)
5500             PRINT "S";P2;" O";P3;
5600             PRINT " N";P4;" C";P5;
5700             PRINT " H";I1;"=";S
```

```
6000          NEXT I1
6100         NEXT I5
6200       NEXT I4
6300     NEXT I3
6400   NEXT I2
9999 END


RUN
Mass interval ? 27.5 , 28.5
S 0  0 0  N 0  C 2  H 4 = 28.0312
S 0  0 0  N 1  C 1  H 2 = 28.0187
S 0  0 0  N 2  C 0  H 0 = 28.0062
S 0  0 1  N 0  C 1  H 0 = 27.9949


RUN
Mass interval ? 88.01 , 88.04
S 0  0 0  N 0  C 7  H 4 = 88.03119
S 0  0 0  N 1  C 6  H 2 = 88.0187
S 0  0 1  N 4  C 1  H 4 = 88.0385
S 0  0 1  N 5  C 0  H 2 = 88.026
S 0  0 2  N 1  C 3  H 6 = 88.03971
S 0  0 2  N 2  C 2  H 4 = 88.0272
S 0  0 2  N 3  C 1  H 2 = 88.0147
S 0  0 3  N 0  C 3  H 4 = 88.01589
S 0  0 5  N 0  C 0  H 8 = 88.03691
S 1  0 0  N 0  C 4  H 8 = 88.0345
S 1  0 0  N 1  C 3  H 6 = 88.02201
S 1  0 3  N 0  C 0  H 8 = 88.01921
S 2  0 0  N 0  C 1  H 12 = 88.03779
S 2  0 0  N 1  C 0  H 10 = 88.0253


RUN
Mass interval ? 234.049 , 234.051
S 0  0 3  N 6  C 8   H 6 = 234.0501
S 1  0 0  N 0  C 16  H 10 = 234.0501
S 1  0 1  N 12 C 1   H 6 = 234.051
S 1  0 5  N 8  C 0   H 10 = 234.0494
S 1  0 6  N 5  C 2   H 12 = 234.0506
S 1  0 10 N 1  C 1   H 16 = 234.049
S 2  0 2  N 2  C 8   H 14 = 234.0494
S 3  0 0  N 8  C 1   H 14 = 234.0503
S 3  0 5  N 1  C 2   H 20 = 234.0499
S 5  0 0  N 1  C 3   H 24 = 234.0508
Ok
```

The program MS can be applied to all compounds which are composed of the five elements H, C, N, O and S. The more elements that have to be considered, the more combinations exist and the longer is the computing time. Generalizing the program and adapting it for special purposes is recommended as an exercise. At the beginning of the program (line 100–500) the exact atomic masses of the five elements (of the most abundant isotopes) are defined. Then the range of the mass of the molar peak is entered by means of the upper and lower limits, the lower limit into the variable A and the upper into the variable B.

The central search program consists of 5 loops nested in each other, because 5 elements are to be considered. For better understanding, the various loops can be distinguished by indenting the program statements within a loop — the corresponding FOR–NEXT lines can now easily be seen. The indentation of lines makes no difference for the execution of a BASIC-program, but it helps us to see the structure of the program. The innermost running variable I1 counts the number of hydrogen atoms. But the outer running variables I2, I3, I4 and I5 contain another quantity, the already 'used' mass.

I2	⟵	mass, used by	S
I3	⟵	mass, used by	S + O
I4	⟵	mass, used by	S + O + N
I5	⟵	mass, used by	S + O + N + C

For this reason the I2-loop starts in line 3000 with 0 and is increased by the increment M2 (the atomic weight of the main isotope of S, see line 200) until the maximum value of M. M was fixed as the highest possible mass (line 2000). I3 starts with I2 and is increased with each loop pass by the atomic weight of oxygen M3; I4 starts with I3 and is increased with the atomic weight of nitrogen M4; I5 starts with I4 and is increased with each loop pass by the atomic weight of carbon M5. The variable I5 now holds the molecular mass which is used by the elements S, O, N and C. This quantity stored in I5 can now be used to calculate the number of hydrogen atoms in order to get the total molecular mass of the molar peak which lies within the specified limits A and B. In line 4000 the minimum number of hydrogen atoms which are necessary to exceed the lower limit A of the molecular mass interval is calculated. This number is contained in the variable J1. In line 4100 the maximum possible number of hydrogen atoms which do not exceed the upper limit B of the molecular mass interval is calculated and this number is transferred into the variable J2. If the maximum number of possible hydrogen atoms J2 is smaller than the necessary minimum number of hydrogen atoms J1, then the specified mass interval for the molecular peak cannot be obtained with the atom combination under consideration and a new element combination has to be calculated. This determination and the conditional jump back to the outer loops occurs in line 4200.

In case the number of maximum possible hydrogen atoms J2 is higher than the lowest number of necessary hydrogen atoms J1, all atom combinations (molecular formula) containing hydrogen atoms between J1 and J2 are in the mass interval of interest. The loop counter, the variable I1, which contains the number of hydrogen atoms, runs from J1 until J2. In lines 5100 and 5200 an elementary check is performed as to whether the result is chemically significant. If the number of hydrogen atoms is negative (line 5100) — this occurs seldom and only if big mass intervals are chosen — then the IF-THEN-statement causes the program to branch to the end of the inner loop, I1 is increased by 1 and the senseless formula will not be printed. Line 5200 serves to exclude atom combinations with too many hydrogen atoms. The number of hydrogen atoms is limited to a quarter of the molecular weight plus 2. If too many hydrogen atoms are necessary then the program jumps directly back from line 5200 into the outer I5-loop.

Line 5300 calculates the molecular weight of the just generated molecular formula. Subsequently the calculated numbers of the atoms S, O, N and C are assigned to the variables P2, P3, P4 and P5, respectively. For that, the difference of the mass portion is determined and used to calculate the mass portion of one element (e.g. I4-I3 is the mass portion of nitrogen). This mass portion is divided by the corresponding atomic weight. The number of atoms of this element is obtained. As truncation errors are always possible, the number 2.999997 for nitrogen atoms can occur. To make the integer number of 3 out of the calculated number, 0.1 is added and the digits after the decimal point are dropped by the standard function INT (lines 5310–5340).

The output of a molecular formula, which is in the specified mass interval for the molecular peak, is done in lines 5500, 5600, and 5700. The missing semicolon at the end of line 5700 produces a carriage return and a line feed in the output. The user must decide which of the printed molecular formulas is correct. In our second example the compound O_5H_8 is printed. This, obviously, is chemical nonsense, but how should the computer know?

Problem 30

Modify the program MS so that a molecular formula is evaluated for compounds containing the elements C, H, O, F, and P (atomic mass for F: 18.9984 and for P: 30.9738).

Problem 31

Vary the program MS so that in addition to the elements C, H, N, O and S, the element F is contained in the search for the molecular formula (You will need an additional loop).

Problem 32

In the output of the program an additional marker — e.g. an asterisk — should indicate whether the printed chemical compound is a free radical with an unsaturated valency. This is another way to reduce the number of possible molecular formulas.

5 Integration

In solving mathematical problems in the natural sciences it is often necessary to perform integrations. Obviously, integrals cannot always be solved analytically. Only in certain cases can the primitive function (antiderivative) of a given function be determined. A general method to integrate any function between integration limits (definite integrals) is numerical integration. The methods of numerical integration are usually easy to understand and simple to program. We will not deal with the details of numerical analysis here, but rather will become acquainted with these methods in principle. If the definite integration is represented as the determination of the area below a function $f(x)$ between the points $x = A$ and $x = E$, an inspection of the graph of Fig. 5.1 might be helpful.

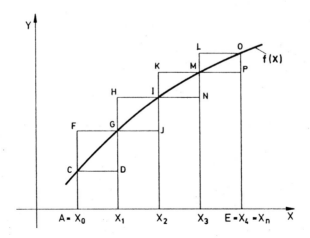

Fig. 5.1. Graph of a function $f(x)$ and the approximation of the definite integral by Euler's method

5.1   Integration According to Euler

The function $f(x)$ of the graph in Fig. 5.1 shall be integrated numerically within the interval from A to E. The sizes of two areas, which are given by 'step construction' above and below the function line, are calculated by summing up the areas of the rectangles. Taking the mean value, the true area can be approximated.

The interval from A to E is divided into equally spaced subintervals. In the first subinterval the function $f(x)$ runs from C to G. The lower estimate area is enclosed by the line from $A = (x_0, 0)$ to C — this equals $(x_0, f(x_0))$ i.e. the function value at the left end of the subinterval (x_0, x_1) —, to $D = (x_1, f(x_0))$ to $(x_1, 0)$ and back to $(x_0, 0)$. Accordingly, the upper estimate area runs from $A = (x_0, 0)$ to F — this is $(x_0, f(x_1))$, i.e. the function value at the right end of the subinterval (x_0, x_1) — then to $G = (x_1, f(x_1))$ and back to the x-axis to $(x_1, 0)$ and $(x_0, 0)$.

In the second subinterval from x_1 to x_2 the lower estimate area encloses the points $(x_1, 0), G, J$ and $(x_2, 0)$, and the upper estimate area $(x_1, 0), H, I$ and $(x_2, 0)$. Because the lengths of the subintervals h are always the same,

$$h = x_{i+1} - x_i$$

the total lower estimate which is called S_l and the total upper area S_u are given by

$$S_l = f(x_0) \cdot h + f(x_1) \cdot h + f(x_2) \cdot h + f(x_3) \cdot h$$
$$S_u = \qquad\qquad f(x_1) \cdot h + f(x_2) \cdot h + f(x_3) \cdot h + f(x_4) \cdot h$$

The true value of the area can now be approximated by calculating the algebraic mean

$$\text{Area} \approx S = \frac{(S_l + S_u)}{2}$$

$$S = \left(\frac{f(A)}{2} + f(x_1) + f(x_2) + f(x_3) + \frac{f(E)}{2} \right) \cdot h$$

This approach is called the trapezoidal formula, because the entire area can be represented as the sum of the four trapezoids T_1–T_4.

$$T_1 = (x_0 \leftrightarrow C \leftrightarrow G \leftrightarrow x_1 \leftrightarrow x_0) \quad = h \cdot \frac{f(x_0) + f(x_1)}{2}$$

$$T_2 = (x_1 \leftrightarrow G \leftrightarrow I \leftrightarrow x_2 \leftrightarrow x_1) \quad = h \cdot \frac{f(x_1) + f(x_2)}{2}$$

$$T_3 = (x_2 \leftrightarrow I \leftrightarrow M \leftrightarrow x_3 \leftrightarrow x_2) \quad = h \cdot \frac{f(x_2) + f(x_3)}{2}$$

$$T_4 = (x_3 \leftrightarrow M \leftrightarrow O \leftrightarrow x_4 \leftrightarrow x_3) \quad = h \cdot \frac{f(x_3) + f(x_4)}{2}$$

$$S = T_1 + T_2 + T_3 + T_4$$

This equation for S is identical to the formula above.

Obviously, for all 'well-behaved' functions the following statement is true: the smaller the intervals h that are chosen, — i.e. the larger the number of subintervals — the more accurate the approximated value. In the trapezoidal formula the error F is approximately proportional to h^2.

$$F \sim h^2$$

For carrying out the evaluation of the definite integral of a given function, the interval A to E is divided into subintervals and the trapezoidal sum S is determined. Then the number of subintervals is increased, the calculation of the trapezoidal sum is repeated, and the result is compared with the value of the preceding calculation. This procedure is carried on until the result is 'sufficiently accurate'. The definition of 'sufficiently accurate' is the toughest problem in developing a suitable computer program.

In the integration code which is listed below, the number of subintervals is doubled after each loop. This is done by an infinite loop. The values of the actual estimate sum can be shown on the monitor screen and, by comparison of consecutive values, it can be decided whether the latest result is sufficiently accurate. Accordingly then the program can be terminated 'by hand'.

The flow diagram in Fig. 5.2 for this computation consists basically of a FOR-NEXT-loop and a few additional statements in which the estimate sum for the integral with n subintervals is calculated. In the external infinite loop the number of subintervals is doubled after each loop and the result of the summation, which is used for the approximation of the integral, is printed on the monitor screen.

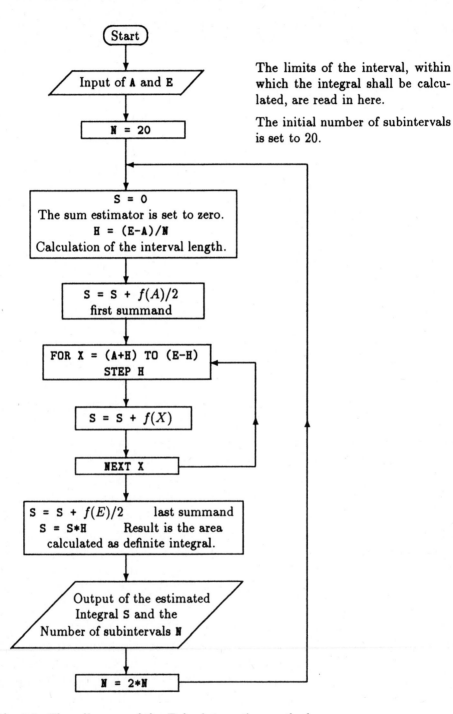

The limits of the interval, within which the integral shall be calculated, are read in here.

The initial number of subintervals is set to 20.

Fig. 5.2. Flow diagram of the Euler integration method

The computer code which corresponds to the flow diagram in Fig. 5.2 is listed below.

```
0 REM "EULER    "          EBERT/EDERER        880518
1 REM *********************************************
2 REM ***              Integration               ***
3 REM *** of any function according to EULER.    ***
5 REM *** The function has to be programmed in   ***
6 REM *** line  2100, 3100 and 4000.             ***
9 REM *********************************************
50     INPUT "Lower integration limit "; A
60     INPUT "Upper integration limit "; E
300    N=20
1000   S=0
1100     H=(E-A)/N
2000     X=A
2100     Y=EXP(-X*X)
2200     S=S+Y/2
3000     FOR X=A+H TO E-H STEP H
3100       Y=EXP(-X*X)
3200       S=S+Y
3300     NEXT X
3900     X=E
4000     Y=EXP(-X*X)
4100     S=S+Y/2
4200     S=S*H
5000     PRINT:PRINT "Integral from";A;"to";E
6000     PRINT "with";N;"subintervals =";S
7000     N=N*2
8000   GOTO 1000
10000 END

RUN
Lower integration limit ? 0
Upper integration limit ? 3

Integral from 0 to 3
with 20 subintervals = .8861613

Integral from 0 to 3
with 40 subintervals = .8861926

Integral from 0 to 3
with 80 subintervals = .8862068
```

```
Integral from 0 to 3
with 160 subintervals = .8862077

Integral from 0 to 3
with 320 subintervals = .8862072

Integral from 0 to 3
with 640 subintervals = .8862068

Integral from 0 to 3
with 1280 subintervals = .8862041
^C
Break in 3200
```

The function, which is integrated in this program, is

$$f(x) = e^{-x^2}$$

This is a non-trivial problem. If the antiderivative of $f(x)$ is multiplied by a constant factor, a function is obtained which is called the error function.

$$\text{erf}(x) = \frac{2}{\pi} \cdot \int_{t=0}^{t=x} e^{-t^2} \, dt$$

Because of its importance for statistical calculations the error function is tabulated in most textbooks on statistics.

In our computer code the function to be integrated $f(x)$ appears at three different positions in the lines 2100, 3100 and 4000. The input of the integration limits follows the REM headings in lines 50 and 60. Then, the number of subintervals n is set to 20 (the number 20 is arbitrary and can be replaced by any other number) and the variable for the sum S is set to zero. In line 1100 the length of a subinterval is calculated; it equals the distance of interval (A, E) divided by the number n of subintervals. In line 2000 the independent variable X is set to the initial value A of the interval and in the next line the function value $f(A)$ is calculated. One-half of this function value is summed up in the variable S. Between the lines 3000 and 3300 there is the FOR-NEXT-loop for the x-values with the step size h of the subinterval. The first and last x-values are neglected, therefore X runs from (A+H) until (E-H). For every x-value the function $f(x)$ is calculated in line 3100 and added to S in the next line. After the loop has been terminated, half of the value of the function at the end of the interval $\frac{f(E)}{2}$ is added to S. In line 4200 the sum S is multiplied by the length of the subinterval H and stored in S. Therefore S designates the approximation sum for this definite integral at n subintervals. In lines 5000 and 6000 the integration interval, the number of subintervals and the corresponding result are printed on the monitor screen. Then the number of subintervals is doubled and the program jumps to line 1000, starting the loop again.

In the example shown, the definite integral is calculated within the limits of integration of 0 to 3. It can be seen that with 80 subintervals 6 digit accuracy is already obtained.

An important disadvantage of this program is obvious: the function is needed at three different positions and has to be programmed three times. With more complicated functions this would be annoying. However a new BASIC-statement, the call of a predefined function, which is also called a user-defined function or function subprogram, can be used to eliminate the difficulty of repeated programming of the same equations. These functions can be used along with BASIC-supplied library functions.

Definition of a Function

$$\boxed{\text{DEF FN}}$$

Syntax: z DEF FNa$(x) = b$

z   :   line number
FNa :   is the name of the function to be defined
a   :   any letter
x :   dummy variable or dummy argument
b :   arithmetic expression using the dummy variable x

Most versions of BASIC allow the definition of a function to be written in any position of the program, but the program must run through the line of definition before this function can be used. (Some forms of BASIC don't have this restriction). However, it is good programming practice to group together all function definitions and place them near the beginning of the program. As a dummy variable any variable name may be used, even if this symbol has already appeared in the program. Modern BASIC-installations allow more than one dummy argument.

It should be clear that the presence of a DEF FN-statement serves only to define a function. In order to evaluate the function it is necessary to refer to the function name elsewhere in the program, just as we would do with a library function.

Calling a Function

FN

Syntax: z ... FNa(b)

z   : line number
...   : any statement in which a function may be called
FNa   : is the name of a user defined function
b : any arithmetic expression

A function is referred (evaluated) by specifying the name of the function within a BASIC-statement, as though the function name were an ordinary variable. The function name is followed by the argument b enclosed in parentheses. If this argument is an arithmetic expression, this expression is evaluated and the value is transferred to the dummy variable in the function definition. Now the function is evaluated and the value of the function is given back to the calling statement.

Of course it is possible to define more than one function in the same program, they only need to have different names (i.e. a different letter after DEF FN). These functions can then be called in any arbitrary order by the main program.

```
Examples:   5   DEF FNC(D) = EXP(-SIN(D*D))
           10   DEF FNY(X) = X^2 + 1/X
           20   INPUT I
           30   I1 = FNY(I)
           40   I2 = FNY(I*I)
           50   I3 = FNY(SQR(I))
           60   PRINT I, I1 , I2 , I3
          100   PRINT FNC(7.35)
          110   PRINT FNC(13.23*FNY(1/I))
          130   GOTO 20
```

```
10 DEF FNA(X) = X*X
20 X = 11
30 FOR I=1 TO 10
40 PRINT FNA(I) , FNA(FNA(I-1)) , X
50 NEXT I
```

If the function definition statement DEF FN is used together with the function call FN in the integration program EULER, then we have the following program called EULER2.

```
0 REM "EULER2  "          EBERT/EDERER          880518
1 REM ***************************************************
2 REM ***                Integration                 ***
3 REM *** of any function according to EULER.        ***
5 REM *** The function has to be programmed in       ***
6 REM *** line # 100.                                ***
9 REM ***************************************************
50    INPUT "Lower integration limit "; A
60    INPUT "Upper integration limit "; E
100   DEF FNF(X)=EXP(-X*X)
300   N=20
1000  S=0
1100    H=(E-A)/N
2000    S=FNF(A)/2
3000    FOR I=1 TO N-1:X=A+I*H
3200      S=S+FNF(X)
3300    NEXT I
4000    S=S+FNF(E)/2
4200    S=S*H
5000    PRINT:PRINT "Integral from";A;"to";E
6000    PRINT "with";N;"subintervals =";S
7000    N=N*2
8000  GOTO 1000
20000 END


RUN
Lower integration limit ? 0
Upper integration limit ? 3

Integral from 0 to 3
with 20 subintervals = .8862058

Integral from 0 to 3
with 40 subintervals = .8862069
```

```
Integral from 0 to 3
with 80 subintervals = .8862071

Integral from 0 to 3
with 160 subintervals = .8862074

Integral from 0 to 3
with 320 subintervals = .8862079

Integral from 0 to 3
with 640 subintervals = .8862078
^C
Break in 3200
```

Line 100, in which the function to be integrated is defined, is new. Therefore, only this line must be modified if another function is to be integrated. Lines 2000, 2100 and 2200 of the program EULER are now combined in line 2000 of EULER2 which contains a function call. Similarly, lines 3900, 4000 and 4100 of the original code are combined in line 4000 of the new code. Also, line 3100 is dropped, and in line 3200 Y is replaced by the function call FNF(X). The parts of the program concerning the output as well as the infinite loop remain the same as in the original EULER code.

Problem 33
Integrate functions of your choice. Proposals

$$\int x^3 \sin x \, dx \qquad \int \frac{1}{\log x + x} \, dx$$

Problem 34
Try to integrate functions which can also be solved analytically and compare the numerical results with the analytical results, e.g.

$$f(x) = \frac{1}{x} \qquad f(x) = x^2 \qquad f(x) = \sin x$$

The antiderivatives are

$$\int \frac{1}{x} \, dx = \log x \qquad \int x^2 \, dx = \frac{x^3}{3} \qquad \int \sin x \, dx = -\cos x$$

Problem 35

The temperature dependence of the molar heat of a hypothetical substance is given by

$$C_p = \frac{3 \log(T + 1)}{1 + \log(T + 1)}$$

C_p is expressed in this formula in units of the gas constant R. In the above formula C_p equals $3R$ for temperatures approaching infinity. Let, arbitrarily, $H = 0$ at $T = 0$, then it follows

$$\Delta H = \int_{T_1}^{T_2} C_p(T)\, dT$$

Set up a table for C_p and H as a function of temperature. To calculate H set the lower limit of temperature in the integral to zero. Plot both functions on graph paper.

5.2 Calculation of the Circumference of an Ellipse

A non-trivial application from mathematics is the calculation of the length L of the circumference of an ellipse. Whereas the area of an ellipse with the half axes b and g can be expressed analytically by

$$\text{area} = b \cdot g \cdot \pi$$

the circumference has to be calculated by an integration or an infinite series. For that purpose some mathematical relations should be reviewed. A circle can be expressed in a parametric form by

$$\begin{pmatrix} x \\ y \end{pmatrix} = \begin{pmatrix} r \cos t \\ r \sin t \end{pmatrix} \qquad 0 \le t < 2\pi$$

where r is the radius and t is the angle, which runs from 0 to 2π in a single rotation. (This is valid in radian units; in degrees it runs from 0 to 360.) If you have problems with this parametric form just calculate the x- and y-values for a number of t-values and plot them.

The ellipse can be described in this parametric form by

$$\begin{pmatrix} x \\ y \end{pmatrix} = \begin{pmatrix} g \cos t \\ b \sin t \end{pmatrix} \qquad 0 \le t < 2\pi$$

g and b are half of the length of the two main axes. For practice you should assume different numbers for g and b, calculate x and y for different parameters t, and again plot the results.

In mathematical textbooks [1] a formula for the length of such parametric curves in the parameter interval (t_1, t_2) is given. If you set $t_1 = 0$ and $t_2 = 2\pi$ then the curve is closed to form an ellipse. The formula for the length of this curve then becomes the formula for the circumference L

$$L = \int\limits_{t_1=0}^{t_2=2\pi} \sqrt{\left(\frac{dx}{dt}\right)^2 + \left(\frac{dy}{dt}\right)^2}\; dt$$

Introducing the equation for the ellipse into this formula one obtains

$$L = \int\limits_{t_1=0}^{t_2=2\pi} \sqrt{g^2 \sin^2 t + b^2 \cos^2 t}\; dt$$

We now modify the program EULER2 to solve this integral; arbitrary values for g and b should be used as input:

```
0 REM "ELLIPSE "          EBERT/EDERER          880601
1 REM *************************************************
2 REM *** Calculation of the circumference of an   ***
3 REM *** ellipse using the integration program    ***
4 REM *** EULER2.                                   ***
9 REM *************************************************
50     INPUT "Length of the 1. half axis "; G
60     INPUT "Length of the 2. half axis "; B
100    DEF FNF(X)=SQR(G*G*SIN(X)^2+B*B*COS(X)^2)
150    PI=4*ATN(1)
200    A=0 : E=2*PI
300    N=10
1000   S=0
1100     H=(E-A)/N
2000     S=FNF(A)/2
3000     FOR I=1 TO N-1
3100       X=A+I*H
3200       S=S+FNF(X)
```

---

[1] E. Kreyszig, *Advanced Engineering Mathematics*, 6.ed., Wiley, New York (1988)

```
3300      NEXT I
4000      S=S+FNF(E)/2
4200      S=S*H
5000      PRINT:PRINT "Circumference of the ellipse"
6000      PRINT "with";N;"subintervals =";S
7000      N=N*2
8000  GOTO 1000
20000 END
```

```
RUN
Length of the 1. half axis ? 1
Length of the 2. half axis ? 1

Circumference of the ellipse
with 10 subintervals = 6.283186

Circumference of the ellipse
with 20 subintervals = 6.283186

Circumference of the ellipse
with 40 subintervals = 6.283186

Circumference of the ellipse
with 80 subintervals = 6.283186
^C
Break in 3200

RUN
Length of the 1. half axis ? 1
Length of the 2. half axis ? 3

Circumference of the ellipse
with 10 subintervals = 13.38405

Circumference of the ellipse
with 20 subintervals = 13.36469

Circumference of the ellipse
with 40 subintervals = 13.3649

Circumference of the ellipse
with 80 subintervals = 13.3649
^C
Break in 3200
```

The function to be integrated (line 100) is replaced by the integrand for the calculation of the ellipse. The upper and lower integration limits are not given in the input anymore, but they are defined in line 200 of the program as these limits are always 0 and 2π respectively.

The program asks for values of both half axes in lines 50 and 60 in the input region. The text "Integral ..." is replaced by "Circumference of the ellipse". The examples show that the circumference of a circle is obtained if identical values for both half axes are chosen and that the integration converges relatively fast in calculating the circumference of an ellipse.

Problem 36

Modify the program ELLIPSE and determine the circumference of the following curves

$$\begin{pmatrix} x \\ y \end{pmatrix} = \begin{pmatrix} 2\cos t - \cos 2t \\ 2\sin t - \sin 2t \end{pmatrix} \quad \text{and} \quad \begin{pmatrix} x \\ y \end{pmatrix} = \begin{pmatrix} u\cos^3 t \\ v\sin^3 t \end{pmatrix}$$

Try to imagine what these curves look like, or plot the result as a graph.

5.3   Simulation of the Sequence of a Copolymer

At this point a problem from polymer chemistry shall be treated: the simulation of the sequential structure of a copolymer.

For the formation of a copolymer the monomers **A** and **B** are present in equal concentrations and in excess. The polymer chain which ends at the active end (radical or ion) with an **A** molecule has a different reactivity to the monomers **A** or **B**. The same is true for a polymer chain ending with a **B**. It is evident that a strictly alternating copolymer is formed if the **A**-end reacts with **B** only, and the **B**-end with **A** only. But what does a copolymer look like, if the reaction probabilities are not the same? This question will be answered by a simulation method based on a Monte-Carlo calculation.

First it has to be specified which monomer starts the chain growth. Then the probability p_1 for the reaction of an end group **A** with the monomer **A** should be known as well as the probability p_3 for a reaction of the endgroup **B** with monomer **B**. The probability $(1 - p_1)$ is then the reactivity of the endgroup **A** with monomer **B** and $(1 - p_3)$ the reactivity of the endgroup **B** with monomer **A**.

The computer must remember whether the active end of the polymer chain is an **A** or **B** molecule. Then it has to choose, following the assumed probabilities, which monomer is next attached to the chain. This monomer is shown on the screen and is stored as the new active end. Then the selection cycle starts anew.

```
0 REM "CO-POL  "           EBERT/EDERER           880511
1 REM **********************************************
2 REM *** Monte-Carlo calculation of the sequence ***
3 REM *** of a copolymer. Given are the probabili- ***
4 REM *** ties A + A and B + B and the concentra- ***
5 REM *** tions of A and B.                       ***
9 REM **********************************************
50      RANDOMIZE (TIMER/3) MOD 32767
100     PRINT "What is the probability for a polymer"
200     PRINT "with an end group A to react with "
300     PRINT "the monomer molecule A";
400     INPUT P1
550     PRINT
600     PRINT "What is the probability for a polymer"
700     PRINT "with an end group B to react with "
800     PRINT "the monomer molecule B";
900     INPUT P3
1100    PRINT
2000    INPUT "Do you want to start with A or B ";E$
2100    PRINT E$;
2200      X=RND(X)
2300      IF E$="b" THEN GOTO 3000
2400      IF X<P1 THEN GOTO 2100
2500      E$="b":GOTO 2100
3000      IF X<P3 THEN GOTO 2100
3100    E$="a":GOTO 2100
4000    END


RUN
What is the probability for a polymer
with an end group A to react with
the monomer molecule A ? 0.9

What is the probability for a polymer
with an end group B to react with
the monomer molecule B ? 0.9
```

```
Do you want to start with A or B ? a
aaaabbaaabbbbbbbbaaaaaaaaaaaabbbbbbbbbbbaabbbbbbbbb
aaaabaaaabbbbbbbbbbbbbbbbbbbbaaaabbbbbbbbbbbbbbbbbbb
bbabbbbbbbbbbbbbbaaaaaabbaabbbbbbbbbbbbbbbbbbbbbbbbb
bbbbbbbaaaaaaaaaaaaaaaaaaabbbaaaaaaaaaaaaaaaaaaabbbb
bbbbbbaaaaabbbbaaaaaabbbabbaaaaaaaaaaaaabbbbbbbbaaa
aaaabbbbaaaaaaaaaaaaaaaaaaaaaaaaaaaaaabbbbbbbbbbbbaa
aaaaaaaaaaabbbbbbbbbbbbbbbbbbbbbbbbbbbbbbbbbbbbbbbbaa
bbbbbbbbbbbabbbaaaaaabbbbbbbbbbbbbbbbbbaaaaaaaaaaaaa
aaabbbbbbbaaaaabbbbbbbbbbbbbbaaaabbbbbbbbbbbbbbbbbaab
bbbbbbbbbbbbbbbbbbbbbbbbbbbbbbbbbb
^C
Break in 2300

RUN
What is the probability for a polymer
with an end group A to react with
the monomer molecule A ? 0.1

What is the probability for a polymer
with an end group B to react with
the monomer molecule B ? 0.1

Do you want to start with A or B ? a
ababababababaababababababababaababababaabaababababababababababab
abababababababababababababababbababababababababababababababbab
babababababbababababbababababaaabababababababbabbabaababababaabab
abababbababababababbabbababababababababababbababababababaaab
ababbababababbabaabababababababababababaabababaabaabbaabba
babaaabababababbababababbbababababbababababababaabaabababaabba
bababababababbababababababaaabababababababababababababababababab
baababbababababababaababababababaababababababaababaabbabbabbb
abababababababababababababababbbabababababababababababababababab
aababababababababbbaabbab
^C
Break in 2200

RUN
What is the probability for a polymer
with an end group A to react with
the monomer molecule A ? 0.2

What is the probability for a polymer
with an end group B to react with
the monomer molecule B ? 0.8
```

```
Do you want to start with A or B ? a
aabbbbbbbbbabbbbbbbbbbbbababbbbbabbbbbbaabbbbbbbbbbbabb
bbbbbbbbbbbabbbbbbabbbbbabbbbbbbabbaabbbbaaabbbbaa
babbbabbbabababbbbbbbabbabbabbbbbaabbbbaabbbbbaabbb
bbbababbbbabbbbbaababababbbbabbbbbabbabbbbabbbbbbbbb
bbbababbbbbbaabbbabababbbabbbbbbbbbbaabbbbabaabbbbbb
bbbbbbababbbbbabbbbbabbbbbbbbbabbbbbbbbbbbbbbbbbbbaba
^C
Break in 2100
```

The input of the probabilities takes place in lines 400 and 900. The starting monomer is written in the variable E$ in line 2000. In E$ the active end is stored during the whole simulation. The active end is printed on the screen in line 2100. This is an example for INPUT and PRINT with the so-called text or string variable. In line 2200 a pseudo random number between 0 and 1 is determined. Lines 2400 and 2500 are appropriate, if the active end of the polymer chain is **A**, and lines 3000 and 3100 are executed if **B** is the active end. This is decided by the IF-statement in line 2300. If E$ is **A**, i.e. the active chain end is **A**, then the program proceeds to line 2400. Here it is determined whether the random number is smaller than the probability p_1 (the probability of reaction of the chain end **A** with monomer **A**). If this is true, then the chain end **A** reacts with **A** and the new chain end is again an **A**-molecule and the character of the chain end does not change. If E$ is not changed, the program jumps back to line 2100, where E$ is printed out and the cycle starts again. In case the random number is not smaller than the probability p_1, then the program proceeds from line 2400, where the decision is made to go to line 2500, in which the variable E$, the polymer end, is changed to **B**. After the colon ':', which is nearly the same as a new line, the program jumps back to the output and the cycle starts again. The sequence in the program for the case that the active end is **B**, lines 3000 and 3100, is completely analogous; the change of the polymer end means in this case the alteration of E$ to **A**. In the first example shown, a block polymer is formed if the probabilities for the monomers to react with themselves are high. An alternating polymer is formed if these probabilities are very small. If one probability is high and the other low, a polymer is formed consisting of blocks of one sort of monomers spiked with few monomers of the other kind.

The Colon in BASIC

:

Syntax: z stmt1 : stmt2

z   :   line number
stmt :   any correct BASIC-statement

The colon is in principle no new statement but it separates different BASIC-statements on the same line. In standard BASIC a new line with a line number is usually written for each statement. However, a new line can be saved by putting a colon at the end of a statement followed directly by the next statement. Therefore colons can be used to keep programs short.

Examples: 10 FOR I=1 TO 50
            20 PRINT LOG(I+1/I)
            30 NEXT I

can be shortened to:

10 FOR I=1 TO 50 : PRINT LOG(I+1/I) : NEXT I

and   10 FOR I=1 TO 10
           20 FOR J=1 TO 10
           30 PRINT I*J;
           40 NEXT J
           50 PRINT
           60 NEXT I

can be shortened to:

10 FOR I=1 TO 10:FOR J=1 TO 10:PRINT I*J;:NEXT J:PRINT:NEXT I

Whether the short forms are used is an individual decision of the programmer. Using the colon can result in more clearly arranged programs, but the contrary can also happen. The short programs need less storage space and run faster. An important restriction exists for the GOTO- and IF-statements, as program jumps always go to the whole line and not only to a part of it.

Problem 37
Modify the program CO-POL in such a way that a copolymerization sequence of exactly 100 monomer units is printed out in blocks of 25 units in 4 lines.

Problem 38

Allow different initial concentrations of the monomers **A** and **B**, and modify the program CO-POL or the program of Problem 37.

The following remarks may assist you in solving this problem. Table 5.1 applies for equal concentrations of the monomers **A** and **B** ($c_A = c_B$). The more general expression for the reaction probabilities with different concentrations of the monomers is shown in Table 5.2.

Table 5.1. Reaction probabilities in copolymerization

Polymer end	Monomer	Probability
A	A	p_1
A	B	$1 - p_1$
B	B	p_3
B	A	$1 - p_3$

Table 5.2. Reaction probabilities in copolymerization with different concentrations c_A and c_B of the monomers — the N_i's are normalizing factors

Polymer end	Monomer	Probability
A	A	$p_1 \cdot c_A / N_1$
A	B	$(1 - p_1) \cdot c_B / N_1$
B	B	$p_3 \cdot c_B / N_2$
B	A	$(1 - p_3) \cdot c_A / N_2$

N_1 is the normalization divisor to guarantee that the sum of the first two probabilities is always unity. The same applies to N_2.

$$N_1 = p_1 \cdot c_A + (1 - p_1) \cdot c_B$$
$$N_2 = p_3 \cdot c_B + (1 - p_3) \cdot c_A$$

Problem 39

Write a program for simulating the copolymerization in which the assumptions of excess concentrations of the monomers has been dropped.
Hint: Two additional variables are needed for the concentrations c_A and c_B. During each polymerization step the concentration of the reacting species is reduced by a small but constant amount and this new concentration is used for the next step.

Problem 40

Write a program for simulating a copolymerization for three different monomers reacting and present in excess.

5.4   Integration According to Simpson

After an excursion to polymerization kinetics, we return to numerical integration. The method of Euler calculates the area of a trapezoid as the integral approximation over a subinterval. If two subintervals are combined, and the area is not approximated with two trapezoids but with the area under a parabola, one obtains the integration method of Simpson (Simpson's barrel rule). Further proof of this method will not be given here, but can be looked up in numerical mathematics textbooks.

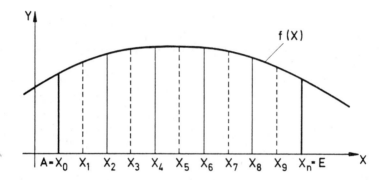

Fig. 5.3. Graph of a function whose definite integral between $A = x_0$ and $E = x_n$ will be determined using Simpson's method

In this example the integration interval (A, E) is divided into 10 subintervals. The length h of a subinterval is

$$h = \frac{E - A}{n} = \frac{x_n - x_0}{n} \qquad n = \text{number of subintervals}$$

In case the curve in the interval (x_0, x_2) is the graph of a parabola, the underlying area S can be expressed exactly by

$$S = \frac{h}{3} \cdot \left(f(x_0) + 4f(x_1) + f(x_2) \right)$$

Summing up all subintervals, the following formula is obtained

$$S = \frac{h}{3} \cdot \Big(f(x_0) + 4f(x_1) + 2f(x_2) + 4f(x_3) + 2f(x_4)$$
$$+ 4f(x_5) + 2f(x_6) + 4f(x_7) + 2f(x_8) + 4f(x_9) + f(x_{10}) \Big)$$

If the definite integral over the interval (A, E) of an arbitrary function $f(x)$ is to be calculated, the Simpson method requires the interval to be divided into $2n$ subintervals.

$$S_{2n} = \frac{E - A}{6n} \cdot \left(f(x_0) + f(x_{2n}) + 4 \sum_{i=1}^{n} f(x_{2i-1}) + \sum_{i=1}^{n-1} f(x_{2i}) \right)$$

The error involved in the Simpson integration becomes smaller with the fourth power of the length h of the subinterval, i.e. if the number of subintervals is increased by the factor 2, the error involved decreases by the factor 16.

$$\text{error} \sim h^4$$

This integration method is applied to the function $f(x) = x^2 \cdot e^{-x^2}$ and the corresponding program is listed in SIMPSON1:

```
0  REM "SIMPSON1"          EBERT/EDERER          880601
1  REM ***************************************************
2  REM ***           Integration                   ***
3  REM *** of an arbitrary function with the method ***
4  REM *** of Simpson.                              ***
5  REM *** The function which will be integrated    ***
6  REM *** has to be programmed in line # 100.      ***
9  REM ***************************************************
50     INPUT "Lower integration limit "; A
60     INPUT "Upper integration limit "; E
100    DEF FNF(X)=X*X*EXP(-X*X)
300    N=4
1000   S=0
1100     H=(E-A)/N
2000     S=FNF(A)
3000     FOR I=1 TO N-1 STEP 2
3100       X=A+I*H
3200       S=S+4*FNF(X)
3300     NEXT I
3500     FOR I=2 TO N-2 STEP 2
3600       X=A+I*H
3700       S=S+2*FNF(X)
3800     NEXT I
4000     S=S+FNF(E)
4200     S=S*H/3
5000     PRINT:PRINT "Integral from";A;"to";E
```

```
6000    PRINT "with";N;"subintervals =";S
7000    N=N*2
8000  GOTO 1000
10000 END

RUN
Lower integration limit ? 0
Upper integration limit ? 1

Integral from 0 to 1
with 4 subintervals = .189512

Integral from 0 to 1
with 8 subintervals = .1894745

Integral from 0 to 1
with 16 subintervals = .1894725

Integral from 0 to 1
with 32 subintervals = .1894723

Integral from 0 to 1
with 64 subintervals = .1894723

Integral from 0 to 1
with 128 subintervals = .1894724

Integral from 0 to 1
with 256 subintervals = .1894724
^C
Break in 3200
```

The program SIMPSON1 is similar to EULER2; both programs contain an infinite external loop in which the number of the subintervals is doubled at each run. The program has to be stopped by hand. The INPUT lines determine the integration intervals that are identical, as are the lines up to 1100. But the defined function is different in this program, and the initial number of subintervals is considerably smaller, namely only 4. In line 2000 the integration is started with the calculation of the function value $f(A)$ of the general function. The N in the program corresponds to the $2n$ of the formula. The loop from line 3000 to 3300 calculates the first sum, the loop from 3500 to 3800 the second sum of the above formula. Both loops contain two statements: in the first one the x-value is calculated, which corresponds to the running variable I, and in the second statement the function value is multiplied by 2 and 4, respectively, and added to the sum S. In line 4000 the function value $f(E)$ is added as the last summand to S. After the multiplication of the sum with one third of the length of the subintervals, the estimated sum of the integral is calculated according to Simpson's method. The output, the increasing of the number of subintervals, and jumping back are again identical with the program EULER2.

Problem 41

Use program SIMPSON1 to integrate the frequency distribution function of Maxwell. As this function is rather extended and does not fit in one program line, use several function definitions. This is a general practice when one has to deal with long formulas.

Problem 42

Integrate the same function $f(x)$ with the programs EULER2 and SIMPSON1 and compare how fast the different programs approach the final values. You will recognize that it pays off to choose the right method of numerical integration. SIMPSON1 is only a little more complicated but considerably more effective than EULER2. For integral transformation, e.g. the Fourier transformation, a great number of integrals have to be calculated for one transformation. In such cases a higher efficiency may allow the calculation to be carried out on a personal computer.

5.5   Integrations with Termination Criteria

If a program is to be written in which many integrals have to be calculated, then the program cannot be terminated by hand for each single integral. The computer has to be programmed in such a way that it decides whether the accuracy of the result is good enough. For that purpose consider the following general statement:

> There are no simple and universal criteria for the termination of a program which loops to get an accurate result. For each example the special termination criterion used has to be tested critically.

In the following example we shall use a method which is easy to program. In this method we compare the most recently calculated approximation of the integral S with the previously calculated approximation S1. If the difference of both values is smaller than a given small number the calculation shall be discontinued and the value S printed as result for the definite integral. If the result of a calculation corresponds to a physical quantity then the user very often can give an absolute value for the accuracy requirement, e.g. a temperature has to be calculated to an accuracy of 0.1°. But it may also happen that a relative accuracy is required, i.e. the result has to be accurate to *n* significant digits.

```
0 REM "SIMPSON2"          EBERT/EDERER          880601
1 REM ***********************************************
2 REM ***              Integration              ***
3 REM *** of an arbitrary function by the method ***
4 REM *** of Simpson.                           ***
5 REM *** The function to be integrated has to be ***
6 REM *** programmed in line # 100.             ***
9 REM ***********************************************
50      INPUT "Lower integration limit "; A
60      INPUT "Upper integration limit "; E
70      INPUT "Accuracy in % ";G
100     DEF FNF(X)=X*X*EXP(-X*X)
300     N=4
500     S1=0
1000    S=0
1100      H=(E-A)/N
2000      S=FNF(A)
3000      FOR I=1 TO N-1 STEP 2
3100        X=A+I*H
3200        S=S+4*FNF(X)
3300      NEXT I
3500      FOR I=2 TO N-2 STEP 2
3600        X=A+I*H
3700        S=S+2*FNF(X)
3800      NEXT I
4000      S=S+FNF(E)
4200      S=S*H/3
4500      IF ABS(S-S1)<ABS(S)*G/100 THEN 9000
4600      IF N>2500 THEN 8000
5000    S1=S : N=N*2 : GOTO 1000
8000    PRINT "Required accuracy is not possible "
8050    PRINT "Last estimation of the integral =";S1;S
```

```
8100   GOTO 70
9000   PRINT:PRINT "Integral from";A;"to";E
9100   PRINT "with";N;"subintervals =";S
9200   PRINT:PRINT "Accuracy required:";ABS(S)*G/100
9300   PRINT "Accuracy calculated:";ABS(S-S1)
10000  END

RUN
Lower integration limit ? 0
Upper integration limit ? 1
Accuracy in % ? 0.1

Integral from 0 to 1
with 8 subintervals = .1894745

Accuracy required: 1.894745E-04
Accuracy calculated: 3.758073E-05
Ok
```

In program SIMPSON2 this extention for the automatic termination is introduced. The first new line is line 70 in which the desired accuracy in percent is put in variable G. As two sums shall be compared, a second variable $S1$ is introduced and set to 0 in the new line 500. The central integration program up to line 4200 is identical with SIMPSON1. In line 4500 the criterion for ending the calculation is defined: if $|S-S1|$ is smaller than the given percentage G from the value of the integral estimate sum S, then the integration is terminated. The IF-statement causes the program to jump to line 9000 and to print the results of the integration.

If the requirement of accuracy in line 4500 is not fulfilled, then in line 4600 it is examined whether the number of the subintervals has exceeded a certain maximum number. This maximum number has been arbitrarily set to 2500 in this program and if exceeded, the IF-statement in 4600 branches the program to line 8000 and a note will be printed that the required accuracy cannot be obtained.

If neither the requirement of accuracy in line 4500 is fulfilled nor the maximum number of intervals exceeded then the program continues with line 5000. There are three statements which are separated by colons. First $S1$ is set to the actual value of S (because subsequently a new S is calculated), then the number of subintervals is doubled and the program jumps back to line 1000 where the sum variable S is set to 0 and the integration restarts from the beginning.

Problem 43

Modify the program SIMPSON2 to calculate within the interval (A, E) a primitive function $F(x)$ numerically from the function $f(x)$. Print it in the form of a table and plot the graph of the function $f(x)$ and its primitive function $F(x)$.

Hint: The primitive function $F(x)$ is given by

$$F(x) = \int_A^x f(t)\, dt \qquad \text{because} \qquad \frac{dF(x)}{dx} = f(x)$$

Set the lower integration limit to A and use an external loop with the upper integration limit as running variable. You can choose the increment k and the end point of the outer loop at will. Then print the upper integration limit together with the calculated integral.

A more effective program for this problem may be written if you use the following sequence

$$F(A) = 0$$

$$F(A + k) = F(A) \qquad + \int_A^k f(t)\, dt$$

$$F(A + 2k) = F(A + k) \qquad + \int_k^{2k} f(t)\, dt$$

$$F(A + 3k) = F(A + 2k) \qquad + \int_{2k}^{3k} f(t)\, dt$$

$$F(A + 4k) = F(A + 3k) \qquad + \int_{3k}^{4k} f(t)\, dt$$

$$\cdots$$

$$F(A + (i + 1) \cdot k) = F(A + i \cdot k) \; + \int_{i \cdot k}^{(i+1) \cdot k} f(t)\, dt$$

Problem 44

The program SIMPSON2 can easily be modified to calculate the Fourier transform of a function $f(x)$. Details on the Fourier transformation and the recommended example should be looked up in the literature.[2] If only symmetric functions are considered — symmetric means: $f(x) = f(-x)$ — then the Fourier transform of $f(x)$ is also a real function (for nonsymmetric functions it is complex) and the formula for this transformation can be written as

$$T(y) = \frac{1}{\sqrt{2\pi}} \cdot \int_{-\infty}^{+\infty} f(x) \cdot \cos(xy) \, dx$$

$f(x)$ is the function to be transformed; $T(y)$ is the Fourier transform. Calculate from a symmetric function $f(x)$ of your choice the Fourier transform $T(y)$ and print it in the form of a table or draw a graph of the functions. Try functions which fade away for large positive or negative arguments, i.e. the function values $f(x)$ become very small if x is large. Then the integration limits can be replaced by finite values.

For example, test the Fourier transformation program with the following function first

$$f(x) = \begin{cases} 1 & \text{if } |x| \leq 1 \\ 0 & \text{else} \end{cases}$$

The equation for the Fourier transform becomes

$$T(y) = \frac{1}{\sqrt{2\pi}} \cdot \int_{-1}^{+1} 1 \cdot \cos(xy) \, dx$$

This can also be solved analytically, and the result is

$$T(y) = \frac{2 \sin y}{y\sqrt{2\pi}}$$

Compare the analytical solution with the numerical result.

Fourier transformations are needed to calculate the usual IR-spectrum from a Fourier IR-spectrometer. The Fourier IR-spectrometer has a considerably better signal to noise ratio than a normal IR-spectrometer and therefore it is far more sensitive. For practical calculations it is recommended not to use the method shown here, but the so-called 'Fast Fourier Transform' method (FFT).[3,4] For the simple method, the transformation of n points needs $n \cdot (n - 1)$ computer operations such as additions or multiplications; the FFT on the other hand needs only $n \log n$ operations. This means that for 4096 points only 50000 operations are used for the FFT method instead of 16 700 000 for the conventional method.

[2] E. Kreyszig, *Advanced Engineering Mathematics*, 6.ed., Wiley, New York (1988)

[3] J.W. Cooley, J.W. Turkey, *Math. Comp.*, **19**, 297 (1965)

[4] E.D. Brigham, *The Fast Fourier Transform*, Prentice Hall, Englewood Cliffs, New York (1974)

Two more examples from physical chemistry, which contain simple applications of the integration procedures introduced, should be treated in more detail.

5.6   Specific Molar Heat of Metals According to Debye

The general theory for the specific molar heat of metals was developed by Debye. The result of this theory is a formula for the specific molar heat C_v, which is dependent on the temperature T and the so-called Debye temperature θ. The Debye temperature θ is a specific characteristic constant for each metal. The formula contains an integral which cannot be solved analytically. Therefore the calculation has to be carried out with a numerical integration program. Strictly speaking, this formula is dependent only on the ratio θ/T; introducing $\theta/T = z$ the Debye formula reads as follows

$$C_v(z) = \frac{9R}{z^3} \cdot \int_0^z x^4 \frac{e^x}{(e^x - 1)^2} \, dx$$

The program DEBYE for the calculation of the Debye formula for the specific heat of a metal is listed as follows:

```
0 REM "DEBYE   "          EBERT/EDERER          880601
1 REM **************************************************
2 REM *** Calculation of Cv according to the      ***
3 REM *** Debye formula. The integration is done  ***
4 REM *** with Simpson's method.                  ***
5 REM *** Cv = 9R/(@/T)^3 *                        ***
6 REM *** Integral(x^4*exp(x)/(exp(x)-1)^2)        ***
7 REM *** @ = Debye temperature                    ***
9 REM **************************************************
50     A=.00001
60     INPUT "@/T ";E
70     G=.01
100    DEF FNF(X)=X^4*EXP(X)/(EXP(X)-1)^2
300    N=4
500    S1=0
1000   S=0
1100     H=(E-A)/N
2000     S=FNF(A)
```

```
3000    FOR I=1 TO N-1 STEP 2
3100       X=A+I*H
3200       S=S+4*FNF(X)
3300    NEXT I
3500    FOR I=2 TO N-2 STEP 2
3600       X=A+I*H
3700       S=S+2*FNF(X)
3800    NEXT I
4000    S=S+FNF(E)
4200    S=S*H/3
4300    S=S*9/E/E/E
4500    IF ABS(S-S1)<ABS(S)*G/100 THEN 9000
4600    IF N>2500 THEN 8000
5000    S1=S : N=N*2 : GOTO 1000
8000    PRINT "The required accuracy is not possible "
8050    PRINT "Last estimation for the integral";S1;S
8100    STOP
9000    PRINT:PRINT "Spec.molar heat Cv of a metal"
9100    PRINT "with @/T of ";E;" = ";S;"* R"
9999    END


RUN
@/T ? 3

Spec.molar heat Cv of a metal
with @/T of  3  =  1.988268 * R
Ok
RUN
@/T ? 1

Spec.molar heat Cv of a metal
with @/T of  1  =  2.855196 * R
Ok
RUN
@/T ? 0.1

Spec.molar heat Cv of a metal
with @/T of  .1  =  2.9985 * R
Ok
RUN
@/T ? 0.01

Spec.molar heat Cv of a metal
with @/T of  .1  =  2.999977 * R
Ok
```

The program DEBYE is based on the integration program SIMPSON2. The function $f(x)$ to be integrated is programmed in line 100. The input of the integration limits was modified. According to the formula, the lower limit is 0. As the numerical calculation for $x = 0$ of the function $f(x)$ causes difficulties (see the next paragraph), the lower limit is set to the very small number of A=0.00001. For the input of the upper integration limit into the variable E the quotient $z = \theta/T$ is needed. The required accuracy is set to 0.01 percent in this program. The central integration program is unchanged. In the new line 4300 the estimate integral is multiplied with the factor 9/E/E/E which is placed in front of the integral. Note that the output is in units of R. A program run shows that C_v converges at higher temperatures against the classical value of $3R$.

Another example in which difficulty arises in calculating the function value at distinct singular points is

$$f(x) = \frac{x^2 - 100}{x - 10} = x + 10$$

This function can be calculated for all x. If the unreduced form is programmed, the computer responds at $x = 10$ with an error statement. Such singular points must either be calculated separately using IF-statements or one should try, as in our DEBYE program, to avoid them by certain tricks.

Problem 45

Use the program DEBYE to calculate the specific heat C_v as a function of θ/T, and plot the results in a diagram.

5.7   Calculation of the Virial Coefficient From a Known Intermolecular Potential

A real gas is often described by adding to the ideal gas law a power series containing terms which are dependent on pressure, with coefficients called virial coefficients. If this series is terminated after the second term the following formula is obtained.

$$pV = RT + B(T) \cdot p$$

The second virial coefficient $B(T)$ can be determined experimentally using the following equation

$$B(T) = \frac{\partial(pV)}{\partial p}$$

The dimension of $B(T)$ is volume/mole. Real gases deviate from the ideal gas law because of the intermolecular forces. The second virial coefficient $B(T)$ shall be calculated from the intermolecular potential $E_p(r)$. For simplification, two spherical atoms or molecules shall be considered. The potential E_p is then dependent only on the distance r and not on the orientation of the two particles. The intermolecular potential shall have the form of a Lennard-Jones potential

$$E_p = 4\epsilon \cdot \left[\left(\frac{\sigma}{r}\right)^m - \left(\frac{\sigma}{r}\right)^n \right]$$

r : Distance between the atoms
σ : Distance, at which the attracting and repulsing forces are equal
ϵ : Depth of the potential
n : Power of the attractive term (usually set to 6)
m : Power of the repulsive term (usually set to 12)

The intermolecular power $K(r)$ can generally be expressed as

$$K(r) = -\frac{dE_p(r)}{dr}$$

The second virial coefficient $B(T)$ can be calculated using the following integral

$$B(T) = 2\pi N_A \cdot \int_0^\infty \left(1 - e^{-\frac{E_p(r)}{kT}} \right) \cdot r^2 \, dr$$

The program to calculate the virial coefficient is listed below:

```
0 REM "VIRIAL  "         EBERT/EDERER         880318
1 REM ************************************************
2 REM *** Calculation of the second virial coeff.  ***
3 REM *** using Simpson's integration routine.     ***
4 REM *** The Lennard-Jones potential is used.     ***
9 REM ************************************************
15    PRINT "First parameter of the Lennard-Jones pot. "
20    INPUT "Sigma ";S9
25    PRINT "Second parameter of the Lennard-Jones pot. "
30    INPUT "Epsilon/k(Boltzmann) ";E9
40    INPUT "Temperature in K ";T9
50    A=S9/1000
60    E=S9*10
70    INPUT "Accuracy in % ";G
100 DEF FNF(X)=(1-EXP(-4*E9/T9*((S9/X)^12-(S9/X)^6)))*X*X
300   N=4
500   S1=0
1000  S=0
1100   H=(E-A)/N
2000   S=FNF(A)
3000   FOR I=1 TO N-1 STEP 2
3100     X=A+I*H
3200     S=S+4*FNF(X)
3300   NEXT I
3500   FOR I=2 TO N-2 STEP 2
3600     X=A+I*H
3700     S=S+2*FNF(X)
3800   NEXT I
4000   S=S+FNF(E)
4200   S=S*H/3
4300   S=S*2*3.14159*6.022E+23*1E-24
4500   IF ABS(S-S1)<ABS(S)*G/100 THEN 9000
4600   IF N>2500 THEN 8000
5000  S1=S : N=N*2 : GOTO 1000
8000  PRINT "The required accuracy is not possible"
8050  PRINT "Last estimation of the integral ";S1;S
8100  STOP
9000  PRINT:PRINT "Integral ";A;"-";E;"Angstroem"
9100  PRINT "with";N;"subintervals =";S
9200  PRINT:PRINT "Accuracy required ";ABS(S)*G/100
9300  PRINT "Accuracy calculated ";ABS(S-S1):PRINT
9400  PRINT "Sigma                 = ";S9;"Angstroem"
9500  PRINT "Epsilon/k(Boltzmann) = ";E9;" K"
9600  PRINT "Temperature           = ";T9;" K"
9700  PRINT "Second virial coeff. = ";S;" cm^3/mol"
9999  END
```

```
RUN
First parameter of the Lennard-Jones pot.
Sigma ? 3.698
Second parameter of the Lennard-Jones pot.
Epsilon/k(Boltzmann) ? 95.05
Temperature in K ? 773
Accuracy in % ? 0.1


Integral   .003698 - 36.98 Angstroem
with 512 subintervals = 26.6624


Accuracy required  .0266624
Accuracy calculated  4.959107E-05


Sigma                = 3.698 Angstroem
Epsilon/k(Boltzmann) = 95.05  K
Temperature          = 773  K
Second virial coeff. = 26.6624  cm^3/mol
Ok
```

Numerical values for the second virial coefficient and the corresponding parameters for the Lennard-Jones potentials can be looked up in physical
chemistry textbooks[5] or handbooks. The program VIRIAL is identical to
the program SIMPSON2 from line 70 up to line 9300 with two exceptions.
The jump in line 8100 was replaced by a STOP-statement and in line 9000
the word Ångström was added to the output statement. Line 4300 has been
added to carry out a normalization and a redimensioning of the output.
The function to be integrated is given by the above formula and is defined
in line 100. It is programmed so that the input of the parameter σ is in
Ångström, the input of the parameter ϵ in Kelvin (i.e. the energy ϵ is to
be divided by the Boltzmann constant k) and the input of the temperature
should also be in Kelvin. These inputs happen in lines 15 through 40. Line
4300 serves for the transformation of the virial coefficients into the units
cm^3/mol. Furthermore, in this line a multiplication with the prefactor of
the above integrals takes place. The output is extended by four lines; the
three input parameters and the second virial coefficient are written on the
screen. In the formula the integration limits run from 0 through ∞. In

---

[5] R.S. Berry, S.A. Rice, J. Ross, *Physical Chemistry*, John Wiley & Sons, (1980),
p.761 and 784ff

numerical methods it is not possible to integrate to an infinite upper integration limit. The limits must be set to finite values. Therefore the limits are simply moved to values where the integrand is already 'very small'. In line 60 the upper integration limit is set to $10\,\sigma$. If one remembers the shape of the potential curves this seems to be reasonable. (A calculation with a higher upper limit can be carried out and the results compared; the computing time becomes longer if the interval is increased.) The lower integration limit of 'zero' leads, as mentioned above, to an error so it will be set to $\sigma/1000$. The example shows the calculation of the second virial coefficient at 773 K with the parameters of nitrogen for the Lennard-Jones potential. (Comment to the unpatient: in this example one usually has to wait for some minutes for the result. Additional lines can be programmed to write intermediate results on the screen.)

Table 5.3. Parameters of the Lennard-Jones potentials for different gases

Substance	σ in [Å]	$\frac{\varepsilon}{k}$ [K]
He	2.556	10.221
H_2	2.928	37.00
Ne	2.749	35.60
Ar	3.405	119.8
Kr	3.60	171
Xe	4.100	221
O_2	3.58	117.5
CO	3.763	100.2
N_2	3.698	95.05
CH_4	3.817	148.2

With the program VIRIAL the second virial coefficient can be calculated from the intermolecular potential. Usually the problem is the other way round, one determines the second virial coefficients experimentally and wants to calculate the intermolecular forces. This is called an inverse problem. We can try to solve it by trial and error using the program VIRIAL.

Problem 46

Calculate directly with VIRIAL or by using a modified program a table which contains the temperature dependence of the second virial coefficients of different substances. Plot the results as a graph against the temperature T and against $1/T$.

Problem 47

Integrate the spectral energy density according to Planck over all wavelengths to get the total energy. Derive the rule that the total energy is proportional to the fourth power of temperature T.

Problem 48

Calculate the difference of the internal energy of two different temperatures from the C_v-values using the following formula

$$\Delta U = U(T_2) - U(T_1) = \int_{T_1}^{T_2} C_v(T)\, dT$$

The C_v-values shall be calculated using the Debye formula. For direct calculation of ΔU you have to write a program for a double integration.

Problem 49

The temperature dependence of C_p shall be calculated with the following formula

$$C_p(T) = 3R \cdot \frac{1 - e^{-\frac{T^2}{10000}}}{1 + \frac{100}{T}}$$

Calculate ΔH and ΔS between two temperatures of your choice (e.g. 78 K and 532 K). Use one of the previous integration programs for this calculation.

Problem 50

An important exercise in polymer analytics is the determination of the mean values of the molecular weight from molecular weight distributions of polymers. Let us assume that the distribution of a polymer can be expressed by the so-called logarithmic normal distribution

$$w(M) = \frac{1}{M} \exp\left(-\frac{(\log M - \log M_0)^2}{b^2}\right)$$

$w(M)$: Weight density distribution function
M : Molecular weight
M_0 : Median of the distribution $w(M)$, viz. half of the polymer sample has a molecular weight smaller than M_0
b : Parameter which describes the width of the distribution

Calculate from a logarithmic normal distribution the number average $\overline{M}_n$, the weight average $\overline{M}_w$, and the centrifuge average $\overline{M}_z$ which are given by the following formulas

$$\overline{M}_n = \frac{\int_0^\infty w(M)\, dM}{\int_0^\infty M^{-1} w(M)\, dM}$$

$$\overline{M}_w = \frac{\int_0^\infty M \cdot w(M)\, dM}{\int_0^\infty w(M)\, dM}$$

$$\overline{M}_z = \frac{\int_0^\infty M^2 \cdot w(M)\, dM}{\int_0^\infty M \cdot w(M)\, dM}$$

Try to explain the relations between the parameters M_0, b and the average values (or their ratios).

Problem 51

Determine the following definite integrals and compare the results of the numerical integration with those of the analytical integration.

$$\int_{x=a}^{x=b} \frac{1}{x^{1.5}}\, dx \qquad \int_{x=a}^{x=b} x^3\, dx$$

$$\int_{x=a}^{x=b} \sin x\, dx \qquad \int_{x=a}^{x=b} x \cdot e^{-x^2}\, dx$$

Problem 52

Write an integration program to calculate the fugacity f as a function of pressure p and temperature T assuming that the van der Waals equation can be applied.

$$\log \frac{f}{p} = \frac{1}{RT} \cdot \int_0^p \left(V - \frac{RT}{p} \right)\, dp$$

The van der Waals equation is given by

$$p = \frac{RT}{V - b} - \frac{a}{V^2}$$

Calculate examples of your choice; the following table should help you to formulate real examples.

Substance	$a/\left[l^2\text{atm} \cdot \text{mol}^{-2}\right]$	$100 \cdot b/\left[\text{lmol}^{-1}\right]$
Benzene	18.000	1.154
Xe	4.194	5.105
Water	5.464	3.049
CO_2	3.592	4.261

Problem 53

The following chemical reaction is given

$$A \xrightarrow{\;k\;} \text{Products}$$

Obeying first order kinetics

$$\frac{d[A]}{dt} = -k \cdot [A]$$

where t is the time and k the reaction rate constant. The Arrhenius equation describes the temperature dependence of k.

$$k(T) = k_0 \cdot e^{-\frac{E_a}{RT}}$$

If the reaction temperature is changed from an initial temperature of 300 K with time by

$$T(t) = 300 + (T_E - 300)(1 - e^{-\frac{t}{q}})$$

T_E is the final temperature and q is the time constant which determines the temperature gradient; then the reaction rate constant is also dependent on time t

$$k\big(T(t)\big) = k(t) = k_0 \cdot e^{-\frac{E_a}{RT(t)}}$$

The concentration of A for a constant reaction temperature T is given by

$$[A] = [A]_0 \cdot e^{-k(T) \cdot t}$$

On the other hand one obtains for the concentration for a time dependent reaction temperature

$$[A] = [A]_0 \cdot e^{-\int_0^t k(T(t))\, dt}$$

Write a program to calculate the dependence of the concentration of A on $[A]_0$, E_a, k_0, T_E, q and t.

Problem 54

In quantum chemistry, overlapping integrals must be calculated frequently. In the following an example is shown in which the principle of the calculation of overlapping integrals is explained. Given is a one-dimensional wave function $w(x)$.

$$w(x) = N \cdot e^{-|x|}$$

First the normalization factor N has to be calculated. N is chosen in such a way that the following equation is satisfied.

$$\int_{-\infty}^{+\infty} w(x) \cdot w(x)\, dx = 1$$

With the normalized wave function overlapping integrals are calculated which have the following form

$$S = N^2 \cdot \int_{-\infty}^{+\infty} w(x - x_a) \cdot w(x - x_b)\, dx$$

x_a and x_b are x-coordinates of the atoms A and B from an arbitrary origin. $(x_a - x_b)$ is the distance between the two atoms. Calculate the overlapping integrals S as a function of the distance of the atoms.

6 Equations

This chapter is limited to solving equations with only one unknown and to providing appropriate examples from chemistry and physics which can be described by these equations. Simple linear but also quadratic equations can easily be solved with a computer by programming the adequate formulas. Algebraic equations of third and fourth degree can be solved analytically; however, the corresponding formulas become so complicated that general numeric methods are usually preferred. Some of them will be treated in this section.

The following equation cannot be solved analytically:

$$x^6 + 4x^5 - 5x^4 + x^3 - 3x^2 - 9x + 11 = 0$$

The same is true for

$$x^2 = \sin x$$

an equation, which we already were concerned with in Problem 3. The following types of equations will be treated in this chapter

Algebraic equations $\qquad x^7 + 3x^4 - 8x^2 - 17x + 1 = 0$

Transcendental equations $\qquad 10x^2 = e^{x^2}$

Implicit functions $\qquad e^{-(x^2+z^2)} = x^2 z + z^2 x^3$

Solving implicit functions seems to be more complicated, as they contain two unknowns. Numerical solutions may be found by setting up a table which includes the (x, z)-pairs which satisfy the implicit equation. To get an idea of the function $z = f(x)$ that solves the implicit equation, the (x, z)-pairs are plotted in a diagram. In general, the three types of equations can be transformed into a normalized form

$$f(x) = 0$$

The example algebraic equation is already in this form. The transcendental equation reads in this normalized form as follows

$$f(x) = 10x^2 - e^{x^2} = 0$$

The same is true for the implicit function

$$f(x) = e^{-(x^2+z^2)} - x^2z - z^2x^3 = 0$$

The expression is considered as a function $f(x)$; z is treated as a parameter to which a certain value has been assigned. Then the expression for the implicit equation is only a function of x. The function can have one or more zero values or none.

Our general problem was reduced by the simple transformation to the determination of the roots of the function $y = f(x)$.

6.1   The Bisection Method

In the following, details such as continuity, differentiability, etc., which are important for mathematicians, are not considered. It is simply assumed that the functions are 'well-behaved'.

How would we try to find the zero positions in the example of the transcendental function if we had time, patience and a pocket calculator? First we would try to find out whether there is a zero value at all. This has to be the case if we find an x-value at which $f(x)$ is positive and another x-value, at which $f(x)$ is negative. Fig. 6.1 illustrates this method.

In our example $f(x) = 10x^2 - e^{x^2}$ for a transcendental equation there is

$$f(0) = -1 < 0$$

$$f(1) = 10 - e > 0$$

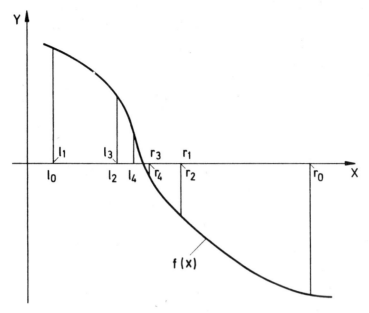

Fig. 6.1. Diagram demonstrating the bisection method

Therefore, there must be a zero value between $x = 0$ and $x = 1$. Naturally this does not mean that there is only one zero value in this interval. Besides, there may be further zero values beyond this interval.

In searching for the zero value between $x = 0$ and $x = 1$ it is logical to calculate first the function value for $x = 0.5$

$$f(0.5) = 1.2156 > 0$$

Therefore a solution value must lie between $x = 0$ and $x = 0.5$. The next trial at $x = 0.25$ results in

$$f(0.25) = -0.43949 < 0$$

Therefore a zero value must be between $x = 0.25$ and $x = 0.5$. If this process is continued the interval in which a zero value must lie is halved (bisected) with each new function calculation. When the interval is small enough — this is decided by the user — then the bisection method is terminated. In our example the first four digits of the zero value are 0.3344. The other roots, which can be calculated the same way are −0.3344, +1.8913 and −1.8913.

This method is called the bisection method because the interval in which the solution of the equation lies is halved at each calculation step.

given: $\qquad l_0 \qquad\qquad\qquad\qquad r_0$

$\qquad\qquad f(l_0) > 0 > f(r_0) \qquad f((l_0 + r_0)/2) < 0$

therefore $\quad l_1 = l_0 \qquad\qquad$ and $\quad r_1 = (l_0 + r_0)/2$

$\qquad\qquad f(l_1) > 0 > f(r_1) \qquad f((l_1 + r_1)/2) > 0$

therefore $\quad l_2 = (l_1 + r_1)/2 \quad$ and $\quad r_2 = r_1$

$\qquad\qquad f(l_2) > 0 > f(r_2) \qquad f((l_2 + r_2)/2) < 0$

therefore $\quad l_3 = l_2 \qquad\qquad$ and $\quad r_3 = (l_2 + r_2)/2$

$\qquad\qquad f(l_3) > 0 > f(r_3) \qquad f((l_3 + r_3)/2) > 0$

therefore $\quad l_4 = (l_3 + r_3)/2 \quad$ and $\quad r_4 = r_3$

$\qquad$ and so on ...

The program for this numerical method must be able to do two things: first, it has to find two different values at which the function values have different signs. Second, it has to perform the bisections. The flow diagram for an appropriate program is shown in Fig. 6.2.

The flow diagram is constructed in a way that the flow jumps to the lower part (line 1000) if the function values at the beginning and the end of the interval have identical signs. The interval (A, E) is divided into a maximum of 1000 subintervals and it will be investigated whether somewhere the sign changes. If this is not the case, it is assumed that no zero value (root) exists within this interval. In case a change of sign was found, this point is taken as the right endpoint of the new interval in which the zero value must lie and the program jumps back into the interval bisecting routine. This point could also have been taken as the new left interval starting point, then the interval bisection would have come to another solution. If the beginning A and the end E of an interval have the same sign for the function value, then there are either none or an even number of zero values in this interval. Of course this is not true if the function has poles like

$$f(x) = \tan(x)$$

or discontinuities like

$$f(x) = \frac{|x|}{x} \cdot \exp(-|x|)$$

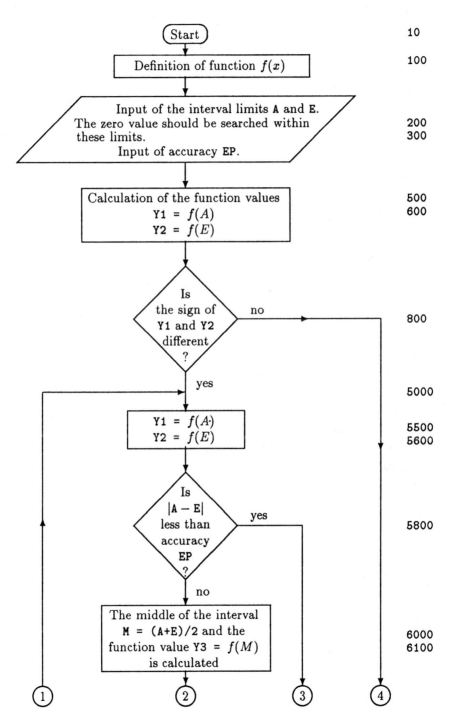

Fig. 6.2. Flow diagram for the bisection method

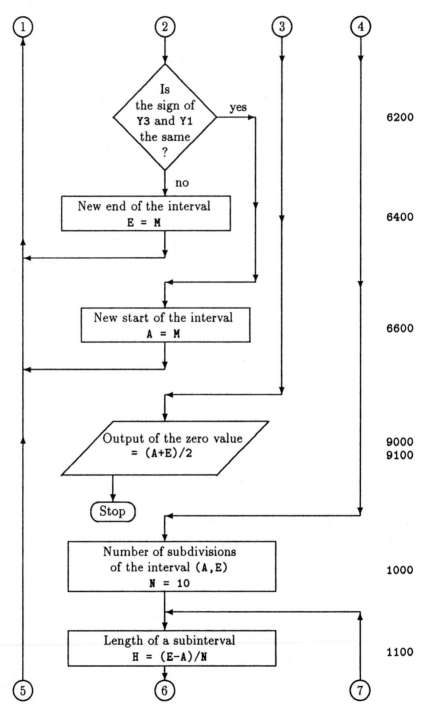

Fig. 6.2. Flow diagram for the bisection method

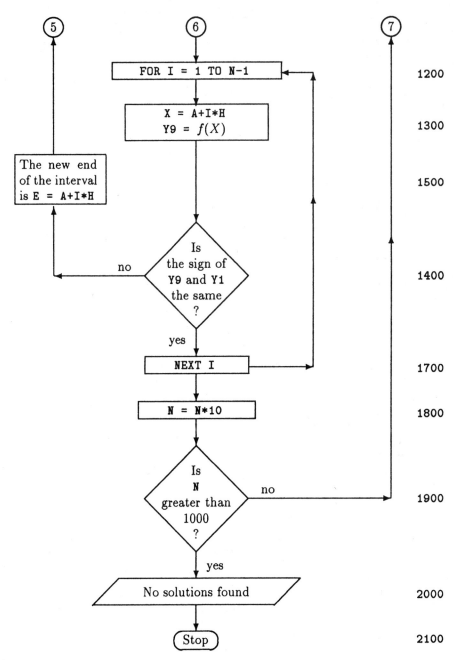

Fig. 6.2. Flow diagram for the bisection method to find a root of a function. It is assumed that the function is continuous which means that two different x-values at which the function values have different signs bracket at least one root of the function.

The BASIC-program BISECT which corresponds to the flow diagram in Fig. 6.2 is listed below. The numbers on the right hand of the flow diagram correspond to the line numbers of the BASIC-program BISECT.

```
0 REM "BISECT  "          EBERT/EDERER          880609
1 REM ***********************************************
2 REM *** This program searches for a root of an   ***
3 REM *** arbitrary function by bisection of the   ***
4 REM *** starting and the following intervals.    ***
5 REM *** The function for which a root shall be    ***
6 REM *** determined has to be programmed in line   ***
7 REM *** # 100.                                    ***
9 REM ***********************************************
100    DEF FNF(X)=23+X*(-17+X*(2+X*(-3+X*(1+.1*X))))
200    INPUT "Boundaries of the searching interval ";A,E
300    INPUT "Absolute accuracy of the root ";EP
500    Y1=FNF(A) : PRINT "y(";A;") = ";Y1
600    Y2=FNF(E) : PRINT "y(";E;") = ";Y2
800    IF   SGN(Y1)=-SGN(Y2) THEN 5000
900 REM ***********************************************
910 REM *** Searching for different signs of the   ***
920 REM *** function values.                       ***
930 REM ***********************************************
1000   N=10
1100   H=(E-A)/N
1200     FOR I=1 TO N-1
1300       Y9=FNF(A+I*H)
1400       IF SGN(Y9)=SGN(Y1) THEN 1700
1500       E=A+I*H
1600       GOTO 5000
1700     NEXT I
1800     N=N*10
1900   IF N<=1000 THEN 1100
2000   PRINT "No root of the function has been found "
2100   STOP
5000 REM ***********************************************
5010 REM *** Start of the bisection routine         ***
5020 REM ***********************************************
5500   Y1=FNF(A):PRINT "y(";A;") = ";Y1
5600     Y2=FNF(E):PRINT "y(";E;") = ";Y2
5800     IF ABS(A-E)<ABS(EP) THEN 9000
6000     M=(A+E)/2
6100     Y3=FNF(M)
6200     IF SGN(Y3)=SGN(Y1) THEN 6600
```

```
6400    E=M : GOTO 5000
6600    A=M : GOTO 5000
9000    PRINT:PRINT "Root =";
9100    PRINT(E+A)/2;"+/-";ABS((E-A)/2)
9999    END


RUN
Boundary of the searching interval ? 0,10
Absolute accuracy of the root ? 0.0001
y( 0 ) =   23
y( 10 ) =   17053
y( 0 ) =   23
y( 2 ) = -7.799999
y( 1 ) =   6.100001
y( 2 ) = -7.799999
y( 1 ) =   6.100001
y( 1.5 ) = -2.303124
y( 1.25 ) =   1.762207
y( 1.5 ) = -2.303124
y( 1.25 ) =   1.762207
y( 1.375 ) = -.3266258
y( 1.3125 ) =   .7068882
y( 1.375 ) = -.3266258
y( 1.34375 ) =   .1870365
y( 1.375 ) = -.3266258
y( 1.34375 ) =   .1870365
y( 1.359375 ) = -7.061768E-02
y( 1.351563 ) =   5.801201E-02
y( 1.359375 ) = -7.061768E-02
y( 1.351563 ) =   5.801201E-02
y( 1.355469 ) = -6.351471E-03
y( 1.353516 ) =   2.581787E-02
y( 1.355469 ) = -6.351471E-03
y( 1.354492 ) =   9.729386E-03
y( 1.355469 ) = -6.351471E-03
y( 1.354981 ) =   1.688004E-03
y( 1.355469 ) = -6.351471E-03
y( 1.354981 ) =   1.688004E-03
y( 1.355225 ) = -2.332688E-03
y( 1.354981 ) =   1.688004E-03
y( 1.355103 ) = -3.223419E-04
y( 1.355042 ) =   6.828308E-04
y( 1.355103 ) = -3.223419E-04

Root = 1.355072 +/- 3.051758E-05
Ok
```

The function for which zero values should be found is defined in line 100 as `DEF FNF(X)`. The function which has been chosen in the example is a polynomial of 5th degree.

$$f(x) = 0.1x^5 + x^4 - 3x^3 + 2x^2 - 17x + 23$$

If this function is transformed so that the high powers are on the right hand side one obtains

$$f(x) = 23 - 17x + 2x^2 - 3x^3 + x^4 + 0.1x^5$$

or

$$f(x) = 23 + x\left(-17 + x\left(2 + x\left(-3 + x\left(1 + 0.1x\right)\right)\right)\right)$$

In this unusual notation no powers are contained any more but only multiplications with x and additions of numerical constants. This polynomial notation, which is also called a Horner scheme is considerably faster to compute.

The actual bisection starts in line 5000. The function values for the beginning and the end of the interval are calculated in lines 5500 and 5600 and printed. This data print-out serves only as intermediate information. The decision whether the calculation is to be terminated or continued is made in line 5800. The value of the actual length of the interval is compared with the value of the user prescribed accuracy `EP` in line 300. This may be critical in some cases. One has to decide at the beginning how much accuracy is required. Other criteria for accuracy can be prescribed as well, e.g. that the accuracy should be one millionth of the initial interval. Another possibility would be to specify the number of interval divisions, e.g. by means of a `FOR-NEXT`-loop. What should not be done is to ask for identical values for the beginning and the end of the interval. Try out what happens in this case, by replacing line 5800 with:

```
5800 IF A=E THEN 9000
```

You will soon find out the reason why the loop will run forever for most examples.

The calculation of the interval center and its function value occurs in lines 6000 and 6100. In lines 6200, 6400, and 6600 it is decided whether the new interval is the left or the right half of the old interval. In printing out the zero value (the root) in lines 9000 and 9100, the center of the interval is the most probable solution (with an error width up to the limits of the interval). There exist more clever estimations, which also consider the function values at both interval ends. In our example the zero value is at 1.355469; this, however, is not the only root of the polynomial.

In line 800 it is decided whether the program jumps to carry out the bisection or whether the first two positions with different signs have to be found. For that purpose the built-in function SGN (abbreviation for sign), the so-called sign function, is applied. The interrogation in line 800 means: if the sign of Y1 is different from the sign of Y2, then jump to line 5000. In reality the function SGN(X) results in $+1$ if x is a positive number and in -1, if x is a negative number, and in 0, if x equals zero. Thus, in line 800 the number $+1$ or -1 is compared with the number $+1$ or -1. This sign function is also used in lines 1400 and 6200. If the signs are the same, then the program investigates the specified interval with a maximum of 1000 subdivisions for a sign change.

Problem 55

Test the difference in computing time necessary for the bisection method, if a polynomial is programmed in the Horner scheme and in the usual notation.

Problem 56

The program of bisection can be speeded up by a factor of two if at each run the function value is calculated at the new interval end only. Rewrite the program BISECT for that purpose. You have to delete two lines and extend two other lines with an additional statement.

Problem 56a

Write a program analogue to BISECT to find a maximum (or a minimum) of a function. You have to be sure that there is exactly one maximum within your initial interval from x_a to x_e, from this follows that every x within the interval has a larger function value than the function values at the interval limits.

$$x \in (x_a, x_e) \Rightarrow f(x) \geq f(x_a) \quad \text{and} \quad f(x) \geq f(x_e)$$

Test two points within the interval, e.g. $x_1 = 0.7x_a + 0.3x_e$ and $x_2 = 0.3x_a + 0.7x_e$. $f(x_1)$ and $f(x_2)$ are larger than both $f(x_a)$ and $f(x_e)$. If $f(x_1)$ is smaller than $f(x_2)$ then substitute x_1 for x_a. If $f(x_2)$ is smaller than $f(x_1)$ then substitute x_2 for x_e. This bracketing method guarantees that the maximum remains within the interval (x_a, x_e) and the length of the interval is decreasing by a factor of 0.7 with every iteration.

Problem 56b

Write a program to find all real roots of a polynomial of second and third order. Generalize this program to polynomials of arbitrary order. If one root x_1 of the polynomial is found, you have to divide the polynomial by $(x - x_1)$ and search for the next zero value x_2. You should try to solve this problem later, when you know more about BASIC's field variables and subroutines.

6.2   Newton's Method

The Newton method for root finding is perhaps the most celebrated method to search zero values of a function. For a better understanding this method shall be explained by means of the diagram Fig. 6.3.

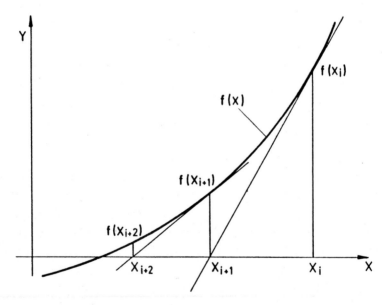

Fig. 6.3. Newton's method extrapolates with the local tangent to find the next estimate of the root of the function.

Let us assume that in the search for a zero point a value x_i is reached (this can also be the estimated initial value). First, the corresponding function value $f(x_i)$ is calculated. Then the tangent at this position to the function graph is calculated and the point of intersection of this tangent with the x-axis is determined. This point of intersection is taken as the new x-value x_{i+1} in the search for zero values and the calculation is repeated with this new estimate of the root as before. As can be seen from the graph, the zero value is found rather quickly. If the method works well — like in the example — then Newton's method converges quadratically. However, functions exist and initial guesses can be chosen for which this tangent method is not successful.

The derivative $df(x_i)/dx$ at the position x_i is the same as the tangent at x_i to the function graph. If the triangle between $f(x_i)$, x_i and x_{i+1} is considered one obtains

$$\frac{df}{dx}(x_i) = f'(x_i) = \frac{f(x_i)}{x_i - x_{i+1}}$$

A simple transformation results in the instruction for the iteration procedure of the Newton method

$$x_{i+1} = x_i - \frac{f(x_i)}{f'(x_i)}$$

This formula shows that for each iteration the function $f(x_i)$ as well as its derivative $f'(x_i)$ is needed. Moreover, an initial x-value has to be indicated. How fast and how certain the Newton method converges depends on this initial value.

The flow diagram for the Newton method can be seen in Fig. 6.4. In this flow diagram not all details are included, rather it shows in principle the operation of the program. So, ahead of the iteration formula in the flow diagram the condition is omitted whether the derivative $f'(x)$ equals zero. If this is the case, then the program generates a mathematical error — a division by zero. Clearly this would mean that the tangent runs parallel to the x-axis and therefore does not intersect. Of similar character is the interrogation at the end of the program whether the specified accuracy has been obtained. It is known from the previous program **BISECT** that there are different methods for determining the desired accuracy. The corresponding BASIC-program **NEWTON** is listed after the corresponding flow diagram (Fig. 6.4).

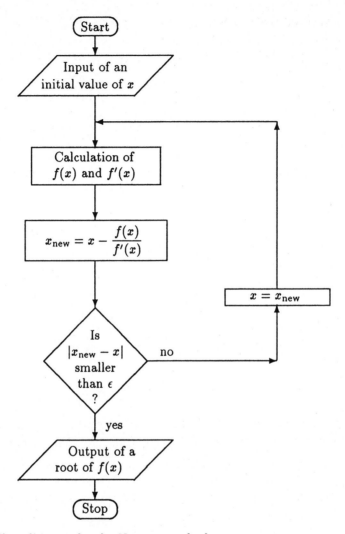

Fig. 6.4. Flow diagram for the Newton method

```
0 REM "NEWTON  "          EBERT/EDERER        880609
1 REM ********************************************
2 REM ***        Root finding by the          ***
3 REM ***          NEWTON method              ***
4 REM *** Function   in line # 100            ***
5 REM *** Derivative in line # 200            ***
9 REM ********************************************
```

```
100    DEF FNF(X)=10*X^5-EXP(X*X)
200    DEF FND(X)=50*X^4-2*X*EXP(X*X)
1100   INPUT "Estimation of the root ";T
1200   Y=FNF(T)
1300     Y1=FND(T)
1400     IF ABS(Y1)<9.999999E-21 THEN 1100
1500     T1=T-Y/Y1
1600     IF ABS(T-T1)<ABS(T*.0000001) THEN 3000
1700   T=T1 : GOTO 1200
3000   PRINT "Root            = "; (T+T1)/2
3100   PRINT "Function value = "; Y
9999   END

RUN
Estimation of the root ? 1
Root            =  .694941
Function value =  0
Ok

RUN
Estimation of the root ? 5
Root            =  2.694024
Function value = -4.882813E-04
Ok
```

The program is very short. In line 100 the function $f(x)$ is defined for which the zero value shall be found, and line 200 contains the derivative of the function $f(x)$ which is called $d(x)$. It is a disadvantage of the Newton method that one has to program the derivative; with complicated functions this can be very time consuming. Besides this is a possible source of error, which often is not recognized quickly.

In line 1100 the estimated initial value is put in. In line 1200 the corresponding function value is assigned to the variable Y and in line 1300 its derivative $d(x)$ is assigned to the variable Y1. Next it will be asked whether the value of the derivative is too small. This would mean a very flat tangent. If this is the case, then the program requests from the user a new initial value for x by jumping back to line 1100. Line 1500 contains the appropriate iteration formula. Next follows the interrogation whether the consecutive x-values agree within the first seven digits. If this is the case, then the program jumps to the output of the zero value (lines 3000 and 3100). If the required accuracy has not yet been obtained, then the new

x-value is stored in the variable T and the program jumps back to line 1200 for another iteration.

   If you test this program you will realize that depending on the initial value the time required to obtain a result can be very different. It may also happen that an overflow error stops the program. To find out what has happened try to solve Problem 57.

Problem 57

Modify program **NEWTON** so that the intermediate results of the single iteration runs are shown on the screen.

Problem 58

As the computing time for the program **NEWTON** may be very long with unfavorable initial values, another improvement of the program may be programmed. Specify a maximum number of iterations. If this number is reached, the program should be stopped and an appropriate message should be shown on the screen.

Problem 58a

Try to write a new program which combines the programs **NEWTON** and **BISECT** in order to obtain a program which is robust like **BISECT** but which uses Newton's method if the root converges fast.

6.3   Regula Falsi or Secant Method

Another frequently used method for determining zero values is the regula falsi or false position method. In this method the function is assumed to be approximately linear in the region of interest, i.e. near the root position. The method is very similar to Newton's method. An advantage however, is that the derivative must not be calculated analytically and is not needed for the program. This method should be explained by means of the diagram in Fig. 6.5. In searching for the zero value, the function values of both last x-values are connected with a straight line. The intersection with the x-axis results in a new x-value which replaces an old one. As can be seen from the diagram in Fig. 6.5 one obtains according to analytical geometry

$$\frac{f(x_i) - 0}{x_i - x_{i+1}} = \frac{f(x_{i-1}) - 0}{x_{i-1} - x_{i+1}}$$

If this equation is solved with respect to x_{i+1} the iteration instruction for the regula falsi is obtained

$$x_{i+1} = x_i - \frac{x_i - x_{i-1}}{f(x_i) - f(x_{i-1})} \cdot f(x_i)$$

This iteration formula differs from the Newton formula in that the differential quotient of the Newton iteration formula is replaced by a difference quotient in the regula falsi. The fact that the derivative of the function does not have to be programmed has its price for the regula falsi. Two initial values for x are required, and frequently more computing time is needed because the false position method converges less rapidly than Newton's method.

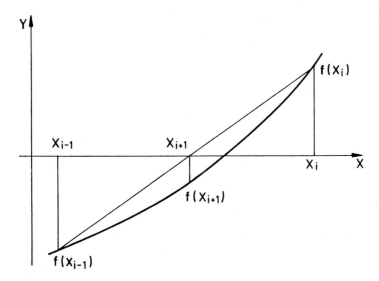

Fig. 6.5. Regula falsi: the interpolation line is drawn through the most recent points to get a new better one.

Fig. 6.6 shows the flow diagram for the regula falsi. The program **REG-FAL** for root finding with the false position method is listed after the corresponding flow diagram.

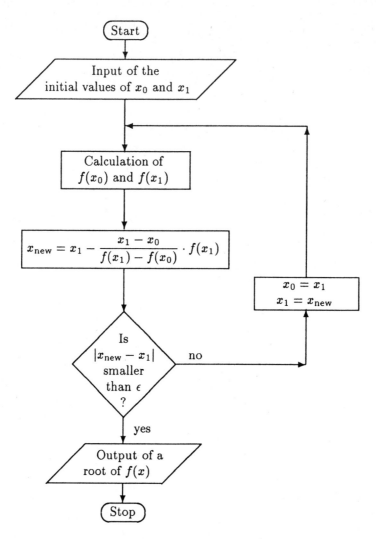

Fig. 6.6. Flow diagram for the false position or the regula falsi method to find roots of a function without using its derivative

```
0 REM "REG-FAL "         EBERT/EDERER         880609
1 REM ************************************************
2 REM *** Finding a root using the regula falsi    ***
3 REM *** method.                                  ***
4 REM *** Program the function in line # 100.      ***
9 REM ************************************************
```

```
100    DEF FNF(X)=(X-4)*(X-2.3)*(X+2.1)*(X+3.34)
1000   INPUT "Starting interval (a,e) ";A,E
1100   INPUT "Intermediate results on the screen ";A$
1200   X0=A : X1=E
1300   Y0=FNF(X0)
1400   Y1=FNF(X1)
1500   IF ABS(X0-X1)<ABS(X0*.0000001) THEN 3000
1600   IF ABS(Y0-Y1)<ABS(Y0*1E-25) THEN 1000
1700   X=X1-(X1-X0)/(Y1-Y0)*Y1
1800   IF A$<>"yes" THEN 2000
1900   PRINT X0;Y0;X1;Y1
2000   X0=X1 : X1=X : GOTO 1300
3000   PRINT "Root            = "; (X0+X1)/2
3100   PRINT "Function value = "; (Y0+Y1)/2
9999   END

RUN
Starting interval (a,e) ? 3,5
Intermediate results on the screen ? n
Root            = 4
Function value =  0
Ok


RUN
Starting interval (a,e) ? 2,3
Intermediate results on the screen ? n
Root            =  2.3
Function value =  0
Ok


RUN
Starting interval (a,e) ? 1.8,1.9
Intermediate results on the screen ? n
Root            =  2.3
Function value =  0
Ok


RUN
Starting interval (a,e) ? -4,-2
Intermediate results on the screen ? n
Root            = -2.1
Function value =  0
Ok
```

```
RUN
Starting interval (a,e) ? -4,-3
Intermediate results on the screen ? n
Root          = -3.34
Function value =  0
Ok
```

This program is also short; the function, for which the zero value should be determined, has been put in line 100 in a form suitable for demonstration purposes. There the roots can be observed directly. Compared to the program NEWTON, line 200, the definition of the derivative, is missing. However, in line 1000 two initial values must be entered, which are stored in the variables A and E. The zero value must not necessarily lie between these two points. It is altogether possible to select close but different x-values.

The corresponding function values are evaluated in lines 1300 and 1400 for both recently determined x-values. The iteration loop runs from line 1300 until 2000. Line 1500 contains the interrogation for the accuracy. If the two latest x-values are close enough, the program jumps to the output in line 3000. In line 1600 it is decided whether both function values are very similar. If this is the case (a nearly horizontal secant) new initial values are requested from the user by jumping back to line 1000. The new x-value is calculated in line 1700 using the iteration formula of the regula falsi. In line 2000 the last but two x-value which was stored in variable X0 is deleted by assigning the next to the last x-value to the variable X0 using the statement X0=X1. Then the value of the last x-calculation assigned to the variable X1. In this way the variable X becomes free for another iteration and the program jumps back to line 1300.

In the INPUT-line 1100 the user is asked for an answer "yes" or "no" to the question for intermediate results. The answer is written with the INPUT-statement in the text variable A$. In line 1800 the IF-statement asks whether the contents of A$ is different from the text "yes". If this is confirmed the program jumps over line 1900 in which the intermediate results are printed on the screen. In case the IF-statement is denied, which means that A$ contains the text "yes", then the intermediate results of each iteration are printed out in line 1900.

Conditional Branching Using Strings

$$\boxed{\text{IF}}$$

Syntax: z1 IF a1\$ ◇ a2\$ THEN z2

            z1 IF a1\$ ◇ a2\$ THEN GOTO z2

            z1 IF a1\$ ◇ a2\$ THEN any statement

z : line numbers

a : string expressions, string variables, or string constants

◇ : comparison operator

In programming with text constants or text variables, the IF-statement is used in the same way as the general IF-statement discussed earlier. However, one has to know the significance of the different comparison operators. The comparison of texts with '=' (equal) or '<>' (unequal) is unequivocal. The comparisons '<' (smaller) or '>' (greater) are related to the lexicographical notation. 'CHEMISTRY' is, in this sense, smaller than 'PHYSICS' but greater than 'CHEMIST'.

Problem 59

Use the programs **NEWTON** or **REG-FAL** or **BISECT** to solve the following implicit equation

$$e^{-(x^2+y^2)} = x^2 y + y^2 x^3$$

Complete the following table

x	4	2	1	0.9	0.8	0.6	0.5	0.4	0.2	0.1	0.05
y											

and draw a graph of this implicit function.

Problem 60

Draw a plot and print a table for the following implicit function.

$$\sin(xy) = \log\left(\frac{x+y}{10}\right)$$

Problem 60a

Write a program to find all the roots between 0 and 2π for the functions

$$f(x) = \sin x + \sin 2x + \sin 3x + \sin 4x$$

and — which is much easier —

$$f(x) = \sin x \cdot \sin 2x \cdot \sin 3x \cdot \sin 4x$$

Problem 60b

Use Planck's equation from Chapter 2.14 and write a program to find its maximum. How does ν_{max} depend on the temperature T? You can solve this problem by writing a program to search the maximum, but you can also calculate the derivative of Planck's equation analytically and look for the root of the derivative function.

Problem 60c

Write a program to find a root, a minimum or a maximum, and a point of inflection of a given function.

Below problems from the natural sciences are discussed in which equations have to be solved by numerical methods.

6.4   Thermoelectric Voltage of a Ni-CrNi-Thermocouple

The measurement of thermoelectric voltage is usually accomplished with two opposing thermocouples, e.g. a pair of Ni-CrNi-thermoelements. One of the thermoelements is kept at $0\,°C$ and the other thermoelement is at the temperature to be measured. Usually the experimental relation between temperature and thermoelectric voltage within a certain temperature range is described by a polynomial. This is done by a so-called linear regression which is discussed in Chapter 7 and 8 in this book.

The temperature can be calculated from the voltage of a Ni-CrNi-thermocouple using the following polynomial, where C is the temperature in $°C$ and V the thermoelectric voltage in mV.

$$\begin{aligned}
C = \ & 25.4497\,V - 0.559195\,V^2 + 0.10452439\,V^3 - 8.776154 \cdot 10^{-3}\,V^4 \\
& + 3.76041 \cdot 10^{-4}\,V^5 - 8.64943 \cdot 10^{-6}\,V^6 + 1.021005 \cdot 10^{-7}\,V^7 \\
& - 4.89101 \cdot 10^{-10}\,V^8
\end{aligned}$$

The specification of dimensions is necessary in reproducing experimental data to a polynomial form. Then one can easily derive the dimensions of the single coefficients and write them down.

If the question is asked the other way round, viz. 'which thermoelectric voltage is to be expected at a certain temperature', then the solution must be found by means of a table or with a programmed search routine. For this purpose the bisection method, which was discussed in Chapter 6.1 will be used.

```
0 REM "NI-CRNI "          EBERT/EDERER          880610
1 REM ************************************************
2 REM *** The polynomial which describes the tem-  ***
3 REM *** perature in C of a Ni-CrNi thermocouple  ***
4 REM *** for an arbitrary voltage is given in line***
5 REM *** # 100. (Reference is assumed to be 0 C.) ***
6 REM *** This program calculates the voltage of a ***
7 REM *** Ni-CrNi thermocouple for a given         ***
8 REM *** temperature.                             ***
9 REM ************************************************
51    K1=  25.4497
52    K2= -  .559195
53    K3=    .10452439
54    K4= - 8.776154E-03
55    K5=   3.76041E-04
56    K6= - 8.64943E-06
57    K7=   1.021005E-07
58    K8= - 4.89101E-10
100   DEF FNF(X)=K1*X+K2*X*X+K3*X*X*X+K4*X^4
               +K5*X^5+K6*X^6+K7*X^7+K8*X^8-TE
150   INPUT "Temperature in Celsius ";TE
200   A=-6:E=55
300   EP=.001
500   Y1=FNF(A):REM print "y(";a;") = ";y1
600   Y2=FNF(E):REM print "y(";e;") = ";y2
800   IF   SGN(Y1)=-SGN(Y2) THEN 5000
900 REM ************************************************
910 REM *** Searching for different signs of the  ***
920 REM *** function values.                      ***
930 REM ************************************************
1000  N=10
1100  H=(E-A)/N
1200    FOR I=1 TO N-1
1300      Y9=FNF(A+I*H)
```

```
1400      IF SGN(Y9)=SGN(Y1) THEN 1700
1500      E=A+I*H
1600      GOTO 5000
1700    NEXT I
1800    N=N*10
1900  IF N<=1000 THEN 1100
2000  PRINT "No root of the function has been found "
2100  STOP
5000 REM ********************************************
5010 REM *** Start of the bisection routine      ***
5020 REM ********************************************
5500  Y1=FNF(A):REM print "y(";A;") = ";y1
5600    Y2=FNF(E):REM print "y(";E;") = ";y2
5800    IF ABS(A-E)<ABS(EP) THEN 9000
6000    M=(A+E)/2
6100    Y3=FNF(M)
6200    IF SGN(Y3)=SGN(Y1) THEN 6600
6400    E=M : GOTO 5000
6600  A=M : GOTO 5000
9000  PRINT "The voltage of the Ni-CrNi TC"
9100  PRINT "at";TE;" Celsius"
9200  PRINT "is";INT(50*(A+E)+.5)/100;"mV"
9999  END


RUN
Temperature in Celsius ? 200
The voltage of the Ni-CrNi TC
at 200  Celsius
is 8.18 mV
Ok


RUN
Temperature in Celsius ? 500
The voltage of the Ni-CrNi TC
at 500  Celsius
is 20.64 mV
Ok
```

The program NI-CRNI corresponds with the program BISECT from line 1000 until 6600 — however the output of intermediate results is deleted. The above polynomial must be rewritten so that the zero value of the function can be searched. This means that for this case only the temperature C has to be transferred to the right hand side of the equation. This polynomial is programmed in the usual notation in the function definition in line 100. The eight coefficients are defined ahead of the function definition (lines 51–58). The temperature is then asked for in an INPUT-statement from the user. If the value for the temperature is put into variable TE, then the problem for the zero value search of the function in line 100 is completed and well-defined and the determination of the root can start. For this purpose the interval limits within which the zero value shall be searched are defined in line 200. As potentials are only physically significant between −6 and +55 mV, the mathematical search is limited to this range. (The formula could easily calculate a thermoelectric voltage for an impossible temperature of 5000 °C.) The required accuracy EP of the results is also limited by the physical realities. Few experimental devices allow higher resolutions than $1\,\mu V$, therefore EP in line 300 is set to the value 0.001.

In an application program such as this, the output should be in the usual physical dimensions. Line 9200 truncates the actual result (A+E)/2 to two digits after the decimal point. This is achieved using the standard function INT. It should be remembered that this function cuts off the digits after the decimal point of any positive real number. If the number is negative, −1 is added after the truncation. In line 9200 the result is first multiplied by 100, then 0.5 is added, this is done for rounding, then the digits after the desired point are cut off, and finally it is divided by 100.

Problem 61

Alter the program NI-CRNI so that one can choose whether, for a given temperature, the thermoelectric voltage in mV is obtained or whether, for a given thermoelectric voltage, the temperature in °C is obtained.

Problem 62

Try to write a new program NI-CRNI using Newton's method or the regula falsi.

6.5   The Persistence Length of a Polymer Molecule

If a polymer molecule in solution has no concrete structure, such as a helix, then it exists as a so-called 'random coil'. To describe such a polymer molecule, models of a 'random coil' are often used.

Whereas in a real macromolecule the single bonds and their angles are substantially rigid and there exists only one more or less free rotation, for the random coil the total molecule is divided into small pieces of the same size, joined together, and the neighboring pieces can be directed at will and randomly. The length of these units is called persistence length. The question arises: how big should the persistence length be in order that the model coil obtains the same physical properties as the real coil? The persistence length characterizes the stiffness of a polymer molecule. A relation exists between the contour length l of a polymer molecule, the persistence length a, and the radius of gyration r.[1]

$$r^2 = a^2 \left(\frac{l}{3a} - 1 + \frac{2a}{l} - 2 \left(1 - e^{-l/a} \right) \cdot \frac{a^2}{l^2} \right)$$

This implicit function is transformed to a function $f(a)$ with the parameters r and l by putting r^2 on the right hand side of the equation. The radius of inertia r of a polymer can be experimentally determined from the angular dependence of the scattering of visible light. The contour length can be directly derived from low angle X-ray scattering. It can also be estimated from the molecular weight and the length of a monomer unit. To solve the above equation the program BISECT is again applied. The program PERSIST, including examples, is listed below:

```
0 REM "PERSIST "          EBERT/EDERER          880325
1 REM ***********************************************
2 REM *** Calculation of the persistence length of ***
3 REM *** a polymer. Given are the radius of gyra- ***
4 REM *** tion r and the contour length l of the   ***
5 REM *** polymer molecule in Angstroem.           ***
6 REM *** The program BISECT is used for finding   ***
7 REM *** the root of the equation.                ***
9 REM ***********************************************
50    INPUT "Radius of gyration in Angstroem "; R
60    INPUT "Contour length in Angstroem ";      L
100   DEF FNF(A)=A*A*(L/3/A-1+2*A/L-2*(1-EXP(-L/A))*A/L*A/L)-R*R
200   A=.0001:E=100000!
300   EP=.0001
500   Y1=FNF(A)
600   Y2=FNF(E)
800   IF    SGN(Y1)=-SGN(Y2) THEN 5000
```

[1] S.K. Garg, S.S. Stivala, *J. Pol. Sci. Polym. Phys.*, **16**, 1419–1434 (1978)

```
900  REM ********************************************
910  REM *** Searching for different signs of the   ***
920  REM *** function values.                        ***
930  REM ********************************************
1000  N=10
1100  H=(E-A)/N
1200    FOR I=1 TO N-1
1300      Y9=FNF(A+I*H)
1400      IF SGN(Y9)=SGN(Y1) THEN 1700
1500      E=A+I*H
1600      GOTO 5000
1700    NEXT I
1800    N=N*10
1900  IF N<=1000 THEN 1100
2000  PRINT "No root of the function has been found "
2100  STOP
5000  REM ********************************************
5010  REM *** Start of the bisection routine         ***
5020  REM ********************************************
5800    IF ABS(A-E)<ABS(EP) THEN 9000
6000      M=(A+E)/2
6100      Y3=FNF(M)
6200      IF SGN(Y3)=SGN(Y1) THEN 6600
6400      E=M:Y2=Y3:GOTO 5000
6600    A=M:Y1=Y3:GOTO 5000
9000  PRINT:PRINT "Persistence length in Angstroem = "
9100  PRINT (E+A)/2;"+/-";ABS((E-A)/2)
9999  END

RUN
Radius of gyration in Angstroem ? 8940
Contour length in Angstroem ? 8900000

Persistence length in Angstroem =
 26.9408 +/- 4.673004E-05
Ok


RUN
Radius of gyration in Angstroem ? 200
Contour length in Angstroem ? 10000

Persistence length in Angstroem =
 12.04345 +/- 4.673004E-05
Ok
```

This program is a slight variation of the program BISECT. The formula is programmed in line 100 as function definition F(A). The user has to input the specifications about the radius of inertia r and contour length l in lines 50 and 60. Ångström has been chosen as the unit for the length. The interval in which the zero value shall be searched has been chosen so big (in line 200) that each physically significant value of the persistence length will be included. The requirement for the accuracy in line 300 is somewhat exaggerated. The output in line 9000 and 9100 relates again to the actual physical chemical problem.

Problem 63

Calculate different persistence lengths as a function of the radius of gyration r at a fixed contour length l and draw this relationship in a diagram.

6.6   Calculation of the Binary Distillation

Let us consider this problem under the assumption that liquids are ideal mixtures. That means that August's formula for vapor pressure can be applied, the distillation column is operated at total reflux, and the distillation stages work ideally.

For the vapor pressures p_i^0 of both pure substances as a function of the temperature T the following formulas can be applied

$$\text{Substance 1:} \qquad \log p_1^0 = \frac{a_1}{T} + b_1$$

$$\text{Substance 2:} \qquad \log p_2^0 = \frac{a_2}{T} + b_2$$

The constants a_1, a_2, b_1 and b_2 are characteristic values of the substances and they can be looked up in appropriate tabular compilations. The numerical values are dependent upon the dimensions applied.

According to Raoult's law, the vapor pressures p_i of the substances above the liquid mixture are given by

$$p_1 = x_1^l \cdot p_1^0 \qquad\qquad p_2 = x_2^l \cdot p_2^0$$

In this equation x_i^l is the mole fraction of substance i in the liquid phase.

The total pressure p_G in the gas phase is accordingly

$$p_G = p_1 + p_2 = x_1^l p_1^0 + x_2^l p_2^0 = x_1^l p_1^0 + (1 - x_1^l) p_2^0.$$

The mole fraction x_i^g of the substance i in the gas phase is given by

$$x_1^g = \frac{p_1}{p_1 + p_2} \qquad\qquad x_2^g = \frac{p_2}{p_1 + p_2}$$

Considering the temperature dependence of the vapor pressures according to August's equation the following function is obtained, where a zero value shall be evaluated to get the boiling temperature T for a total pressure p_G.

$$f(T) = p_G - x_1^l \cdot e^{a_1/T + b_1} - (1 - x_1^l) \cdot e^{a_2/T + b_2}$$

To calculate the zero value of this formula we need the characteristic values of both substances a_1, a_2, b_1 and b_2. For the total pressure we assume 760 torr. If a value for the mole fraction x_1^l in the liquid phase in the first distillation stage is given, then the boiling temperature of the mixture can be calculated from the above equation. Next, the composition of the gas phase x_1^g can be calculated from the formula for the vapor pressure and Raoult's law. The mole fraction of the substance 1 in the gas phase over the first stage is equivalent to the mole fraction of the liquid phase in the next distillation stage. If one carries out the calculation in this way, then one can calculate values for the whole distillation column. The flow diagram for this problem is given in Fig. 6.7 and the corresponding program follows.

```
0 REM "DISTILL "          EBERT/EDERER          880610
1 REM ************************************************
2 REM *** Distillation of an ideal binary mixture. ***
3 REM *** The vapor pressure of each substance is  ***
4 REM *** expressed by the formula:                ***
5 REM ***        log(P) = A/T + B                  ***
6 REM *** The parameters A and B have to be read   ***
7 REM *** for each substance by an input-statement.***
9 REM ************************************************
100 DEF FNF(T)=X1*EXP(A1/T+B1)+(1-X1)*EXP(A2/T+B2)-760
200    INPUT "a1,b1 ";A1,B1
300    INPUT "a2,b2 ";A2,B2
500    A9=A1/(LOG(760)-B1)
```

```
600    E9=A2/(LOG(760)-B2)
700    IF A9<E9 THEN 1000
800      H=E9:E9=A9+1:A9=H-1
900      H=A1:A1=A2:A2=H
920      H=B1:B1=B2:B2=H
1000   PRINT "Mole fraction of the lower"
1100   PRINT "boiling substance  ";
1200   INPUT X1
1300   INPUT "How many theoretical plates ";I6
1500   PRINT:PRINT "Plate  x-liq      x-vap       Bp in C"
2000   I5=1
2100   A=A9:E=E9
5000 REM ********************************************
5010 REM *** Start of the bisection routine      ***
5020 REM ********************************************
5500     Y1=FNF(A)
5600      Y2=FNF(E)
5800      IF ABS(A-E)<ABS(A)/1E+07 THEN 9000
6000      M=(A+E)/2
6100      Y3=FNF(M)
6200      IF SGN(Y3)=SGN(Y1) THEN 6600
6400      E=M : GOTO 5000
6600      A=M : GOTO 5000
9000      X3=X1*EXP(A1/A+B1)/760
9100      PRINT I5;TAB(8);
9110      PRINT INT(X1*10000)/10000;TAB(19);
9125      PRINT INT(X3*10000)/10000;TAB(29);
9130      PRINT INT((A-273.15)*1000)/1000
9200      IF I6=I5 THEN 9999
9300   I5=I5+1 : X1=X3 : GOTO 2100
9999   END


RUN
a1,b1 ? -3809.9 , 17.785
a2,b2 ? -3330.0 , 17.403
Mole fraction of the lower
boiling substance  ? 0.1
How many theoretical plates ? 6

Plate  x-liq      x-vap       Bp in C
  1      .1         .2398      63.395
  2      .2398      .4792      57.285
  3      .4792      .736       48.877
  4      .736       .8972      41.844
  5      .8972      .9652999   38.165
  6      .9652999   .9889      36.746
```

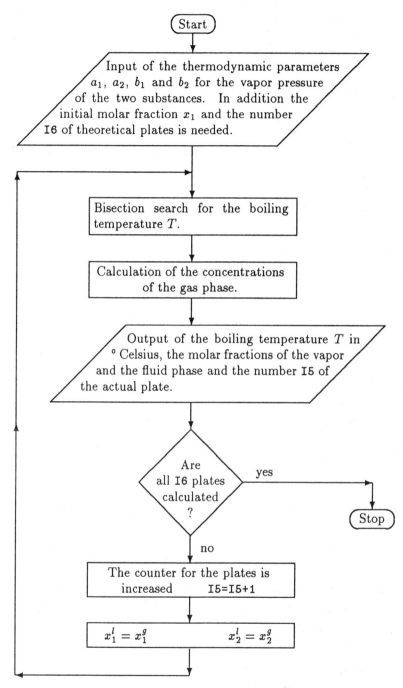

Fig. 6.7. Flow diagram for the calculation of the separation of binary mixtures in a distillation column

The function for the zero value search, i.e. the determination of the boiling temperature, is programmed in line 100. The total pressure is set to 760 torr. Then the input of the physical parameters, a_1, a_2, b_1 and b_2, is required. In lines 500 and 600 the boiling temperatures of both substances A9 and E9 at 760 torr are calculated. In line 700 it is asked whether substance 1 has the lower boiling point. If this is confirmed, the program jumps to line 1000. In case substance 2 has the lower boiling point, the substances 1 and 2 and their parameters are exchanged in lines 800, 900 and 920. The temperature that is less than or equal to that of the lower-boiling component is stored as variable A9. The temperature that is greater than or equal to that of the higher-boiling component is stored as variable E9. Then (in lines 1000 to 1200) the user is asked for the mole fraction of the component with the lower boiling point, which is stored as variable X1. The next input in line 1300 requires the number of the theoretical plates to be calculated. The number of the theoretical plate, for which the current calculation is made is set to an initial value of 1 in line 2000.

The limits of the interval for the search of the boiling point of the mixture are set in line 2100 to a temperature A9, which is less than the boiling point of the more volatile substance, and to a temperature E9 which is higher than the temperature of the less volatile substance. Lines 5000 through 6000 contain the core of the program BISECT which calculates the boiling temperature of the mixture. As soon as the iteration is finished, the program jumps from line 5800 to line 9000, where the mole fraction of the more volatile substance in the gas phase is calculated. In lines 9100–9130 the mole fraction of the more volatile component in the gas phase and in the liquid phase, the corresponding boiling point, and the number of the plate in the distillation column are printed. Line 9200 tests whether all considered plates have been calculated. If this is confirmed the program jumps to the end in line 9999. In case this is not confirmed, the mole fraction of the liquid phase in the next plate is set equal to the mole fraction of the gas phase in the present plate by the command X1=X3 in line 9300. At the same time the plate counter I5 is raised by 1 and the program jumps back to line 2100, the start of a new bisection. In the example given with the program the characteristic data for n-hexane and n-pentane are used.

Problem 64

Generalize the program DISTILL so that the total pressure is not necessarily set to 760 torr, but so that any other total pressure can be chosen.

Problem 65

Modify the generalized program DISTILL so that it can be used for a finite reflux. An amplification line has to be considered. Two additional inputs are needed, one for the reflux ratio v and the other for the mole fraction x_d for the composition of the destillate. The formula in line 9300 should be altered as follows

$$x_{i+1,1}^l = \left(x_{i,1}^g - \frac{x_d}{v+1}\right) \cdot \frac{v+1}{v}$$

The new line 9300 then reads

```
9300 I5=I5+1:X1=(X3-XD/(V+1))*(V+1)/V:GOTO 2100
```

The program should be terminated if $x_{i,1}^l$ exceeds the value of x_d.

Problem 66

Modify the program DISTILL for the calculation of the separation of ternary mixtures.

6.7   The pH-Value of a Weak Acid

The weak acid HA dissociates in water according to the following equations

$$\text{HA} \quad \rightleftharpoons \quad \text{H}^\oplus + \text{A}^\ominus \tag{1}$$

$$\text{H}_2\text{O} \quad \rightleftharpoons \quad \text{H}^\oplus + \text{OH}^\ominus \tag{2}$$

The corresponding equilibrium constants are

$$K_A = \frac{[\text{H}^\oplus][\text{A}^\ominus]}{[\text{HA}]} \quad \text{Equilibrium constant of the acid} \tag{3}$$

$$K_W = [\text{H}^\oplus][\text{OH}^\ominus] \quad \text{Dissociation constant of water} \tag{4}$$

The mole balance for the undissociated acid and the anion is related to the concentration of the total acid $[\text{HA}]_E$ which is put in the beaker

$$[\text{HA}]_E = [\text{HA}] + [\text{A}^\ominus] \quad \text{Particle balance of the acid} \tag{5}$$

As the electrolyte solution is electrically neutral, the number of positive ions has to be equal to the number of negative ions.

$$[H^{\oplus}] = [A^{\ominus}] + [OH^{\ominus}] \qquad \text{Balance of charge} \qquad (6)$$

If the concentration of the undissociated acid [HA] is calculated from equation (3) and is put into equation (5), one obtains

$$[HA]_E = \frac{[H^{\oplus}][A^{\ominus}]}{K_A} + [A^{\ominus}] \qquad (7)$$

If the $[OH^{\ominus}]$-ion concentration is calculated from equation (4) and is put into equation (6) the following equation is obtained

$$[A^{\ominus}] = [H^{\oplus}] - \frac{K_W}{[H^{\oplus}]} \qquad (8)$$

If equation (8) is put into equation (7) the final formula is obtained, in which the equilibrium constants, the concentration of the total acid and the hydrogen ion concentration are related.

$$[HA]_E = \frac{[H^{\oplus}]^2 - K_W}{K_A} + [H^{\oplus}] - \frac{K_W}{[H^{\oplus}]} \qquad (9)$$

$$pH = -\log_{10}[H^{\oplus}]$$

The BASIC-program for the determination of the pH-value is listed as follows:

```
0 REM "ACID    "        EBERT/EDERER          880610
1 REM ***********************************************
2 REM *** Determination of the pH-value of a weak  ***
3 REM *** acid. The resulting cubic equation is    ***
4 REM *** solved by the bisection method.          ***
5 REM *** All dimensions are in mole and liter.    ***
9 REM ***********************************************
100    DEF FNF(X)=X*(X-KW/X)/KA+X-KW/X-EW
200    A=1E-10 : E=1
320    INPUT "Concentration of the acid in mol/l ";EW
340    INPUT "Dissociation constant Ka ";KA
360    INPUT "Dissociation constant Kw of water ";KW
```

```
500    Y1=FNF(A)
600    Y2=FNF(E)
800    IF SGN(Y1)=-SGN(Y2) THEN 5000
900 REM *********************************************
910 REM *** Searching for different signs of the   ***
920 REM *** function values.                       ***
930 REM *********************************************
1000   N=10
1100   H=(E-A)/N
1200     FOR I=1 TO N-1
1300       Y9=FNF(A+I*H)
1400       IF SGN(Y9)=SGN(Y1) THEN 1700
1500       E=A+I*H
1600       GOTO 5000
1700     NEXT I
1800     N=N*10
1900   IF N<=1000 THEN 1100
2000   PRINT "No root of the function has been found "
2100   STOP
5000 REM *********************************************
5010 REM *** Start of the bisection routine         ***
5020 REM *********************************************
5500   Y1=FNF(A)
5600     Y2=FNF(E)
5700     EP=(A+E)/10000
5800     IF ABS(A-E)<ABS(EP) THEN 9000
6000     M=(A+E)/2
6100     Y3=FNF(M)
6200     IF SGN(Y3)=SGN(Y1) THEN 6600
6400     E=M : GOTO 5000
6600   A=M : GOTO 5000
9000   PRINT:PRINT "pH =";-LOG((E+A)/2)/LOG(10)
9999   END


RUN
Concentration of the acid in mol/l ? 0.1
Dissociation constant Ka ? 1.85E-5
Dissociation constant Kw of water ? 1E-14

pH = 2.869345

RUN
Concentration of the acid in mol/l ? 0.001
Dissociation constant Ka ? 1.85E-5
Dissociation constant Kw of water ? 1E-14

pH = 3.895907
```

The formula (9) is written in line 100 of the program ACID as a function definition. The bisection method is used for determining the hydrogen ion concentration. The two limiting values in line 200 of 10^{-10} and 1 are certainly acceptable for a weak acid. Then, the program asks for the values of the three parameters in the formula: the total concentration of the acid in mol/l, the dissociation constant of the acid, and the dissociation constant of water. Through line 6600, the program is identical with the program BISECT. Only line 5700 has been added for the automatic request of the accuracy. In our program a relative accuracy of at least four digits is required. The output in line 9000 is adapted to the problem. The pH-value is the negative logarithm to the base 10 of the hydrogen ion concentration. As a number of computer languages can handle only natural logarithms, the transformation can be carried out using the following formula

$$\log_{10} x = \frac{\log x}{\log 10}$$

Problem 67
Use the program ACID to calculate two tables or plots which show the dependence of the pH-value on the weighed amount of acid at a given acid constant, and the dependence on the acid constant at a given amount of acid.

Problem 68
Rewrite the program ACID and use the program NEWTON to solve the equation. As the program NEWTON needs an initial value, replace the above equations (5) and (6) by the following approximations

$$[HA]_E \approx [HA] \qquad\qquad (5')$$
$$[H^{\oplus}] \approx [A^{\ominus}] \qquad\qquad (6')$$

If these approximations are put into equation (3) then a value for an adequate initial value is obtained
$$[H^{\oplus}] = \sqrt{K_A \cdot [HA]_E}$$

Problem 69
Write a program to calculate the pH-values of a buffer solution considering also the addition of a strong acid.

A buffer solution consists of a weak acid and a salt of this acid. The pH-value of such a solution changes relatively little in adding moderate amounts of a strong acid or strong base. Quantitatively this can be expressed as follows

$$HA \quad \rightleftharpoons \quad H^{\oplus} + A^{\ominus} \tag{1}$$

$$H_2O \quad \rightleftharpoons \quad H^{\oplus} + OH^{\ominus} \tag{2}$$

$$KA \quad \rightarrow \quad K^{\oplus} + A^{\ominus} \quad \text{Complete dissociation of the salt KA} \tag{3}$$

As a strong acid we choose HCl which is completely dissociated.

$$HCl \quad \rightarrow \quad H^{\oplus} + Cl^{\ominus} \tag{4}$$

$$K_A = \frac{[H^{\oplus}][A^{\ominus}]}{[HA]} \quad \text{Equilibrium constant of the acid} \tag{5}$$

$$K_W = [H^{\oplus}][OH^{\ominus}] \quad \text{Dissociation constant of water} \tag{6}$$

The particle balance for the anion results from the initial amount of weak acid and the salt of this acid

$$[HA]_E + [KA]_E = [HA] + [A^{\ominus}] \quad \text{Balance for anions} \tag{7}$$

$$[KA]_E = [K^{\oplus}] \quad \text{Complete dissociation of KA} \tag{8}$$

The charge balance reads

$$[H^{\oplus}] + [K^{\oplus}] = [A^{\ominus}] + [OH^{\ominus}] + [Cl^{\ominus}] \tag{9}$$

If the concentration of the undissociated acid [HA] from equation (5) is put into equation (7), the following formula is obtained

$$[HA]_E = \frac{[H^{\oplus}][A^{\ominus}]}{K_A} + [A^{\ominus}] - [KA]_E \tag{10}$$

The $[OH^{\ominus}]$-ion concentration is calculated from equation (6) and put into equation (9) for the charge balance. Provided that the strong acid HCl and the salt KA are completely dissociated, one obtains by rearrangement

$$[A^{\ominus}] = [H^{\oplus}] - \frac{K_W}{[H^{\oplus}]} + [KA]_E - [HCl]_E \tag{11}$$

A combination of the equations (10) and (11) gives ultimately the final formula, in which the equilibrium constants, the initial concentrations of the weak acid, the salt and the strong acid are related as well as the hydrogen ion concentration

$$[HA]_E + [KA]_E = \left(\frac{[H^\oplus]}{K_A} + 1 \right) \left([H^\oplus] + [KA]_E - [HCl]_E - \frac{K_W}{[H^\oplus]} \right)$$

Test the program with the following values: the initial concentrations of weak acid and its salt shall be 1 mol/l respectively, the equilibrium constant of the weak acid should have the value of $1.85 \cdot 10^{-5}$, and the dissociation constant of water should be 10^{-14}. Vary the initial amount of the strong acid and calculate the changes of the pH-value. Draw this dependence as a plot.

Problem 70

Write a program for the calculation of the pH-value of a buffer solution considering the addition of a strong base.

Problem 71

Write a program for the calculation of the pH-value of a mixture of two weak acids.

Problem 72

A weak electrolyte K_2A dissociates into three ions

$$K_2A \quad \rightleftharpoons \quad 2K^\oplus + A^{2\ominus}$$

The corresponding equilibrium constant K_c can be written as follows

$$K_c = \frac{4\alpha^3 c_0^2}{1 - \alpha}$$

Calculate the degree of dissociation α for a given K_c and initial concentration c_0 (e.g. $K_c = 0.01$ and $c_0 = 0.001$).

Problem 73

Given is the equilibrium reaction of first order

$$A \quad \underset{k_2}{\overset{k_1}{\rightleftharpoons}} \quad B$$

Evaluating the time law and integrating it, one obtains

$$\log \left(\frac{k_1[A] - k_2[B]}{k_1[A]_0 - k_2[B]_0} \right) = -(k_1 + k_2) \cdot t$$

$[A]_0$ and $[B]_0$ are the initial concentrations of the substances A and B; t stands for the reaction time. Additionally a conversion parameter x is introduced which connects the actual concentrations $[A]$ and $[B]$ with the initial concentrations $[A]_0$ and $[B]_0$ in the following way

$$[A] = [A]_0 - x \qquad \text{and} \qquad [B] = [B]_0 + x$$

Given are initial concentrations of substances $[A]_0$ and $[B]_0$ and the reaction rate constant k_1. Further the value for x shall be known for a given reaction time t. Write a program for the calculation of the reaction rate constant k_2.

Problem 74

For the temperature dependence of the reaction rate constant, Eyring's theory (activated complex theory) provides the following formula

$$k = \frac{k_B T}{h} \cdot e^{\frac{\Delta S^{\ddagger}}{R}} \cdot e^{-\frac{\Delta H^{\ddagger}}{RT}}$$

where k_B is Boltzmann's constant and h is Planck's constant.

Choose values for the activation entropy $\Delta S^{\ddagger}$ (e.g. –11.8 cal/(mol K)), the activation enthalpy $\Delta H^{\ddagger}$ (e.g. 25000 cal/mol), and the reaction rate constant k (e.g. $0.001\,\text{s}^{-1}$) and calculate the temperature T which corresponds to the value of k. Compare the numerical result with the analytical approximation which is obtained if the temperature dependence of the preexponential factor is neglected. Then you can use Arrhenius' formula

$$k = k_0 \cdot e^{-\frac{E_a}{RT}}$$

to transfer Eyring's formula into the Arrhenius equation, an average temperature T_m is assumed and the following relations are considered

$$k_0 = \frac{k_B T_m}{h} \cdot e^{\frac{\Delta S^{\ddagger}}{R}} \qquad \text{and} \qquad E_a = \Delta H^{\ddagger} + RT_m$$

6.8   Iterative Methods

Before finishing the chapter on equations a problem shall be treated which was mentioned earlier in Chapter 2.

```
10    INPUT X
20    X=SQR(ABS(SIN(X)))
30    PRINT X,
40    GOTO 20
```

If we write the program a little more in detail one can see that the solution of the equation

$$x^2 = \sin x$$

is calculated by an iterative method.

```
10    INPUT X
20    Y=SQR(ABS(SIN(X)))
22    IF X=Y THEN STOP
24    X=Y
30    PRINT X,
40    GOTO 20
```

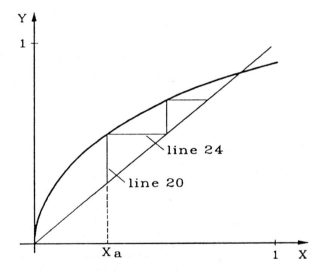

Fig. 6.8. Iterative procedure to find the root of an equation which is given in the form $x = g(x)$. The graphs $y = g(x)$ and $y = x$ are drawn and their intersection points are roots of the equation. A solution point can be either attractive or repulsive to the iteration process: $x_{i+1} = g(x_i)$.

As soon as an x-value is obtained which is identical with the y-value calculated in line 20, this x-value is considered as the solution of the above equation. The question why the iterative method converges to this value can be answered if one plots in a (x, y)-diagram the graph of the equation used in line 20 together with the angle bisector in the first quadrant. (The angle bisector is the graph of the equation in line 24.) If the above mathematical method is reconstructed by means of this graph, one can recognize that the solution point $x = 0.876726216$ is an 'attractive or converging point' and the solution point $x = 0$, however, is a 'repulsive or diverging point'. It is obvious that this property of a solution point is dependent on the local value of the derivative of function $g(x)$. It should be mentioned here that Newton's method is one special case of the iterative methods.

7 Arrays

In mathematics and applications in natural sciences indexed variables are frequently used, e.g. in the production of tables from experimental data, in the formation of sums, in characterization of energy states and in many more problems.

These variables can be defined in the different programming languages. They are also called fields, arrays, vectors, lists, tables, matrices or subscripted variables. In carrying out calculations with indexed variables the computer must know how many different indices are required and how much memory area has to be reserved. For that purpose the memory requirement must be programmed first. For arrays the same variable name conventions are valid as for ordinary variables. The indices are written in parentheses after the variable name. The comma is used as separator if more than one index is needed.

Dimensioning of Indexed Variables

DIM

Syntax: z1 `DIM` V(n)
          z2 `DIM` Y(n,m)
          z3 `DIM` Z(n,m,l)

z   : line numbers
`V`, `Y`, `Z` : variable names
n, m, l   : positive integer numbers

The keyword `DIM` is followed by one or more field names separated by commas. Each field name must be followed by one or more integer constants in parentheses. These integers indicate the maximum value of each index that is permitted in an indexed variable. It is assumed that the BASIC used can work with variables up to three dimensions, that means at least three subscripts can be used. (In BASICA or GWBASIC up to 255 dimensions are allowed.) A DIM-statement can appear anywhere in a program. However it is good practice to place these statements at the beginning of the program. BASIC would automatically assign 11 elements (X(0) – X(10)) to an indexed variable X if X is used without the dimension statement, but this is usually bad programming style. Subscripted variables can be used within a program in the same manner as ordinary variables. An index need not necessarily be written as a constant. Variables and expressions can also be used. Example programs which illustrate the use of arrays are listed below:

```
10 DIM X(10),Y(10)
20 FOR I=1 TO 10
30   X(I)=I*I
40   Y(I)=EXP(-X(I))
50   PRINT I,X(I),Y(I)*X(I)
60 NEXT I

10 DIM F(3,3,3)
20 FOR I = 0 TO 3
30   FOR J = 0 TO 3
40     FOR K = 0 TO 3
50       F(I,J,K) = SQR(I*I+J*J+K*K)
60     NEXT K
70   NEXT J
80 NEXT I
```

The next program concerns the calculation of the arithmetic mean value of experimental data and the deviations of each single datum from the average value. This program using arrays is considerably simpler and more clearly structured than the program for determining mean value which was explained earlier in Chapter 4.3. One can call upon each value directly, whereas in the earlier program one had to work with RESTORE- and READ-commands.

```
 0  REM MEAN
10  DIM X(100),Y(100)
20  INPUT "How many data"; N
```

```
30   FOR I = 1 TO N
40     INPUT X(I)
50   NEXT I
60   S = 0
70   FOR I = 1 TO N
80     S = S + X(I)
90   NEXT I
100 PRINT "Mean Value = ";S/N
200 FOR I = 1 TO N
210    PRINT I,X(I),X(I)-S/N
220 NEXT I
```

There are problems in which factorial functions are frequently required. It is useful to calculate first the values of the factorial functions and to store them as arrays. Then they can be directly called when they are required.

Recall: $0! = 1,$ $1! = 1,$ $n! = n(n-1)!$

The program for the calculation of factorial functions reads as follows:

```
10 DIM G(33)
20 G(0) = 1
30 G(1) = 1
40 FOR I = 2 TO 33
50   G(I) = G(I-1)*I
60 NEXT I
```

Another example using arrays is the calculation of the Fibonacci numbers F_i which are often used in solving problems of optimization. They are defined as follows

$$F_1 = 1 \qquad F_2 = 1 \qquad F_i = F_{i-2} + F_{i-1} \qquad \text{for } i > 2$$

A program for the calculation of these numbers reads as follows:

```
1000 DIM F(100)
1100 F(1) = 1 : F(2) = 1
1200 FOR I = 3 TO 100
1300   F(I) = F(I-2) + F(I-1)
1400 NEXT I
1500 FOR I = 1 TO 100
1600   PRINT I , F(I)
1700 NEXT I
```

By means of the statement DIM F(100) memory space was reserved for variables F(0) through F(100). The highest possible indices are either limited by syntax rules or by the available memory space. Details should be looked up in the computer or BASIC-handbook or, which is frequently simpler, should be determined by experiment.

Problem 75

Write the program MOL-VEL (Chapter 4.2) for printing out a table of molecular velocities of different substances using arrays.

Problem 76

Rewrite the program MEAN so that arrays can be used; then it is much easier to test for outliers.

Problem 76a

Use subscripted variables to solve Problem 43 and store the function values and the respective values of the antiderivative in a vector. Use this program later as a basis for plotting the graph of the function and of the integral when you have become acquainted with computer graphics.

7.1   Sorting of Numbers and Words

A problem which frequently appears in processing data is the sorting of data. The program which we will develop shall follow a method which is used in sorting filing cards — this method is often called 'bubble sort'.

The flow diagram in Fig. 7.1 explains this method of sorting experimental data. If, to make it clearer, the sorting of n filing cards is referred to, then the method can be described as follows: first, the card with the highest value is searched and found and it is exchanged with the first card. Then the card with the next highest value is found from the remaining $n-1$ cards and exchanged with the second card. This is repeated until all cards are in the right order. The program for this sorting method is listed after the flow diagram (Fig. 7.1).

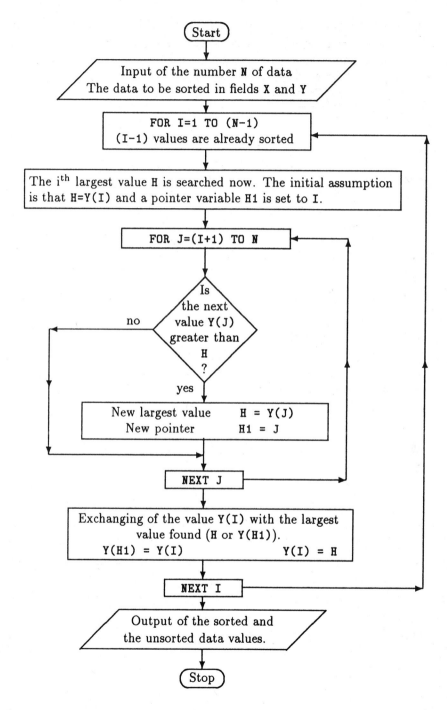

Fig. 7.1. Flow diagram for the 'bubble sort' method

```
0 REM "SORTNUM "          EBERT/EDERER          880620
1 REM ************************************************
2 REM *** Descending sort of numerical values which***
3 REM *** are read in with an INPUT statement.     ***
9 REM ************************************************
1000  DIM X(100),Y(100)
1100  INPUT "Number of values ";N
1200  FOR I=1 TO N
1300    INPUT X(I) : Y(I)=X(I)
1400  NEXT I
2000  FOR I=1 TO N-1
2100    H=Y(I) : H1=I
3000    FOR J=I+1 TO N
3100      IF H>Y(J) THEN 3400
3200      H=Y(J):H1=J
3400    NEXT J
4000    Y(H1)=Y(I) : Y(I)=H
4100  NEXT I
5000  PRINT:PRINT "Original  values "
5100  FOR I=1 TO N:PRINT X(I);:NEXT I:PRINT
5200  PRINT:PRINT "Sorted  values"
5300  FOR I=1 TO N:PRINT Y(I);:NEXT I:PRINT
9999  END

RUN
Number of values ? 10
? 4
? 2
? -6
? 6
? 3
? 0
? 2
? 9
? 1
? -3

Original  values
 4  2 -6  6  3  0  2  9  1 -3

Sorted  values
 9  6  4  3  2  2  1  0 -3 -6
Ok
```

Dimensioning of the vectors X and Y takes place in line 1000. It is established as 100. If a greater number of values must be sorted, the dimensioning must be increased accordingly.

Input occurs in lines 1100 through 1400. The values are stored simultaneously in Y(I) and in X(I). The field X serves to save the unsorted data.

The external I-loop runs from line 2000 through 4100 and indicates at any time the number of the value being searched (the Ith highest value). The inner J-loop — which carries out the search — is written in lines 3000 through 3400. Before entering this inner loop the value of Y(I) is assigned to the auxiliary variable H. The pointer variable H1 holds the position of H which is I at the beginning of the inner loop. The inner loop is seeking for a higher value. If such a value is found, and therefore the IF-interrogation in line 3100 is denied, this new higher value is assigned to the variable H and the position at which this value has been found is saved in variable H1. The search for the highest value of the remaining data is continued, and at the end of the J-loop the highest value is stored in H and the position in which it was found in H1. In line 4000, immediately after the end of the inner loop, the Ith value is exchanged for the highest value that was found. The output of the sorted and unsorted data follows in lines 5000 through 5300.

Problem 77

Write a program for optional ascending or descending sorting.

Problem 77a

Given are data pairs (x_i, y_i) which should be sorted in their x-values. Write a program to sort ascending the x-values, but the correspondence between x_i and y_i should be maintained.

With the following program, SORTTEXT, words or strings can be sorted. It is nearly identical to the program SORTNUM, however, the variables in which words are stored are text variables as characterized by the $.

```
0 REM "SORTTEXT"          EBERT/EDERER          880620
1 REM ************************************************
2 REM *** Descending sort of string values which  ***
3 REM *** are read in with an INPUT statement.     ***
9 REM ************************************************
1000  DIM X$(100),Y$(100)
1100  INPUT "Number of strings ";N
1200  FOR I=1 TO N
1300    INPUT X$(I) : Y$(I)=X$(I)
1400  NEXT I
2000  FOR I=1 TO N-1
2100    H$=Y$(I) : H1=I
3000    FOR J=I+1 TO N
3100      IF H$>Y$(J) THEN 3400
3200      H$=Y$(J) : H1=J
3400    NEXT J
4000    Y$(H1)=Y$(I) : Y$(I)=H$
4100  NEXT I
5000  PRINT:PRINT "Original strings "
5100  FOR I=1 TO N:PRINT X$(I);" ";:NEXT I:PRINT
5200  PRINT:PRINT "Sorted strings "
5300  FOR I=1 TO N:PRINT Y$(I);" ";:NEXT I:PRINT
9999  END


RUN
Number of strings ? 6
? sorting
? may
? be
? a
? total
? nonsense

Original strings
sorting  may  be  a  total  nonsense

Sorted strings
total  sorting  nonsense  may  be  a
Ok
```

Problem 78

Assume you have data pairs, e.g. the data: 'temperature / reaction rate' or 'average pH-value of a lake / calendar year' or 'telephone number / name' and you must sort these data pairs by the first value (e.g. telephone number) or by the second value (e.g. name). Write a program which can do both. Generalize this problem from data pairs to data triples and multiple data points (data tuples).

Problem 78a

Given is a rectangular matrix with n rows and m columns. Write a program to sort the matrix by an arbitrary column or an arbitrary row.

7.2   Monte-Carlo Calculation of First Order Kinetics

In a reaction of first order (examples: radioactive decay, disintegration reactions or isomerizations) the number of the reacting particles per time interval is proportional to the number of the particles still available. This results in an exponential time relationship, if the number of particles is very large.

$$N_t = N_0 \cdot e^{-kt}$$

N_t : Number of particles at time t
N_0 : Number of particles at time 0
k : Rate constant

The simulation of a first order reaction by a random method can be clearly explained: At the beginning there are N drawers and in each drawer is a particle which is able to react. A drawer is selected randomly; if this drawer contains a particle which has not yet reacted, this particle will react now. If the particle has already reacted, nothing will happen. This selection is done in a time interval Δt. In this way the reaction probability is proportional to the number of particles which have not yet reacted. The following program KIN-1ORD is based on this method.

```
0 REM "KIN-1ORD"        EBERT/EDERER        880611
1 REM ***********************************************
2 REM *** Monte-Carlo simulation of first order   ***
3 REM *** kinetics.                               ***
9 REM ***********************************************
50    RANDOMIZE (TIMER/3) MOD 32767
100   DIM X(100)
200   FOR I=1 TO 100 : X(I)=1 : NEXT I
220   T=0
```

```
250    N=100
260    PRINT T,N,EXP(-.01*T)
300      FOR I=1 TO 10
350        T=T+1
400        J=INT(RND*100+1)
500        IF X(J)=0 THEN GOTO 1000
600        X(J)=0
700        N=N-1
1000     NEXT I
1200   GOTO 260
9999   END

RUN
 0                100              1
 10                91              .9048375
 20                83              .8187308
 30                74              .7408182
 40                67              .6703201
 50                63              .6065307
 60                57              .5488117
 70                53              .4965853
 80                48              .449329
 90                45              .4065697
100                40              .3678795
110                40              .3328712
120                39              .3011942
130                35              .2725318
140                33              .246597
150                30              .2231302
160                26              .2018965
170                23              .1826835
180                21              .1652989
190                19              .1495686
200                17              .1353353
210                15              .1224565
220                14              .1108032
230                11              .1002589
240                10              9.071796E-02
250                10              .082085
260                10              .0742736
270                 7              6.720553E-02
280                 6              6.081007E-02
290                 6              5.502323E-02
300                 6              4.978708E-02
^C
Break in 260
```

The number of drawers is defined in line 100 as 100 by the statement DIM X(100). In line 200 the whole array X is filled with the number 1. The 1 represents a non-reacted particle. The number of the random shots is proportional to the time, which in line 220 obtains the initial value zero. In the variable N the number of the available, not yet reacted, particles is stored. Therefore this variable is set to 100 at the beginning of the run. In lines 300 through 1000 ten random experiments are carried out without showing the results on the screen. The time counter is increased by 1 for each random experiment. In line 400 a random number between 1 and 100 is generated and stored in variable J. In case the variable X(J) in the Jth drawer is already a zero — a particle which already has reacted — nothing more happens and the program jumps to the end of the FOR-NEXT-loop (line 1000) and starts the next random experiment. In case the number one is in the drawer X(J), the particle disintegrates, X(J) changes to zero and the number of the reactive particles N is decreased by one (line 700). After ten random experiments the program jumps back to line 260. There, the number of experiments — which corresponds to the reaction time — the number of the remaining non-reacted particles and the analytical result (for comparison) are printed out. The infinite loop continues in line 1200 by jumping to line 260 for the start of the next ten random experiments.

Problem 79

Write a Monte-Carlo program for simulating a consecutive first order reaction.

$$A \xrightarrow{k_a} B \xrightarrow{k_b} C$$

Problem 80

Write a Monte-Carlo program for simulating the kinetics of a second order reaction.

$$A + B \xrightarrow{k} C + D$$

Problem 81

Given are n different (e.g. 100) drawers. Arrange m spheres randomly into these drawers. How does the frequency distribution depend on n and m? Write a Monte-Carlo program to demonstrate this problem.

7.3　Multiplication of Two Square Matrices

Of the matrix operations (addition, subtraction and multiplication) multiplication is the most complicated and laborious. In the following, matrices
are represented by capital letters.

Addition:　　　　　　$C = A + B$

$$c_{i,j} = a_{i,j} + b_{i,j}$$

The corresponding matrix elements are added.

Subtraction:　　　　$C = A - B$

$$c_{i,j} = a_{i,j} - b_{i,j}$$

The corresponding matrix elements are subtracted.

Multiplication:　　　$C = A \cdot B$

$$c_{i,j} = \sum_{k=1}^{n} a_{i,k} \cdot b_{k,j}$$

In multiplication, the ith row of the matrix A is made to correspond with
the jth column of matrix B and the corresponding elements are multiplied.
The n products are then added to obtain the element $c_{i,j}$ of the product
matrix. The number n is the number of rows and columns.

　　　The program MATMULT for the multiplication of two square matrices
reads as follows:

```
0 REM "MATMULT "          EBERT/EDERER           880620
1 REM ************************************************
2 REM *** Multiplication of two square matrices.   ***
9 REM ************************************************
200   DIM A(15,15),B(15,15),C(15,15)
300   INPUT "Dimension of the matrices ";N
400   PRINT
500   PRINT "Input of the elements of matrix A and B "
600   FOR I=1 TO N : FOR J=1 TO N
700      PRINT I;"'th row ";J;"'th column ";
800      INPUT A(I,J),B(I,J)
900   NEXT J : NEXT I
```

```
1000   FOR I=1 TO N
1100     FOR J=1 TO N
1200       C(I,J)=0
1300       FOR K=1 TO N
1400         C(I,J)=C(I,J)+A(I,K)*B(K,J)
1500       NEXT K
1600     NEXT J
1700   NEXT I
2000   FOR I=1 TO N : PRINT : FOR J=1 TO N
2100       PRINT A(I,J); : NEXT J : NEXT I
2200   PRINT : FOR I=1 TO N : PRINT : FOR J=1 TO N
2300       PRINT B(I,J); : NEXT J : NEXT I
2400   PRINT : FOR I=1 TO N : PRINT : FOR J=1 TO N
2500       PRINT C(I,J); : NEXT J : NEXT I
9999   END


RUN
Dimension of the matrices ? 2

Input of the elements of matrix A and B
 1 'th row   1 'th column ? 1,2
 1 'th row   2 'th column ? 3,4
 2 'th row   1 'th column ? 8,7
 2 'th row   2 'th column ? 2,6


 1   3
 8   2


 2   4
 7   6


 23  22
 30  44
Ok
```

The formula for the calculation of one element $c_{i,j}$ of the new matrix C is programmed in lines 1200 through 1500. This is the innermost loop of the whole program. As this calculation has to be done for each element of matrix C, the inner loop is surrounded by two further loops. The I-loop counts the number of rows and the J-loop counts the number of columns of the matrix C. The entire matrix multiplication program is found in lines 1000 through 1700. The lines with higher numbers serve for the output of the three matrices A, B, and C. Line 200 contains the dimensioning of the three matrices. The INPUT-statement in line 300 requires information about the actual size of the matrices. If the matrices contain more than 15 lines or columns the dimensioning in line 200 has to be modified. Line 500 contains additional information for the following input loops. The input of the elements of both matrices A and B has to proceed in the right order simultaneously. The example shown as computer output after the program listing may be more clearly written as

$$\begin{pmatrix} 1 & 3 \\ 8 & 2 \end{pmatrix} \cdot \begin{pmatrix} 2 & 4 \\ 7 & 6 \end{pmatrix} = \begin{pmatrix} 23 & 22 \\ 30 & 44 \end{pmatrix}$$

Problem 82

Rewrite the program MATMULT so that the matrices A and B are not read in with an INPUT-command but by DATA- and READ-statements.

Problem 83

An example of the multiplication of non-square matrices follows: The normalized mass spectra of the following four hydrocarbons at 70 eV ionization potential are:

Table 7.1. Normalized mass spectra of four different hydrocarbons. The highest peak of each substance is normalized to 100.

Substance	\multicolumn{12}{c}{Mass}
	12	13	14	15	16	24	25	26	27	28	29	30
CH_4	1	3	12	100	35							
C_2H_6			2	80		3	4	13	37	24	100	70
C_2H_4	4	7	48	13		2	7	21	18	100		
C_2H_2	5	23	57	2		40	45	100				

Write a program which calculates the mass spectra of mixtures. A necessary physical assumption is the additivity of mass spectra. Use the above data and calculate the mass spectra for the following mixtures: 1:1:1:1, 1:2:3:4 or 4:5:2:8 of CH_4, C_2H_6, C_2H_4 and C_2H_2 respectively.

Extend your program to include the relative ionization sensitivities of the four hydrocarbons, for example: $I_{CH_4} = 0.8$, $I_{C_2H_6} = 0.8$, $I_{C_2H_4} = 1.8$ and $I_{C_2H_2} = 0.9$.

Problem 84

Write a program for optional addition, subtraction and multiplication of matrices, and generalize this program for non-square matrices.

For addition and subtraction the two matrices must correspond in their number of rows and columns. For multiplication the number of columns of the first matrix has to correspond to the number of rows of the second matrix. If the first matrix has the dimension $(n \times m)$ and the second matrix dimension $(m \times l)$ then the product matrix has dimension $(n \times l)$.

Problem 85

Write a program which raises a square matrix A to a higher power. The exponent n should be a positive integer, e.g.

$$A^7 = A \cdot A \cdot A \cdot A \cdot A \cdot A \cdot A$$

7.4   Algebra of Complex Numbers

A complex number z contains both a real part x and an imaginary part y. They are frequently written as follows

$$z = x + iy \qquad \text{where} \quad i = \sqrt{-1}$$

More suitable for computers is, however, the vectorial notation

$$z = (x, y)$$

For that purpose an array Z of length 2 is defined. In Z(1) the x-value, the real part, is written and in Z(2) the y-value, the imaginary part. The four fundamental algebraic operations with complex numbers $z_1 = x_1 + iy_1$ and $z_2 = x_2 + iy_2$ are

$$z_1 + z_2 = (x_1 + x_2) + i(y_1 + y_2)$$

$$z_1 - z_2 = (x_1 - x_2) + i(y_1 - y_2)$$

$$z_1 z_2 = (x_1 x_2 - y_1 y_2) + i(x_1 y_2 + x_2 y_1)$$

$$\frac{z_1}{z_2} = \frac{(x_1 x_2 + y_1 y_2) + i(x_2 y_1 - x_1 y_2)}{x_2 x_2 + y_2 y_2}$$

The general program for the calculation of complex quantities contains all four operations.

```
0 REM "COMPLEX "         EBERT/EDERER         880620
1 REM ************************************************
2 REM *** Algebra of complex numbers. The        ***
3 REM *** fundamental arithmetic expressions can  ***
4 REM *** be calculated ( + - * / ). The complex  ***
5 REM *** numbers are written as:                 ***
6 REM ***     real part , imaginary part          ***
9 REM ************************************************
200    DIM A(2),B(2),C(2)
300    PRINT "Input of the complex number"
400    PRINT "a + b*i     as     a,b"
500    INPUT "1st complex number "; A(1),A(2)
600    INPUT "2nd complex number "; B(1),B(2)
700    INPUT "Which operation (+ - * /) ";A$
1000    IF A$="+" THEN 2000
1100    IF A$="-" THEN 3000
1200    IF A$="*" THEN 4000
1300    IF A$="/" THEN 5000
1400  GOTO 700
2000    FOR I=1 TO 2 : C(I)=A(I)+B(I) : NEXT I
2100      GOTO 6000
3000    FOR I=1 TO 2 : C(I)=A(I)-B(I) : NEXT I
3100      GOTO 6000
4000    C(1)=A(1)*B(1)-A(2)*B(2)
4100      C(2)=A(1)*B(2)+A(2)*B(1) : GOTO 6000
5000    H=B(1)*B(1)+B(2)*B(2)
5100      C(1)=(A(1)*B(1)+A(2)*B(2))/H
5200      C(2)=(A(2)*B(1)-A(1)*B(2))/H
6000  PRINT
6100  PRINT "(";A(1);"+";A(2);"*i) ";A$;
6200  PRINT " (";B(1);"+";B(2);"*i) = ";
6300  PRINT "(";C(1);"+";C(2);"*i)"
9999  END


RUN
Input of the complex number
a + b*i     as     a,b
1st complex number ? 2,3
2nd complex number ? 1,1
Which operation (+ - * /) ? +

( 2 + 3 *i) + ( 1 + 1 *i) = ( 3 + 4 *i)
```

```
RUN
Input of the complex number
a + b*i     as     a,b
1st complex number ? 2,3
2nd complex number ? 1,1
Which operation (+ - * /) ? /

( 2 + 3 *i) / ( 1 + 1 *i) = ( 2.5 + .5 *i)
Ok
```

First, three complex quantities A, B and C are defined in the form of arrays of length 2. The first coordinate contains the real part and the second one the imaginary part. After the definition in line 200, two complex numbers are put in the vectors A and B. In line 700 the program asks for information about the calculation operation and the operational sign is stored in the text variable A$. In the following four lines the program compares A$ with each of the four operations. If all questions are denied the program jumps back to the input of the calculation operation. The user should then put in a new appropriate operational sign. The program COMPLEX is organized in four parts — each for one basic operation — to which the program jumps alternatively. At the end of each of these program parts, in which the complex number C is calculated, the program jumps to line 6000. There follows the output of the complex quantities A and B in the usual notation together with the operational sign and the result of the calculation. The first arithmetic program part which is run through if the operational sign is 'plus' is contained in line 2000 through 2100. Here the sums of the real parts and the imaginary parts are formed and stored in C. This sum formation could be programmed explicitly, but for demonstration purposes a loop was chosen in this program instead. In the section from line 3000 through 3100 the same calculation follows for the subtraction. In section 4000 through 4100 the products are calculated according to the above formula. The quotient is calculated in the section from line 5000 through 5200. The example shows an addition and a division of two complex numbers.

Problem 86

A complex number z is given. Write a program for calculating the following complex valued functions

$$f(z) = z^5 = z \cdot z \cdot z \cdot z \cdot z$$

$$f(z) = \frac{z^2 + z^3}{i + z}$$

$$f(z) = \frac{(5 + 3i)z^5}{z - 3}$$

Try to keep the program general and write the input part of the program for the complex functions (quotients of polynomials) so simple that it can be used by someone who does not know the program.

Problem 87

Write a program for calculation of the complex valued function

$$f(z) = \sqrt[n]{z}$$

n should be an integer quantity. If z is given as $z = x + iy$ you have to solve the equations

$$x = r \cos \phi \qquad y = r \sin \phi$$

for r and ϕ. After you have done this transformation into polar coordinates you can write the complex number z as

$$z = r(\cos \phi + i \sin \phi)$$

The function $f(z)$ can now be written in a form which is easy to program

$$f(z) = \sqrt[n]{z} = \sqrt[n]{r} \left(\cos \frac{\phi}{n} + i \sin \frac{\phi}{n} \right)$$

Problem 88

Write a program for the calculation of e^z, $\sin z$, $\cos z$ and $\log z$.

$$e^z = e^{x+iy} = e^x \cdot e^{iy}$$

$$e^{iy} = \cos y + i \sin y \qquad \text{Euler's equation}$$

From Euler's equation, the following equations can be derived directly

$$\cos z = \frac{1}{2}\left(e^{iz} + e^{-iz}\right)$$

$$\sin z = \frac{1}{2}\left(e^{iz} - e^{-iz}\right)$$

The logarithm function is given by

$$\log z = \log r + i\phi$$

The polar coordinates have to be determined as indicated in problem 87.

Problem 89

For readers who already have some background in calculus: write a program for the integration of complex valued functions. Start with the integration along the real axis. For this purpose the integration programs which have already been presented can be rewritten. Next, try to integrate over an arbitrary curve in the complex plane.

The integration in the complex plane can be done analogously to integration along an interval of the real axis. A sequence of points z_k is chosen and the following sum is formed whose limiting value is defined as the respective definite integral.

$$\int_C f(z)\,dz = \lim_{n\to\infty} \sum_{k=1}^{n} f(z_k)(z_k - z_{k-1})$$

These points z_k lie on curve C, along which the integration should be performed. The points must not lie too far apart in order to get an approximation of the unknown integral which is defined as the limiting value.

7.5   Bragg's Scattering Angle of X-Ray Diffraction

More details on X-ray diffraction from crystals can be found in textbooks on physical chemistry.[1]

For simplicity we consider a cristallographic elementary cell with identical centers of scattering. Such an elementary cell is completely characterized by three lengths a, b and c and three angles α, β and γ.

---

[1] R.S. Berry, S.A. Rice, J. Ross, *Physical Chemistry*, John Wiley & Sons (1980), p.427ff

Through the crystal lattice, which is defined by elementary cells that are joined together, an infinite number of planes can be constructed which touch certain scattering centers. These planes are characterized by the so-called Miller indices h, k, and l. The calculation of the distance d of two neighboring planes with the Miller indices hkl of an elementary cell with lengths a, b, and c and angles α, β and γ can be performed using the following formulas (α_a, β_a, γ_a, a_a, b_a and c_a, are auxiliary quantities.)

$$\cos \alpha_a = \frac{\cos \beta \cos \gamma - \cos \alpha}{\sin \beta \sin \gamma}$$

$$\cos \beta_a = \frac{\cos \gamma \cos \alpha - \cos \beta}{\sin \gamma \sin \alpha}$$

$$\cos \gamma_a = \frac{\cos \alpha \cos \beta - \cos \gamma}{\sin \alpha \sin \beta}$$

$$a_a = \frac{1}{a \sin \beta \sin \gamma_a}$$

$$b_a = \frac{1}{b \sin \gamma \sin \alpha_a}$$

$$c_a = \frac{1}{c \sin \alpha \sin \beta_a}$$

Using these formulas, a further auxiliary variable q can be calculated

$$q = h^2 a_a^2 + k^2 b_a^2 + l^2 c_a^2 + 2kl b_a c_a \cos \alpha_a + 2hl c_a a_a \cos \beta_a + 2hk a_a b_a \cos \gamma_a$$

From q the distance d between two planes can be obtained

$$d = \sqrt{\frac{1}{q}}$$

Bragg's condition for a scattering angle θ for X-rays of wavelength λ at which no destructive interference takes place is

$$d \sin \theta = \frac{\lambda}{2}$$

This equation can be rearranged to obtain Bragg's condition:

$$\theta = \arcsin \frac{\lambda}{2d}$$

```
0 REM "BRAGG    "         EBERT/EDERER         880621
1 REM ********************************************
2 REM *** The Bragg equation is used for calcula- ***
3 REM *** ting the angles of incidence as a func- ***
4 REM *** tion of the wavelength of the X-rays and ***
5 REM *** the Miller indices hkl. The necessary in-***
6 REM *** put is the wavelength of the X-rays and  ***
7 REM *** the geometry of the primitive unit cell. ***
9 REM ********************************************
100    PI=4*ATN(1)
200    DIM D(3),D1(3),A(3)
300    DIM C(3),C1(3),S(3),S1(3)
400    PRINT "All lengths in Angstroem"
500    PRINT "Wavelength of the X-rays"
520    INPUT L9
600    PRINT "Length of the 3 axes of the unit cell"
620    INPUT D(1),D(2),D(3)
700    PRINT "The three angles of the unit cell"
720    INPUT A(1),A(2),A(3)
1000   FOR I=1 TO 3
1100     S(I)=SIN(A(I)/180*PI)
1200     C(I)=COS(A(I)/180*PI)
1300   NEXT I
2000   C1(1)=(C(2)*C(3)-C(1))/(S(2)*S(3))
2100   C1(2)=(C(3)*C(1)-C(2))/(S(3)*S(1))
2200   C1(3)=(C(1)*C(2)-C(3))/(S(1)*S(2))
2500   FOR I=1 TO 3
2600     S1(I)=SQR(1-C1(I)^2)
2700   NEXT I
3000   D1(1)=1/(D(1)*S(2)*S1(3))
3100   D1(2)=1/(D(2)*S(3)*S1(1))
3200   D1(3)=1/(D(3)*S(1)*S1(2))
5000   J=1 : J1=0
5100   FOR L=0 TO J
5200     FOR K=0 TO J-L
5300       H  = J - L  - K
6000       Q=H*H*D1(1)^2+K*K*D1(2)^2+L*L*D1(3)^2
6100       Q=Q+2*K*L*D1(2)*D1(3)*C1(1)+2*L*H*D1(3)*D1(1)*C1(2)
6200       Q=Q+2*H*K*D1(1)*D1(2)*C1(3)
6300       D=SQR(1/Q)
7000       IF J1/12<>INT(J1/12) THEN 7800
7500         PRINT"***"
7520         A$="":A$=INKEY$:IF A$="" THEN 7520
7600         PRINT" h  k  l       incidence angle"
7800       IF L9/2/D>1 THEN 8500
7900       S=L9/2/D:S=S/SQR(1-S*S):S=ATN(S)*180/PI
```

```
8000      PRINT H;K;L;TAB(15);S
8100      J1=J1+1:GOTO 9000
8500      J1=J1+1:PRINT H;K;L;" no incidence angle"
9000    NEXT K
9200   NEXT L
9500   J=J+1
9700   GOTO 5100
9999   END
```

RUN
All lengths in Angstroem
Wavelength of the X-rays
? 1.537
Length of the 3 axes of the unit cell
? 5.64 , 5.64 , 5.64
The three angles of the unit cell
? 90 , 90 , 90
***

h	k	l	incidence angle
1	0	0	7.831419
0	1	0	7.831419
0	0	1	7.831419
2	0	0	15.81414
1	1	0	11.11034
0	2	0	15.81414
1	0	1	11.11034
0	1	1	11.11034
0	0	2	15.81414
3	0	0	24.128
2	1	0	17.73916
1	2	0	17.73916

***

h	k	l	incidence angle
0	3	0	24.128
2	0	1	17.73916

^C
Break in 7520

RUN
All lengths in Angstroem
Wavelength of the X-rays
? 1.537
Length of the 3 axes of the unit cell
? 3.962 , 13.858 , 3.697
The three angles of the unit cell
? 90 , 90 , 90

```
***
h  k  l        incidence angle
1  0  0        11.18442
0  1  0        3.178987
0  0  1        11.99763
2  0  0        22.82609
1  1  0        11.63869
0  2  0        6.367814
1  0  1        16.51779
0  1  1        12.42381
0  0  2        24.56608
3  0  0        35.5845
2  1  0        23.07146
1  2  0        12.91106
***
h  k  l        incidence angle
0  3  0        9.576595
2  0  1        26.11144
1  1  1        16.83825
0  2  1        13.62753
1  0  2        27.30744
^C
Break in 7520
```

In program BRAGG the above formulas are transformed into computer instructions. For clarification Table 7.2 contains the assignment of the symbols of the above formulas to the BASIC-variables in the program.

Table 7.2. Symbols used in the equations for Bragg's condition and the respective BASIC-variables

Equation Symbols	Program Variables
$h \quad k \quad l$	H K L
λ	L9
$a \quad b \quad c$	D(1) D(2) D(3)
$\alpha \quad \beta \quad \gamma$	A(1) A(2) A(3)
$\sin\alpha \quad \sin\beta \quad \sin\gamma$	S(1) S(2) S(3)
$\cos\alpha \quad \cos\beta \quad \cos\gamma$	C(1) C(2) C(3)
$\sin\alpha_a \quad \sin\beta_a \quad \sin\gamma_a$	S1(1) S1(2) S1(3)
$\cos\alpha_a \quad \cos\beta_a \quad \cos\gamma_a$	C1(1) C1(2) C1(3)
$a_a \quad b_a \quad c_a$	D1(1) D1(2) D1(3)
$q \quad d$	Q D

For clarification the method of this program is represented in two flow diagrams. The first one in Fig. 7.2 shows the procedure in rough outline, the second flow diagram in Fig. 7.3 shows the central part of Fig. 7.2 in more detail.

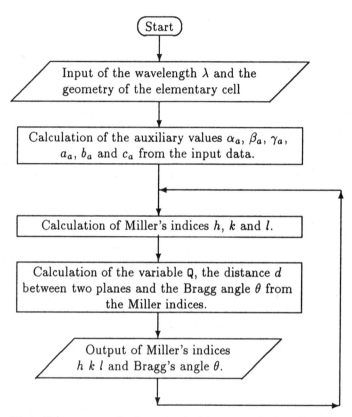

Fig. 7.2. Flow diagram to calculate the incidence angle for all Miller indices using Bragg's condition

All arrays which are used in this program are dimensioned in lines 200 and 300. In the input part from line 400 to 720 the information about the wave length λ of the X-rays and the lengths and angles of the elementary cell are required. Because the trigonometric functions sine and cosine of a certain angle are frequently used in the above formulas, they are calculated in a loop (line 1100 through 1200) and stored in the variables $S(I)$ and $C(I)$. The input of the angles is required in degrees, but the functions require the arc measure. A transformation must be performed in the argument

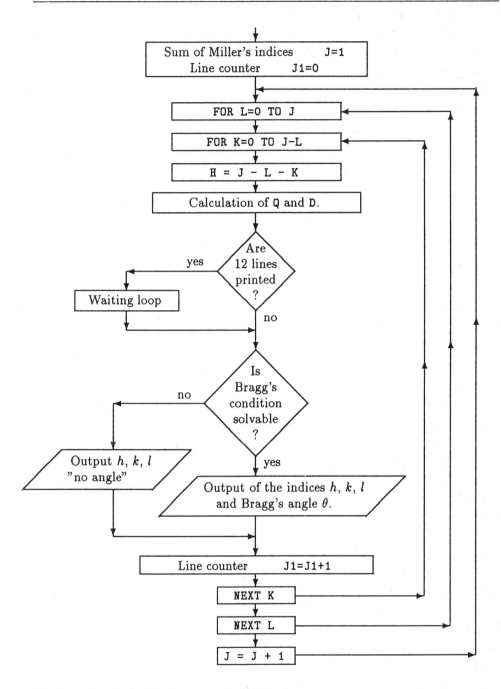

Fig. 7.3. Detailed flow diagram of the loops of program BRAGG, which is roughly outlined in flow diagram in Fig. 7.2

of the sines and cosines. In lines 2000, 2100 and 2200 the cosines of the auxiliary angles α_a, β_a and γ_a are calculated. In the following loop the sines of the auxiliary angles are calculated from the cosines using the well-known transformation formula. The auxiliary quantities a_a, b_a, c_a are calculated using the corresponding formulas in lines 3000, 3100 and 3200.

To understand the following loops one has to observe that the physical estimation of scattering planes is the more important the smaller the sum of the Miller indices is.

In the program, the sum of the Miller indices is accumulated in the variable J. The first value of J is set to unity in line 5000. The variable J1 is used as a counter for the output lines. After 12 output lines the program stops; the corresponding test is contained in line 7000. The quantity J1/12 is only then equal to the quantity INT(J1/12) if J1 is a multiple of 12. If this is the case, the program does not jump to line 7800 but proceeds to run through the next three lines. These three lines serve to print out the three asterisks *** in line 7500, to enter a waiting loop in line 7520, and to print out the heading for the next 12 Bragg angles in 7600. (The waiting loop technique shall be explained after the description of the program.) The next 12 lines are printed out as soon as the user has pressed any key.

If the wavelength λ of the incident light is greater than twice the distance d between the planes under consideration, then no scattering angle exists. This can be formally recognized from the above formula; the arcsine of a quantity greater than unity is not defined. Then Bragg's condition has no solution. This test with the corresponding branching takes place in line 7800. In line 7900 Bragg's angle θ is calculated in degrees in three separate statements. The first statement is the calculation of the sine of this angle θ. As most BASIC-dialects do not possess the arcsine function as a built-in standard function, in the second command the tangent is calculated from the sine. The third command calculates, with help of the arctangent function, the angle, which is transformed simultaneously from arc measure into degrees.

$$\arcsin x = \arctan \frac{x}{\sqrt{1 - x^2}}$$

This Bragg angle θ is then printed together with the corresponding Miller indices. In the next line the counter J1 is increased by one and the program jumps by means of the command GOTO 9000 to the statement NEXT K. The infinite J-loop, which counts the sum of the Miller indices, runs from line 5100 through line 9700. Immediately before the jump in 9700, the J-counter is increased by one. The L-loop runs from zero to J. This has been programmed in the FOR-NEXT-loop from line 5100 through 9200. The K-loop also starts from zero, the index K however can reach only a maximum quantity of J-L, as the sum of all indices cannot become greater than J. This is programmed in the inner loop, the K-loop from line 5200 through 9000. The third Miller index h must have the value J-(K+L) because H+K+L=J must be obeyed (line 5300).

Now the purpose of the waiting loop will be explained, which is programmed in line 7520. First, the text variable A$ is made blank by the command A$="". The third statement in this line asks whether A$ is empty. If this is the case, the program jumps back to the beginning of this line. This looks like a dead loop, which is certainly true; but the INKEY$-statement allows to quit this loop.

Character Input During Program Execution

INKEY$

Syntax: z Z$ = INKEY$

z   : line number

Z$: string variable

The INKEY$-statement makes possible the input of any character during the running of a program. The INKEY$-command does not interrupt the program. The computer asks whether the keyboard buffer contains a character. This character is then stored in the assigned text variable. The keyboard buffer is filled with the character if a key is pressed, even if the program is running. In case you use a BASIC which does not possess the INKEY$-statement, the waiting loop can be replaced by a pause with the help of the command INPUT A$.

Problem 90

Determine the Bragg's angle spectrum of KCl with Mo-X-rays (0.708 Å). KCl has a cubic lattice and a lattice length of 3.1 Å.

7.6   Simulation of a Gas-Chromatographic Column

In this chapter we will simulate not a real gas-chromatographic column but rather an ideal model. This model can also be treated by mathematical-analytical methods. The numerical simulation shown here can be achieved by mathematical methods with much less work.

The model divides a chromatographic column into n ideal equal distribution chambers, and in each of these chambers a given substance will be distributed between the liquid phase and the gas phase by a constant distribution coefficient. The flow of the gas through the column is also ideal. The gas phase is transported in steps from one distribution chamber to the next. This step-wise gas transport is also the unit of the time scale.

For better understanding a short numerical example will be discussed. The distribution coefficient between the liquid and the gas phase is assumed to be 1. A certain amount of substance (in this example 2048 units) is injected into the first distribution chamber and is distributed between the two phases immediately. As the distribution coefficient equals unity, we only have to note in this example the concentration in one of the two phases. At time $t = 1$ the gas phase is transported into the next mixing chamber. The first chamber is filled with pure carrier gas and the distributions are adjusted in all chambers. Table 7.3 shows how the concentrations in each of the two

Table 7.3. Amount of a substance in the gas phase (or the liquid phase, because the distribution coefficient is assumed to be 1) of the first eleven distribution chambers of an ideal column as a function of time

Time	Chamber										
	1	2	3	4	5	6	7	8	9	10	11
t=0	1024	0	0	0	0	0	0	0	0	0	0
t=1	512	512	0	0	0	0	0	0	0	0	0
t=2	256	512	256	0	0	0	0	0	0	0	0
t=3	128	384	384	128	0	0	0	0	0	0	0
t=4	64	256	384	256	64	0	0	0	0	0	0
t=5	32	160	320	320	160	32	0	0	0	0	0
t=6	16	96	240	320	240	96	16	0	0	0	0
t=7	8	56	168	280	280	168	56	8	0	0	0
t=8	4	32	112	224	280	224	112	32	4	0	0
t=9	2	18	72	168	252	252	168	72	18	2	0
t=10	1	10	45	120	210	252	210	120	45	10	1

phases vary in the different chambers with time. The following program GC
calculates the retention time and the half-width of a substance after flowing
through a gas-chromatographic column.

```
0 REM "GC       "          EBERT/EDERER          880524
1 REM ***********************************************
2 REM *** Simulation of the operation of a gaschro-***
3 REM *** matographic column with n theoretical   ***
4 REM *** plates. N may be arbitrary but < 250.    ***
9 REM ***********************************************
100    DIM G(250),F(250)
200    INPUT "Number of theoretical plates ";N
300    INPUT "Partition coefficient ";A
400    P1=A/(1+A) : P2=1/(1+A)
420    REM P1/P2=A and P1+P2=1
600    G(1)=1 : REM Injection
1000   T=0 : T5=0 : D5=0
1200   M=0 : REM Maximum
1300   REM T5 is the retention time of the maximum
2000   T=T+1
3000     REM print t,g(n)
3050     IF T5>0 THEN GOTO 4000
3100     IF G(N)<=G(N-1) THEN GOTO 5000
3200     T5=T+1 : M=G(N)
3300     GOTO 5000
4000     IF D5>0 THEN GOTO 9000
4100     IF G(N)>M/2 THEN GOTO 5000
4200     D5=T-T5
5000     REM Partition equilibrium is established
5100     FOR I=1 TO N
5200       S=G(I)+F(I)
5300       G(I)=S*P1
5400       F(I)=S*P2
5500     NEXT I
6000     REM The gas is flowing now
6100     FOR I=N TO 2 STEP -1
6200       G(I)=G(I-1)
6300     NEXT I
6400     G(1)=0
7000   GOTO 2000
9000   PRINT "Retention time =";T5-N
9100   PRINT "Width at half peak height =";D5
9999   END
```

```
RUN
Number of theoretical plates ? 20
Partition coefficient ?  0.2
Retention time = 91
Width at half peak height = 29
Ok
```

In the program GC, two arrays, G for the gas phase and F for the liquid phase, are dimensioned to 250 in line 100. This is the maximum possible number of theoretical plates (caution: high numbers of plates increase the calculation time!). The number of plates (variable N) is obtained by the input command as well as the distribution coefficient α (variable A). A is defined as the ratio of the amount in the gas phase to the amount in the liquid phase. From the distribution coefficient in line 400 the 'amounts' p_1 and p_2 of the two phases, are calculated obeying the two conditions

$$\frac{p_1}{p_2} = \alpha \qquad p_1 + p_2 = 1$$

As the phases after each transport are not in equilibrium, the equilibrium distribution must be calculated for every chamber. This occurs in the program from line 5000 through 5500. The I-loop runs over all mixing chambers. In line 5200 the sum of both amounts is calculated and in lines 5300 and 5400 the new equilibrium distribution is obtained. The transportation of the gas phase is simulated in section from line 6000 to 6400. The $(i-1)^{th}$ gas phase becomes the ith gas phase (line 6200); the gas phase of the first mixing chamber is set to zero (line 6400). Why does this loop run backwards?

When the equilibrium distribution and the gas flow calculation are finished, the program jumps back from line 7000 to line 2000. There the time counter T is increased by one and printed out together with the contents of the gas phase of the last chamber. (In our printed example this output is deleted.) As long as the maximum concentration in the last mixing chamber G(N) has not been attained, the program jumps from line 3100 to 4000 where the simulation cycle is continued. The time where the maximum is obtained is the retention time which is stored in variable T5 along with the concentration of the maximum in variable M. In the following cycles the program jumps from line 3050 directly to line 4000. In line 4100 the concentration of the last chamber G(N) is compared with half of the maximum concentration

M/2. If G(N) drops below M/2, the half-width is calculated and in the next run the program jumps to the final output in lines 9000 and 9100.

Problem 91

There is a simple mathematical relation between the number of plates, the retention time and the half-width. Try to discover this relation by experimenting. (Experimenting with the computer to solve certain problems is sometimes considerably easier than a theoretical evaluation of the mathematical relations. You can look up the theory in textbooks of physical chemistry.)

Problem 92

Write a Monte-Carlo simulation program for anionic polymerization. For this purpose we assume a certain number of starting anions which are equally distributed in an array, in which the number 1 is written for each element of the array. For a polymerization step, one anion should be randomly chosen and a monomer should be added. Calculate the mean values of the distribution each time as soon as the average degree of polymerization is increased by 1.

Simulate these mean values for a polymer distribution which would result if the number of anions decreases because of contamination during an anionic polymerization (living polymers). The decrease of the anions should also be simulated with the help of a random number generator.

7.7   Linear Regression

In the experimental determination of parameters (e.g. the difference of enthalpies or activation energies) the correlation of experimental data (e.g. equilibrium constants or rate constants with temperature) are often linearized by plotting the data pairs in a graph and connecting the points by a straight line. From the intercept and the slope of the straight line the physical parameters can be determined. To carry out this procedure — to approximate a number of points by a best fit straight line — with a computer, one has to specify a mathematical method.

The method most used tries to put the straight line in a position that will minimize the sum of the squared errors of the y-values. This is called the least squares method (Fig. 7.4). Given are n data pairs (x_i, y_i) and a model which predicts a linear relationship between the measured independent quantity x and the dependent variable y which can be described by a straight line. In this most simple case there are two adjustable parameters a and b.

$$y = a + bx$$

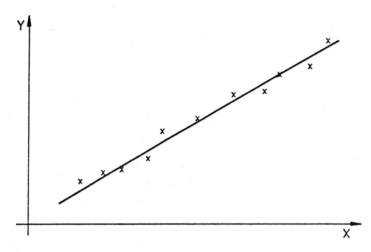

Fig. 7.4. Data points and the resulting straight line as determined with the least squares method

The least squares sum S between the experimental and the simulated y-values is represented by the following formula

$$S = \sum_{i=1}^{n}(y_{\text{exp}} - y_{\text{theory}})^2 = \sum_{i=1}^{n}\left(y_i - (a + bx_i)\right)^2$$

Both parameters a and b of the straight line should be determined so that the least squares sum S is a minimum. For that purpose the partial derivatives of S with respect to a and b are formed and set equal to zero. This is a necessary condition form minimizing S.

$$\frac{\partial S}{\partial a} = -2\sum_{i=1}^{n}(y_i - a - bx_i) = 0$$

$$\frac{\partial S}{\partial b} = -2\sum_{i=1}^{n}(y_i - a - bx_i)\cdot x_i = 0$$

If both equations are divided by -2 and the expressions expanded and rearranged, after factoring the parameters a and b one obtains

$$\sum_{i=1}^{n} y_i = a \cdot n \quad + b \cdot \sum_{i=1}^{n} x_i$$

$$\sum_{i=1}^{n} x_i y_i = a \cdot \sum_{i=1}^{n} x_i + b \cdot \sum_{i=1}^{n} x_i^2$$

Defining

$$S_1 = n \qquad S_2 = \sum_{i=1}^{n} x_i \qquad S_3 = \sum_{i=1}^{n} y_i$$

$$S_4 = \sum_{i=1}^{n} x_i^2 \qquad S_5 = \sum_{i=1}^{n} x_i y_i$$

one obtains the following system of equations for the parameters a and b

$$S_3 = S_1 \cdot a + S_2 \cdot b$$
$$S_5 = S_2 \cdot a + S_4 \cdot b$$

The sums S_1, S_2, S_3, S_4 and S_5 can be calculated from the data pairs and used to solve for a and b

$$a = \frac{S_3 S_4 - S_5 S_2}{S_1 S_4 - S_2 S_2}$$

$$b = \frac{S_1 S_5 - S_2 S_3}{S_1 S_4 - S_2 S_2}$$

The problem is now solved mathematically. In Fig. 7.5 a flow diagram for a linear regression program using the method described above is shown. The respective program with the name REGLIN is listed below.

```
0 REM "REGLIN  "          EBERT/EDERER          880628
1 REM ***********************************************
2 REM ***          Linear Regression            ***
3 REM *** Linear least squares fit to a straight ***
4 REM *** line is calculated.                    ***
5 REM ***          y = a + b*x                   ***
9 REM ***********************************************
```

```
1000   DIM X(100),Y(100)
1100   INPUT "How many data points (y,x) ";N
1200   FOR I=1 TO N
1300     PRINT "y(";I;") , x(";I;")  ";
1400     INPUT Y(I),X(I)
1700   NEXT I
1800   PRINT
2000   S1=N
2100   S2=0:S3=0:S4=0:S5=0
2200   FOR I=1 TO N
2300     S2=S2+X(I)
2400     S3=S3+Y(I)
2500     S4=S4+X(I)*X(I)
2600     S5=S5+Y(I)*X(I)
2700   NEXT I
3000   D1=S1*S4-S2*S2
3100   D2=S3*S4-S5*S2
3200   D3=S1*S5-S2*S3
4000   A=D2/D1
4100   B=D3/D1
5000   PRINT "The regression line for the data is"
5100   PRINT "   y = ";A;"  +  ";B;"* x"
6000   S=0
6100   FOR I=1 TO N
6200     S=S+(Y(I)-(A+B*X(I)))^2
6300   NEXT I
6500   PRINT:PRINT "Sum of squared deviations =";S
9999   END

RUN
How many data points (y,x) ? 7
y( 1 ) , x( 1 )  ? 1 , 0
y( 2 ) , x( 2 )  ? 3.1 , 0.9
y( 3 ) , x( 3 )  ? 4.8 , 1.98
y( 4 ) , x( 4 )  ? 7.2 , 3.05
y( 5 ) , x( 5 )  ? 8.9 , 4.2
y( 6 ) , x( 6 )  ? 11 , 5.02
y( 7 ) , x( 7 )  ? 12.7 , 6.3

The regression line for the data is
   y =  1.229611   +   1.869125 * x

Sum of squared deviations = .4533441
Ok
```

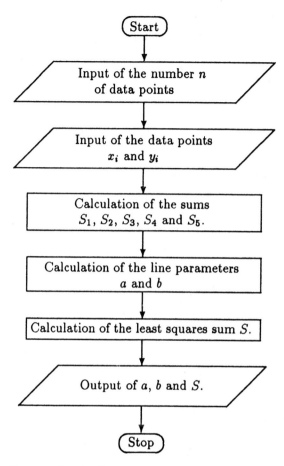

Fig. 7.5. Flow diagram for the linear regression to a straight line

In line 1000 the X- and Y-vectors are dimensioned. In the next line the number of data pairs N is entered. N is required as the end value for the following input loop. This input loop for the x- and y-values runs from line 1200 through 1700. At line 2000 the calculation of the sums begins. First, the definition S1=N is made. In the FOR-NEXT-loop from line 2200 through 2700 the sums which are needed in the formula are generated. The summing variables have to be set to zero before the beginning of the loop. From these sums the denominator is formed in line 3000, and in lines 3100 and 3200 the corresponding nominators are calculated according to the formulas for a and b. Then a and b are calculated and printed out. The loop at the end of the program (6100 through 6300) calculates the sum of the least squares, which are written on the screen in the last executable line of the program.

Problem 93

Extend the program REGLIN in order to:

- generate a table of the input values together with the theoretical y-values from the regression line,
- calculate the y-values for certain x-values which don't have to be experimental data, and print them out,
- carry out a regression with weighted data pairs.

Problem 94

The kinetics of the following reaction was investigated at $190\,°C$

$$\text{Tricyclo-}(5.3.0^{2,6}\text{-deca-3.9-diene}) \xrightarrow{k} 2\text{ Cyclopentadiene}$$

The concentration of the reacting substance was experimentally determined at different reaction times, and the order of the reaction is assumed to be 1.

$$-\frac{d[C]}{dt} = k[C]$$

From this follows by integration and rearrangement

$$-\log[C] = -\log[C]_0 + kt$$

The following experimental data were obtained (concentrations $[C]$ in mol/l and time t in seconds):

$1/[C]$	1.85	2.04	2.34	2.70	3.83	5.28
t	524	620	752	876	1188	1452

Determine from this experimental data the values for $[C]_0$ and k by linear regression.

Problem 95

The reaction rate constants k of the first order decomposition reaction

$$CH_3CHF_2 \xrightarrow{k} CH_2CHF + HF$$

were determined as a function of the temperature. The temperature dependence is as usual described by the Arrhenius law

$$k = k_0 \cdot e^{-\frac{E_a}{RT}}$$

The experimental results are as follows:

$k \cdot 10^7 [s^{-1}]$	7.9	26	52	58	69	230	250	620	1400
T [°C]	429	447	460	462	463	483	487	507	521

Determine by linear regression the pre-exponential factor k_0 and the activation energy E_a. Linearize the nonlinear problem by using $\log k$ and $1/T$ as the dependent and independent variables.

7.8 Linear Regression with Error Estimation

In the following a little more useful linear regression program will be discussed which includes the option of weighting and has different output variations. Furthermore the errors of the parameters a and b are calculated. The mathematics required to solve these problems can be found in books on statistics.[2] The appropriate formulas using D and T as auxiliary functions are

$$D = n \sum_{i=1}^{n} x_i^2 - \left(\sum_{i=1}^{n} x_i \right)^2 = n \cdot S_4 - S_2^2$$

$$T = \sqrt{\sum_{i=1}^{n} \frac{(y_i - a - bx_i)^2}{n-2}} = \sqrt{\frac{S}{n-2}}$$

$$\text{Error of } a = T\sqrt{\frac{S_4}{D}}$$

$$\text{Error of } b = T\sqrt{\frac{n}{D}}$$

The corresponding program REGLINW reads as follows:

---

[2] e.g. L. Sachs, *Applied Statistics: A Handbook of Techniques*, 2.ed., Springer, New York (1984)

```
0 REM "REGLINW "         EBERT/EDERER         880506
1 REM ***********************************************
2 REM ***         Weighted Linear Regression      ***
3 REM *** The linear least squares fit to a       ***
4 REM *** straight line is calculated with weighted***
5 REM *** data points. In addition the errors     ***
6 REM *** in the parameters a and b are obtained.  ***
9 REM ***********************************************
1000  DIM X(100),Y(100),W(100)
1010  A$="                          "
1100  INPUT "How many data points (y,x) ";N
1120  FOR I=1 TO N : W(I)=1 : NEXT I : PRINT
1140  PRINT "Do you want to weight your data points ";
1150  INPUT B$
1200  FOR I=1 TO N
1300     PRINT "  y(";I;") = ";
1310     INPUT Y(I)  : LOCATE CSRLIN-1,1
1330     PRINT A$;A$ : LOCATE CSRLIN-1,1
1350     PRINT "  y(";I;") =";Y(I);
1400     PRINT "  x(";I;") = ";
1410     INPUT X(I)  : LOCATE CSRLIN-1,1
1450     IF B$<>"yes" THEN GOTO 1615
1490     PRINT A$;A$ : LOCATE CSRLIN-1,1
1500     PRINT "  y(";I;") =";Y(I);
1600     PRINT "  x(";I;") =";X(I);
1610     PRINT "  w(";I;") = ";
1613     INPUT W(I)  : LOCATE CSRLIN-1,1
1615     PRINT A$;A$ : LOCATE CSRLIN-1,1
1620     PRINT "  y(";I;") =";Y(I);
1630     PRINT "  x(";I;") =";X(I);
1640     PRINT "  w(";I;") =";W(I)
1700  NEXT I
2000  S1=0
2100  S2=0 : S3=0 : S4=0 : S5=0
2200  FOR I=1 TO N
2250     S1=S1+W(I)
2300     S2=S2+X(I)*W(I)
2400     S3=S3+Y(I)*W(I)
2500     S4=S4+X(I)*X(I)*W(I)
2600     S5=S5+Y(I)*X(I)*W(I)
2700  NEXT I
3000  D1=S1*S4-S2*S2
3100  D2=S3*S4-S5*S2
3200  D3=S1*S5-S2*S3
4000  A=D2/D1
4100  B=D3/D1
```

```
5000   PRINT "The regression line for the data is"
5100   PRINT " y  = ";A;"  +  ";B;"* x"
6000   S=0
6100   FOR I=1 TO N
6200     S=S+W(I)*(Y(I)-(A+B*X(I)))^2
6300   NEXT I
6500   PRINT:PRINT "Sum of squared deviations =";S
6600   D=SQR(S/(N-2))
6605   PRINT:PRINT "Standard deviation  =";D
6610   DA=D*SQR(S4/D1) : DB=D*SQR(S1/D1)
6620   PRINT:PRINT "Number of data points =";N : PRINT
6630   PRINT "a = ";A;" +/- ";DA:PRINT
6640   PRINT "b = ";B;" +/- ";DB:PRINT
7000   PRINT:PRINT "Do you want to compare the input "
7100   PRINT "values with the regression values ";
7200   INPUT A$
7300   IF A$="yes" THEN GOTO 8000
7500   GOTO 8400
8000   PRINT
8050   PRINT " x            y(input)   y(regression) "
8100   FOR I=1 TO N
8200     PRINT X(I);TAB(12);Y(I);TAB(23);A+B*X(I)
8300   NEXT I
8400   PRINT:PRINT "Do you want to interpolate data"
8500   PRINT "using the regression line ";
8600   INPUT A$
8700   IF A$="yes" THEN GOTO 9000
8800   GOTO 9999
9000   PRINT "x-value";:INPUT X
9050     PRINT A$;A$;: LOCATE CSRLIN-1,1
9100   PRINT "y(";X;") = ";A+B*X
9200   GOTO 9000
9999   END


RUN
How many data points (y,x) ? 4

Do you want to weight your data points ? yes
   y( 1 ) = 4.1    x( 1 ) = 2.1    w( 1 ) = .5
   y( 2 ) = 8.1    x( 2 ) = 3.9    w( 2 ) = 3
   y( 3 ) = 9.9    x( 3 ) = 5.1    w( 3 ) = 10
   y( 4 ) = 12.1   x( 4 ) = 5.8    w( 4 ) = 1
The regression line for the data is
 y  = .5361733   +   1.864496 * x

Sum of squared deviations = 1.090794
```

```
Standard deviation  = .7385099

Number of data points = 4

a =  .5361733  +/-  1.269276

b =  1.864496  +/-  .2615152


Do you want to compare the input
values with the regression values ? yes

   x           y(input)   y(regression)
   2.1         4.1        4.451614
   3.9         8.100001   7.807707
   5.1         9.899999   10.0451
   5.8         12.1       11.35025

Do you want to interpolate data
using the regression line ? yes
y(-1 ) = -1.328322
y( 2 ) =  4.265165
y( 17 ) =  32.2326
y( 55 ) =  103.0834
y( 0 ) =  .5361733
y( 7.23 ) =  14.01648
x-value?
Break in 9000
```

In this program the input part is enhanced for the possibility of weighting the experimental data. The weights are set to one before the input of the data. If the input of the weight is ignored, it is automatically equal to 1. The least squares sum S is now given by

$$S = \sum_{i=1}^{n} w_i (y_{\text{exp}} - y_{\text{theory}})^2 = \sum_{i=1}^{n} w_i \Big(y_i - (a + bx_i) \Big)^2$$

where w_i is the weight for the data point (x_i, y_i). The central loop from line 2200 through 2700 is extended in each line by the multiplication of the weight. Line 2250 was added in which the sum of the weights is calculated. This sum is then used instead of the number of the experimental values in the regression formula. In the further lines through 6500 there is only one variation from the original program, the least squares are multiplied by the weight. In the following lines the overall standard deviation (line 6600) and the errors of the parameters *a* and *b* (line 6610) are calculated using the above formulas and printed on the screen. In line 7200 information is requested as to whether the table should contain the experimental data together with the theoretical *y*-values. If this is affirmed, a print-out follows in lines 8000 through 8300. Finally, the possibility is offered to obtain for any *x*-value the corresponding theoretical *y*-value, which is also printed out (line 8400 through 9200). The special statements in the lines 1330, 1490, 1615, and 9100 imply cursor movements for a better arrangement of the print-out on the screen.

Problem 96

The program REGLINW is very useful for many practical applications, but it still has a rather serious practical disadvantage. If a mistake occurs in typing in experimental data there is no other choice than starting the program from the beginning. As these mistakes occur frequently in typing a great number of experimental points, this is indeed very annoying.

Try to extend the program REGLINW to allow corrections in the input data. The simplest method is to replace the input part by DATA- and READ-commands. A more elegant method is the addition of statements correcting the input data between lines 1700 and 2000; this might be more difficult to program. After finishing with the input, the data should be written on the screen again, one after the other. The user has to confirm each point or type in a correction.

7.9   Mass Distribution of the Molecular Peak

In the mass spectroscopy of halogenated organic substances there is usually no unique molecular peak, because, for example, Br and Cl have different isotopes. Using the natural abundances of the isotopes we will find for the molecule HBr two molecular peaks of almost the same height, one with mass 80 and the other with mass 82. In this discussion we only consider integer values for the atomic masses of the isotopes.

The distribution of molecular peaks — using the natural abundances of the isotopes of an element — will be calculated in the program ISOMASS. The frequency distribution is calculated by adding one atom after the other to get the molecule under consideration. If we already have a molecule fraction with distribution

$$d = (d_1, d_2, \ldots, d_m, \ldots, d_{m_{\max}})$$

$$\text{where} \qquad \sum_{j=1}^{m_{\max}} d_j = 1$$

and the index m stands for the mass, and if we add an atom with isotope distribution

$$i = (i_1, i_2, \ldots, i_n, \ldots, i_{n_{\max}})$$

we will get for the new molecule fraction a new mass distribution given by the following equation.

$$d_{m,\text{new}} = \sum_{\substack{m_d, m_i \\ m_d + m_i = m}} d_{m_d} \cdot i_{m_i}$$

This means that the sum of the frequency products of all combinations of masses m_d and m_i which add up to the mass m is calculated in order to get the new frequency d_m.

   This calculation is done in program ISOMASS which is listed below:

```
0 REM "ISOMASS "          EBERT/EDERER         880909
1 REM ***********************************************
2 REM *** Determination of the mass distribution   ***
3 REM *** of the molecular peak, considering        ***
4 REM *** the isotope abundance of the elements.    ***
9 REM ***********************************************
100    DIM FREQ(512) , FRAUX(512) , FR(512)
400    REM Name,lowest mass,number of isotopes,frequencies
500    DATA "C" , 12 , 2 , 0.989  , 0.011
510    DATA "H" ,  1 , 2 , 0.99985, 0.00015
520    DATA "N" , 14 , 2 , 0.9963 , 0.0037
530    DATA "O" , 16 , 3 , 0.99762, 0.00038 , 0.00200
600    DATA "Cl", 35 , 3 , 0.7577 , 0.0      , 0.2423
650    DATA "Br", 79 , 3 , 0.5069 , 0.0      , 0.4931
900    DATA"FIN", 0 , 0
1000   NMAX = 1 : NAM$=""
1020   FREQ(1) = 1
1040   MMIN = 0 : MMAX = 0
1200   READ NAM$ , AMIN , ANUM
1300   IF NAM$="FIN" THEN GOTO 20000
```

```
1340   FOR I=1 TO ANUM
1350     READ FR(I)
1380   NEXT I
1400   PRINT "How many  *** ";NAM$;" *** atoms ";
1420   INPUT NA
1440   IF NA=0 THEN GOTO 1200
1500   FOR J=1 TO NA
1600     FOR I=1 TO NMAX
1620        FRAUX(I) = FREQ(I)
1640     NEXT I
1660     NMAX2=NMAX
1670     MMIN = MMIN + AMIN
1680     MMAX = MMAX + AMIN + ANUM - 1
1690     NMAX = NMAX + ANUM - 1
1700     FOR I=1 TO NMAX : FREQ(I)= 0 : NEXT I
1720     FOR I2=1 TO NMAX2
1740       FOR I=1 TO ANUM
1760         FREQ(I2+I-1) = FREQ(I2+I-1) + FRAUX(I2)*FR(I)
1780       NEXT I
1800     NEXT I2
1900   NEXT J
2300   GOTO 1200
20000 REM output of the frequency distribution
20050 PRINT : MW=0
20100 FOR I=1 TO NMAX
20110   MW = MW + FREQ(I)*(MMIN+I-1)
20120   IF FREQ(I)<1E-10 THEN GOTO 20140
20130   PRINT "Frequency of mass ";MMIN+I-1;" = "; FREQ(I)
20140 NEXT I
20200 PRINT : PRINT "Average molecular mass = "; MW
30000 END

RUN
How many  *** C *** atoms ? 0
How many  *** H *** atoms ? 1
How many  *** N *** atoms ? 0
How many  *** O *** atoms ? 0
How many  *** Cl *** atoms ? 0
How many  *** Br *** atoms ? 1

Frequency of mass  80  =  .506824
Frequency of mass  81  =  7.603501E-05
Frequency of mass  82  =  .493026
Frequency of mass  83  =  7.3965E-05

Average molecular mass =  80.98635
Ok
```

```
RUN
How many  *** C *** atoms ? 8
How many  *** H *** atoms ? 8
How many  *** N *** atoms ? 1
How many  *** O *** atoms ? 1
How many  *** Cl *** atoms ? 2
How many  *** Br *** atoms ? 1

Frequency of mass  283  =  .2644364
Frequency of mass  284  =  2.492937E-02
Frequency of mass  285  =  .4279346
Frequency of mass  286  =  4.026985E-02
Frequency of mass  287  =  .1941001
Frequency of mass  288  =  .0181805
Frequency of mass  289  =  2.744861E-02
Frequency of mass  290  =  2.53445E-03
Frequency of mass  291  =  1.582187E-04
Frequency of mass  292  =  7.516612E-06
Frequency of mass  293  =  2.461351E-07
Frequency of mass  294  =  5.405402E-09

Average molecular mass =  285.0527
Ok
```

In line 100 three arrays are defined. The first array **FREQ** holds the total frequency distribution, the second one **FRAUX** is an auxiliary vector for the frequency distributions and the third one **FR** is used for the isotope distribution of an element. **FREQ(I)** is the frequency of mass (MMIN+I-1). In the lines beginning with line 500 the isotope distribution for each element used is written in one DATA-statement. The first item of this DATA-statement is the name of the element, the second item is the lowest mass of an isotope of this element and the third item is the number of isotopes followed by their frequencies. All isotopes between the isotope with the lowest mass and the isotope with the highest mass have to be included, which means that some of the frequencies may be zero. In line 1000 the variable **NMAX** is set to one; it always holds the total number of possible masses for the distribution. **MMIN** and **MMAX** are the variables for the minimum mass and maximum mass, respectively, and they are initially set to zero.

In line 1200 the name of an element is read into **NAM\$**, the lowest mass of the isotopes into **AMIN**, and the number of different isotopes into **ANUM**. In the following loop the frequencies of all isotopes are read into the array

FR. The next statement asks the user for the number NA of atoms of this element within the molecule. If there are none, the program branches back to line 1200 to read the items for the next element.

If there are one or more atoms of this element, the J-loop is entered which counts the number of atoms. The body of this loop determines the new frequency distribution for the masses. From line 1600 through 1640 the old distribution is stored in the auxiliary vector FRAUX. After this loop the new minimum mass and maximum mass and the number of different masses is calculated. NMAX2 is used to hold the number of the different masses of the old distribution, which is now in the vector FRAUX.

In line 1700 the vector FREQ is set to zero. In the following two nested loops each old frequency FRAUX(I2) is multiplied with each isotope frequency FR(I) and the products are added to the new frequency FREQ(I2+I-1). After the new mass distribution is calculated, the NEXT J statement for the next atom is executed. If there is no further atom the program jumps back to line 1200 for the next chemical element. Now the same calculations are done for the atoms of this element. The last element of the DATA lines has the name "FIN". If this is read in line 1200 the program jumps to line 20000, the output and the end of the program.

In the output the masses and their frequencies are written on the screen. If the frequency is too low then the output is suppressed by the IF-statement in line 20120. For control, an average of the molecular mass is calculated and printed. This average is not the correct average molecular weight because only integer masses for the isotopes are used.

Problem 96a

Given is not only the molecular peak but also one or more additional fragments of the molecule together with their appearance probabilities.

Calculate the mass distribution of the total spectrum if the different isotopes of the elements are taken into account.

8 Linear Systems

Linear dependences often occur in the natural and engineering sciences. They describe what often is called the 'additive superposition of effects'. Frequently an ideal behavior is in the mathematical sense a linear behavior. Concrete examples are:

- Concentration dependence of thermodynamical functions in ideal mixtures,
- Superposition of mass spectra, if mixtures are analyzed,
- Analysis of mixtures by gas chromatography,
- Reaction rates of reaction systems of first order with respect to the concentrations of the educts (the dependence on the temperature is obviously nonlinear).
- Mass and heat transport at stationary states are dependent on concentration and temperature gradients,
- The dependence of fluxes on potentials in irreversible thermodynamics (Onsager-relations).
- Electrical current in a net of resistances,
- Vapor pressures and partial pressures as functions of the composition of liquids (in ideal systems).
- Cost of chemicals as a function of the required quantities (this is only valid for the ideal case and not if the salesman smells big business and raises the prices).
- LCAO-methods in quantum chemistry,
- and many more problems.

Frequently nonlinear behavior is linearized by dividing the range of interest into intervals, within which linear behavior is assumed. For extremely nonlinear behavior — e.g. the temperature dependence of the reaction rate —

the intervals chosen must be very small. In thermodynamics, real behavior can frequently be approximated by a linear dependence if the region of validity is limited (ideal limiting behavior). Raoult's and Henry's limiting laws for small concentrations are appropriate examples. Linearizing by dividing in validity regions is a general principle which is not only used in natural and engineering sciences but also in mathematics — the derivative is a local linear approximation.

8.1   Subroutines

As the programs and problems become more extensive (but not necessarily more complicated) at this time a very important structural element of the BASIC-languages, the subroutine, will be introduced. Subroutines help to improve the structure of programs.

In the integration program EULER we had to write the functions to be integrated at three different positions in the program; in program EULER2 a so-called function subroutine was used instead, which, however, required mathematical expressions to be written on a single line. General subroutines serve a similar purpose; they are program parts which can be used as often as required from a main program or from other subroutines at different positions. Subroutines can consist of entire programs and they are not restricted to one line or to one mathematical expression as in the DEF FN-statements. Subroutines are called with the statement GOSUB.

Calling a Subroutine

$$\boxed{\text{GOSUB}}$$

Syntax: z1 GOSUB z2

z : line number

---

z2 is the line number of the first line of the subroutine called. The beginning of a subroutine is not further indicated in BASIC. Subroutines have to be terminated by the command RETURN.

End of a Subroutine

RETURN

Syntax: z RETURN

z : line number

A subroutine does not have an explicit start statement. The GOSUB-command is very similar to the GOTO-command. Both commands let the program jump to the line number z2. The difference is that the GOTO-jump forgets where it comes from whereas the GOSUB-statement keeps its origin — that is the line number z1 — in memory and by the command RETURN the program jumps back to the line following the GOSUB-statement.

A program may contain more than one reference to the same subroutine. Control will always be returned from the subroutine to the statements following the particular GOSUB-statement that referred the subroutine. It is possible for one subroutine to contain a reference to another subroutine. Subroutines that are structured in this manner are said to be nested. Nested subroutines must maintain a strict hierarchical ordering.

```
Example:  10      DIM  X(100)
          20      FOR I=1 TO 100
          30         X(I) = I*I
          40      NEXT I
          50      GOSUB 1000
          60      GOSUB 2000
          70      FOR I=1 TO 100
          80         X(I) = I*I*I
          90      NEXT I
          100     GOSUB 1000
          110     GOSUB 2000
          120     STOP
          1000    S=0
          1010    FOR I=1 TO 100
          1020       S = S + X(I)
          1030    NEXT I
          1040    S = S/100
          1050    RETURN
          2000    PRINT "Mean  = ";S
```

```
2010    RETURN
2020    END
```

In the beginning of this example the first hundred squared integers are calculated and assigned to the array X (line 20–40); in line 50 the subroutine is called which carries out the calculation of the mean value. The next subroutine call in line 60 serves for the output of data on the screen. Then the cubed integer numbers are calculated and assigned to the vector X by the main program (lines 70 through 90). The next two subroutine calls serve again for the calculation of the mean value and the output. The program can be extended so that the variables X(I) in the main program can be filled with any possible values. By means of the subroutine call GOSUB 1000 the average value is calculated.

A further example calculates the magnitude of π by a Monte-Carlo method. However this program looks very different to that which was used earlier in Chapter 3.5.

```
5       RANDOMIZE
10      N1 = 0
20      N2 = 0
30      GOSUB 1000: REM Calculation of a Random Vector
40      GOSUB 1200: REM Decission of hitting
50      GOSUB 1400: REM Output
60      GOTO 30
1000    X1 = RND
1010    X2 = RND
1020    RETURN
1200    H = X1*X1 + X2*X2
1210    IF H>1 THEN GOTO 1230
1220    N1 = N1 + 1
1230    N2 = N2 + 1
1240    RETURN
1400    PRINT N1,N2,N1/N2*4
1500    RETURN
2000    END
```

This program was separated into subroutines, not because they are often used, but to improve the organization of the program. The main program (lines 10 through 60) is short and is composed of subroutine or block calls. The blocks (subroutines) contain the different mathematical operations and the output.

The following short demonstration program carries out the smoothing experimental data, a procedure often needed in processing experimental data:

```
0       REM 1 3 7 3 1 Smoothing of Data
100     DIM X(100),Y(100)
200     GOSUB 2000 : REM Input of Data
300     GOSUB 3000 : REM Smoothing of Data and Output
500     INPUT "Smoothing the Data again"; A$
600     IF A$="YES" THEN GOTO 300
700     STOP

2000    N = 10
2010    DATA 1,1.8,2.8,4.3,5,6,6.9,7.8,9.2,10
2020    FOR I=1 TO N: READ X(I): Y(I)=X(I): NEXT I
2040    RETURN

3000    Y(1) = X(1) : Y(N) = X(N)
3010    Y(2) = (X(1)+3*X(2)+X(3))/5
3020    Y(N-1) = (X(N-2)+3*X(N-1)+X(N))/5
3100    FOR I=3 TO (N-2)
3110      Y(I)=(X(I-2)+3*X(I-1)+7*X(I)+3*X(I+1)+X(I+2))/15
3120    NEXT I
3200    GOSUB 4000: REM Output of the smoothed data
3900    RETURN

4000 FOR I=1 TO N:PRINT I,X(I),Y(I):X(I)=Y(I):NEXT I
4100 RETURN
```

This method of smoothing experimental data is very widely used. It must be decided beforehand, how extensive the smoothing should be, that is how many neighboring points should be considered in the smoothing algorithm. The decision to use two neighboring values on each side with the weights 1 3 7 3 1, as done in the above program, can obviously be changed at any time. The general equation for neighboring smoothing is given by

$$x_{i,\text{smoothed}} = \frac{\sum\limits_{j=-n}^{n} w_j \cdot x_{i+j}}{\sum\limits_{j=-n}^{n} w_j}$$

where $w = (w_{-n}, \ldots, w_0, \ldots, w_n)$ are the weights of the $2n + 1$ points used to smooth the data. The data at the beginning and the end of the data list have to be smoothed with a separate algorithm or they have to remain unchanged.

The frequent use of subroutines in the example program serves again to improve the structure of the program. In this example nested subroutines are used, that is, a subroutine is called from another subroutine. The maximum possible number of nesting levels of subroutines depends on the computer and software used. The necessary information can be looked up in the language handbook but it can also be determined by testing. Sometimes the level of nesting is limited only by the memory available.

> Warning: avoid calling a subroutine directly or indirectly by itself (recursive call).

The simplicity with which BASIC-program parts can be arranged as subroutines, and the simple call of subroutines (in comparison to other higher programming languages) have one disadvantage. A program (e.g. a main program) including all its subroutines is one BASIC unit. Variable names apply to the whole program. Variables of a subroutine are not independent from other subroutines or from the main program (such as the dummy variable in the DEF FN-statement) in standard BASIC. This is a great disadvantage in large programs, in which parts have to be programmed independently from others (possibly even by different authors) and which are then combined into one program. If, nevertheless, large programs are to be produced and combined in this way, the use of different variables for the different subroutines has to be established beforehand.

8.2   Systems of Linear Equations

The Gauss-Jordan method will be discussed here; for a better understanding, however, a system of linear equations should be solved 'by hand'. The following three equations containing three unknown quantities are given.

$$
\begin{aligned}
4x &+ 3y + z &= 13 \\
2x &- y - z &= -3 \\
7x &+ y - 3z &= 0
\end{aligned}
$$

These equation systems can also be written deleting the symbols x, y, z and $=$ and then the so-called extended coefficient matrix is obtained

$$\begin{pmatrix} 4 & 3 & 1 & 13 \\ 2 & -1 & -1 & -3 \\ 7 & 1 & -3 & 0 \end{pmatrix} \quad \times \tfrac{1}{4}$$

In such an equation system each equation can be divided by any quantity without changing the solution vector. In order that the first coefficient of the first equation becomes unity, the first equation is divided by 4.

$$\begin{pmatrix} 1 & \tfrac{3}{4} & \tfrac{1}{4} & \tfrac{13}{4} \\ 2 & -1 & -1 & -3 \\ 7 & 1 & -3 & 0 \end{pmatrix} \quad \begin{array}{l} [\text{2}^{\text{nd}}\ \text{row}] - 2 \cdot [\text{1}^{\text{st}}\ \text{row}] \\ [\text{3}^{\text{rd}}\ \text{row}] - 7 \cdot [\text{1}^{\text{st}}\ \text{row}] \end{array}$$

As the different equations can also be multiplied by any quantity and the equations may be added or subtracted from each other, the other coefficients in the first column can be converted into zeroes. Therefore we multiply the first row by 2 and subtract it from the second row. In the same way we multiply the first line by 7 and then subtract it from the third line

$$\begin{pmatrix} 1 & \tfrac{3}{4} & \tfrac{1}{4} & \tfrac{13}{4} \\ 0 & -\tfrac{5}{2} & -\tfrac{3}{2} & -\tfrac{19}{2} \\ 0 & -\tfrac{17}{4} & -\tfrac{19}{4} & -\tfrac{91}{4} \end{pmatrix} \quad \times \left(\tfrac{-2}{5} \right)$$

Now the next diagonal element should be converted into unity. Therefore the second equation is divided by the diagonal element $(-5/2)$ and the following equations are obtained

$$\begin{pmatrix} 1 & \tfrac{3}{4} & \tfrac{1}{4} & \tfrac{13}{4} \\ 0 & 1 & \tfrac{3}{5} & \tfrac{19}{5} \\ 0 & -\tfrac{17}{4} & -\tfrac{19}{4} & -\tfrac{91}{4} \end{pmatrix} \quad \begin{array}{l} [\text{1}^{\text{st}}\ \text{row}] - \tfrac{3}{4} \cdot [\text{2}^{\text{nd}}\ \text{row}] \\ \\ [\text{3}^{\text{rd}}\ \text{row}] + \tfrac{17}{4} \cdot [\text{2}^{\text{nd}}\ \text{row}] \end{array}$$

Again, we try to convert the coefficients in the same column to zero. Therefore the second equation $\times 3/4$ is subtracted from the first equation, and the $17/4$'s of the second equation is added to the third equation

$$\begin{pmatrix} 1 & 0 & -\tfrac{1}{5} & \tfrac{2}{5} \\ 0 & 1 & \tfrac{3}{5} & \tfrac{19}{5} \\ 0 & 0 & -\tfrac{11}{5} & -\tfrac{33}{5} \end{pmatrix} \quad \times \left(\tfrac{-5}{11} \right)$$

The third diagonal element becomes 1 by dividing the third equation by the third diagonal element

$$\begin{pmatrix} 1 & 0 & -\frac{1}{5} & \frac{2}{5} \\ 0 & 1 & \frac{3}{5} & \frac{19}{5} \\ 0 & 0 & 1 & 3 \end{pmatrix} \qquad \begin{array}{l} [1^{st} \text{ row}] + \frac{1}{5} \cdot [3^{rd} \text{ row}] \\ [2^{nd} \text{ row}] - \frac{3}{5} \cdot [3^{rd} \text{ row}] \end{array}$$

The coefficients of the third column are converted to zero by adding or subtracting parts of the third equation

$$\begin{pmatrix} 1 & 0 & 0 & 1 \\ 0 & 1 & 0 & 2 \\ 0 & 0 & 1 & 3 \end{pmatrix}$$

Now, the 'forgotten' symbols can be added again to the equation set, and one obtains

$$\begin{array}{ccccccc} 1x & + & 0y & + & 0z & = & 1 \\ 0x & + & 1y & + & 0z & = & 2 \\ 0x & + & 0y & + & 1z & = & 3 \end{array}$$

from which the result is directly obtained

$$x = 1$$
$$y = 2$$
$$z = 3$$

The flow diagram for the Gauss-Jordan method is shown in Fig. 8.1. To understand the flow diagram for programming this method, some definitions and commands must be given. The extended coefficient matrix is called A, its elements $a_{i,j}$. The dimension of the equation system, i.e. the number of equations, is called n. Normalization of the ith row means: dividing the ith equation by the ith diagonal element $a_{i,i}$. If this diagonal element is equal to zero then the ith equation is exchanged with the one following it. If this does not help, the exchange operation is carried out with one of the next rows until the ith diagonal element becomes unequal to zero. If this is not possible, the equation system is not uniquely solvable. Equations can be exchanged ad libitum without changing the result. The corresponding BASIC-program GJ is listed after the flow diagram (Fig. 8.1).

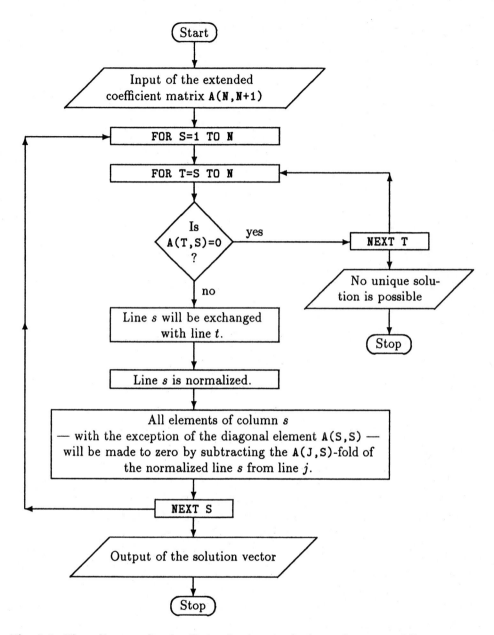

Fig. 8.1. Flow diagram for the Gauss-Jordan method to solve a set of linear equations

```
0 REM "GJ       "        EBERT/EDERER        880628
1 REM ***********************************************
2 REM *** Solving a system of n linear equations  ***
3 REM *** with n unknowns using the method of     ***
4 REM *** Gauss-Jordan                            ***
9 REM ***********************************************
50000 DIM A(20,21)
50100 INPUT "Number of equations ";N
50200 PRINT "Input of the extended coefficient matrix"
50300 FOR I=1 TO N
50400    FOR J=1 TO N
50500       PRINT "A(";I;",";J;")=";:INPUT A(I,J)
50600    NEXT J
50700    PRINT "B(";I;")=";:INPUT A(I,N+1)
50800 NEXT I : PRINT
50900 FOR S=1 TO N
51000    FOR T=S TO N
51100       IF A(T,S)<>0 THEN 51300
51200    NEXT T
51210    PRINT "No solution is possible" : GOTO 63999
51300    GOSUB 53000 : REM EXCHANGE lines
51400    C=1/A(S,S)
51500    GOSUB 54000 : REM NORMALIZE the actual line
51600    FOR T=1 TO N
51700       IF T=S THEN 52000
51800       C=-A(T,S)
51900       GOSUB 55000 : REM make zeros in the column
52000    NEXT T
52100 NEXT S
52200 GOSUB 56000 : GOTO 63999 :REM OUTPUT and END
52210 REM #########################################
52220 REM ### Main program ends here.          ###
52230 REM #########################################
53000 REM ********************************************
53002 REM *** Subroutine: EXCHANGE                ***
53004 REM *** Lines S and T are exchanged.        ***
53010 REM ********************************************
53100    FOR J=1 TO N+1
53200       B=A(S,J) : A(S,J)=A(T,J) : A(T,J)=B
53300    NEXT J
53400 RETURN : REM end of subroutine EXCHANGE
54000 REM ********************************************
54002 REM *** Subroutine: NORMALIZE               ***
54004 REM *** Normalizing the actual line S.      ***
54010 REM ********************************************
```

```
54100    FOR J=1 TO N+1
54200      A(S,J)=C*A(S,J) : NEXT J
54400 RETURN : REM end of subroutine NORMALIZE
55000 REM *******************************************
55002 REM *** Subroutine: MAKEZERO              ***
55004 REM *** The elements of line S are multi-  ***
55006 REM *** plied with C and subtracted from the ***
55008 REM *** elements of line T.               ***
55010 REM *******************************************
55100    FOR J=1 TO N+1
55200      A(T,J)=A(T,J)+C*A(S,J)
55300    NEXT J
55400 RETURN : REM end of subroutine MAKEZERO
56000 REM *******************************************
56002 REM *** Subroutine: OUTPUT               ***
56004 REM *** Output of the solution.          ***
56010 REM *******************************************
56100    FOR T=1 TO N
56200      PRINT "X(";T;")=";A(T,N+1)
56300    NEXT T
56400 RETURN : REM end of subroutine OUTPUT
63999 END

RUN
Number of equations ? 3
Input of the extended coefficient matrix
A( 1 , 1 )=? 4
A( 1 , 2 )=? 7
A( 1 , 3 )=? 2
B( 1 )=? 38
A( 2 , 1 )=? 6
A( 2 , 2 )=? 3
A( 2 , 3 )=? 2
B( 2 )=? 26
A( 3 , 1 )=? 5
A( 3 , 2 )=? -2
A( 3 , 3 )=? -3
B( 3 )=? -1

X( 1 )= 2
X( 2 )= 4
X( 3 )= 1.000001
Ok
```

```
RUN
Number of equations ? 3
Input of the extended coefficient matrix
A( 1 , 1 )=? 1
A( 1 , 2 )=? 2
A( 1 , 3 )=? 3
B( 1 )=? 6
A( 2 , 1 )=? 1
A( 2 , 2 )=? 1
A( 2 , 3 )=? 1
B( 2 )=? 3
A( 3 , 1 )=? 2
A( 3 , 2 )=? 3
A( 3 , 3 )=? 4
B( 3 )=? 9

No solution is possible
Ok
```

The program has four subroutines. The first, which starts in line 53000, exchanges two matrix lines, the sth line with the tth one. The next subroutine multiplies an entire matrix row with the constant c (line 54000–54400). The subroutine which starts in line 55000 subtracts the c-fold of the sth line of the extended coefficient matrix from the tth matrix line. The subroutine starting in line 56000 serves to output the n x-values.

The input part of the main program runs from line 50000 through 50800. The dimension of the extended coefficient matrix A is set to a maximum of 20 unknown variables (line 50000); therefore the matrix needs 20 rows and 21 columns. The input of the actual number of unknowns and equations follows in the next line. In the following two nested loops the elements of the coefficient matrix and the vector of the right hand side of the equation system are read in and are written into the extended coefficient matrix A.

The central calculation loop runs from line 50900 through 52100. In this S-loop the diagonal elements of the matrix lines are set to 1, one after the other and the elements above and below these 1's are converted to zero. But first one must be sure that the diagonal element itself is not zero. This is done in the T-loop from 51000 through 51200. In this loop an element in the sth row beginning with the sth matrix line is sought that is unequal to zero. In case no such element is found, the T-loop runs to the end, a message is shown on the screen that no unique solution of the equation system is possible, and the program run is terminated. If, however, an element $a_{t,s}$ unequal to zero is found in matrix line t and column s the program jumps — because of the IF-statement in line 51100 — to line 51300. The T-loop is now discontinued and the variable T still holds the number t. In this line the subroutine 53000 is called for exchanging the tth matrix line with the sth one. Thus the element $a_{s,s}$ becomes unequal to zero. In the next line 51400 the normalization factor $c = 1/a_{s,s}$ is calculated and the subroutine starting with line 54000 is called which multiplies the matrix line s by the normalization factor c and thus makes the sth diagonal element $a_{s,s}$ equal to unity.

Then follows a loop (lines 51600–52000) that makes the other elements in the sth column equal to zero. As this loop runs again with the variable T, the 'out of order' jump from the preceding T-loop is 'forgotten'. In this T-loop the $a_{t,s}$-fold of the sth row is subtracted from the tth line and the element $a_{t,s}$ of the matrix becomes zero. If this T-loop has done its work for all lines except the sth one, the next diagonal element which is unequal to zero can be sought, and the S-loop is increased by 1. If the S-loop has finished, all n diagonal elements are equal to 1, and in the $(n+1)^{\text{th}}$ column the results appear; the rest of the matrix contains only zeros. In the next line 52200 the subroutine for printing out the results is called, and then the program jumps to the end in line 63999.

In the above example the solution of three equations containing three unknown quantities was calculated. The calculation time required for the solution of linear equation systems increases roughly by the square of the number of equations.

Problem 97

Given are four tanks containing diluted acids in four different concentrations.

a mixture of the four tanks in the relation 1:1:1:1 results in 25% acid

a mixture of the four tanks in the relation 4:3:2:1 results in 20% acid

a mixture of the four tanks in the relation 4:1:1:4 results in 25% acid

a mixture of the four tanks in the relation 4:1:4:1 results in 22% acid

What are the concentrations of acid in the four containers?

Problem 98

Given are the mass spectra of Problem 83 of four different hydrocarbons and there are four unknown mixtures of these hydrocarbons. An analysis gives the heights of four different mass peaks for each of the four different mixtures, each containing four substances. Calculate the composition of the different mixtures. (The calibration factors, which relate the mass peaks and the concentrations, should be set to 1, and can therefore be forgotten.)

Table 8.1. The height of four mass peaks obtained from mass spectroscopy of four mixtures is given. The mass spectra of the pure substances should be taken from Problem 83.

Peak heights	Mixture 1	Mixture 2	Mixture 3	Mixture 4
Mass 14	119	169	326	60
Mass 15	195	284	501	113
Mass 26	134	168	433	21
Mass 28	124	148	124	100

Problem 99

As you may have noticed in solving the above problem, a weak point of the program GJ is the input of the equation system in the form of an extended coefficient matrix. You must not make any mistakes in entering the numbers and, moreover, you cannot prove what you have put in.

Try to write the input part to be more user-friendly. The first thing to do should be to build in the possibility of correcting input data. Moreover, the original equation system should remain in memory even after the calculation of the unknown quantities so that you can call upon it afterwards. A user-friendly input part, once written, is applicable, in principle, to all programs. To such an input part belongs: the original data input, the possibility of listing data (singly and one after the other) with the opportunity to correct mistakes and to add or delete data.

Problem 100

w is a quantity much smaller than 1, given is the equation system

$$x + wy = 1$$
$$wx + y = 1$$

The result can easily be calculated. The values of x and y are very near to 1; they are exactly

$$x = \frac{1}{1+w} \qquad y = \frac{1}{1+w}$$

If GJ and BASICA is used, which internally calculates to eight decimal digits, and $w = 10^{-10}$ is set into the above equation system, the result $x = 1$, $y = 1$ is obtained. If both equations are interchanged to

$$wx + y = 1$$
$$x + wy = 1$$

the program GJ produces the result $x = 0$, $y = 1$. This is evidently wrong. If the interchanged equations are both divided by w and if w is assumed to be $5 \cdot 10^{-8}$, then the result $x = 2$, $y = 1$ is obtained. Using mathematical operations that normally should not change a result produces arbitrarily wrong answers. The reason for this lies in the limited accuracy of the computer and the resulting rounding errors. An equation system which reacts as in the above example is called a 'weakly conditioned equation system'. In equation systems with more than two equations it is frequently difficult to recognize whether it is a weakly conditioned system or not. Therefore, it must be tested whether the calculated results for the unknown quantities do truly satisfy the equation system.

Test the above equations system and try to find out when (with which w) the abnormal behavior starts with your computer. As the wrong results depend on the computer accuracy, they are dependent on the hardware and software used.

Modify the program GJ by an additional subroutine that sets the calculated values for the unknowns into the equation system, compares the results of this revision with the right hand sides of the orginal equations, and calculates their differences.

Problem 100a

In textbooks on numerical analysis you often find the sentence: never perform the Gauss-Jordan elimination without pivoting! What is pivoting? Nothing more than interchanging rows (partial pivoting) or rows and columns (total pivoting). Partial pivoting is easier than full pivoting, because we don't have to keep the permutations for the solution vector. Partial pivoting is almost as good as full pivoting.

The simplest way to implement partial pivoting into GJ is to pick the largest (in magnitude) available element as the pivot element in the search for a suitable diagonal element $a_{i,i}$. We must not only look for an element in column i which is unequal to zero, but also for the largest absolute value. The row where the pivot element is found is interchanged with the ith row.

Modify GJ to a Gauss-Jordan elimination with partial pivoting. The algorithm is now numerically stable — test Problem 100.

8.3   General Linear Regression

We have become acquainted with linear regression following a straight line in Chapters 7.7 and 7.8. Frequently, experimental data pairs (x_i, y_i) are not directly related by a linear relation but the quantity y is linearly dependent on functions $f_j(x)$. An example of this is the dependence of the height of fall y on the time of fall t for a free-falling object. The general dependence is expressed by the following polynomial

$$y = a \cdot t^2 + b \cdot t + c$$

y is in this case linearly dependent on three functions of t

$$f_1(t) = t^2 \qquad f_2(t) = t \qquad f_3(t) = 1$$

The parameters which should be determined by regression are a, b and c. Linear regression following a straight line, which has been treated earlier, is a special case of the generalized linear regression. This is easily recognized if in the above example the first function is taken away. Then the relation remains

$$y = b \cdot t + c$$

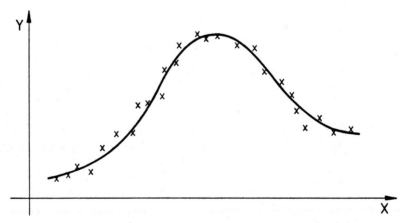

Fig. 8.2. A general least squares function and the appropriate experimental points

Another example of general linear regression is the following simple system of first order reactions

$$A \xrightarrow{k_1} P$$

$$B \xrightarrow{k_2} P$$

The resulting equation for the concentration of the product P for given rate constants k_1 and k_2 reads as follows

$$[P] = [P]_0 + [A]_0 \cdot \left(1 - e^{-k_1 t}\right) + [B]_0 \cdot \left(1 - e^{-k_2 t}\right)$$

If experimental values of the concentration of P in dependence of the time t are given and if the values of the rate constants k_1 and k_2 are known, again a general linear regression problem is obtained in which the concentration $[P]$ depends linearly upon three functions of t. The unknown constants that can be determined by regression of the experimental data are $[P]_0$, $[A]_0$ and $[B]_0$.

To obtain the mathematical relations for the general linear regression we shall, to begin with, limit it to two general functions. The extension to more functions will be done later by analogy.

Given are n experimental data pairs (x_i, y_i) for the quantities x and y. Further, the mathematical relation between x and y is known

$$y = a \cdot f_1(x) + b \cdot f_2(x)$$

The parameters a and b will be determined under the condition that the sum of the deviation squares is minimized. In a mathematical formulation the problem reads as follows

$$\text{Minimize} \quad S = \sum_{i=1}^{n} \left(y_i - a \cdot f_1(x_i) - b \cdot f_2(x_i)\right)^2$$

As S, the least squares sum, is to be minimized, its partial derivatives with respect to a and b must be set to zero.

$$\frac{\partial S}{\partial a} = -2 \cdot \sum_{i=1}^{n} \left(y_i - a \cdot f_1(x_i) - b \cdot f_2(x_i) \right) \cdot f_1(x_i) = 0$$

$$\frac{\partial S}{\partial b} = -2 \cdot \sum_{i=1}^{n} \left(y_i - a \cdot f_1(x_i) - b \cdot f_2(x_i) \right) \cdot f_2(x_i) = 0$$

If both equations are solved and the factors which belong to the parameters a and b are summarized, one obtains the following system of linear equations

$$a \cdot \sum_{i=1}^{n} f_1(x_i) f_1(x_i) + b \cdot \sum_{i=1}^{n} f_1(x_i) f_2(x_i) = \sum_{i=1}^{n} y_i f_1(x_i)$$

$$a \cdot \sum_{i=1}^{n} f_2(x_i) f_1(x_i) + b \cdot \sum_{i=1}^{n} f_2(x_i) f_2(x_i) = \sum_{i=1}^{n} y_i f_2(x_i)$$

It again consists of two linear equations with two unknown quantities a and b, which, as described in the last chapter, can be solved simply.

If the above set of two functions is further generalized — m different functions $f_i(x)$ are then given — the general linear least squares problem can now be solved.

For n experimental data pairs (x_i, y_i) there exists a theoretical relation between y and x with m functions $f_i(x)$

$$y = \sum_{j=1}^{m} k_j \cdot f_j(x) = k_1 f_1(x) + k_2 f_2(x) + \ldots + k_m f_m(x)$$

The coefficients k_j are determined by regression of the experimental data. The sum of the squared errors S is minimized.

$$S = \sum_{i=1}^{n} \left(y_i - \sum_{j=1}^{m} k_j \cdot f_j(x_i) \right)^2$$

The partial derivatives of S with respect to all k_j's are formed and all these derivatives are set equal to zero. By rearrangement one obtains a linear equation system. Using the abbreviated notation

$$f_{jk} = \sum_{i=1}^{n} f_j(x_i) \cdot f_k(x_i) \qquad \text{and} \qquad b_j = \sum_{i=1}^{n} f_j(x_i) \cdot y_i$$

the linear equation system can be written as follows

$$f_{11}k_1 + f_{12}k_2 + f_{13}k_3 + \ldots + f_{1m}k_m = b_1$$

$$f_{21}k_1 + f_{22}k_2 + f_{23}k_3 + \ldots + f_{2m}k_m = b_2$$

$$f_{31}k_1 + f_{32}k_2 + f_{33}k_3 + \ldots + f_{3m}k_m = b_3$$

$$\ldots$$

$$f_{m1}k_1 + f_{m2}k_2 + f_{m3}k_3 + \ldots + f_{mm}k_m = b_m$$

The same equation system written in matrix notation results in

$$\begin{pmatrix} f_{11} & f_{12} & f_{13} & \cdots & f_{1m} \\ f_{21} & f_{22} & f_{23} & \cdots & f_{2m} \\ f_{31} & f_{32} & f_{33} & \cdots & f_{3m} \\ \cdots & \cdots & \cdots & \cdots & \cdots \\ f_{m1} & f_{m2} & f_{m3} & \cdots & f_{mm} \end{pmatrix} \cdot \begin{pmatrix} k_1 \\ k_2 \\ k_3 \\ \vdots \\ k_m \end{pmatrix} = \begin{pmatrix} b_1 \\ b_2 \\ b_3 \\ \vdots \\ b_m \end{pmatrix}$$

This system of equations can be solved for the unknown k_j's with the Gauss-Jordan method of program GJ. The BASIC-program and an example are given in the following:

```
0 REM "REGLIN-G"          EBERT/EDERER          880628
1 REM ****************************************************
2 REM ***         General Linear Regression      ***
3 REM *** The regression can be done for the     ***
4 REM *** coefficients of n arbitrary functions. ***
5 REM *** Those functions have to be programmed   ***
6 REM *** in the subroutine function beginning at ***
7 REM *** line # 40000.                           ***
9 REM ****************************************************
600    DIM X(100),Y(100),A(20,21)
700    DIM K(20),F(20)
1000   GOSUB 10000 : REM INPUT of data points
2000   GOSUB 20000 : REM The REGRESSION is calculated
3000   GOSUB 30000 : REM OUTPUT of the COEFFicients
4000   END
4100   REM ###########################################
4200   REM ### Main program ends here.          ###
4300   REM ###########################################
```

```
10000 REM ********************************************
10050 REM *** Subroutine: DATAINPUT            ***
10100 REM ********************************************
10200   INPUT "How many functions ";N
10400   INPUT "How many data points ";N1
10600   FOR I=1 TO N1
10700     PRINT "x(";I;")  ";:INPUT X(I)
10800     PRINT "y(";I;")  ";:INPUT Y(I)
10900   NEXT I
10950 RETURN : REM end of subroutine DATAINPUT
20000 REM ********************************************
20050 REM *** Subroutine: REGRESSION           ***
20100 REM ********************************************
20200   FOR I=1 TO N : FOR J=1 TO N+1
20300     A(I,J)=0 : NEXT J : NEXT I
20400   FOR I=1 TO N1:GOSUB 40000 : REM FUNCTIONS
20700     FOR I1=1 TO N
20800       FOR I2=1 TO I1
20900         A(I1,I2)=A(I1,I2)+F(I1)*F(I2)
21000         A(I2,I1)=A(I1,I2) : NEXT I2
21200       A(I1,N+1)=A(I1,N+1)+Y(I)*F(I1)
21300     NEXT I1
21400   NEXT I : GOSUB 50900 : REM GJ
21500 RETURN : REM end of subroutine REGRESSION
30000 REM ********************************************
30050 REM *** Subroutine: COEFF_OUTPUT         ***
30100 REM ********************************************
30400   PRINT:PRINT "Coefficients of the regression"
30500   FOR I=1 TO N
30600     PRINT "k(";I;") = ";A(I,N+1)
30700   NEXT I
30900   S=0 : FOR I=1  TO N1
31000     GOSUB 40000:S1=0
31100     FOR I1=1 TO N:S1=S1+A(I1,N+1)*F(I1):NEXT I1
31200   S=S+(Y(I)-S1)^2
31300   NEXT I:PRINT
31500   PRINT "Sum of squared deviations =";S
31800 RETURN : REM end of subroutine COEFF_OUTPUT
40000 REM ********************************************
40002 REM *** Subroutine: FUNCTIONS            ***
40004 REM *** The user defined functions have to   ***
40006 REM *** be programmed here.                  ***
40010 REM ********************************************
40400   F(1)=X(I)*X(I)
40500   F(2)=EXP(X(I))
40600   F(3)=1/X(I)
42000 RETURN : REM end of subroutine FUNCTIONS
```

```
50000 REM ********************************************
50002 REM *** Subroutine: GJ                      ***
50004 REM *** Solving a system of n linear equa-  ***
50006 REM *** tions with n unknowns using the     ***
50008 REM *** method of Gauss-Jordan.             ***
50010 REM*********************************************
50100   INPUT "Number of equations ";N
50200   PRINT "Input of the extended coeff. matrix"
50300   FOR I=1 TO N
50400     FOR J=1 TO N
50500       PRINT "A(";I;",";J;")=";:INPUT A(I,J)
50600     NEXT J
50700     PRINT "B(";I;")=";:INPUT A(I,N+1)
50800   NEXT I : PRINT
50900   FOR S=1 TO N
51000     FOR T=S TO N
51100       IF A(T,S)<>0 THEN 51300
51200     NEXT T
51210     PRINT "No solution is possible" : GOTO 63999
51300     GOSUB 53000 : REM EXCHANGE lines
51400     C=1/A(S,S)
51500     GOSUB 54000 : REM NORMALIZE the actual line
51600     FOR T=1 TO N
51700       IF T=S THEN 52000
51800       C=-A(T,S)
51900       GOSUB 55000 : REM  zeros in the column
52000     NEXT T
52100   NEXT S
52200 RETURN : REM end of subroutine GJ
53000 REM ********************************************
53002 REM *** Subroutine: EXCHANGE                ***
53004 REM *** Lines S and T are exchanged.        ***
53010 REM ********************************************
53100   FOR J=1 TO N+1
53200     B=A(S,J) : A(S,J)=A(T,J) : A(T,J)=B
53300   NEXT J
53400 RETURN : REM end of subroutine EXCHANGE
54000 REM ********************************************
54002 REM *** Subroutine: NORMALIZE               ***
54004 REM *** Normalizing the actual line S.      ***
54010 REM ********************************************
54100   FOR J=1 TO N+1
54200     A(S,J)=C*A(S,J) : NEXT J
54400 RETURN : REM end of subroutine NORMALIZE
```

```
55000 REM ********************************************
55002 REM *** Subroutine: MAKEZERO              ***
55004 REM *** The elements of line S are multi-  ***
55006 REM *** plied with C and subtracted from the ***
55008 REM *** elements of line T.              ***
55010 REM ********************************************
55100   FOR J=1 TO N+1
55200     A(T,J)=A(T,J)+C*A(S,J)
55300   NEXT J
55400 RETURN : REM end of subroutine MAKEZERO
56000 REM ********************************************
56002 REM *** Subroutine: OUTPUT               ***
56004 REM *** Output of the solution.          ***
56010 REM ********************************************
56100   FOR T=1 TO N
56200     PRINT "X(";T;")=";A(T,N+1)
56300   NEXT T
56400 RETURN : REM end of subroutine OUTPUT
63999 END

RUN
How many functions ? 3
How many data points ? 8
x( 1 )  ? 1
y( 1 )  ? 2
x( 2 )  ? 2
y( 2 )  ? 4.5
x( 3 )  ? 3
y( 3 )  ? 9.3333
x( 4 )  ? 4
y( 4 )  ? 16.25
x( 5 )  ? .5
y( 5 )  ? 2.25
x( 6 )  ? .1
y( 6 )  ? .9
x( 7 )  ? 5.1
y( 7 )  ? 25
x( 8 )  ? 1.8
y( 8 )  ? 3.9

Coefficients of the regression
k( 1 ) =  1.097955
k( 2 ) = -2.196133E-02
k( 3 ) =  .1367191

Sum of squared deviations = 4.155648
Ok
```

From line 50000 through 63999 the program contains the Gauss-Jordan program for solving arbitrary linear equation systems which is built in as a subroutine. For that, the last statement END in the original program has been replaced by RETURN. This is an example of how larger programs can be constructed from smaller programs. The entire original program GJ was used, along with those parts that are not necessary for general regression. A minor difference to the program GJ exists at the beginning of the subroutine in line 50000, which now contains a REM-statement. The dimensioning of the extended coefficient matrix, which was in this line before, was shifted to the front of the main program.

The three functions, which are used in this example, are contained in the subroutine from line 40000 through 42000. In this subroutine the N functions at the position X(I) are calculated and the results are stored in the variables F(1), F(2) and F(3).

The main program is short; it consists of the REM-statements for the description of the program, the dimension declaration of the vectors X and Y for the data, the vectors F and K for the function values and the regression parameters, of the extended coefficient matrix A and the call of three subroutines. First the input subroutine is called, then the subroutine for calculating the regression and finally the output. The main program ends with the statement STOP so that it does not run into the subroutines which follow.

The input subroutine from line 10000 through 10950 asks the user for the number of functions in the regression equation and stores this number in variable N. The number of experimental data pairs is requested and stored in N1. A loop follows for the input of the N1 x- and y-values.

The subroutine which carries out the general regression is contained in lines 20000 through 21500. This subroutine calls further subroutines and therefore is relatively short and clear. It contains essentially three imbedded loops which serve to calculate the extended coefficient matrix with the above formulas. First, the elements of this matrix are set to zero (double loop from line 20200 through 20300). The external loop, the I-loop, runs from line 20400 through line 21400 over all N1 data pairs. In the second statement of line 20400 the subroutine for calculating the N function values at the position X(I) is called. In the following two nested loops (I1 and I2) every function value is multiplied by each other and the product F(I1)*F(I2) is added in the corresponding matrix element A(I1,I2). As these products are symmetrical, A(I2,I1) does not need to be calculated, if A(I1,I2) is known. In the same way the b-values of the above formulas are calculated and added as the $(N+1)^{th}$ column of the matrix A. For that all function values F(I1) are multiplied with the corresponding Y(I) data value.

After finishing these three loops only two more statements remain in the regression subroutine. The first is the call of the subroutine GJ for calculating the linear equation system which is contained in matrix A (by GOSUB 50900 the calculating part of GJ is called directly), the second is the statement RETURN, which jumps back to the main program.

Line 3000 of the main program jumps to the output subroutine with GOSUB 30000. As in our Gauss-Jordan method, the results of the linear equation system are contained in the $(N+1)^{th}$ column; the elements of this column are printed on the screen as the calculated regression coefficients. This is done in lines 30000 through 30700.

In the second half of the output subroutine the sum of the least squares is calculated. Again, a loop over all experimental data is required (lines 30900–31300). For each experimental value the N functions at this position are computed by calling the function subroutine. If each of these function values is multiplied with the corresponding regression coefficient and the products are added, one obtains the theoretical y-value which is stored in variable S1. The difference to the experimental y-value is formed, squared and added to the least squares sum S. After finishing with this calculation over all experimental data, the least squares sum S is printed out in line 31500. After the jump back from this output subroutine into the main program, the program is terminated by the STOP-statement.

In all our examples the experimental data scatter only very little from the theoretical values. This can be seen from the small value of the least squares sum.

Problem 101

What do we have to change in program REGLIN-G to be able to use it for a linear regression following a straight line?

Problem 102

Improve the input part of the program REGLIN-G by adding a suitable editor and extend the input and regression part to allow the use of weighted experimental data.

Problem 103

Use twenty artificial data pairs, which describe a sine curve, and try to find different polynomials (beginning with third degree) which run through these points. For this problem extend your output subroutine with a table which compares the 'experimental' and 'theoretical' y-values.

Problem 104

If Ostwald's law of dilution is combined with the degree of dissociation, which is defined as the ratio of the equivalent conductivity Λ_c at concentration c and at infinite dilution Λ_0, the following equation is obtained

$$c = K_c \Lambda_0^2 \cdot \left(\frac{1}{\Lambda_c}\right)^2 - K_c \Lambda_0 \cdot \left(\frac{1}{\Lambda_c}\right)$$

c : Concentration
K_c : Equilibrium constant of the acid
Λ_c : Equivalent conductivity at concentration c
Λ_0 : Equivalent conductivity at concentration $c = 0$

For acetic acid, the following experimental data at 25 °C are given:[1]

c [10^{-3} mol/l]	Λ_c [Ω^{-1} cm^{-1} val^{-1} l]
0.028014	210.38
0.15321	112.05
1.02831	48.146
2.41400	32.217
5.91153	20.962
12.829	14.375
50.000	7.358
52.303	7.202

Using these experimental data and the program REGLIN-G calculate the equilibrium constant K_c and the limiting value of equivalent conductivity at infinite dilution Λ_0.

Compare the results using unweighted experimental data with a regression calculation in which you weight the data with $(\Lambda_c/c)^2$. The correct results for Λ_0 and K_c are 390.71 [Ω^{-1} cm^{-1} val^{-1} l] and $1.8 \cdot 10^{-5}$ [mol/l], respectivly.

---

[1] G. Kortüm, *Lehrbuch der Elektrochemie*, Verlag Chemie, Weinheim (1972), p.149

Problem 105

Modify program REGLIN-G so that problems from the so-called multilinear regression can be calculated. Multilinear means that there are $(x_{1,i}, x_{2,i}, x_{3,i}, \ldots, x_{m,i}, y_i)$ data sets instead of experimental data pairs (x_i, y_i) and the unknown parameters k_1, k_2 etc. are related to y by the following equation.

$$y = \sum_{j=1}^{m} k_j \cdot f_j(x_j) = k_1 \cdot f_1(x_1) + k_2 \cdot f_2(x_2) + \ldots + k_m \cdot f_m(x_m)$$

Even the more general case with n different functions $f_j(x_1, \ldots, x_m)$, depending all on x_i and on n unknown parameters k_j, may be treated. The resulting regression function is written as the following equation.

$$y = \sum_{j=1}^{n} k_j \cdot f_j(x_1, x_2, \ldots, x_m)$$

Such dependences frequently occur in the natural and engineering sciences as approximations for a limited range or as pure empirical relations. This will be explained in more detail in the following example.

The yield or conversion U of a product of a certain chemical reaction is dependent upon temperature T and pH-value. The reaction has been investigated experimentally in a pH-range from 3 through 7 and in a temperature range from $10\,°C$ to $40\,°C$. The following experimental data were obtained:

Conversion	°C	pH-value
66.8	38.6	4.36
47.1	23.6	5.31
64.8	30.6	3.15
50.0	22.4	4.68
63.8	31.0	3.44
44.3	25.3	6.21
33.6	11.2	5.51
56.2	38.3	6.43

A linear relation between yield, temperature and pH-value was assumed.

$$U = a + b \cdot T + c \cdot \text{pH}$$

Try to calculate with the above data the coefficients a, b and c using a multilinear regression program.

Problem 106

Choose three series of data through which three parallel straight lines should be constructed. The three different axis intercepts a_i and the common slope b should be determined, under the condition that the common least squares sum of all data should be minimized. You can solve this program by writing down the least squares sum and developing the partial derivatives of the four parameters a_1, a_2, a_3 and b.

$$S = \sum_{i=1}^{n_1}(y_i - a_1 - bx_i)^2 + \sum_{i=n_1+1}^{n_2}(y_i - a_2 - bx_i)^2 + \sum_{i=n_2+1}^{n}(y_i - a_3 - bx_i)^2$$

You can also consider this problem as a special case of the multilinear regression, which was described in the last problem.

Problem 107

Given are the calibration mass spectra from Problem 83. In addition the spectra of two mixtures are supplied. They are assumed to be a linear superposition of the spectra of the pure substances.

Table 8.2. Normalized mass spectra of four different hydrocarbons (the largest peak of each substance is normalized to 100) and two spectra of mixtures of the four hydrocarbons.

Substance	Mass											
	12	13	14	15	16	24	25	26	27	28	29	30
CH_4	1	3	12	100	35							
C_2H_6			2	80		3	4	13	37	24	100	70
C_2H_4	4	7	48	13		2	7	21	18	100		
C_2H_2	5	23	57	2		40	45	100				
Mixture 1	11	32	120	193	37	44	57	136	53	124	102	68
Mixture 2	10	33	110	104	17	44	52	117	35	113	52	33

As the spectra of the mixtures, as well as the calibration spectra, contain experimental inaccuracies, the spectra of the mixtures are not exact linear combinations of the calibration spectra from Problem 83. There exists no combination of concentrations of the four pure substances which exactly corresponds to the spectra of the mixtures. A procedure should be developed to calculate the concentration of the substances so that the sum of the squares of the deviations between the experimental (spectra of the mixtures) and the simulated spectra of the mixtures (gained by linear combination of the calibration spectra) is minimized.

An indication is given here how this problem can be solved. Achieving this on the computer is recommended as an exercise. We use the symbols:

M : Vector of the spectrum of a mixture (12 rows, 1 column)
E : Calibration matrix of the four substances (12 rows, 4 columns)
E^T : Transposed calibration matrix E (4 rows, 12 columns) (transposed means reflected at the main diagonal)
C : Vector of the concentrations (4 rows, 1 column)

If the experimental spectrum of the mixture were an exact linear combination of the spectra of the pure substances, then the relation could be expressed by the following matrix equation

$$M = E \cdot C \qquad (*)$$

Because of the usual inaccuracy of the experiments this over-determined equation system is not solvable. However, a solution for the vector C can be obtained for which $|M - E \cdot C|$ is minimized. The above equation is multiplied on both sides with the transposed matrix E^T

$$E^T \cdot M = E^T \cdot E \cdot C$$

$E^T \cdot M$ is a vector with 4 rows and 1 column that can easily be calculated from the transposed calibration matrix and the spectrum of the mixture. Similarly the 4×4 matrix $E^T \cdot E$ can be calculated from the calibration matrix. Then a simple linear equation system is obtained, which can be solved with the program GJ.

This simple method for a relatively complicated data processing problem is given without proof. This can be proven analogously to the demonstration of the proof for the linear regression. The basic equation is

$$S = \sum_{i=1}^{n} (h_{i,\exp} - h_{i,\text{calc}})^2$$

$h_{i,\exp}$ is the measured height of the peak with mass i and $h_{i,\text{calc}}$ is the calculated height of the peak with mass i.

If in this equation, the above linear equation $(*)$ is inserted for the simulated peak heights, a partial derivation with respect to the unknown concentration is carried out and the resulting system of equations rearranged, the operation mentioned above is obtained.

For problems of this kind it is frequently reasonable to add weights to the single peaks. Try to write the program so that weights can be assigned to the individual input values.

8.4   Inversion of a Square Matrix

Although in solving linear problems and exercises one should always choose a solution using linear equation systems, it may be necessary or advantageous to calculate the inverse of a matrix. The method used to invert a matrix is also called Gauss-Jordan. The corresponding program is very similar to the program GJ. The extended coefficient matrix A, which in program GJ had the dimension $n \times (n+1)$, is extended to a $n \times 2n$ matrix for the inversion.

The matrix to be inverted is extended by the n-dimensional unity matrix, i.e. the equation system for n right hand sides should be solved simultaneously, and the same example is considered as for solving the equation system according to Gauss-Jordan

$$\begin{array}{rrrrrrr} 4x & + & 3y & + & z & = & 1 \quad 0 \quad 0 \\ 2x & - & y & - & z & = & 0 \quad 1 \quad 0 \\ 7x & + & y & - & 3z & = & 0 \quad 0 \quad 1 \end{array}$$

This can be written in matrix form as a coefficient matrix

$$\begin{pmatrix} 4 & 3 & 1 & 1 & 0 & 0 \\ 2 & -1 & -1 & 0 & 1 & 0 \\ 7 & 1 & -3 & 0 & 0 & 1 \end{pmatrix} \qquad \begin{array}{l} \times \frac{1}{4} \\ [2^{\text{nd}}\text{ row}] - \frac{1}{2} \cdot [1^{\text{st}}\text{ row}] \\ [3^{\text{rd}}\text{ row}] - \frac{7}{4} \cdot [1^{\text{st}}\text{ row}] \end{array}$$

The first line is divided by 4 and the remaining elements $a_{i,1}$ of the first column are made to zero by appropriate additions and subtractions. One obtains

$$\begin{pmatrix} 1 & \frac{3}{4} & \frac{1}{4} & \frac{1}{4} & 0 & 0 \\ 0 & -\frac{5}{2} & -\frac{3}{2} & -\frac{1}{2} & 1 & 0 \\ 0 & -\frac{17}{4} & -\frac{19}{4} & -\frac{7}{4} & 0 & 1 \end{pmatrix} \qquad \begin{array}{l} [1^{\text{st}}\text{ row}] + \frac{3}{10} \cdot [2^{\text{nd}}\text{ row}] \\ \times -\frac{2}{5} \\ [3^{\text{rd}}\text{ row}] - \frac{17}{10} \cdot [2^{\text{nd}}\text{ row}] \end{array}$$

The second diagonal element becomes 1 after multiplying the second equation by $-\frac{2}{5}$. Again, the elements above and below shall be made to zero by suitable additions or subtractions

$$\begin{pmatrix} 1 & 0 & -\frac{1}{5} & \frac{1}{10} & \frac{3}{10} & 0 \\ 0 & 1 & \frac{3}{5} & \frac{1}{5} & -\frac{2}{5} & 0 \\ 0 & 0 & -\frac{11}{5} & -\frac{9}{10} & -\frac{17}{10} & 1 \end{pmatrix} \qquad \begin{array}{l} [1^{\text{st}}\text{ row}] - \frac{1}{11} \cdot [3^{\text{rd}}\text{ row}] \\ [2^{\text{nd}}\text{ row}] + \frac{3}{11} \cdot [3^{\text{rd}}\text{ row}] \\ \times -\frac{5}{11} \end{array}$$

The third diagonal element becomes 1, and the other elements of column 3 become zero by appropriate matrix line operations

$$\begin{pmatrix} 1 & 0 & 0 & \frac{2}{11} & \frac{5}{11} & -\frac{1}{11} \\ 0 & 1 & 0 & -\frac{1}{22} & -\frac{19}{22} & \frac{3}{11} \\ 0 & 0 & 1 & \frac{9}{22} & \frac{17}{22} & -\frac{5}{11} \end{pmatrix}$$

The last three columns of this matrix are the inverse matrix of the original matrix. The reader should try to prove this evaluation as an exercise.

The program INV-GJ for inverting matrices reads as follows:

```
0 REM "INV-GJ  "          EBERT/EDERER          880628
1 REM **********************************************
2 REM *** Inversion of a Matrix with the Algorithm ***
3 REM ***              of Gauss-Jordan              ***
4 REM *** The dimension of the matrices and the     ***
5 REM *** numerical values of the original matrix   ***
6 REM *** are written as DATA-statements beginning  ***
7 REM *** with line # 60000.                        ***
9 REM **********************************************
50000 DIM A(20,40)
50100 READ N :REM dimension of the matrix
50300 FOR I=1 TO N
50400   FOR J=1 TO N
50500     READ A(I,J)
50550     A(I,N+J)=0
50600   NEXT J
50700   A(I,N+I)=1
50800 NEXT I
50900 FOR S=1 TO N
51000   FOR T=S TO N
51100     IF A(T,S)<>0 THEN 51300
51200   NEXT T
51210   PRINT "No solution is possible" : GOTO 63999
51300   GOSUB 53000 : REM EXCHANGE lines
51400   C=1/A(S,S)
51500   GOSUB 54000 : REM NORMALIZE the actual line
51600   FOR T=1 TO N
51700     IF T=S THEN 52000
51800     C=-A(T,S)
51900     GOSUB 55000 : REM make zeros in the column
52000   NEXT T
```

```
52100 NEXT S
52200 GOSUB 56000 : GOTO 63999 :REM OUTPUT and END
52210 REM #########################################
52220 REM ### Main program ends here.          ###
52230 REM #########################################
53000 REM *****************************************
53002 REM *** Subroutine: EXCHANGE             ***
53004 REM *** Lines S and T are exchanged.     ***
53010 REM *****************************************
53100   FOR J=1 TO 2*N
53200     B=A(S,J) : A(S,J)=A(T,J) : A(T,J)=B
53300   NEXT J
53400 RETURN : REM end of subroutine EXCHANGE
54000 REM *****************************************
54002 REM *** Subroutine: NORMALIZE            ***
54004 REM *** Normalizing the actual line S.   ***
54010 REM *****************************************
54100   FOR J=1 TO 2*N
54200     A(S,J)=C*A(S,J) : NEXT J
54400 RETURN : REM end of subroutine NORMALIZE
55000 REM *****************************************
55002 REM *** Subroutine: MAKEZERO             ***
55004 REM *** The elements of line S are multi- ***
55006 REM *** plied with C and subtracted from the ***
55008 REM *** elements of line T.              ***
55010 REM *****************************************
55100   FOR J=1 TO 2*N
55200     A(T,J)=A(T,J)+C*A(S,J)
55300   NEXT J
55400 RETURN : REM end of subroutine MAKEZERO
56000 REM *****************************************
56001 REM *** Subroutine: OUTPUT               ***
56002 REM *** The original matrix, the inverted ***
56003 REM *** matrix and, as a control, the product***
56004 REM *** of the matrices are printed out.  ***
56010 REM *****************************************
56030   RESTORE:READ N
56100   FOR I=1 TO N
56200     FOR J=1 TO N
56220       READ A(I,J)
56300       PRINT TAB(1+15*(J-1));A(I,J);
56400     NEXT J:PRINT
56500   NEXT I:PRINT
57100   FOR I=1 TO N
57200     FOR J=N+1 TO 2*N
57300       PRINT TAB(1+15*(J-N-1));A(I,J);
```

```
57400     NEXT J:PRINT
57500    NEXT I:PRINT
58100    FOR I=1 TO N
58200      FOR J=1 TO N
58210        C=0
58220        FOR K=1 TO N
58230          C=C+A(I,K)*A(K,J+N)
58240        NEXT K
58300        PRINT TAB(1+15*(J-1));C;
58400      NEXT J:PRINT
58500    NEXT I:PRINT
59400 RETURN : REM end of subroutine OUTPUT
59900 REM #########################################
59910 REM ### Data section:                    ###
59920 REM ### The first datum is the dimension of  ###
59930 REM ### the matrix, followed by the elements.###
59940 REM #########################################
60000 DATA 4           : REM Dimension of the matrix
60100 DATA 7,5,3,1
60200 DATA 0,1,1,1
60300 DATA 4,2,1,3
60400 DATA 8,1,9,5
63999 END
```

```
RUN
 7              5              3              1
 0              1              1              1
 4              2              1              3
 8              1              9              5


 5.263162E-02  -.4999999      .1052634       2.631579E-02
 .1403508       .4999995     -5.263233E-02  -9.649123E-02
 3.508759E-02   .2500005     -.2631574       .1008772
-.1754386       .25           .3157895      -4.385963E-03


 .9999996      -5.960465E-08  -8.344651E-07  -1.490116E-08
-1.788139E-07   .9999999      -2.384186E-07   0
-5.960465E-08  -1.788139E-07   .9999998      -1.490116E-08
-8.940697E-07   4.529953E-06   5.483628E-06   .9999999
```

In comparing this program with the program GJ one notices that the sub-routines for exchanging, normalizing and subtracting equations differ in a single point: the running variable J runs to 2*N instead of N+1. The input part has been simplified. Instead of input statements, the dimension and the elements of the matrix are put in by READ-statements (50100–50800). Simultaneously, the columns from N+1 through 2*N were set to zero and only the elements A(I,N+I) are set to 1. At the end of the program the corresponding data of the original matrix are contained in the DATA-statements. The rest of the main program is identical with the program GJ.

The output subroutine is new. First, the given matrix is again read into the left half of A (lines 56030–56500) and printed. The output of the inverted matrix occurs in lines 57100 through 57500. In the following lines both halves of A are multiplied according to the laws of matrix multiplication and written on the screen. If the inversion was carried out correctly, this control matrix must be identical to the corresponding unity or identity matrix. It might be noted from the example that small rounding errors may lead to small deviations from this unity matrix.

An improved program for inverting matrices, also based on the method of Gauss-Jordan, was programmed according to an example given in the literature.[2] The reader is advised to analyze this program, possibly with help from this literature. The program INVERS reads as follows:

```
0 REM "INVERS  "          EBERT/EDERER          881013
1 REM ****************************************************
2 REM *** Inversion of a matrix by the method of  ***
3 REM *** Gauss-Jordan using a pivot technique.   ***
4 REM *** The dimension of the matrices and the   ***
5 REM *** numerical values of the original matrix ***
6 REM *** are written as DATA-statements beginning ***
7 REM *** with line # 50000.                       ***
9 REM ****************************************************
100    DIM A(20,20),B(20,20),U(20,20)
110    DIM V(20,20),W(20,20)
300    READ N : REM dimension of the matrix
400    FOR I=1 TO  N
420      FOR J=1 TO  N
440        READ A(I,J)
460        B(I,J)=0
480        V(I,J)=0
500        IF I<>J THEN GOTO 600
520        B(I,J)=1
540        V(I,J)=1
600      NEXT J
```

---

[2] W. Törnig, *Numerische Mathematik für Ingenieure*, Springer, Heidelberg (1979)

```
620    NEXT I
1000   FOR Z=1 TO N
1050     S=0
1100     REM ****************************************
1105     REM *** Searching for the pivot element.   ***
1110     REM ****************************************
1120     FOR I=Z TO N
1150       IF S>ABS(A(I,Z)) THEN GOTO 1300
1200       S=ABS(A(I,Z)) : T=I
1300     NEXT I
2000     REM ****************************************
2001     REM *** Lines Z and T will be exchanged.   ***
2005     REM ****************************************
2050     FOR I=1 TO N
2100       S=A(Z,I) : A(Z,I)=A(T,I) : A(T,I)=S
2200     NEXT I
2300     IF ABS(A(Z,Z))>1E-30 THEN GOTO 2400
2350     PRINT "No inversion is possible " : END
2400     V(Z,Z)=0 : V(T,T)=0
2450     V(Z,T)=1 : V(T,Z)=1
3000     REM ****************************************
3020     REM *** Gauss-Jordan elimination           ***
3040     REM ****************************************
3100     FOR I=1 TO N
3200       FOR J=1 TO N
3300         IF I=Z THEN GOTO 4000
3350         IF J=Z THEN GOTO 4500
3400         U(I,J)=A(I,J)-A(Z,J)*A(I,Z)/A(Z,Z)
3500         GOTO 5000
4000         IF I=J THEN GOTO 4350
4050         U(I,J)=-A(I,J)/A(Z,Z)
4100         GOTO 5000
4350         U(Z,Z)=1/A(Z,Z)
4400         GOTO 5000
4500         U(I,Z)=A(I,Z)/A(Z,Z)
5000       NEXT J:NEXT I
5100     REM ****************************************
5102     REM *** Multiplication  B = V * B          ***
5110     REM ****************************************
5200     FOR I=1 TO N
5250       FOR J=1 TO N
5300         W(I,J)=0
5350         FOR K=1 TO N
5400           W(I,J)=W(I,J)+V(I,K)*B(K,J)
5450     NEXT K : NEXT J : NEXT I
5500     FOR I=1 TO N : FOR J=1 TO N
```

```
5550         B(I,J)=W(I,J)
5600    NEXT J : NEXT I
6000    FOR I=1 TO N
6050      FOR J=1 TO N
6100        A(I,J)=U(I,J):V(I,J)=0
6250        IF I=J THEN V(I,J)=1
6300      NEXT J : NEXT I
6500    NEXT Z
7000    REM *******************************************
7005    REM *** The result is obtained by multiply- ***
7010    REM *** ing matrix A with the permutation   ***
7020    REM *** matrix B.                           ***
7030    REM *******************************************
7100    FOR I=1 TO N
7200      FOR J=1 TO N
7300        W(I,J)=0
7400        FOR K=1 TO N
7500          W(I,J)=W(I,J)+A(I,K)*B(K,J)
7600    NEXT K : NEXT J : NEXT I
8000    REM *******************************************
8010    REM *** Checking of the result.             ***
8020    REM *******************************************
8040    RESTORE : READ N
8050    FOR I=1 TO N
8060      FOR J=1 TO N
8070        READ A(I,J)
8080    NEXT J:NEXT I
8100    FOR I=1 TO N
8200      FOR J=1 TO N
8300        B(I,J)=0
8400        FOR K=1 TO N
8500          B(I,J)=B(I,J)+A(I,K)*W(K,J)
8600    NEXT K : NEXT J : NEXT I
9000    REM*********************************************
9050    REM *** Output of the original matrix A, the ***
9100    REM *** inverted matrix W and the product    ***
9200    REM *** A*W of both.                         ***
9250    REM *******************************************
9300    FOR I=1 TO N : FOR J=1 TO N
9310        PRINT TAB(1+15*(J-1));A(I,J);
9320    NEXT J : PRINT : NEXT I: PRINT
9400    FOR I=1 TO N : FOR J=1 TO N
9410        PRINT TAB(1+15*(J-1));W(I,J);
9420    NEXT J : PRINT : NEXT I: PRINT
9500    FOR I=1 TO N : FOR J=1 TO N
9510        PRINT TAB(1+15*(J-1));B(I,J);
```

```
9520  NEXT J : PRINT : NEXT I: PRINT
49900 REM ##########################################
49910 REM ### Data section:                      ###
49920 REM ### The first datum is the dimension of  ###
49930 REM ### the matrix, followed by the elements.###
49940 REM ##########################################
50000 DATA 4         : REM Dimension of the matrix
50100 DATA 7,5,3,1
50200 DATA 0,1,1,1
50300 DATA 4,2,1,3
50400 DATA 8,1,9,5
60000 END
```

```
RUN
  7               5               3               1
  0               1               1               1
  4               2               1               3
  8               1               9               5


   .0526316       -.5              .1052632        2.631579E-02
   .1403509        .5            -5.263157E-02    -9.649123E-02
  3.508772E-02     .25            -.2631579        .1008772
 -.1754386         .25             .3157895       -4.385962E-03


  1               2.980232E-08    8.940697E-08    1.583248E-08
 -1.490116E-08    1               0               8.381903E-09
  0               1.192093E-07    1               2.793968E-09
  1.192093E-07    1.192093E-07    0               1
```

Problem 108

Given is a fictional chemical reaction system in which two reactions are running at the same time. After a fixed reaction time they are characterized by two conversions U_1 and U_2. The conversions, which are measured in %, are linearly dependent on four physical quantities according to the following formulas

$$U_1 = 10 + 1 \cdot c_1 - 2 \cdot c_2 + 2.0 \cdot \text{pH} + 0.5 \cdot (T - 300)$$

$$U_2 = 15 - 1 \cdot c_1 - 1 \cdot c_2 + 1.2 \cdot \text{pH} + 0.9 \cdot (T - 300)$$

pH : pH-value
T : Temperature in Kelvin
c_1 : Concentration of substance 1 in [mol/l]
c_2 : Concentration of substance 2 in [mol/l]

Using reaction conditions $c_1=2\,[\text{mol/l}]$ and $c_2=2\,[\text{mol/l}]$, a temperature of 340 K and a pH-value of 4, we get conversions of $U_1 = 36\%$ and $U_2 = 51.8\%$, respectively, if the formulas above are used.

The problem is to find conditions at which the above yields are both increased by 1%. But in addition, the sum of the squared changes of the parameter values, which govern the reaction conditions, are minimized. Generally the parameter variations are also weighted. Problems of this kind exist if the number of parameters is higher than the number of quantities which can be adjusted (over-determined systems). In other words — we want to change the quantities U_1 and U_2 by a given amount, but we want to change the parameters as little as possible.

Examples of this are complex chemical reaction mechanisms which consist of a great number of chemical species which react in even more chemical reactions. The kinetics of each of these reactions is characterized by two kinetic parameters, which influence the dynamics of the concentrations of the species.

The following method for solving such problems is given without proof. The method can be derived either with the usual pattern (writing down a least squares sum) or it can be evaluated with Lagrange's method for searching extreme values with additional conditions. This method is contained in most textbooks on applied mathematics.[3]

If the variations of yields with parameter variations are considered, we get nearly the same two equations, but all quantities have been replaced by the variation of the quantities.

$$\Delta U_1 = 1 \cdot \Delta c_1 - 2 \cdot \Delta c_2 + 2.0 \cdot \Delta \text{pH} + 0.5 \cdot \Delta T$$

$$\Delta U_2 = -1 \cdot \Delta c_1 - 1 \cdot \Delta c_2 + 1.2 \cdot \Delta \text{pH} + 0.9 \cdot \Delta T$$

Arranged in matrix form one obtains

$$\begin{pmatrix} \Delta U_1 \\ \Delta U_2 \end{pmatrix} = \begin{pmatrix} 1 & -2 & 2.0 & 0.5 \\ -1 & -1 & 1.2 & 0.9 \end{pmatrix} \cdot \begin{pmatrix} \Delta c_1 \\ \Delta c_2 \\ \Delta \text{pH} \\ \Delta T \end{pmatrix}$$

or in a short notation where S is the sensitivity matrix.

$$\Delta U = S \cdot \Delta k$$

The parameter variation Δk, which will be calculated, should be as small as possible. The variation vector ΔU of the yield variation and the sensitivity matrix S are given. In matrix notation the solution reads as follows

$$\Delta k = S^T \cdot \left(S \cdot S^T\right)^{-1} \cdot \Delta U$$

---

[3] C.E. Pearson (ed.), *Handbook of Applied Mathematics*, 2.ed., Reinhold, New York, (1983)

To obtain the unknown vector Δk, the matrix S must be multiplied with its transposed matrix S^T and from this the inverse matrix must be calculated. The inverse matrix must be multiplied from the left side with the transposed sensitivity matrix S^T, and then a new matrix is obtained. After multiplying this auxiliary matrix with the vector ΔU, the unknown variation vector Δk is obtained. Use one of the matrix inversion programs to solve this problem.

Problem 109

The following principle holds: the inverse of an inverse matrix leads to the original matrix. Write a program to test this principle numerically.

8.5   Determination of Eigenvalues of a Matrix

Chemists need to calculate eigenvalues for example in problems of reaction dynamics (e.g. reaction kinetics and reactor theory etc.), in the analysis of normal modes in molecular spectroscopy and — probably most frequently — in quantum chemical approximation calculations. Before we go into the practical treatment of such problems some mathematical explanations will be given.

A should be a square matrix, whose eigenvalues λ_i and eigenvectors $\vec{v}_i$ shall be determined. As we are interested only in real eigenvalues, we only need to treat symmetrical matrices (this means the elements $a_{i,j} = a_{j,i}$), because only these have exclusively real eigenvalues. The eigenvalue λ_i and the corresponding eigenvector $\vec{v}_i$ are defined by the following equation

$$A \cdot v_i = \lambda v_i$$

This means that if the matrix A is multiplied with an eigenvector $\vec{v}_i$ one obtains in principle again the eigenvector $\vec{v}_i$, but it is multiplied with a scalar, the eigenvalue λ_i. To determine the eigenvalues of a symmetrical matrix we use a method which is known in numerical analysis textbooks as the Jacobi method. In this method the original matrix A is consecutively transformed until the elements, apart from the diagonal elements, consist exclusively of very small quantities (possibly becoming zero). The diagonal elements are then the eigenvalues. This method to determine the eigenvalues is given in the flow diagram of Fig. 8.3. The corresponding BASIC-program EIGEN is listed below.

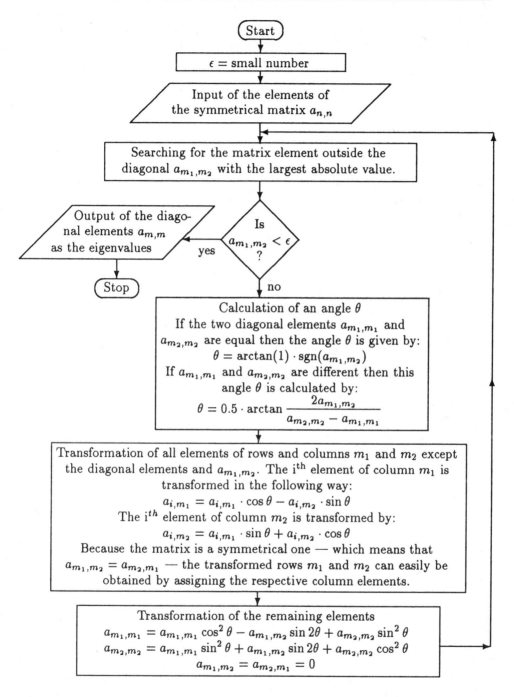

Fig. 8.3. Flow diagram of the Jacobi method for the determination of eigenvalues of a symmetrical matrix

```
0 REM "EIGEN    "        EBERT/EDERER        880511
1 REM ***********************************************
2 REM *** Determination of the eigenvalues of a   ***
3 REM *** symmetrical matrix by Jacobi's method.  ***
4 REM *** The dimension of the matrix and the num- ***
5 REM *** erical values of the matrix elements    ***
6 REM *** are written as DATA-statements beginning ***
7 REM *** with line # 10000.                       ***
9 REM ***********************************************
10    DIM A(20,20)
20    E9=1E-08 : REM Accuracy eps
30    READ N : REM dimension of the symm. matrix
34    REM ***********************************************
35    REM *** Reading the symmetrical matrix.        ***
36    REM ***********************************************
40    FOR I=1 TO N
50      FOR J=1 TO N
60        READ A(I,J)
80      NEXT J
90    NEXT I
100   FOR Z9=1 TO 400
200     REM ***********************************************
202     REM *** Searching for the pivot element,      ***
204     REM *** the highest absolute values.          ***
206     REM ***********************************************
210     M=0
220     FOR I=1 TO N-1
230       FOR J=I+1 TO N
240         IF M>ABS(A(I,J)) THEN GOTO 280
250         M=ABS(A(I,J))
260         M1=I : M2=J
280       NEXT J
290     NEXT I
300     IF M<E9 THEN GOTO 9000
400     REM ***********************************************
402     REM *** Calculation of angles.                ***
406     REM ***********************************************
410     IF A(M1,M1)<>A(M2,M2) THEN GOTO 500
420     P=ATN(1)*SGN(A(M1,M2))
430     GOTO 600
500     P=2*A(M1,M2)/(A(M2,M2)-A(M1,M1))
520     P=.5*ATN(P)
600     REM ***********************************************
602     REM *** Matrix is now modified.               ***
606     REM ***********************************************
620     FOR I=1 TO N
```

```
630        IF I=M1 THEN GOTO 700
635        IF I=M2 THEN GOTO 700
640        A5=A(I,M1)
660        A(I,M1)=A(I,M1)*COS(P)-A(I,M2)*SIN(P)
670        A(M1,I)=A(I,M1)
680        A(I,M2)=A5*SIN(P)+A(I,M2)*COS(P)
690        A(M2,I)=A(I,M2)
700      NEXT I
720      A6=A(M1,M1)
730      A7=A(M2,M2)
740      A8=A(M1,M2)
750      A(M1,M2)=0
760      A(M2,M1)=0
770      A(M1,M1)=A6*COS(P)^2-A8*SIN(2*P)+A7*SIN(P)^2
780      A(M2,M2)=A6*SIN(P)^2+A8*SIN(2*P)+A7*COS(P)^2
8000   NEXT Z9
9000   REM *******************************************
9010   REM *** Output of the eigenvalues.          ***
9020   REM *******************************************
9100   FOR I=1 TO N
9200     PRINT "Eigenvalue = "; A(I,I)
9210   NEXT I
9900   REM #########################################
9910   REM ### Data section:                    ###
9920   REM ### The first datum is the dimension of  ###
9930   REM ### the matrix, followed by the elements.###
9940   REM #########################################
10000 DATA 6 : REM dimension of the matrix
10100 DATA 0,1,0,0,1,0
10200 DATA 1,0,1,0,0,0
10300 DATA 0,1,0,1,0,0
10400 DATA 0,0,1,0,1,0
10500 DATA 1,0,0,1,0,1
10600 DATA 0,0,0,0,1,0
63999 END

RUN
Eigenvalue = -1.618035
Eigenvalue =  .6180345
Eigenvalue = -1.860807
Eigenvalue =  2.114909
Eigenvalue = -.2541019
Eigenvalue =  1.000001
Ok
```

In line 10 the matrix A, whose eigenvalues are to be determined, is dimensioned. In line 20 the small quantity ϵ is defined as E9=1E-8. For each new class of problems this magnitude ϵ has to be redefined. In line 30 the real dimension n of matrix A is put in by a READ-statement and in the following lines 40 through 90 the elements of the matrix A are entered.

Lines 100 through 8000 contain the program part for the matrix transformation. This section stops running if after 400 transformations the required accuracy has not been obtained.

In the program section from line 200 through 290 the non-diagonal element with the highest absolute value is determined. As the matrix is symmetrical, it is sufficient to consider only the elements that lie above the diagonal. The characteristics of the maximum element are stored, by putting in M the absolute value, in M1 the row number, and in M2 the column number.

In line 300 the absolute value of the maximum element is interrogated. If it is already sufficiently small (less than ϵ), the program jumps to line 9000 and the diagonal elements are printed out as the unknown eigenvalues.

From line 400 through 780 the matrix A is transformed according to the scheme which is given in Fig. 8.3 in the flow diagram.

The matrix, which is given in our example and which consists of zeroes and ones only (line 10100 and the following) is the Hückel matrix of the substance fulvene. The eigenvalues are quantities which are proportional to the energy levels of the MO-orbitals.

Problem 110

Calculate the eigenvalues of the following matrices

$$
\begin{pmatrix} 2 & -1 & 0 \\ -1 & 2 & -1 \\ 0 & -1 & 2 \end{pmatrix}
\qquad
\begin{pmatrix} 7 & -3 & 1 & 9 \\ -3 & 0 & -2 & 2 \\ 1 & -2 & -9 & -1 \\ 9 & 2 & -1 & -1 \end{pmatrix}
$$

Problem 111

Determine the eigenvectors $\vec{v}_i$ which are associated with the eigenvalues λ_i of the Hückel matrix for fulvene. The eigenvectors define how the molecular orbitals, with their corresponding energy values, are composed as a linear combination of the atomic orbitals.

For the calculation of these eigenvectors $\vec{v}_i$ you can use the program GJ for solving linear equation systems. The following system of linear equations must be solved

$$(A - \lambda_i \cdot I) \cdot v_i = 0$$

I : Identity matrix
λ_i : i^{th} eigenvalue
$\vec{v}_i$: i^{th} eigenvector, associated to the eigenvalue λ_i
0 : Zero vector

Systems of equations of this kind do not possess a unique solution. Try to diagonalize the matrix as completely as possible using a modified program GJ. For that, substitute line 51200 by:

```
51200 NEXT T : S9=S : GOTO 52100
```

For the same reason the test for zero in line 51100 must be replaced by the test for the small quantity ϵ.

```
51100 IF ABS (A(T,S))>1E-6 THEN 51300
```

In this way the variable X(S9) is determined which can be defined arbitrarily after finishing the GJ-diagonalization. (In the S9th column the elements in and below the diagonal contain only zeroes. It is logical to choose X(S9)=-1. If this S9th column is completed by -1 in the S9th line, this column contains the unknown eigenvector. Frequently this eigenvector is normalized to length 1. For that reason the length of the vector (root of the square sum of the vector elements) is formed and each vector element is divided by this length.

Difficulties arise, if a matrix contains more than one equal eigenvalue. Then, in determining the eigenvectors using the program GJ, two lines are obtained which contain only zeroes. This means that the corresponding vector elements can be chosen arbitrarily and from the remaining system two linearly independent eigenvectors can be determined.

Problem 112

Calculate the eigenvalues of the Hückel matrix for the pentadienyl cation, for benzene, hexatriene, cyclooctatetraene, butadiene, and anthracene. The Hückel matrices for π-systems contain zeroes and ones only. If the C-atoms of the molecules are characterized by numbers and a chemical bond exists between atom i and atom j, then the matrix elements $a_{i,j}$ and $a_{j,i}$ are both unity. More details on HMO calculations can be found in the literature.[4]

---

[4] E. Heilbronner, H. Bock, *The HMO-Modell and its Application*, Wiley & Sons, New York (1976)

8.6   Calculation of the Number of Isomers

The problem to calculating the possible number of structural isomers of a chemical compound probably fits better into Chapter 7 (arrays). But as subroutines are used extensively in the following program, it is presented here.

The solution of this chemical problem lies in mathematical graph theory. Readers, who would like to learn about this in more detail should consult an appropriate textbook.[5] At this point a program will be described which is able to calculate the number of alkyl groups, alkanes, and as an example for more complicated structures, alkenones (ketones or aldehydes containing a C–C double bond but no rings).

The calculation of the number of different alkyl groups or alkyl radicals using a certain number of C-atoms is the basis for the calculation of extended chemical structures of molecules. For the treatment of this problem graph theory considers only the carbon skeleton. The corresponding graph contains only points for the C-atoms and connecting lines for the C–C bonds. The position which holds a free valence is marked and is called the root of the graph. The mathematical problem consists of counting all branched trees with a maximum of four branchings (why?) at a single branching point. The solution of this problem is produced by graph theory in form the of a polynomial $\mathbf{A}(x)$ containing an infinite number of elements

$$\mathbf{A}(x) = A_0 + A_1 x + A_2 x^2 + A_3 x^3 + \ldots = \sum_{i=0}^{\infty} A_i x^i$$

The coefficients A_i of this polynomial in x correspond to the number of isomeric alkyls with i carbon atoms. A_0 is set equal to 1, as the H-radical is considered as an alkyl without a C-atom. The calculation of the other coefficients is performed using the following implicit definition formula for $\mathbf{A}(x)$

$$\mathbf{A}(x) = 1 + \frac{x}{6} \cdot \left(\mathbf{A}^3(x) + 3\mathbf{A}(x)\mathbf{A}(x^2) + 2\mathbf{A}(x^3) \right)$$

This equation can be solved using a recursion. If for example, the first four coefficients (isomers through propyl) are known, then by means of the

---

[5] e.g. A.T. Balaban, *Chemical Application of Graph Theory*, Academic Press, London (1976)

above formula the fifth coefficient can be calculated. This is done through the required number of C-atoms in the compound.

From this polynomial $A(x)$ for the alkyls, the polynomials for homologues of other chemical substances can be obtained by symmetry and permutation considerations. For the alkenones the following polynomial $E1(x)$ is obtained

$$E1(x) = x^3 A^4(x) \cdot \sum_{p=0}^{\infty} \left(\frac{1}{2} x \left(A^2(x) + A(x^2) \right) \right)^p$$

The number of aldehydes within these alkenones can be derived with graph theory as

$$AL1(x) = x^3 A^3(x) \cdot \sum_{p=0}^{\infty} \left(\frac{1}{2} x \left(A^2(x) + A(x^2) \right) \right)^p$$

Finally the polynomial for the ketones $KE1(x)$ within the class of alkenones can be calculated

$$KE1(x) = AL1(x) \cdot \left(A(x) - 1 \right)$$

The corresponding equations for the alkenones, without considering cis-trans isomers, reads

$$E2(x) \quad = x^3 A^2(x) \cdot \sum_{p=0}^{\infty} \left(\frac{1}{2} x \left(A^2(x) + A(x^2) \right) \right)^p \cdot \frac{1}{2} \left(A^2(x) + A(x^2) \right)$$

$$AL2(x) \quad = x^3 A(x) \cdot \sum_{p=0}^{\infty} \left(\frac{1}{2} x \left(A^2(x) + A(x^2) \right) \right)^p \cdot \left(A^2(x) + A(x^2) \right)$$

$$KE2(x) \quad = AL2(x) \cdot \left(A(x) - 1 \right)$$

The construction of the polynomials $AK(x)$ for the alkanes from the polynomial $A(x)$ for the alkyls is a little more complicated. Therefore, one calculates first two auxiliary polynomials $P(x)$ and $Q(x)$ and uses them in the equation for $AK(x)$.

$$P(x) \quad = \frac{1}{24} x \left(A^4(x) + 6 A^2(x) A(x^2) + 3 A^2(x^2) + 8 A(x) A(x^3) + 6 A(x^4) \right)$$

$$\mathbf{Q}(x) \;\; = \frac{1}{2}\left(\Big(\mathbf{A}(x)-1\Big)^2 + \mathbf{A}(x^2) - 1\right)$$

$$\mathbf{AK}(x) = \mathbf{P}(x) - \mathbf{Q}(x) + \mathbf{A}(x^2) - 1$$

The BASIC-program ISOMERS which calculates these polynomials is listed as follows:

```
0 REM "ISOMERS "          EBERT/EDERER          880105
1 REM ***********************************************
2 REM *** Calculation of the number of isomers of  ***
3 REM *** alkyls, alkanes and alkenones as a       ***
4 REM *** function of the number of C-atoms.       ***
5 REM *** Lit.: Balaban; Chemical Application of   ***
6 REM ***        Graph Theory, Academic Press 1976 ***
9 REM ***********************************************
100   DIM A(50),A2(50),A3(50),A4(50),A9(50),AK(50)
110   DIM B2(50),B3(50),B4(50),C(50),C2(50)
120   DIM E(50),E2(50),F(50),H2(50),HY(50),K2(50)
130   DIM KE(50),P(50),Q(50),X(50),Y(50),Z(50)
200   REM ***********************************************
210   REM *** Recursion formula for the series      ***
220   REM *** A(x) which stands for the alkyls.     ***
230   REM *** A(x)=1+x/6*(A(x)^3+                    ***
240   REM ***       3*A(x)*A(x^2)+2*A(x^3)          ***
300   REM ***********************************************
350   INPUT "Maximum number of C-atoms "; N
400   A(0)=1 : A(1)=1 : A(2)=1 : A(3)=2
500   FOR I=3 TO N-1
520     S=0
550     FOR J=I TO 0 STEP -1
600       FOR K=0 TO I-J
620         L=I-J-K
700         S=S+A(J)*A(K)*A(L)
720       NEXT K
740     NEXT J
800     FOR J=0 TO I STEP 2
820       K=I-J
850       S=S+3*A(J/2)*A(K)
870     NEXT J
900     IF INT(I/3)=I/3 THEN S=S+2*A(I/3)
910     S=S/6
920     A(I+1)=S
960   NEXT I
```

```
1000   REM Calculation of A^2
1020   M=N:FOR I=0 TO M:X(I)=A(I):Y(I)=A(I):NEXT I
1060   GOSUB 9000:FOR I=0 TO M:A2(I)=F(I):NEXT I
1200   REM Calculation of A^4
1220   M=N:FOR I=0 TO M:X(I)=A2(I):Y(I)=A2(I):NEXT I
1240   GOSUB 9000:FOR I=0 TO M:A4(I)=F(I):NEXT I
1360   REM Calculation of A^3
1362   M=N:FOR I=0 TO M:X(I)=A2(I):Y(I)=A(I):NEXT I
1364   GOSUB 9000:FOR I=0 TO M:A3(I)=F(I):NEXT I
1400   REM Calculation of (.5*x*(A(x)^2+A(x^2)) in AK()
1420   FOR I=0 TO N:AK(I)=0:NEXT I
1440   FOR I=0 TO N/2:AK(2*I)=A(I):NEXT I
1460   FOR I=0 TO N:AK(I)=AK(I)+A2(I):NEXT I
1480   FOR I=N+1 TO 1 STEP -1:AK(I)=AK(I-1)/2:NEXT I
1490   AK(0)=0
2000   REM **********************************************
2001   REM *** Summing up the AK(x)^m                 ***
2002   REM *** for all m from 0 to n. The result is ***
2003   REM *** in Z(). E() and P() are auxiliary     ***
2004   REM *** vectors.                              ***
2005   REM **********************************************
2010   FOR I=0 TO N:Z(I)=0:NEXT I:Z(0)=1
2020   FOR I=0 TO N:Z(I)=Z(I)+AK(I):P(I)=AK(I):NEXT I
2100   FOR M=2 TO N
2120     FOR I=0 TO N:E(I)=0:NEXT I
2200     FOR J=0 TO N
2220       FOR K=0 TO J
2240         L=J-K:E(J)=E(J)+AK(K)*P(L)
2260     NEXT K:NEXT J
2300     FOR I=0 TO N:P(I)=E(I):Z(I)=Z(I)+P(I):NEXT I
2400   NEXT M
3000   REM Calculation of the alkenones in E(i)
3100   FOR I=0 TO N+3:E(I)=0:NEXT I
3200   FOR J=0 TO N
3220     FOR K=0 TO J
3240       L=J-K:E(J+3)=E(J+3)+Z(K)*A4(L)
3260   NEXT K:NEXT J
3300   REM Calculation of the aldehydes in HY(i)
3310   M=N+3:FOR I=0 TO N:X(I)=A3(I):Y(I)=Z(I):NEXT I
3320   GOSUB 9000:FOR I=0 TO N:HY(I+3)=F(I):NEXT I
3400   REM Calculation of the ketones in KE(i)
3410   M=N+3:FOR I=0 TO N:X(I)=HY(I):Y(I)=A(I):NEXT I
3420   Y(0)=0:GOSUB 9000:FOR I=0 TO M:KE(I)=F(I):NEXT I
4000   REM Alkenones without c-t-isomers
4030   FOR I=0 TO N:X(I)=AK(I+1):Y(I)=A2(I):NEXT I
4040   GOSUB 9000
```

```
4045   FOR I=0 TO N:X(I)=F(I):Y(I)=Z(I):NEXT I
4050   GOSUB 9000:FOR I=0 TO N:E2(I+3)=F(I):NEXT I
4100   REM Aldehydes without c-t-isomers
4130   M=N+3
4135   FOR I=0 TO M:X(I)=AK(I+1):Y(I)= A(I):NEXT I
4140   GOSUB 9000
4145   FOR I=0 TO M:X(I)=F(I):Y(I)=Z(I):NEXT I
4150   GOSUB 9000:FOR I=0 TO M:H2(I+3)=F(I):NEXT I
4200   REM Ketones without c-t-isomers
4230   M=N+3:FOR I=0 TO M:X(I)=H2(I):Y(I)= A(I):NEXT I
4240   Y(0)=0:GOSUB 9000:FOR I=0 TO M:K2(I)=F(I):NEXT I
5000   REM Calculation of A(x^2) in B2()
5010   FOR I=0 TO N:B2(I)=0:NEXT I
5020   FOR I=0 TO N/2:B2(2*I)=A(I):NEXT I
5100   REM Calculation of A(x^3) in B3()
5110   FOR I=0 TO N:B3(I)=0:NEXT I
5120   FOR I=0 TO N/3:B3(3*I)=A(I):NEXT I
5200   REM Calculation of A(x^4) in B4()
5210   FOR I=0 TO N:B4(I)=0:NEXT I
5220   FOR I=0 TO N/4:B4(4*I)=A(I):NEXT I
5300   REM Calculation of A(x^2)^2 in C2()
5310   M=N:FOR I=0 TO M:X(I)=B2(I):Y(I)=B2(I):NEXT I
5320   GOSUB 9000:FOR I=0 TO M:C2(I)=F(I):NEXT I
5500   REM Calculation of p(x) in P()
5510   REM p(x)=x*(a4+6*a2*b2+3*c2+8*a*b3+6*b4)/24
5580   M=N:FOR I=0 TO M:X(I)=A2(I):Y(I)=B2(I):NEXT I
5620   GOSUB 9000:FOR I=0 TO M:P(I)=A4(I)+F(I)*6:NEXT I
5640   FOR I=0 TO N:P(I)=P(I)+3*C2(I):NEXT I
5660   M=N:FOR I=0 TO M:X(I)=A(I):Y(I)=B3(I):NEXT I
5680   GOSUB 9000:FOR I=0 TO M
5690     P(I)=(P(I)+F(I)*8+6*B4(I))/24:NEXT I
5700   FOR I=N TO 0 STEP -1:P(I+1)=P(I):NEXT I:P(0)=0
5800   REM Calculation of q(x) in Q()
5810   REM q(x) = ((a(x)-1)^2+a(x^2)-1)/2
5840   M=N:FOR I=0 TO M:X(I)=A(I):Y(I)=A(I):NEXT I
5850   X(0)=0:Y(0)=0:GOSUB 9000
5860   FOR I=0 TO M:Q(I)=(F(I)+B2(I))/2:NEXT I
5880   Q(0)=Q(0)-.5
6000   REM Calc of the alkanes c(x)=p(x)-q(x)+a(x^2)-1
6040   FOR I=0 TO N:C(I)=P(I)-Q(I)+B2(I):NEXT I
6050   C(0)=C(0)-1
8500   PRINT:PRINT" C-atoms  alkyls  alkenones  ";
8505   PRINT "aldehydes       ketones     alkanes";
8510   PRINT:FOR I=0 TO N
8520   PRINT TAB(3) I;TAB(11) A(I);TAB(19) E(I);
8530   PRINT TAB(32) HY(I);TAB(47) KE(I);
```

```
8540  PRINT TAB(61) C(I):PRINT TAB(21) E2(I);
8550  PRINT TAB(34) H2(I);TAB(49) K2(I)
8560  NEXT I
8999  STOP
9000  REM *********************************************
9001  REM *** Subroutine: MULTIPLICATION          ***
9002  REM *** The polynomials X and Y are multi-   ***
9003  REM *** plied, the result is in F. The       ***
9004  REM *** maximum degree of the polynomial     ***
9005  REM *** which is calculated is m.            ***
9006  REM *********************************************
9010  FOR I9=0 TO M:F(I9)=0
9020    FOR J9=0 TO I9
9030      K9=I9-J9 : F(I9)=F(I9)+X(J9)*Y(K9)
9040  NEXT J9:NEXT I9
9050  RETURN
63999 END
```

```
RUN
Maximum number of C-atoms ? 10
```

C-atoms	alkyls	alkenones	aldehydes	ketones	alkanes
0	1	0	0	0	0
		0	0	0	
1	1	0	0	0	1
		0	0	0	
2	1	0	0	0	1
		0	0	0	
3	2	1	1	0	1
		1	1	0	
4	4	5	4	1	2
		4	3	1	
5	8	16	11	5	3
		12	8	4	
6	17	47	30	17	5
		34	21	13	
7	39	135	82	53	9
		95	56	39	
8	89	380	222	158	18
		262	149	113	
9	211	1056	599	457	35
		718	398	320	
10	507	2921	1621	1300	75
		1965	1068	897	

```
Break in 8999
```

In lines 100 through 130 a number of vectors are dimensioned; they will be used as vectors and auxiliary vectors containing the required results and intermediate values. The only input occurs in line 350. Here, the maximum number of C-atoms is put in, up to which the number of isomers will be calculated.

In the program section from line 400 through 960 the number of alkyls is calculated and stored in the array A. In the starting line 400 the first four A(I) are already assigned. In the main loop (I-loop) from line 500 through 960 the coefficients A_{i+1}, which accompany x^{i+1}, are calculated. The nested loops from line 550 through 740 serve to calculate the coefficient, which accompanies x_i if $\mathbf{A}^3(x)$ is multiplied out. This happens by counting using the variable S. The loop from line 800 through 870 performs this counting. Here that part of the overall coefficient is determined which is contributed by the expression $3\mathbf{A}(x)\mathbf{A}(x^2)$. In the next two lines the counting is completed by considering the contribution of $\mathbf{A}(x^3)$ and dividing by 6.

Before continuing the explanation of the main program, an important and frequently used subroutine 9000 will be described. This subroutine calculates the first m coefficients of a polynomial $\mathbf{F}(x)$ which is formed by the multiplication of two given polynomials $\mathbf{X}(x)$ and $\mathbf{Y}(x)$

$$\mathbf{F}(x) = \mathbf{X}(x) \cdot \mathbf{Y}(x)$$

For this all positive number combinations of K9 and J9 are considered for determining F(I9) (the I9th coefficient of $\mathbf{F}(x)$), whose sum results in I9, and the corresponding products X(J9)*Y(K9) are summed up. If in the following the term 'calculation of polynomials' is used, this is a short term for the 'calculation of the coefficients of polynomials'.

In lines 1000 through 1364 the polynomials $\mathbf{A}^2(x)$, $\mathbf{A}^4(x)$ and $\mathbf{A}^3(x)$ are calculated by the subroutine 9000 and the results are stored in the vectors A2(I), A4(I) and A3(I).

The polynomial $\frac{x}{2}\left(\mathbf{A}^2(x) + \mathbf{A}(x^2)\right)$, which is contained in the above formulas under the sum symbol, is calculated in lines 1400 through 1480 and stored in the vector variable AK(I). The calculation of the sum is performed in lines 2000 through 2400. The results are stored in variable Z(I). Z(I) is obtained by adding the appropriate powers of the polynomials AK(I); the auxiliary vectors P(I) and E(I) are used in raising to a power.

As soon as the fundamental polynomials are calculated, the above formulas for the alkenones, aldehydes and ketones, with and without considering the cis–trans isomers, can be easily calculated. This happens in the program section from line 3000 through 4240. An extensive discussion of this section is not necessary as the calculation is limited in general to a multiplication of different polynomials. Which polynomials are involved in this calculation can be seen from the formulas; the multiplication is carried out by the subroutine 9000.

The calculation of the number of alkanes starts in line 5000. First, auxiliary polynomials are calculated and stored (lines 5000 through 5300). These are: $A(x^2)$, $A(x^3)$, $A(x^4)$ and $A^2(x^2)$, which are stored in the vectors B2(I), B3(I), B4(I), and C2(I). The auxiliary polynomial $P(x)$ for the calculation of the alkanes is computed in the program lines 5500 through 5700. In this section the arithmetic is again limited to additions and multiplications of polynomials carried out by subroutine 9000. The auxiliary polynomial $Q(x)$ is calculated in the following program section from 5800 through 5880. The lines 6000 and 6040, in which the final polynomial for the alkanes is calculated, only add up the auxiliary polynomials which were evaluated earlier.

The output of the calculated numbers of isomers follows from line 8500 onwards. Two numbers are shown for the alkanes, ketones, and aldehydes. The upper figure is the number of isomers, counting cis–trans isomers as different isomers, the lower figure is the corresponding number of isomers without considering cis–trans isomerization.

Problem 113

Write a program or change the existing program ISOMERS to calculate the number of isomers of other chemical structures. Formulas which are derived from the polynomial $A(x)$ for the alkyls are given below

Primary alcohols: $PA(x) = xA(x)$

Secondary alcohols: $SA(x) = \dfrac{x}{2}\left(A^2(x) - 2A(x) + A(x^2)\right)$

Tertiary alcohols: $TA(x) = \dfrac{x}{6}\left(A^3(x) - 3A^2(x) + 3A(x)A(x^2) - 3A(x^2) + 2A(x^3)\right)$

Sum of ketones and aldehydes: $KA(x) = \dfrac{x}{2}\left(A^2(x) + A(x^2)\right)$

The aldehydes, separately, have the same formula as the primary alcohols. The ketones separately, can therefore be calculated simply by subtracting $\mathbf{PA}(x)$ from the above formula for $\mathbf{KA}(x)$.

Olefins: $\mathbf{AY}(x) = \dfrac{1}{8}\left(\mathbf{A}^4(x) + 2\mathbf{A}^2(x)\mathbf{A}(x^2) + 3\mathbf{A}^2(x^2) + 2\mathbf{A}(x^4)\right)$

Acetylenes: $\mathbf{AC}(x) = \dfrac{x^2}{2}\left(\mathbf{A}^2(x) + \mathbf{A}(x^2)\right)$

Esters: $\mathbf{ES}(x) = x\,\mathbf{A}(x)\left(\mathbf{A}(x) - 1\right)$

9   Differential Equations

If dynamical behavior must be described in engineering or physical sciences, this is done most nearly 'in a natural manner' by differential equations (DE's) or by systems of differential equations. This shall be shown explicitly by some examples:

a) Reaction kinetics $\frac{dc}{dt} = f(c, t)$
    c can stand for the concentration of a substance or for an array of concentrations.

b) Transport phenomena
    With the equations for transport phenomena, the transport of heat, mass transport and transport of momentum are described. With stationary treatment ordinary differential equations (first laws) may be used, non-stationary treatment makes the use of partial differential equations (second laws) necessary. If technical problems in heat and mass transfer are treated, often very similar equations, which may be empirical, are used (for example: heat exchange, mixing, drying, extraction, adsorption etc.).

c) Population dynamics
    The interactions of prey and predators can, in principle, be described by two very simple differential equations. In these equations assumptions are made about the birth and death rates of animals as a function of the 'concentration' of the respective animal population.

d) Spreading of epidemics (or of information)
    Under certain conditions differential equations can be used to describe the spreading of epidemics, news, rumors, knowledge and opinions. For such purposes the population is separated into healthy, infected, resistant and eliminated species. The theory is the basis from which the respective time laws (differential equations) may be derived.

Problems which may be solved with differential equations are numerous and can be subdivided into three types.

a) Initial value problems
    These are problems which result, for example, from reaction kinetics. Given are the concentrations of substances at time zero, the time dependence of the concentrations must be calculated.

b) Boundary value problems
    In these problems the function values or the derivative values at certain values of the independent variable are given. The solution between these values is calculated. An example for such a differential equation is the calculation of a beam which is fixed at both ends with a force applied at a certain point. In this case the boundary values for both ends of the beam are set to zero. Another example may be explained by the question: What is the stationary temperature distribution in a cube if three faces are set to 20° C and the other three to 80° C?

c) Eigenvalue problems
    These are often subproblems of boundary value problems which shall be explained with the help of the following example: The total energy is a parameter of the Schrödinger equation for a harmonic oscillator. The question we are interested in is: What are the total energies for which the Schrödinger equation provides physically reasonable results? (Reasonable results are wave functions which converge to zero with increasing distance). Total energies for which the differential equations are of a reasonable form are called eigenvalues; the resulting wave function is called an eigenfunction.
    An even simpler example is the differential equation for a vibrating string. The boundary values of the string are fixed by the physical limit at both ends as in a musical instrument. There is a parameter in the differential equation which has the meaning of a wavelength. The question to be answered is: which eigenvalues have this special parameter? Or otherwise: only at certain values of this parameter is it possible to obtain a solution of the differential equations with the specified boundary conditions.

The first two methods, which we will introduce here, are for solving initial value problems. They are rather simple and they are closely related to the integration routines treated in earlier sections. If more sophisticated algorithms are needed you should consult textbooks on numerical analysis.

9.1   The Euler-Cauchy Method

A differential equation $\frac{dy}{dx} = f(x, y)$ and an initial value $y(a) = y_a$ are given. (The integration routines treated in earlier chapters can be used for solving $\frac{dy}{dx} = f(x)$. In the generalized form the right hand side of the differential equation is a function of both the independent variable x and the dependent variable y.) To solve this differential equation means that a function $y = g(x)$ must be found which fulfills both the differential equation and the initial condition.

$$\text{For example}: \qquad \frac{dy}{dx} = x \cdot y \qquad \text{with initial value } y(0) = 1$$

$$\text{The solution is}: \qquad y = g(x) = e^{\frac{1}{2}x^2}$$

If you want to know the solution for $x = 1$, using $g(x)$ to calculate $g(1)$ you will obtain $y = 1.648721$. An analytical function is obtained as the solution of the differential equation only in very few examples. Generally the solution for $x = 1$ has to be calculated by a numerical procedure (and, of course, also for all the other values of x which are of interest). A numerical treatment of a differential equation doesn't result in an analytical function but you will obtain the solution value y for every x-value given. Then, by a plot of the corresponding (x, y)-values the graph of the solution function can be constructed.

Our first aim is to make the numerical method clear and easily understandable in principle — this differs somewhat from an exact mathematical derivation and critial analysis of the methods. The primary step is the approximation of the differential equation by a difference equation

$$\text{from} \qquad y' = \frac{dy}{dx} = f(x, y)$$

$$\text{we get} \qquad \frac{\Delta y}{\Delta x} \approx f(x, y)$$

By rearrangement of the difference equation the following formula for the numerical calculation is obtained

$$\Delta y \approx f(x, y) \cdot \Delta x$$

The numerical calculation starts at x_0 where $y = y_0$ and proceeds stepwise as shown in the example below.

initial values:	y_0	x_0
step 1:	$y_1 = y_0 + f(x_0, y_0) \cdot \Delta x$	$x_1 = x_0 + \Delta x$
step 2:	$y_2 = y_1 + f(x_1, y_1) \cdot \Delta x$	$x_2 = x_1 + \Delta x$
step 3:	$y_3 = y_2 + f(x_2, y_2) \cdot \Delta x$	$x_3 = x_2 + \Delta x$
step 4:	$y_4 = y_3 + f(x_3, y_3) \cdot \Delta x$	$x_4 = x_3 + \Delta x$
. . .	. . .	. . .

In this way you can calculate, using a relatively small increment Δx, until a result of y for a given x-value is obtained. The numerical error or deviation decreases if the step size Δx for the calculation is decreased. In textbooks on numerical mathematical methods the proof can be found that the resulting error is proportional to the step size.

$$\text{error} \sim \Delta x$$

Similar to the integration procedure of Euler there is no criterion programmed for program termination and an infinite loop is used to calculate more and more accurately through smaller and smaller increments. The flow diagram for this procedure according to Euler-Cauchy is shown in Fig. 9.1. This flow diagram is divided into three parts. The first part is the main program. The main program calls the second part, the subroutine for the Euler-Cauchy algorithm. The Euler-Cauchy subroutine needs in its loop the function value of the right hand side $f(x, y)$ of the differential equation. This is done by calling the third part, the subroutine RHS.

The BASIC-program corresponding to this flow diagram has the name EUL-ODE and reads as follows:

```
0 REM "EUL-ODE "        EBERT/EDERER        880706
1 REM ************************************************
2 REM *** Solving ordinary differential equations  ***
3 REM *** by the method of Euler-Cauchy.           ***
4 REM *** ODE:  dy/dx = f(x,y)                      ***
5 REM *** The right hand side of this differential ***
6 REM *** equation has to be programmed in the     ***
7 REM *** subroutine RHS starting at line # 10000. ***
9 REM ************************************************
```

```
500    INPUT "Initial value of y"; Y1
600    INPUT "x-values  start,end"; A,E
1000   N1=2
1050   PRINT
1100   H=(E-A)/N1
1200     GOSUB 5000 : REM ODE calculation
1300     PRINT N1;TAB(7);"subintervals";
1400     PRINT "    y=";Y
1500     N1=N1*2
1600   GOTO 1100
2000   REM ##########################################
2010   REM ### Main program ends here.          ###
2020   REM ##########################################
5000   REM ******************************************
5002   REM *** Subroutine: ODE                   ***
5004   REM *** The solution of the ODE is approxi- ***
5006   REM *** mated by using N1 subintervals.   ***
5010   REM ******************************************
5100   Y=Y1
5200   FOR X=A TO E-H STEP H
5300     GOSUB 10000 : REM RHS is evaluated
5400     Y=Y+D*H
5500   NEXT X
5600   RETURN
10000  REM ******************************************
10002  REM *** Subroutine: RHS                   ***
10004  REM *** The right hand side of the ODE is ***
10006  REM *** calculated here.                  ***
10010  REM ******************************************
11000  D=X*Y
12000  RETURN
63999  END


RUN
Initial value of y? 1
x-values  start,end? 0,1

   2    subintervals    y= 1.25
   4    subintervals    y= 1.419434
   8    subintervals    y= 1.524006
  16    subintervals    y= 1.583386
  32    subintervals    y= 1.61524
  64    subintervals    y= 1.631768
 128    subintervals    y= 1.640189
 256    subintervals    y= 1.644442
 512    subintervals    y= 1.646578
```

```
    1024 subintervals    y= 1.647649
    2048 subintervals    y= 1.648183
    4096 subintervals    y= 1.648459
    8192 subintervals    y= 1.648592
^C
Break in 10000

PRINT EXP(0.5)
  1.648721
```

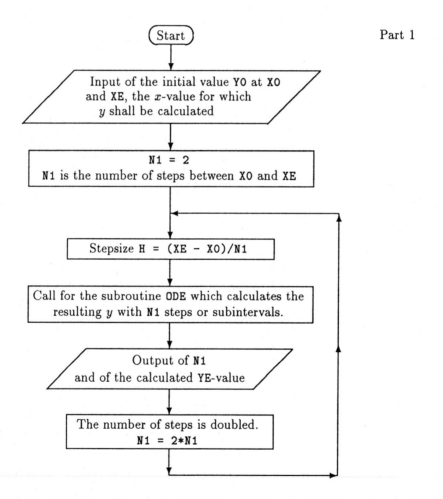

Part 1

Fig. 9.1. Part 1: flow diagram for the Euler-Cauchy method to solve an initial value problem for an ordinary differential equation

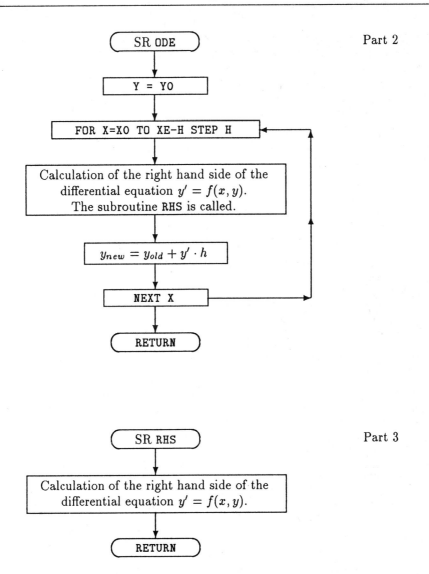

Fig. 9.1. Parts 2 and 3: flow diagram for the Euler-Cauchy method to solve an initial value problem for an ordinary differential equation

It was already mentioned that no criteria of termination were introduced into the program to keep it short and clearly arranged. The program contains an infinite loop and the differential equation will be infinitely recalculated with doubled accuracy after each repetition. The user has to stop the program from the keyboard as soon as he/she is satisfied with the result. That is usually the case if the difference between two consecutive results is sufficiently small. As a consequence of the doubling of the calculation, computing time will also be doubled for each following run and therefore a 'natural' end of the program exists.

The y-value belonging to the starting x-value fixes the solution at the beginning and has to be entered into the computer. This is the reason for the designation 'initial value problems' for this class of problems in differential equations. In addition the x-value, for which the y-value will be calculated, has to be typed in.

If not only one but more points should be calculated, the program EUL-ODE needs a separate run for each point (of course using the same initial conditions). The calculated points (x_i, y_i) together with the initial values (x_0, y_0) can be plotted in a diagram and will then give an impression of the solution curve of the differential equation problem.

As an example for solving a differential equation the following equation has been chosen.

$$\frac{dy}{dx} = x \cdot y$$

The set of solutions of this differential equation can be calculated analytically

$$y = A \cdot e^{\frac{1}{2}x^2}$$

The correctness of this solution is easily proven by setting it into the right hand side and left hand side of the differential equation. The constant A is determined by the initial value. For an initial value $y = 1$ for $x = 0$ the solution equation fits only if $A = 1$. With this result the initial value problem is completely solved. The analytical solution of this special case may be used for comparing it with the numerical solutions which are obtained by a computation with the program EUL-ODE.

In principle every differential equation can be treated with the program EUL-ODE. For doing so the right hand side of the differential equation has to be programmed in line 11000 and assigned to variable D.

Problem 114

Find numerical solutions for the given differential equations; find a solution for different points and draw a graph of the solution curve. Compare the numerical solutions with results you may get by integrating the differential equations analytically.

$$\frac{dy}{dx} = -y \qquad\qquad \frac{dy}{dx} = -k \cdot y^2 + \sin^2 xy$$

$$\frac{dy}{dx} = k - y^2 \qquad\qquad \frac{dy}{dx} = x^y + y^x$$

Start with the initial value problem $y = 1$ for $x = 0$, then solve the differential equation problems for other initial conditions.

Problem 115

The decomposition constant k_d of a substance A during a radiolytic decomposition is assumed to be proportional to the intensity of the radiation D ($k_d = k_0 D$; D may be considered as dimensionless).

$$\frac{d[A]}{dt} = -k_d \cdot [A] = -k_0 \cdot D \cdot [A]$$

The intensity of the radiation D decreases according to an exponential time law

$$D = D_0 \cdot e^{-kt}$$

The values of the parameters are assumed to be

$$k = 0.001\,\mathrm{s}^{-1}$$
$$D_0 \cdot k_0 = 0.001\,\mathrm{s}^{-1}$$

If these values are used, the following differential equation is obtained.

$$\frac{d[A]}{dt} = -0.001 \cdot e^{-0.001 \cdot t} \cdot [A]$$

The variation of the quotient $[A]/[A]_0$ with time can be calculated with the program EUL-ODE. Determine this ratio after infinite time.

The analytical solution of this problem is as follows

$$\log\left(\frac{[A]_0}{[A]}\right) = \frac{k_0 D_0}{k} \cdot (1 - e^{-kt})$$

Compare the numerical result with the analytical solution.

Problem 116

The reaction order of the thermal decomposition of a substance A should be two and the reaction order for the decomposition by UV-radiation should be zero (quantum yield $= 1$). From this the following differential equation for the overall reaction results

$$\frac{d[A]}{dt} = -k_2[A]^2 - u(t)$$

$u(t)$ is assumed to be proportional to the intensity of the UV-radiation of sunlight during the day. Therefore $u(t)$ can be approximated by a sine function.

The reaction starts with the sunrise. What is the concentration of A at sunset if the following parameters are assumed:

sun raise	6:00 am	i.e.: $t = 0$
sun set	6:00 pm	i.e.: $t = 43200$ s
$[A]_0$	1 mol/l	
k_2	0.00002 l/(mol·s)	

$$u(t) = 0.00001 \cdot \sin(t \cdot \pi/43200)\,\text{mol}/(\text{l·s})$$

Problem 117

Generalize the last problem for an arbitrary thermal decomposition order, even non-integer numbers should be allowed. Two different sources of radiation are used. The two sources have different time behavior with reaction orders, referring to the concentration of A, not necessarily 1 or 0.

9.2   The Improved Euler-Cauchy Method

In the simple Euler method differential equations are converted into difference equations and from this the following iteration formula is obtained after an algebraic conversion.

$$y_{i+1} = y_i + h \cdot f(x_i, y_i)$$

The letter h is very often used as a simpler symbol for Δx. The right hand side of the differential equation $f(x_i, y_i)$ is calculated for the given point (x_i, y_i). Because it is desirable for the algorithm for the solution of the differential equation to be the same for calculating in the positive x-direction and in the negative x-direction, it would certainly be an improvement if the right hand side of the differential equation were evaluated somewhere between (x_i, y_i) and (x_{i+1}, y_{i+1}). But this is not possible in a direct way because the value for y_{i+1} is not yet known. But with a calculation step — usually called predictor step — an appropriate value y_{i+1}^* can be calculated. For the predictor step the Euler iteration formula shown above is chosen

$$y_{i+1}^* = y_i + h \cdot f(x_i, y_i)$$

The predictor step is followed by a corrector step in which the right hand side of the differential equation is evaluated between the two points.

$$y_{i+1} = y_i + \frac{h}{2} \cdot \Big(f(x_i, y_i) + f(x_{i+1}, y_{i+1}^*) \Big) \qquad (1)$$

If this simple predictor-corrector procedure is used the calculation error is decreased by reducing the step size h, but this improvement is not only proportional to h, but the error is a function of h^3.

$$\text{error} \sim h^3$$

The sequence of steps can be transformed and this can be programmed according as follows:

```
Y2   = Y + H*F(X  ,Y )
Y3   = Y + H*F(X+H,Y2)
YNEW = (Y2 + Y3)/2
```

If the first and second equations (written in BASIC-notation) are used for substitution in the third one, the above equation (1) results. Included in the program EUL-ODE the program below is obtained:

```
0 REM "EUL-IMPR"         EBERT/EDERER          880706
1 REM ***********************************************
2 REM *** Solving ordinary differential equations  ***
3 REM *** by the improved method of Euler-Cauchy.  ***
4 REM *** ODE:  dy/dx = f(x,y)                      ***
5 REM *** The right hand side of this differential ***
6 REM *** equation has to be programmed in the     ***
7 REM *** subroutine RHS starting at line # 10000  ***
8 REM *** as D = f(X9,Y9) .                         ***
9 REM ***********************************************
```

```
500    INPUT "Initial value of y"; Y1
600    INPUT "x-values  start,end"; A,E
1000   N1=2
1050   PRINT
1100   H=(E-A)/N1
1200     GOSUB 5000 : REM ODE calculation
1300     PRINT N1;TAB(7);"subintervals";
1400     PRINT "    y=";Y
1500     N1=N1*2
1600   GOTO 1100
2000   REM ###########################################
2010   REM ### Main program ends here.          ###
2020   REM ###########################################
5000   REM *******************************************
5002   REM *** Subroutine: ODE                   ***
5004   REM *** The solution of the ODE is approxi- ***
5006   REM *** mated by using N1 subintervals.    ***
5010   REM *******************************************
5100   Y=Y1
5200   FOR X=A TO E-H STEP H
5250     X9=X : Y9=Y
5300     GOSUB 10000 : REM RHS is evaluated
5320     Y2=Y+D*H
5340     X9=X+H : Y9=Y2
5350     GOSUB 10000 : REM RHS is evaluated
5360     Y3=Y+D*H
5400     Y=(Y2+Y3)/2
5500   NEXT X
5600   RETURN
10000 REM *******************************************
10002 REM *** Subroutine: RHS                   ***
10004 REM *** The right hand side of the ODE is ***
10006 REM *** calculated here.                  ***
10010 REM *******************************************
11000 D=X9*Y9
12000 RETURN
63999 END

RUN
Initial value of y? 1
x-values  start,end? 0,1

  2    subintervals    y= 1.617188
  4    subintervals    y= 1.642286
  8    subintervals    y= 1.647355
 16    subintervals    y= 1.648415
```

```
 32    subintervals    y= 1.648649
 64    subintervals    y= 1.648704
128    subintervals    y= 1.648717
256    subintervals    y= 1.64872
512    subintervals    y= 1.648717
^C
Break in 5360

PRINT EXP(0.5)
 1.64872
Ok
```

The structure of the program remains the same — the main program is identical to that of EUL-ODE.

The right hand side of the differential equation is calculated in the subroutine beginning with line 10000. The variables X and Y cannot be used, because they were used already in the subroutine beginning with line 5000. Therefore the variables X and Y in EUL-ODE are substituted by X9 and Y9 in EUL-IMPR. The subroutine ODE beginning at line 5000 has been extended. Up to line number 5320 the same calculations are programmed as in the respective subroutine of the program EUL-ODE. But the result now is no longer considered the final new y-value but the predictor value for y and is therefore stored in variable Y2. The next four program lines (up to 5400) are the corrector step.

The differential equation chosen is the same as in program EUL-ODE. Therefore, by comparing runs of both programs, one can see how much faster the improved version converges to the exact numerical value. A small addition to the programming code with the help of numerical mathematics yields considerable improvement and acceleration of the calculation method.

Problem 118

Solve the following differential equations and plot one or more solution graphs (depending on the initial conditions).

$$\frac{dy}{dx} = 2y + e^x \qquad\qquad \frac{dy}{dx} = \log(x) + \frac{x}{y}$$

$$\frac{dy}{dx} = x^3 + 2xy \qquad\qquad \frac{dy}{dx} = \frac{(1+y) \cdot x \cdot y}{\frac{dy}{dx}}$$

Problem 119

Compare the solution of several differential equations using the program `EUL-ODE` and the program `EUL-IMPR`.

Problem 120

Write a program for the following fundamental and simple model of population dynamics (Verhulst's law). In this model it is assumed that the birth rate is proportional to the actual population p and the death rate is proportional to the square of the population (because of overcrowding and lack of resources). The following differential equation results from these assumptions.

$$\dot{p} = \frac{dp}{dt} = a \cdot p - b \cdot p^2$$

Calculate the variation of the population with time, if $a = 1$ and $b = 0.0001$ are assumed. Starting with a population number (or better: population concentration or density) of 10 is recommended. What is the population after a very long time (stable population)?

Problem 121

Many species need a partner for reproduction but have a death rate proportional to the population. From this follows the differential equation for the population

$$\dot{p} = \frac{dp}{dt} = -a \cdot p + b \cdot p^2$$

Assuming $a = 1$ and $b = 0.001$, a limit for the population p_l can be calculated below which the species are dying out.

Select different initial populations and determine the limit p_l, below which an initial population is dying out.

Problem 122

The following problem treats the spreading of news within a population. The number of the entire population is denoted by P. p individuals possess knowledge of the news at time zero. The respective differential equation can be written as follows

$$\dot{p} = \frac{dp}{dt} = a \cdot p \cdot (P - p) + b \cdot (P - p)^c$$

The first term describes the spreading of the news by contacts of uninformed individuals $(P - p)$ with informed individuals (p). The second term describes the influence of news media like TV or newspapers. Naturally they act on uninformed individuals only.

Try to describe the spreading of the news with time, as an initial value of $p = 1$ is assumed at time 0. At what time is $p = 1000$, if the parameters are as follows: $a = 0.00001$, $b = 0.001$, $c = 0.9$ and $P = 5000$. Determine the development of the spreading if $b = 0$ is assumed, that means that no modern media are involved. At what time is the influence of the news media of most importance for the spreading of news?

9.3   The Runge-Kutta Method

This numerical method, which is very often used to solve scientific problems, will be introduced here without any mathematical proof. It is a method which is reliable, robust and not too difficult to program. The method is self-starting, which is another advantage, and furthermore it is useful for starting more sophisticated numerical methods. The Runge-Kutta method is similar to that of Euler, but for the error it can be shown that

$$\text{error} \sim h^4$$

This means that the error of the results decreases much faster than with the method of Euler; h is again an abbreviation for Δx.

The Runge-Kutta algorithm needs four auxiliary quantities k_1, k_2, k_3 and k_4 (these are different Δy-values). Given is again a differential equation.

$$\frac{dy}{dx} = y' = f(x, y)$$

The calculation of the next solution point (x_{i+1}, y_{i+1}), beginning with (x_i, y_i) is done using the following scheme.

$$k_1 = h \cdot f(x_i, y_i)$$

$$k_2 = h \cdot f(x_i + \frac{h}{2}, y_i + \frac{k_1}{2})$$

$$k_3 = h \cdot f(x_i + \frac{h}{2}, y_i + \frac{k_2}{2})$$

$$k_4 = h \cdot f(x_i + h, y_i + k_3)$$

$$x_{i+1} = x_i + h$$

$$y_{i+1} = y_i + \frac{1}{6} \cdot (k_1 + 2k_2 + 2k_3 + k_4)$$

This calculation procedure is used in the following BASIC-program called R-K-N:

```
0 REM "R-K-N  "           EBERT/EDERER           880706
1 REM *************************************************
2 REM *** Solving ordinary differential equations  ***
3 REM *** by the method of Runge-Kutta-Nystroem.   ***
4 REM *** ODE:  dy/dx = f(x,y)                      ***
5 REM *** The right hand side of the  differential ***
6 REM *** equation has to be programmed in the     ***
7 REM *** subroutine RHS starting at line # 10000 .***
9 REM *************************************************
500    INPUT "Initial value of y"; Y1
600    INPUT "x-values  start,end"; A,E
1000   N1=2
1050   PRINT
1100   H=(E-A)/N1
1200     GOSUB 5000 : REM ODE is evalulated
1300     PRINT N1;TAB(7);"subintervals";
1400     PRINT"    y=";Y
1500     N1=N1*2
1600   GOTO 1100
2000   REM ###########################################
2010   REM ### Main program ends here.          ###
2020   REM ###########################################
5000   REM *************************************************
5002   REM *** Subroutine: ODE                   ***
5004   REM *** The solution of the ODE is approxi- ***
5006   REM *** mated by using N1 subintervals.    ***
5010   REM *************************************************
5100   Y=Y1
5200   FOR X=A TO E-H STEP H
5300     X9=X  : Y9=Y : GOSUB 10000 : K1=H*D
5400     X9=X+H/2 : Y9=Y+K1/2 : GOSUB 10000 : K2=H*D
5500     X9=X+H/2 : Y9=Y+K2/2 : GOSUB 10000 : K3=H*D
5600     X9=X+H  : Y9=Y+K3  : GOSUB 10000 : K4=H*D
5700     Y=Y+(K1+2*K2+2*K3+K4)/6
```

```
5800  NEXT X
5900  RETURN
10000 REM ********************************************
10002 REM *** Subroutine: RHS                    ***
10004 REM *** The right hand side of the ODE is  ***
10006 REM *** calculated here.                   ***
10010 REM ********************************************
11000 D=X9*Y9
12000 RETURN
63999 END

RUN
Initial value of y? 1
x-values   start,end? 0,1

  2    subintervals    y= 1.648528
  4    subintervals    y= 1.64871
  8    subintervals    y= 1.648721
 16    subintervals    y= 1.648721
 32    subintervals    y= 1.648722
 64    subintervals    y= 1.648721
128    subintervals    y= 1.648722
^C
Break in 10006

PRINT EXP(0.5)
 1.64872
Ok
```

You can model experiments by practicing solving differential equations with this program. You will quickly see that R-K-N produces more exact solutions much faster than the two other programs EUL-ODE and EUL-IMPR.

The main program and the subroutine for calculating the right hand side of the differential equation is identical with that of EUL-ODE. And again there is an infinite loop for halving the step size at each new loop run.

In the subroutine beginning with line 5000, some changes were made within the FOR-NEXT-loop compared with EUL-ODE. The above mentioned calculation scheme is programmed in the five lines from 5300–5700. The function $f(x, y)$ is again called up by GOSUB 10000, but not before the new variables X9 and Y9 have been filled with the appropriate values.

Problem 123

Compare the three different methods (EUL-ODE, EUL-IMPR and R-K-N) for solving differential equations by using the following test equations

$$y' = -xy \qquad y' = x + y$$
$$y' = y \qquad y' = \sin y$$
$$y' = -y \qquad y' = \log y$$

Problem 124

Expand the common main program of all three programs for differential equations with criteria for program termination. For doing so, three consecutive results should be compared. The output should be limited to the final result. There may be a rounding problem with the main FOR-NEXT-loop from line 5200–5800. Try to improve it!

Problem 125

Modify the BASIC-program R-K-N to obtain not only the final result but also 20 y-values between x-start and x-end. This may be useful to be able to observe the entire resulting solution function from the user-given starting value until the unknown function value at the end of the x-interval.

Problem 126

Solve some or all of the following differential equations with a numerical procedure and compare the results obtained with the analytical solutions.

$$y' = -0.1y - 0.5y^2 \qquad y' = y \cdot \log x - y$$

$$y' = 1 - y - e^{-x} \qquad y' = xy^3 - y$$

$$y' = \sin xy + \cos \frac{x}{y} \qquad y' = e^x \cdot y^2 - 2y$$

Problem 127

A chemical reaction is carried out in a thermally insulated batch reactor. The differential equation for the concentration c of the reactant is the following

$$\dot{c} = \frac{dc}{dt} = -k \cdot c$$

k can be written according to the Arrhenius' formula

$$k = k_0 \cdot e^{-\frac{E_a}{RT}}$$

The temperature T in the reaction vessel is changing with the conversion and can be expressed as

$$T = T_0 + (c_0 - c) \cdot H$$

The parameter H is proportional to the reaction enthalpy.

Determine the course of the concentration c with reaction time both for an exothermic reaction and for an endothermic reaction. The respective differential equations can easily be obtained from the equations above. Try to solve the same problems if the reaction is of second order.

Problem 128

The growth of a tumor, which is often exponential, can frequently be described with simple differential equations. The growth-constant in most cases is decreasing with time. An example for describing the growth of a tumor is the following differential equation.

$$\frac{dV}{dt} = a \cdot V \cdot e^{-b \cdot t^c}$$

V : Size of the tumor
c : Usually $= 1$
a, b : Characteristic constants

Determine the growth of a tumor if it is growing according to the differential equation above. For certain values of parameter c there is a limiting value for the size of the tumor (for example if $c = 1$); for other values there is no such size limit (for example if $c = 0$). Try to find out by practicing with the computer for which values of c the tumor has a finite size, even in the case where growing time is nearly infinite. a and b may be chosen arbitrarily.

Problem 129

Consider a continuously stirred flow reactor with a volume of V liters. The reaction order for the conversion of a substance C is assumed to be a. The concentration of the substance C at the beginning of the reaction ($t = 0$) is c_0. The reaction is assumed to be constant in volume, therefore the input flow, z, is equal to the product stream. The concentration of the feed is c_z, the product flow has, of course, the same concentration as in the reaction vessel c.

The differential equation for the concentration behavior in this reaction vessel may be written as

$$\frac{dc}{dt} = -k \cdot c^a + \frac{c_z \cdot z}{V} - \frac{c \cdot z}{V}$$

Determine the course of the concentration for the initial stage of the reaction and the stationary state which is established after a reasonably long time. The reaction dynamics is dependent on the parameters c_0, c_z, k, a, z and V. For practice use the following parameters: $c_0 = 3 \, \text{mol/l}$, $c_z = 2 \, \text{mol/l}$, $a = 1$, $k = 0.001 \, \text{s}^{-1}$, $z = 1 \, \text{l/sec}$ and $V = 1000 \, \text{l}$.

9.4   The Euler Method for Systems of Differential Equations

The following simple reaction mechanism for enzyme action is considered:

$$E \ + \ S \ \xrightarrow{k_1} \ ES$$
$$E \ + \ S \ \xleftarrow{k_2} \ ES$$
$$ES \ \xrightarrow{k_3} \ P \ + \ E$$

E   :   Enzyme
S   :   Substrate
ES :   Enzyme-substrate complex
P   :   Product

From the reaction mechanism the following system of coupled differential equations can be derived.

$$\frac{d[E]}{dt} = -k_1 \cdot [E] \cdot [S] + k_2 \cdot [ES] + k_3 \cdot [ES]$$

$$\frac{d[ES]}{dt} = +k_1 \cdot [E] \cdot [S] - k_2 \cdot [ES] - k_3 \cdot [ES]$$

$$\frac{d[S]}{dt} = -k_1 \cdot [E] \cdot [S] + k_2 \cdot [ES]$$

$$\frac{d[P]}{dt} = +k_3 \cdot [ES]$$

If this system is treated analytically, assuming a quasistationary state, the well-known Michaelis-Menten kinetics is obtained.

If an exact solution of this system without using approximations is required, a numerical method has to be used instead. For the sake of clearness Euler's method will be discussed first. For simplifying the description a system of only two coupled differential equations is used as an example. (At the right hand side the Euler notation for only one differential equation is printed.)

$$\left\{\begin{array}{l} \dfrac{dy_1}{dx} = f_1(x, y_1, y_2) \\[2mm] \dfrac{dy_2}{dx} = f_2(x, y_1, y_2) \end{array}\right\} \qquad \dfrac{dy}{dx} = f(x, y)$$

If the differential equations are approximated by difference equations and if the step size in the direction of x is called h, then the following formulas are obtained for stepping forward in x-direction.

$$y_1(x + h) = y_1(x) + f_1\left(x, y_1(x), y_2(x)\right) \cdot h$$

$$y_2(x + h) = y_2(x) + f_2\left(x, y_1(x), y_2(x)\right) \cdot h$$

In comparison the one-dimensional equation reads.

$$y(x + h) = y(x) + f\left(x, y(x)\right) \cdot h$$

The comparison between both evaluations shows that the formulas are analogous. The difference is that we now have two different right hand sides, both of them may depend on x, y_1 and y_2.

The following program SYS-EUL was developed analogously to EUL-ODE:

```
0 REM "SYS-EUL "        EBERT/EDERER        880711
1 REM ************************************************
2 REM *** Solving systems of ordinary differential ***
3 REM *** equations by the method of Euler-Cauchy. ***
5 REM *** The right hand sides of the differential ***
6 REM *** equations have to be programmed in the   ***
7 REM *** subroutine RHS starting at line # 10000; ***
8 REM *** the dimension of the system N in line 200***
9 REM ************************************************
```

```
100    DIM Y(20),Y1(20),D(20)
200    N=2 : REM N is the dimension of the ODE-system
300    FOR I=1 TO N
400      PRINT "Initial value of y(";I;") = ";
500      INPUT Y1(I)
600    NEXT I
700    INPUT "x-values   start,end ";A,E
1000   N1=4
1050   PRINT
1100   H=(E-A)/N1
1200     GOSUB 5000
1300     PRINT N1;TAB(7);"subintervals"
1400     FOR I=1 TO N
1420       PRINT "y(";I;") = ";Y(I)
1440     NEXT I
1500     N1=N1*2
1600   GOTO 1100
2000   REM ##########################################
2010   REM ### Main program ends here.        ###
2020   REM ##########################################
5000   REM ******************************************
5002   REM *** Subroutine: ODE                   ***
5004   REM *** The solution of the ODE is approxi-  ***
5006   REM *** mated by using N1 subintervals.    ***
5010   REM ******************************************
5100   FOR I=1 TO N : Y(I)=Y1(I) : NEXT I
5200   FOR X=A TO E-H STEP H
5300     GOSUB 10000 : REM RHS is evaluated
5400   FOR I=1 TO N : Y(I)=Y(I)+D(I)*H : NEXT I
5500   NEXT X
5600   RETURN
10000 REM ******************************************
10002 REM *** Subroutine: RHS                   ***
10004 REM *** The right hand sides of the ODE's ***
10006 REM *** are calculated here as            ***
10008 REM *** D(I) = f(X,Y(1),Y(2),..Y(N))      ***
10010 REM ******************************************
10100 D(1)=-3*Y(1)*Y(1)+2*Y(2)+EXP(-X)
10200 D(2)=3*Y(1)*Y(1)-12*Y(2)+EXP(-X)
12000 RETURN
63999 END


RUN
Initial value of y( 1 ) = ? 1
Initial value of y( 2 ) = ? 1
x-values   start,end ? 0,1
```

```
  4      subintervals
y( 1 ) = -3.023438
y( 2 ) =  13.42922
  8      subintervals
y( 1 ) =  .5130753
y( 2 ) =  .1167976
 16      subintervals
y( 1 ) =  .5220019
y( 2 ) =  .1070964
 32      subintervals
y( 1 ) =  .5252197
y( 2 ) =  .108302
 64      subintervals
y( 1 ) =  .5268594
y( 2 ) =  .1089248
 128     subintervals
y( 1 ) =  .5276858
y( 2 ) =  .109241
 256     subintervals
y( 1 ) =  .5280998
y( 2 ) =  .1094002
^C
Break in 5400
```

The initial values of y, the actual values of y and the derivatives of y are stored in the vectors Y, Y1 and D. The statement concerning the declaration of their dimensions is contained in line number 100. This is the main difference from the program EUL-ODE which can handle only one-dimensional differential equations. The variables in EUL-ODE (Y, Y1 and D) become vectors in SYS-EUL.

In the subroutine for the calculation of the right hand sides of the differential equations (beginning with line 10000), the variables D and Y are indexed. Both right hand sides of the differential equations are programmed, and their result is stored in the variables D(1) and D(2).

The main program starts after the declaration of the vectors with the statement N=2 which means that the actual system of differential equations is of dimension 2. Instead of only one initial value for Y there have to be read in N initial values, one for each Y(I). This is done in lines 300 to 600. There is another last modification in the main program. The output has to be programmed for N y-values (lines 1400–1440).

In the subroutine for calculation according to the Euler scheme (beginning with line 5000), the assignment statements of the lines 5100 and 5400 are substituted by vector assignments. Vector assignments have to be accomplished by a FOR-NEXT-loop, therefore line 5100 becomes:

```
    SYS-EUL:    5100 FOR I=1 TO N : Y(I)=Y1(I) : NEXT I
instead of
    EUL-ODE:    5100 Y=Y1
```

and line 5400 becomes:

```
    SYS-EUL:    5400 FOR I=1 TO N : Y(I)=Y(I)+D(I)*H : NEXT I
instead of
    EUL-ODE:    5400 Y=Y + D*H
```

If you wish to calculate other systems of differential equations than those contained in the BASIC-program SYS-EUL, then you have to adjust the dimension N of the differential equation system in line 500, and in the subroutine beginning with 10000 you have to program the new right hand sides of the system of differential equations.

Problem 130

Because the program SYS-EUL is analogous to EUL-ODE there is also no criterion for terminating the program. Complete the program SYS-EUL with such a criterion and modify the output part so that there is a choice of printing the final result only or in addition intermediate results as well.

Problem 131

Given is the following rather complicated chemical reaction model. The letters stand for different chemical substances and the k_i's for the appropriate reaction rate constants.

$$A \xrightarrow{k_1 = 1} M + M$$

$$A + M \xrightarrow{k_2 = 10} B + C$$

$$C \xrightarrow{k_3 = 4} M + D$$

$$M + M \xrightarrow{k_4 = 20} E + C$$

From this reaction mechanism a kinetic is derived which can be described by the following system of differential equations

$$\frac{d[A]}{dt} = -[A] - 10[A][M]$$

$$\frac{d[M]}{dt} = 2[A] - 10[A][M] + 4[C] - 40[M][M]$$

$$\frac{d[B]}{dt} = 10[A][M]$$

$$\frac{d[C]}{dt} = 10[A][M] - 4[C]$$

$$\frac{d[D]}{dt} = 4[C]$$

$$\frac{d[E]}{dt} = 20[M][M]$$

In this system of differential equations the mathematical symbols are substituted by indexed variables according to the following scheme

$$\frac{d[A]}{dt} \longrightarrow D(1) \qquad\qquad [A] \longrightarrow Y(1)$$

$$\frac{d[M]}{dt} \longrightarrow D(2) \qquad\qquad [M] \longrightarrow Y(2)$$

$$\frac{d[B]}{dt} \longrightarrow D(3) \qquad\qquad [B] \longrightarrow Y(3)$$

$$\frac{d[C]}{dt} \longrightarrow D(4) \qquad\qquad [C] \longrightarrow Y(4)$$

$$\frac{d[D]}{dt} \longrightarrow D(5) \qquad\qquad [D] \longrightarrow Y(5)$$

$$\frac{d[E]}{dt} \longrightarrow D(6) \qquad\qquad [E] \longrightarrow Y(6)$$

With this transformation we obtain a system of differential equations in a form needed by SYS-EUL beginning with line 10000.

```
D(1) = -Y(1) -10*Y(1)*Y(2)
D(2) = 2*Y(1)-10*Y(1)*Y(2)+4*Y(4)-40*Y(2)*Y(2)
D(3) = 10*Y(1)*Y(2)
D(4) = 10*Y(1)*Y(2)-4*Y(4)
D(5) = 4*Y(4)
D(6) = 20*Y(2)*Y(2)
```

Try to accomplish the derivation of the system of differential equations from the reaction mechanism, and calculate with the help of program SYS-EUL the time behavior of the decomposition of substance A and the formation of the different products B, C, D and E. (It should be noted that this system of differential equations can be solved better and faster with the program SYS-RKN which will be described in the next section.)

Problem 132

Given is – as in earlier examples – a simple chemical decomposition reaction of the order a

$$\frac{dc}{dt} = -k_0 \cdot e^{-\frac{E_a}{RT}} \cdot c^a$$

The reaction has a reaction enthalpy and the reaction vessel is kept constant at a temperature T_b. The heat flow from the reaction vessel to the thermostat is proportional to $(T - T_b)$. The temperature changes with time can be expressed by the following differential equation

$$\frac{dT}{dt} = H \cdot \left(k_0 \cdot e^{-\frac{E_a}{RT}} \cdot c^a \right) + \alpha \cdot (T - T_b)$$

H is proportional to the heat of reaction and α is the coefficient which describes the heat flow through the reactor wall.

Write a program to solve these coupled differential equations and calculate the time behavior of c and T with given constant parameters T_b, k_0, E_a, H, α, a and the initial values $c_{t=0}$ and $T_{t=0}$.

(Note: If you are taking a highly exothermic reaction and a very small heat flow you may get numerical problems, resulting in negative concentrations as intermediate values, or very high oscillations, or much too high concentrations and other nonsense. Then, you must use smaller increments which means many more subintervals to obtain the correct results.)

9.5   The Runge-Kutta Method for Systems of Differential Equations

The program SYS-EUL is an extension of the program EUL-ODE to systems of differential equations. The same can be said about the following program SYS-RKN, which is an extension of the above Runge-Kutta program R-K-N to systems of differential equations. It contains the same two differential equations as SYS-EUL.

```
0 REM "SYS-RKN "          EBERT/EDERER          880620
1 REM *************************************************
2 REM *** Solving systems of ordinary differential ***
3 REM *** equations following the method of        ***
4 REM *** Runge-Kutta-Nystroem.                     ***
5 REM *** The right hand sides of the differential ***
6 REM *** equations have to be programmed in        ***
7 REM *** subroutine RHS starting at line # 10000. ***
8 REM *** The dimension of the system N appears     ***
9 REM *** in line # 200.                            ***
10 REM ************************************************
100    DIM Y(20),Y1(20),D(20),Y9(20)
120    DIM K1(20),K2(20),K3(20),K4(20)
200    N=2 : REM N is the dimension of the ODE-system
300    FOR I=1 TO N
400      PRINT "Initial value of y(";I;") = ";
500      INPUT Y1(I)
600    NEXT I
700    INPUT "x-values   start,end ";A,E
1000   N1=4
1100   H=(E-A)/N1
1200     GOSUB 5000 : REM call for ODE
1300     PRINT N1;TAB(7);"subintervals"
1400     FOR I=1 TO N
1420       PRINT "y(";I;") = ";Y(I)
1440     NEXT I
1500     N1=N1*2
1600   GOTO 1100
2000   REM ###########################################
2010   REM ### Main program ends here.            ###
2020   REM ###########################################
5000   REM *****************************************
5002   REM *** Subroutine: ODE                   ***
5004   REM *** The solution of the ODE is approxi- ***
5006   REM *** mated by using N1 subintervals.    ***
5010   REM *****************************************
5100   FOR I=1 TO N : Y(I)=Y1(I) : NEXT I
5200   FOR I9=0 TO N1-1 : X=A+I9*H
5300     X9=X : FOR I=1 TO N : Y9(I)=Y(I) : NEXT I
5320     GOSUB 10000
5340     FOR I=1 TO N : K1(I)=H*D(I) : NEXT I: X9=X+H/2
5400     FOR I=1 TO N : Y9(I)=Y(I)+K1(I)/2 : NEXT I
5420     GOSUB 10000
5440     FOR I=1 TO N : K2(I)=H*D(I) : NEXT I: X9=X+H/2
5500     FOR I=1 TO N : Y9(I)=Y(I)+K2(I)/2 : NEXT I
5520     GOSUB 10000
```

```
5540     FOR I=1 TO N : K3(I)=H*D(I) : NEXT I : X9=X+H
5600     FOR I=1 TO N : Y9(I)=Y(I)+K3(I) : NEXT I
5620     GOSUB 10000
5640     FOR I=1 TO N : K4(I)=H*D(I) : NEXT I
5700     FOR I=1 TO N
5720        Y(I)=Y(I)+(K1(I)+2*K2(I)+2*K3(I)+K4(I))/6
5740     NEXT I
5800  NEXT I9 : RETURN
10000 REM *******************************************
10002 REM *** Subroutine: RHS                    ***
10004 REM *** The right hand sides of the ODE's  ***
10006 REM *** are calculated here as:            ***
10008 REM *** D(I) = f(X9,Y9(1),Y9(2),..Y9(N))   ***
10010 REM *******************************************
10100 D(1)=-3*Y9(1)*Y9(1)+2*Y9(2)+EXP(-X9)
10200 D(2)=3*Y9(1)*Y9(1)-12*Y9(2)+EXP(-X9)
12000 RETURN
63999 END


RUN
Initial value of y( 1 ) = ? 1
Initial value of y( 2 ) = ? 1
x-values    start,end ? 0,1
  4     subintervals
y( 1 ) = -1.665671
y( 2 ) =  5.698063
  8     subintervals
y( 1 ) =  .5284656
y( 2 ) =  .1096451
 16     subintervals
y( 1 ) =  .5285151
y( 2 ) =  .1095636
 32     subintervals
y( 1 ) =  .5285158
y( 2 ) =  .1095607
^C
Break in 5400
```

Comparing the two Runge-Kutta programs you can see that each statement which uses one of the variables Y, Y1, Y9, D, K1, K2, K3 or K4 is substituted by a FOR-NEXT-loop and the respective array variables. You can convince yourself easily that SYS-RKN is working better and faster than the program SYS-EUL. It obtains the same precision with 8 subintervals for which SYS-EUL needs 128 subintervals.

If you wish to use this program for another system of differential equations you simply have to change the dimension of the system in line 200 and program the right hand side of the new system of differential equations into the subroutine beginning in line 10000. Remember in this subroutine that you must use the variable X9 and the vector variables D and Y9.

Problem 133

Compare the programs SYS-RKN and SYS-EUL by solving the same problems with SYS-RKN, which follow the description of program SYS-EUL.

Problem 134

To make the program SYS-RKN more general, two improvements are proposed:

1. A criterion for terminating the program should be built in which stops the program as soon as an accuracy defined is obtained by the user. Intermediate results should only be printed on the screen if they are explicitely desired by the user.

2. The program should contain an option to print out 20 equidistant values of the solution function between x-start and x-end. Hint: begin with 20 subintervals by writing N1=20 in line 1000; additionally, define a result matrix E(20,20) in line 100, from which only E(N,20) is needed if the dimension of the system of differential equations is N. On the first run through the 20 subintervals, the calculated y-values are assigned to E. This may be done by an additional line 5760 with the following statements:

```
5760 FOR I=1 TO N : E(I,I9+1)=Y(I) : NEXT I
```

During the next loop only every second y-value is written into the result matrix, then every fourth and so on. This should be done in an additional program line after line 5740 with a conditional branching containing an IF-interrogation.

Problem 135

With program SYS-RKN differential equations of higher orders than 1 can also be solved. We take as a simple example the differential equation, whose solutions are sines or cosines

$$y'' = \frac{d^2y}{dx^2} = -y$$

If y is replaced by y_2 and $\frac{dy}{dx}$ by y_1, then the above differential equation of second order can be written as a system of ordinary differential equations of first order

$$\frac{dy_1}{dx} = -y_2$$

$$\frac{dy_2}{dx} = y_1 \qquad \text{(this is the definition of } y_1)$$

The system of DE's can now be solved by the program **SYS-RKN**. The two components of the solution vector are the y-value and the value for $\frac{dy}{dx}$.

Try to solve with this method the second order differential equation given above. Also calculate the following differential equation

$$\frac{d^3y}{dx^3} = -y$$

Use for example

$$\frac{dy_1}{dx} = y_2 \qquad \frac{dy_2}{dx} = y_3 \qquad \frac{dy_3}{dx} = -y_1$$

Compare the numerical results with the analytical solutions. Solve numerically the differential equation for forced oscillation (x is the displacement from the point of equilibrium).

$$m \cdot \frac{d^2x}{dt^2} + k \cdot \frac{dx}{dt} + D \cdot x = K \cdot \cos(a \cdot t)$$

Try to find out the frequency a, which has a maximum amplitude for x at given parameters m, k, D and K.

Problem 136

Consider a simple model for the spread of epidemics. The population density or population number is N. An increase of the population by birth is not assumed.

G is the number of individuals which are healthy, but may become infected
I is the number of infected individuals and
E is the number of individuals which are eliminated, which means they are dead or isolated or immune.

The chance of being infected is proportional to the number of contacts between individuals from G and I. The rate for the elimination of individuals of class I is proportional to the number of I itself. From this consideration the following system of differential equations can be derived

$$\frac{dG}{dt} = -a \cdot G \cdot I$$

$$\frac{dI}{dt} = a \cdot G \cdot I - b \cdot I$$

$$\frac{dE}{dt} = b \cdot I$$

Assume a certain initial value of the populations of G and I and values for the parameters a and b. Calculate from this the process (development course) of the epidemic. Try to verify the famous 'limit value theorem' for the spreading of an epidemic, which says that an infectious desease is not going to be an epidemic if, for the number of people which can be infected, the following relation holds

$$G < \frac{b}{a}$$

Problem 137

There are some clear examples of population dynamics concerning predators and prey (e.g. snow hare and snow fox, predator fish and prey fish). The number of predators (concentration) is expressed by R and the number of prey by B. In addition we assume that there is enough food for the prey. If there is no influence of the predators on the prey the number of prey grows with a rate proportional to their actual number: $\frac{dB}{dt} = k \cdot B$. If the existence of predators is considered, then the number of prey which are eaten is proportional to the number of contacts between these two species. From this the differential equation for the number or concentration of prey is obtained

$$\frac{dB}{dt} = k \cdot B - r \cdot B \cdot R$$

The predators have a natural death rate which is proportional to the number of predators $(-s \cdot R)$. The increase of the number of predators depends on their number and the amount of food available, which is the number of prey $(l \cdot B \cdot R)$. From these considerations the following differential equation results for the number of predators

$$\frac{dR}{dt} = -s \cdot R + l \cdot B \cdot R$$

Try to show that this system of differential equations can result in sustained oscillations for both populations if suitable parameters k, r, s and l are chosen.

If you apply this model for predator fish and prey fish, and if you complete this model by assuming that the number of both fish is reduced by the work of fishermen, then the following system of differential equations is obtained

$$\frac{dB}{dt} = k \cdot B - r \cdot B \cdot R - f \cdot B$$

$$\frac{dR}{dt} = -s \cdot B - l \cdot B \cdot R - f \cdot R$$

Try to show that the average number of predator fish is decreasing and of prey fish is increasing if the rate constant f of fishing is less then the growing rate constant *k* for the prey fish. If there is no fishing at all, the average number of prey fish, which is often the fish humans prefer to eat, is decreasing (Volterra's principle).

Problem 138

A chemical reaction A $\longrightarrow$ B occurs in a continuously stirred tank flow reactor with constant input and output. The kinetical order n of the reaction and the reaction enthalpy are known. From these assumptions the following system of differential equations is derived in which one equation describes the concentration of A with time and the other one the temperature T of the reaction vessel

$$\frac{d[A]}{dt} = -k_0 \cdot e^{-\frac{E_a}{RT}} \cdot [A]^n + ([A]_e - [A]) \cdot q$$

$$\frac{dT}{dt} = H \cdot k_0 \cdot e^{-\frac{E_a}{RT}} \cdot [A]^n + (T_f - T) \cdot u + w \cdot (T_w - T)$$

In both differential equations the first two terms are very similar; they refer to the chemical reaction. The second terms consider the feed and output flow of the reactor. The third term in the equation for the temperature describes the heat transfer through the wall of the reaction vessel.

Calculate the dynamics of this reaction concerning conversion and temperature. Vary the initial values of [A] and T with other fixed parameters and calculate the stationary state which is reached after a long time.

What are the conditions to obtain only one stationary state and for what conditions may different stationary states (multiple steady states) be obtained?

Problem 139

The reaction model OREGONATOR of Fields and Noyes[1] for oscillatory chemical reactions is given by the following scheme:

$$A + Y \xrightarrow{k_1} X$$

$$X + Y \xrightarrow{k_2} P$$

$$B + X \xrightarrow{k_3} 2X + Z$$

$$X + X \xrightarrow{k_4} Q$$

$$Z \xrightarrow{k_5} f \cdot Y \qquad 0.5 < f < 2.4$$

---

[1] R.J. Fields, R.M. Noyes, *Acc. Chem. Res.*, **10**, 221 (1977)

The substances A and B are consumed, and P and Q are produced. The concentration of the intermediates X, Y and Z may have an oscillatory behavior if certain conditions, which are given by the reaction parameters, are satisfied.

Write down the differential equations for all substances which are participating in the chemical reaction system. Make assumptions for the reaction parameters and calculate the reaction dynamics. Try to find values for the reaction parameters which result in an oscillatory behavior. (If difficulties arise with this problem consult the literature given below.)

Write down the differential equations for the reaction system above, when instead of a batch process the reaction takes place in a continuously stirred flow reactor. Compare the results of the calculations of both reactor types.

Problem 140

In order to simulate the concentration of a certain drug in a defined volume of liquid it is necessary to distinguish between the amount of drug not yet dissolved (F) and the dissolved part of the drug (L). In addition, the following assumptions are made:

The rate of consumption of the dissolved substance is proportional to the concentration of the dissolved substance itself. (The degradation law of first order can be substituted by a law of other orders if necessary.) The dissolution rate of the substance F is proportional to $[F]^a$. (If F is a drug in the form of a spherical pill then this assumption is reasonable because dissolving is then proportional to the surface of the pill and $a = \frac{2}{3}$.)

A further assumption should be a dissolving rate proportional to the difference $[L]_{max} - [L]$, where $[L]_{max}$ is the maximum concentration of the drug in the liquid if the liquid is saturated with the drug. The differential equations following from these assumptions read as follows

$$\frac{d[F]}{dt} = -k_1 \cdot [F]^a \cdot ([L]_{max} - [L])$$

$$\frac{d[L]}{dt} = +k_1 \cdot [F]^a \cdot ([L]_{max} - [L]) - k_a [L]^n$$

Calculate the concentration change with time of the dissolved drug, if the following values for the parameters are used:

$$a = \frac{2}{3}, \quad k_1 = 1, \quad [L]_{max} = 0.8, \quad n = 1, \quad k_a = 0.1$$

The initial value for $[F]$ is 1 and for $[L]$ is 0. If you are calculating the differential equation for a long time period then you may get negative concentrations for F (why?). Treat this case with an IF-statement and branch to program statements for assigning the physically reasonable value 0 to F.

Because it is important for practical purposes, difficulties which often arise from kinetical problems shall be mentioned here. Generally it is very simple to generate the system of differential equations from a reaction mechanism which is composed of elementary chemical reactions. If the reaction mechanism becomes very extensive then the setting up of the differential equations should preferably be done with a computer program (for details consult the section on non-numerical data processing).

The attempt to solve such differential equations with the usual numerical methods very often leads to 'impossible results', which are obviously erroneous, e.g. negative concentrations, oscillations with large amplitudes, or on the other hand, no changes at all in the concentrations.

The reason for such behavior is in most cases a difference in the reaction rate constants of many orders of magnitude. Such a system of differential equations is called 'stiff'. For the numerical calculation of stiff systems of differential equations it is necessary to use special computer programs which are usually called 'stiff solvers'. The best solvers now available for stiff differential equations, which are also very suitable for complex chemical reaction systems, were developed by Deuflhard [2,3] and Gear.[4,5] A very elementary introduction in the modeling and simulation of complicated chemical reaction systems can be found in the literature.[6]

In the last chapter of this book a program is given that uses the implicit Euler method to integrate systems of stiff differential equations.

---

[2] P. Deuflhard, G. Bader, U. Novak, *LARKIN - A Software Package for the Numerical Simulation of LARge Systems Arising in Chemical Reaction KINetics*, Proceedings of the Workshop on Modelling of Chemical Reaction Systems, *Springer Series in Chemical Physics*, **18**, 38 (1981). Editors: K.H. Ebert, P. Deuflhard, W. Jäger.

[3] P. Deuflhard, G. Bader, U. Novak, *An Advanced Simulation Package for Large Chemical Reaction Systems*, in Aiken (ed.): *Stiff Computation*, Oxford University Press (1983)

[4] C.W. Gear, *Numerical Initial Value Problems in Ordinary Differential Equations*, Prentice Hall, New Jersey (1971)

[5] C.W. Gear, L.F. Shampine, *SIAM Rev.*, **21**, 1 (1979)

[6] K.H. Ebert, H.J. Ederer, G. Isbarn; *Angew. Chem. (Int. Ed. Eng.)*, **19**, 333 (1980)

9.6 Boundary Value Problems

Up until now we have treated problems and tasks in this chapter using initial values $y_i(x_a)$ for the y_i's in the differential equations. If not all of the initial values are known, but other ones instead of these, then the problem is called a boundary value problem.

A mathematical example for this is the following differential equation

$$\frac{d^2y}{dx^2} = x - y$$

with the boundary conditions: $y(0) = 0$ and $y(1) = 2$.

For illustration an example from reaction kinetics is given consisting of the following reactions and reaction rate constants:

$$A \quad \xrightarrow{\quad k_1 = 0.1 \quad} \quad B$$

$$A \; + \; B \quad \xrightarrow{\quad k_2 = 0.1 \quad} \quad Products$$

The corresponding differential equations are of the following form

$$\frac{d[A]}{dt} = -0.1 \cdot [A] - 0.1 \cdot [A] \cdot [B]$$

$$\frac{d[B]}{dt} = +0.1 \cdot [A] - 0.1 \cdot [A] \cdot [B]$$

If all concentrations and reaction rate constants are considered as dimensionless, the question of the resulting boundary value problem may be formulated as: for time $t = 0$ the concentration of A should be 1 and at time $t = 1$ the concentration of A should be 0.75. How do the concentrations of A and B change with time if these two boundary conditions are fulfilled? The first boundary condition is the initial value of [A]. The initial concentration of B is the variable, which may be chosen freely to fulfil the second boundary condition.

If the above system of differential equations is solved with **SYS-RKN** using different initial conditions for [B], the following results are obtained:

	$[A]_{t=0}$	$[B]_{t=0}$	$[A]_{t=1}$	$[B]_{t=1}$
	1.0	0.0	0.90061	0.09064
	1.0	5.0	0.55774	4.70887

From these results it can be concluded that $[B]_{t=0}$, the initial value for $[B]$ which leads to the boundary condition $[A]_{t=1} = 0.75$, lies in the interval from 0 to 5.

	$[A]_{t=0}$	$[B]_{t=0}$	$[A]_{t=1}$	$[B]_{t=1}$
	1.0	2.5	0.70929	2.37839
	1.0	1.25	0.79941	1.22858

A search routine must start with a given initial value for $[B]$ and the ODE-system is then evaluated. The calculated $[A]$-value is compared with the boundary condition and a new and better initial value for $[B]$ is determined. Because this method corresponds to target shooting it is called a 'shooting method'.

This shooting method consists of a solver for the differential equation and a trial and error algorithm for finding a better initial value for $[B]$. But shooting can also be done using the bisection method, which is shown in the following program BOUNDARY.

```
0 REM "BOUNDARY"          EBERT/EDERER          880621
1 REM ***************************************************
2 REM *** A boundary value problem resulting from   ***
3 REM *** reaction kinetics is solved using the     ***
4 REM *** programs SYS-RKN and BISECT.              ***
9 REM ***************************************************
100   DIM Y(20),Y1(20),D(20)
120   DIM K1(20),K2(20),K3(20),K4(20)
200   N=2 : REM N is the dimension of the ODE-system
300   Y1(1)=1 : A=0 : E=1
400   YA=0 : YE=5
500   YS=.75 : REM This is the concentration Y(1)
510          REM at t=E
1000  N1=16
1100  H=(E-A)/N1
1200    Y1(2)=YA : GOSUB 5000 : F1=Y(1) : GOSUB 6000
1300    Y1(2)=YE : GOSUB 5000 : F9=Y(1) : GOSUB 6000
1350    IF SGN(F1-YS)=SGN(F9-YS) THEN GOTO 1700
1400    YM=(YA+YE)/2 : Y1(2)=YM : GOSUB 5000
```

```
1410    FM=Y(1) : GOSUB 6000
1420    IF SGN(F1-YS)<>SGN(FM-YS) THEN GOTO 1460
1440    YA=YM : F1=FM : GOTO 1400
1460    YE=YM : F2=FM : GOTO 1400
1600  GOTO 1100
1700  PRINT "Make a better choice for the interval"
1800  STOP
1900  END
2000  REM ###########################################
2010  REM ### Main program ends here.            ###
2020  REM ###########################################
5000  REM *******************************************
5002  REM *** Subroutine: ODE                     ***
5004  REM *** The solution of the ODE is approxi- ***
5006  REM *** mated by using N1 subintervals.     ***
5010  REM *******************************************
5100  FOR I=1 TO N : Y(I)=Y1(I) : NEXT I
5200  FOR I9=0 TO N1-1 : X=A+I9*H
5300    X9=X : FOR I=1 TO N : Y9(I)=Y(I) : NEXT I
5320    GOSUB 10000
5340    FOR I=1 TO N : K1(I)=H*D(I) : NEXT I: X9=X+H/2
5400    FOR I=1 TO N : Y9(I)=Y(I)+K1(I)/2 : NEXT I
5420    GOSUB 10000
5440    FOR I=1 TO N : K2(I)=H*D(I) : NEXT I: X9=X+H/2
5500    FOR I=1 TO N : Y9(I)=Y(I)+K2(I)/2 : NEXT I
5520    GOSUB 10000
5540    FOR I=1 TO N : K3(I)=H*D(I) : NEXT I : X9=X+H
5600    FOR I=1 TO N : Y9(I)=Y(I)+K3(I) : NEXT I
5620    GOSUB 10000
5640    FOR I=1 TO N : K4(I)=H*D(I) : NEXT I
5700    FOR I=1 TO N
5720      Y(I)=Y(I)+(K1(I)+2*K2(I)+2*K3(I)+K4(I))/6
5740    NEXT I
5800  NEXT I9 : RETURN
6000  REM *******************************************
6001  REM *** Subroutine: OUTPUT                   ***
6002  REM *** Output of intermediate results.     ***
6010  REM *******************************************
6100  PRINT "t =";A;":";TAB(20);"y(1) = ";Y1(1);
6150  PRINT            TAB(40);"y(2) = ";Y1(2)
6200  PRINT "t =";E;":";TAB(20);"y(1) = ";Y(1);
6250  PRINT            TAB(40);"y(2) = ";Y(2)
6300  PRINT : RETURN
```

```
10000 REM ********************************************
10002 REM *** Subroutine: RHS                     ***
10004 REM *** The right hand sides of the ODE's   ***
10006 REM *** are calculated here as              ***
10008 REM *** D(I) = f(X9,Y9(1),Y9(2),..Y9(N))    ***
10010 REM ********************************************
10100 D(1)=-.1*Y9(1)-.1*Y9(1)*Y9(2)
10200 D(2)=+.1*Y9(1)-.1*Y9(1)*Y9(2)
12000 RETURN
```

```
RUN
t = 0 :            y(1) =  1          y(2) =  0
t = 1 :            y(1) =  .9006101   y(2) =  9.064119E-02

t = 0 :            y(1) =  1          y(2) =  5
t = 1 :            y(1) =  .5577449   y(2) =  4.708875

t = 0 :            y(1) =  1          y(2) =  2.5
t = 1 :            y(1) =  .7092928   y(2) =  2.37839

t = 0 :            y(1) =  1          y(2) =  1.25
t = 1 :            y(1) =  .7994166   y(2) =  1.228578

t = 0 :            y(1) =  1          y(2) =  1.875
t = 1 :            y(1) =  .7530459   y(2) =  1.802079

t = 0 :            y(1) =  1          y(2) =  2.1875
t = 1 :            y(1) =  .7308511   y(2) =  2.089892

t = 0 :            y(1) =  1          y(2) =  2.03125
t = 1 :            y(1) =  .7418678   y(2) =  1.945899

t = 0 :            y(1) =  1          y(2) =  1.953125
t = 1 :            y(1) =  .7474364   y(2) =  1.873967

t = 0 :            y(1) =  1          y(2) =  1.914063
t = 1 :            y(1) =  .7502361   y(2) =  1.838018

t = 0 :            y(1) =  1          y(2) =  1.933594
t = 1 :            y(1) =  .748835    y(2) =  1.855991

t = 0 :            y(1) =  1          y(2) =  1.923828
t = 1 :            y(1) =  .7495352   y(2) =  1.847004

t = 0 :            y(1) =  1          y(2) =  1.918945
t = 1 :            y(1) =  .7498856   y(2) =  1.842511
```

```
t = 0 :          y(1) = 1          y(2) = 1.916504
t = 1 :          y(1) = .7500608   y(2) = 1.840264

t = 0 :          y(1) = 1          y(2) = 1.917725
t = 1 :          y(1) = .7499733   y(2) = 1.841388

t = 0 :          y(1) = 1          y(2) = 1.917114
t = 1 :          y(1) = .7500169   y(2) = 1.840826

t = 0 :          y(1) = 1          y(2) = 1.917419
t = 1 :          y(1) = .749995    y(2) = 1.841107

^C
Break in 5640
```

The BASIC-program BOUNDARY is an extension of the program SYS-RKN, but only the main program has been changed. In the subroutine starting with line 10000 the respective differential equations are written as program statements.

The initial value for [A] (variable Y1(1)) which is set to unity, the starting time (variable A), and the end time (variable E) are written directly into the program by the statements in line 300. In line 400 the upper (YE) and lower (YA) limits of the interval, within which the correct initial value for [B] (variable Y1(2)) has to be searched, are programmed. The interval has to be found by trial and error or by analyzing the problem. In line 500 the second boundary condition YS=0.75 is programmed; YS=0.75 is the value for Y(1) at time $t = 1$. The same trial and error technique, which is used for determining the initial interval for [B], is applied to determine the number of subintervals for the Runge-Kutta solver in order to reach a satisfying accuracy for the solutions of the differential equations. For the example in the program it can be shown that for N1=16 the results are exact to within seven decimal places. Therefore N1 is set to 16 in line 1000.

From line 1100 to 1600 the searching algorithm is programmed which is very similar to the well-known bisection method. In lines 1200 and 1300 the concentration of B for time $t = 1$ is calculated using the boundaries YA and YE for the initial values of [B]. The results are stored in the variables F1 and F2, respectively, and printed on the screen. These calculations are carried out by assigning the initial value to the variable Y1(2) and calling the Runge-Kutta subroutine by GOSUB 5000. GOSUB 6000 calls the subroutine for the output of the calculated shot. This output routine from line 6000 to 6300, which prints the results for [A] and [B] at both boundaries, is an expansion of the program SYS-RKN.

If the limits YA and YE in the main program are not chosen correctly, which means the unknown initial value for [B] (variable Y(2)) cannot be found between the two values, then both the differences of the calculated Y(1)-values with the given boundary values have the same sign. This is asked for in line 1350 and the program jumps to line 1700 if this condition is true; in this case a message is printed on the screen.

The center of the solution interval is calculated in line 1400. Then a new shot is done with GOSUB 5000. The shooting result is stored in variable FM and printed by GOSUB 6000.

In line 1420 it is asked whether the sign of the deviation of the calculated shots from the boundary value given (YS), which is 0.75 in our example, is changing in the interval (YA,YM) or the interval (YM,YE). Depending on the answer, YE=YM or YA=YM is the next statement; this means that the new interval, which includes the correct solution, has half the length of the previous interval (bisection method). The shooting result FM has to be stored in the variable F1 or F2 depending on which side of the new interval YM is situated. In line 1600 the program jumps back to line 1100 starting a new iteration cycle.

Because the searching routine is programmed as an infinite loop, the user stops the run when the result has an accuracy that he or she considers sufficient. Methods in which the program decides such a condition have been presented earlier in this book in many other programs and problems.

For our example, 10 iterations are needed to obtain a shot which is accurate to 4 decimal places. The initial value for [B], which leads to this result, is 1.916.

Problem 141
Improve the program BOUNDARY

- by including an automatic search for an appropriate starting interval from YA to YE
- by an automatic determination of the number of subintervals N1 for the Runge-Kutta subroutine and
- by programming a termination criterion which stops the program if a given accuracy is obtained.

Problem 142

For the reaction system above with the same differential equations, determine the initial concentrations at time $t = 0$ for A and B. Given are the boundary conditions: $[A] = 1.1$ at time $t = 2$ and $[A] = 0.7$ at time $t = 4$. This problem can be illustrated by the following table:

	$t = 0$	$t = 2$	$t = 4$
[A]	$?_1$	1.1	0.7
[B]	$?_2$	$?_3$	$?_4$

Hint: In using **BOUNDARY**, the concentration of B at times $t = 2$ and $t = 4$ should be calculated first by starting at $t = 2$ as the initial time. The value of $[A] = 0.7$ at time $t = 4$ is then considered as the other boundary condition. With this procedure you will obtain the values for $?_3$ and $?_4$.

  If you are using **SYS-RKN** with the initial values $[A] = 1.1$, and $[B] = ?_3$ (the value which has been calculated as the first part of the problem) you will get $?_1$ and $?_2$. The starting time then is $t = 2$ and the final time is $t = 0$. The time proceeds in the negative direction, which is done automatically if the input question is answered in the following way:

```
X-Value  Start,End ? 2,0
```

Thus you will obtain the concentrations of A and B for time $t = 0$.

Problem 143

Calculate the following boundary value problem for a differential equation of second order

$$\frac{d^2y}{dx^2} - \left(\frac{dy}{dx}\right)^2 + x^2 - y = 0$$

Remember: a differential equation of second order can be transformed into a system of ordinary differential equations of dimension two. This is obtained if the substitution $\frac{dy}{dx} = z$ is used

$$\frac{dy}{dx} = z \qquad\qquad \frac{dz}{dx} = z^2 - x^2 + y$$

Then the boundary value problem is the following: How is y changing with x between $x = 0$ and $x = 1$ if the boundary conditions are given by $y_{x=0} = 1$ and $y_{x=1} = 0.5$? To obtain these conditions only $\left(\frac{dy}{dx}\right)_{x=0}$ can be chosen independently; in the transformed system this means that $z_{x=0}$ should be used. For this problem we propose to start the search between -3 and 0. If the initial value of z is determined, it is easy to calculate, with the two initial values for y and z, the course of y between $x = 0$ and $x = 1$ using the program **SYS-RKN**.

Problem 144

A beam has both its ends fixed by a support. The ends should be defined as $x = 0$ and $x = 1$. The bending of the beam is described with the variable y. If an external force is acting in the middle of the beam and if the beam has elastic behavior then the bending of the beam can be described in a dimensionless form by the following differential equation.

$$\frac{d^2 y}{dx^2} = -a \cdot (1 - 2 \cdot |x - 0.5|) + b \cdot y$$

Assuming $a = 0.1$ and $b = 5.0$, calculate the bending of the beam between $x = 0$ and $x = 1$.

9.7   The Harmonic Oscillator – an Eigenvalue Problem

The time independent Schrödinger equation, which is restricted to one dimension, is not a partial but an ordinary differential equation of order two. If this equation is applied to a diatomic molecule one obtains

$$\frac{d^2 W}{dx^2} = -\left(E - V(x)\right) \cdot \frac{8\pi^2 \mu}{h^2} \cdot W$$

W :   Wave function (unknown)
E :   Total energy, the eigenvalue (unknown)
h :   Planck's constant
μ :   Reduced mass of the diatomic molecule
x :   Distance between the two atoms
$V(x)$:   Potential energy as a function of x

The problem is to find values for the total energy E (the eigenvalues of the energy) which yield 'reasonable' wave functions W by using the differential equations above. Because the square of the wave function W has the physical meaning of a probability of residence, a 'reasonable' physical condition is that the amplitude of the wave function approaches zero with increasing x-values.

To be able to solve this problem with the program SYS-RKN it is necessary to divide this differential equation of second order into a system of two first order differential equations. This is done again by using an additional variable U, which is the derivative of the wave function W. The

ODE-system obtained reads as follows

$$\frac{dW}{dx} = U$$

$$\frac{dU}{dx} = -\Big(E - V(x)\Big) \cdot W \cdot f$$

f is the combination of all the factors of the original equation. The assumption of an harmonic oscillator supplies the equation for $V(x)$

$$V(x) = \frac{k}{2} \cdot (x_g - x)^2$$

$$k = \mu \cdot (2\pi\nu)^2$$

ν : Oscillation frequency

k : Force constant

x_g : Equilibrium distance between the two atoms

The problem can be solved with the following program HARM, which contains the Runge-Kutta differential equation solver as a subroutine.

```
0 REM "HARM    "          EBERT/EDERER          880620
1 REM ************************************************
2 REM *** An eigenvalue problem using the Runge-   ***
3 REM *** Kutta-Nystroem method for solving the 1- ***
4 REM *** dimensional Schroedinger equation for the***
5 REM *** harmonic oscillator. The eigenvalues are ***
6 REM *** those energy values which result in      ***
7 REM *** reasonable wave functions.               ***
9 REM ************************************************
100    DIM Y(20),Y1(20),D(20)
120    DIM K1(20),K2(20),K3(20),K4(20)
123    NL=6.022E+23
125    DG=7.6064E-12 : REM for H2,
126    REM:          the depth of the potentials in erg
130    MY=.504/NL : REM reduced mass of H2
135    RG=7.412E-09 : REM equilibrium distance in cm
140    AA=(2* 3.1415)^2*1.3192E+14*MY*1.3192E+14/2
145    HH=6.626176E-27 : REM Planck's constant in erg/hz
195    DEF FNV(R)=AA*(R-RG)*(R-RG)
200    N=2 : REM N is the dimension of the ODE-system
300    PRINT "Wave function at equilibrium distance";
```

```
310    INPUT Y1(1)
400    PRINT "Derivation of the wave function at "
405    PRINT "equilibrium distance ";
410    INPUT Y1(2)
500    INPUT "Total energy in erg ";EG
510    REM Delta(EG) for H2 is about 1E-12
600    A=RG : REM starting value for x
650    E=RG*5 : REM end value for x
1000   N1=100
1100   H=(E-A)/N1
1200   GOSUB 5000
1600   STOP
2000   REM #############################################
2010   REM ### Main program ends here.              ###
2020   REM #############################################
5000   REM *******************************************
5002   REM *** Subroutine: ODE                     ***
5004   REM *** The solution of the ODE is approxi- ***
5006   REM *** mated by using N1 subintervals.     ***
5010   REM *******************************************
5100   FOR I=1 TO N : Y(I)=Y1(I) : NEXT I
5200   FOR I9=0 TO N1-1 : X=A+I9*H
5300     X9=X : FOR I=1 TO N : Y9(I)=Y(I) : NEXT I
5320     GOSUB 10000
5340     FOR I=1 TO N : K1(I)=H*D(I) : NEXT I: X9=X+H/2
5400     FOR I=1 TO N : Y9(I)=Y(I)+K1(I)/2 : NEXT I
5420     GOSUB 10000
5440     FOR I=1 TO N : K2(I)=H*D(I) : NEXT I: X9=X+H/2
5500     FOR I=1 TO N : Y9(I)=Y(I)+K2(I)/2 : NEXT I
5520     GOSUB 10000
5540     FOR I=1 TO N : K3(I)=H*D(I) : NEXT I : X9=X+H
5600     FOR I=1 TO N : Y9(I)=Y(I)+K3(I) : NEXT I
5620     GOSUB 10000
5640     FOR I=1 TO N : K4(I)=H*D(I) : NEXT I
5700     FOR I=1 TO N
5720       Y(I)=Y(I)+(K1(I)+2*K2(I)+2*K3(I)+K4(I))/6
5740     NEXT I
5770     GOSUB 6000 : REM intermediate RESULTS
5800   NEXT I9 : RETURN
6000   REM *******************************************
6002   REM *** Subroutine: RESULTS                 ***
6004   REM *** Output of the wave function         ***
6010   REM *******************************************
6200   PRINT "Psi(";X;") = ";Y(1),EG,Y(2)
6300   RETURN
```

```
10000 REM *********************************************
10002 REM *** Subroutine: RHS                      ***
10004 REM *** The right hand sides of the ODE's    ***
10006 REM *** are calculated here as               ***
10008 REM *** D(I) = f(X9,Y9(1),Y9(2),..Y9(N))     ***
10010 REM *********************************************
10100 D(1)=Y9(2)
10200 D(2)=-(EG-FNV(X9))*78.95201*MY/HH*Y9(1)/HH
12000 RETURN
63999 END
```

The two differential equations which represent the Schrödinger equation are programmed in the subroutine starting in line 10000. The variable EG is used for the total energy E. The potential energy $V(x)$ is calculated by the function call FNV(X9). The respective DEF FN-definition is programmed in line 195.

As an example, the hydrogen molecule is treated in this program. In lines 123 to 145 the physical constants are defined in the cgs-system. The specific constants for hydrogen — they must be changed if a different molecule is to be calculated — are: the reduced mass in line 130, the distance of equilibrium in line 135 and the oscillation frequency within the formula for the force constant in line 140.

Because the potential energy of the harmonic oscillator is symmetrical at the equilibrium distance, only two types of wave functions are possible:

1. symmetrical wave functions, which means $W'(x_g) = 0$ at the point of equilibrium x_g and
2. antisymmetrical wave functions, which results in $W(x_g) = 0$ for the equilibrium distance.

The calculation of the system of differential equations is begun at the equilibrium distance, therefore A=RG is programmed in line 600. In addition, the final value for x is chosen as the 5-fold equilibrium distance (E=RG*5) and the number of subintervals is fixed as N1=100 in line 1000.

To be able to calculate the ODE-system, the two initial values Y1(1) and Y1(2) are needed. The INPUT-statements in lines 300 and 400 ask for them from the user. The physical interpretation of these initial values is the wave function and its derivative at equilibrium. If you want to describe a symmetrical wave function, you should answer with 1 and 0. The reverse answer, 0 and 1, should be given for the calculation of an antisymmetrical wave function. We don't have to care about the normalization factors in this problem. In line 500 the program asks the user for the total energy EG. Then the subroutine 5000 calculates the wave function beginning from the equilibrium distance RG up to 5*RG in 100 steps. Line 5770 in the ODE-subroutine is added for the output of the intermediate results.

The determination of a 'correct' value for the total energy (the eigenvalue of the Schrödinger equation) is done by trial and error. It is useful to know that these energy values are expected to be about 10^{-12} erg.

An example for the search of the zero-point energy is shown in more detail. The program is started with initial values Y1(1)=1 and Y1(2)=0. If, in addition, it is started with a total energy of EG= $0.3 \cdot 10^{-12}$ the wave function is rapidly increasing towards positive infinity; if EG= $0.5 \cdot 10^{-12}$ is used, the result for the wave function diverges to negative infinity. The procedure for the determination of the zero-point energy is shown in Table 9.1.

Table 9.1. Trial and error runs with program **HARM** to bracket the total energy EG. The divergence characteristics ($+\infty$ or $-\infty$) lead to the value of the total energy for the next trial. In the last column of the table the distance is given where the overflow error occurs.

Y1(1)	Y1(2)	EG [10^{-12} erg]	Diverges to	Distance [10^{-8} cm]
1	0	0.300	$+\infty$	1.90
1	0	0.500	$-\infty$	1.90
1	0	0.400	$+\infty$	1.93
1	0	0.450	$-\infty$	1.93
1	0	0.430	$+\infty$	1.93
1	0	0.440	$-\infty$	1.96
1	0	0.436	$+\infty$	1.96
1	0	0.438	$-\infty$	1.96
1	0	0.437	$+\infty$	1.99

Table 9.1 shows that the zero-point energy is determined to an accuracy of three decimal places. If you need a more accurate result just continue this searching strategy. However, it is recommended to increase the calculation accuracy by increasing the number of subintervals in line 1000 for the ODE solver. With this program the first four eigenvalues were calculated (Table 9.2).

Table 9.2. The first four energy levels of the harmonic oscillator for the hydrogen
molecule, calculated with HARM

Type	Y1(1)	Y1(2)	Total Energy $[10^{-12}\text{erg}]$
sym	1	0	0.43708
anti	0	1	1.31140
sym	1	0	2.18620
anti	0	1	3.06176

From the results shown in Table 9.2 you can derive the well known fact that
the energy differences between two adjacent energy levels are constant and
that the zero-point energy is exactly half of this difference.

In addition to the eigenvalues, the program produces the course of the
the wave function starting from the equilibrium distance. Therefore it is
easy to generate a graph from these data. One should know, however, that
the divergence of the wave function is caused by small numerical errors.
Depending on small changes of the total energy, the wave function diverges
either to $-\infty$ or $+\infty$ after being very close to zero. In order to obtain correct
graphs of the wave functions, this divergence should be neglected and the
graph should smoothly approach the x-axis.

Problem 145

Determine additional energy eigenvalues using the program HARM. Calculate the
energy levels of the following diatomic molecules with the physical constants given
in the table.

Molecule	Frequency $[10^{13}\text{ s}^{-1}]$	Equiblibrium Distance [Å]	Reduced Mass [amu]
H_2	1.31920	0.7412	0.5039
HD	1.14290	0.7412	0.6717
D_2	0.93445	0.7412	1.00705
N_2	0.70693	1.0940	7.00154
F_2	0.27551	1.4090	9.49910

Problem 146

Modify the program HARM and introduce an automatic search for the eigenvalues.
We propose a method which is analogous to the bisection method.

Problem 147

Vary the function definition in line 195 of program **HARM** to obtain an anharmonic oscillator. Use the following Morse potential

$$V(x) = D_e \cdot \left(1 - e^{a \cdot (x - x_g)^2}\right)$$

D_e is the depth of the potential; the numeric value for hydrogen is $7.6064 \cdot 10^{-12}$ erg. The parameter a of the Morse potential can be calculated according to the following formula

$$a = \pi \cdot \sqrt{\frac{2\mu}{D_e}}$$

Because the symmetry is broken in the anharmonic case the wave functions are not exactly symmetric or antisymmetric.

Now you have not only to search for the total energy but also for the correct initial value for the derivation of the wave function, if you have chosen the initial value at the point of equilibrium arbitrarily. This means a substantial increase in calculation and searching effort. It is also recommended to calculate to a higher numerical precision. Use as the starting values for the anharmonic oscillator the eigenvalues of the harmonic oscillator.

9.8   Partial Differential Equations

As an example we will treat a thermal conductivity problem using the second Fourier law

$$\frac{\partial T}{\partial t} = c \cdot \frac{\partial^2 T}{\partial x^2}$$

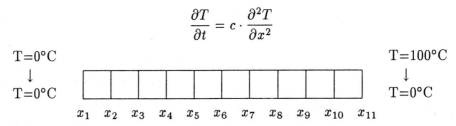

A rod is homogeneous and has a length of 100 cm. The initial temperature distribution is linear, increasing from the left end to the right end from 0 °C to 100 °C, according to the formula

$$T_{t=0}(x) = x$$

Starting from time $t = 0$, both ends of the rod are set to $T = 0$ °C.

If the length of the rod is divided into n equal subintervals, then the first partial derivative of T to x at the point x_i can be approximated by

$$\frac{\partial T}{\partial x}(x_i) \approx \frac{T(x_{i+1}) - T(x_{i-1})}{x_{i+1} - x_{i-1}} \stackrel{\text{def}}{=} T'(x_i)$$

The second partial derivative of T to x can be expressed by the following approximation

$$\frac{\partial^2 T}{\partial x^2}(x_i) \approx \frac{\partial T'}{\partial x}(x_i) \approx \frac{T'(x_{i+1}) - T'(x_{i-1})}{x_{i+1} - x_{i-1}}$$

If the approximation above for $T'(x_i)$ is inserted in the last equation and if the length of the interval $x_{i+1} - x_i$ is written as h, then the following formula is obtained

$$\frac{\partial^2 T}{\partial x^2}(x_i) \approx \frac{T(x_{i-2}) - 2T(x_i) + T(x_{i+2})}{4h^2}$$

If the interval is halved and the same substitutions are carried out, the final approximation formula is obtained ($2h$ is substituted by h)

$$\frac{\partial^2 T}{\partial x^2}(x_i) \approx \frac{T(x_{i-1}) - 2T(x_i) + T(x_{i+1})}{h^2}$$

If this approximation for the second derivative is used with the original second Fourier law then we obtain

$$\frac{\partial T}{\partial t}(x_i) \approx c \cdot \frac{T(x_{i-1}) - 2T(x_i) + T(x_{i+1})}{h^2}$$

As proposed in the scheme above, the rod is subdivided into 10 intervals leading to 11 points of interest which are considered separately. If, for the sake of briefness, $T(x_i)$ is written as T_i, $\frac{\partial T(x_i)}{\partial t}$ as $\frac{dT_i}{dt}$, and $\frac{c}{h^2}$ as b, and if we remember that the temperatures at the ends of the rod are defined as T_1 and T_{11}, which are kept constant at $0\,^\circ\text{C}$, the following equations are obtained

$$\frac{dT_1}{dt} = 0$$

$$\frac{dT_2}{dt} = b \cdot (T_1 - 2 \cdot T_2 + T_3)$$

$$\frac{dT_3}{dt} = b \cdot (T_2 - 2 \cdot T_3 + T_4)$$

$$\frac{dT_4}{dt} = b \cdot (T_3 - 2 \cdot T_4 + T_5)$$

$$\frac{dT_5}{dt} = b \cdot (T_4 - 2 \cdot T_5 + T_6)$$

$$\frac{dT_6}{dt} = b \cdot (T_5 - 2 \cdot T_6 + T_7)$$

$$\frac{dT_7}{dt} = b \cdot (T_6 - 2 \cdot T_7 + T_8)$$

$$\frac{dT_8}{dt} = b \cdot (T_7 - 2 \cdot T_8 + T_9)$$

$$\frac{dT_9}{dt} = b \cdot (T_8 - 2 \cdot T_9 + T_{10})$$

$$\frac{dT_{10}}{dt} = b \cdot (T_9 - 2 \cdot T_{10} + T_{11})$$

$$\frac{dT_{11}}{dt} = 0$$

So the partial differential equation has been approximated by a system of ordinary differential equations. If smaller intervals are used, the number of intervals and therefore the dimension of the system of differential equations increases. Of course the computational effort increases too, but the results become more precise.

The system of differential equations obtained can now be solved very easily using the routine SYS-RKN. This is done in the following program PDE:

```
0 REM "PDE    "         EBERT/EDERER        880624
1 REM ************************************************
2 REM *** Solving a partial differential equation ***
3 REM *** using the program SYS-RKN as a core.    ***
4 REM *** The example shown is a rod whose tempera-***
5 REM *** ture distribution is described by the   ***
6 REM *** second Fourier law.                     ***
9 REM ************************************************
100    DIM Y(20),Y1(20),D(20),Y9(20)
120    DIM K1(20),K2(20),K3(20),K4(20)
200    N=11 : REM N number of points of support
250    Y1(1)=0 : REM Temperature at the left end
```

```
260    Y1(N)=0 : REM Temperature at the right end
270    REM Temperature distribution at time=0
300    FOR I=2 TO N-1
400      Y1(I)=(I-1)*10
600    NEXT I
650    A=0
680    PRINT "At which time do you wish to calculate"
690    PRINT "the temperature distribution";
700    INPUT E
1000   N1=4
1100   H=(E-A)/N1
1200     GOSUB 5000 : PRINT
1210     PRINT "Temperature distribution at time = ";E
1220     PRINT "in degree C  with";N1;"time intervals "
1230     FOR I=1 TO N
1240       PRINT TAB(I*6-6);(I-1)*10;"cm";
1250     NEXT I : PRINT
1400     FOR I=1 TO N
1410       PRINT TAB(I*6-6);(INT(Y(I)*10+.5))/10;
1420     NEXT I : PRINT
1500     N1=N1*2
1600   GOTO 1100
2000   REM #############################################
2010   REM ### Main program ends here.             ###
2020   REM #############################################
5000   REM *********************************************
5002   REM *** Subroutine: ODE                      ***
5004   REM *** The solution of the ODE is approxi-  ***
5006   REM *** mated by using N1 subintervals.      ***
5010   REM *********************************************
5100   FOR I=1 TO N : Y(I)=Y1(I) : NEXT I
5200   FOR I9=0 TO N1-1 : X=A+I9*H
5300     X9=X : FOR I=1 TO N : Y9(I)=Y(I) : NEXT I
5320     GOSUB 10000
5340     FOR I=1 TO N : K1(I)=H*D(I) : NEXT I: X9=X+H/2
5400     FOR I=1 TO N : Y9(I)=Y(I)+K1(I)/2 : NEXT I
5420     GOSUB 10000
5440     FOR I=1 TO N : K2(I)=H*D(I) : NEXT I: X9=X+H/2
5500     FOR I=1 TO N : Y9(I)=Y(I)+K2(I)/2 : NEXT I
5520     GOSUB 10000
5540     FOR I=1 TO N : K3(I)=H*D(I) : NEXT I : X9=X+H
5600     FOR I=1 TO N : Y9(I)=Y(I)+K3(I) : NEXT I
5620     GOSUB 10000
5640     FOR I=1 TO N : K4(I)=H*D(I) : NEXT I
5700     FOR I=1 TO N
5720       Y(I)=Y(I)+(K1(I)+2*K2(I)+2*K3(I)+K4(I))/6
```

```
5740    NEXT I
5800  NEXT I9 : RETURN
10000 REM ******************************************
10002 REM *** Subroutine: RHS                   ***
10004 REM *** The right hand sides of the ODE's ***
10006 REM *** are calculated here as            ***
10008 REM *** D(I) = f(X9,Y9(1),Y9(2),..Y9(N))  ***
10010 REM ******************************************
10100 FOR D9=2 TO 10
10200   D(D9)=(Y9(D9+1)-2*Y9(D9)+Y9(D9-1))*.0023
10300 NEXT D9
10400 D(1)=0
10500 D(11)=0
12000 RETURN
63999 END


RUN
At which time do you wish to calculate
the temperature distribution? 1000
Temperature distribution at time =  1000
in degree C  with 4 time intervals
 0 cm 10 cm 20 cm 30 cm 40 cm 50 cm 60 cm 70 cm 80 cm 90 cm 100 cm
 0     10    19.9  29.8  39.2  47.6  53.1  53.5  44.8  26.3  0

Temperature distribution at time =  1000
in degree C  with 8 time intervals
 0 cm 10 cm 20 cm 30 cm 40 cm 50 cm 60 cm 70 cm 80 cm 90 cm 100 cm
 0     10    19.9  29.8  39.2  47.6  53.2  53.4  44.9  26.1  0
^C
Break in 10200
```

The system of equations is programmed in the subroutine starting in line 10000. The first and the last time derivative are set to zero as in the table above. The time derivatives from $\frac{dT_2}{dt}$ until $\frac{dT_{10}}{dt}$ are written as a loop. For problems of this kind, this can usually be done in this way because the 9 differential equations have a very similar form — only the indices are shifted.

The dimension 11 is fixed in line 200, the temperature for both ends of the rod is set to 0 °C in lines 250 and 260. The remaining points of the initial distribution at time $t = 0$ are calculated in the loop from line 300 to 600. The initial time is set to zero in line 650. The final time E, for which the temperature distribution shall be calculated, is required from the user by the INPUT-statement in line 700.

With the exception of the output, the rest of the program is identical with program SYS-RKN. The output has been modified to obtain a printout which is adapted to the physical problem. In the example the temperature distribution for time $t = 1000$ is calculated.

Problem 148

Modify program PDE to calculate the above problem with a subdivision of 20 intervals. Compare the results with that of 10 intervals. The following changes have to be made:

- the dimension of the system of equations
- the programming of the formula for the initial distribution
- the output
- the end value for the loop in the subroutine for the calculation of the right hand sides of the differential equations
- the factor b in the same subroutine, which is a combination of the physical constants and the interval length h (see formula above)

Problem 149

For treating diffusion problems a completely analogous equation can be used which is called Fick's law. If only one dimension in space is considered then this equation has the following form

$$\frac{\partial c}{\partial t} = a \cdot \frac{\partial^2 c}{\partial x^2}$$

Let us try to solve the following problem: There are two parallel glass tubes which are connected vertically by another tube resulting in an arrangement of the form of an H.

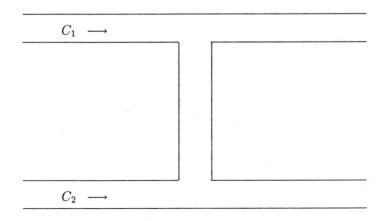

A chemical substance having different concentrations c_1 and c_2, respectively, flows through the two parallel tubes. At time $t = 0$ the concentration in the connecting tube is assumed to be zero everywhere. Determine the concentration distribution in the connecting tube at given times.

In a mathematical sense this problem is analogous to the example which was solved with the program PDE. If the time, the length of the vertical tube, and the concentrations are converted into dimensionless quantities, then it is easy to choose a numerical value for the factor a to obtain correct solutions for the concentrations at different times.

What is the stationary concentration distribution? To answer this question do calculations for increasing times until the results no longer change.

Problem 150

Given is a tube whose diameter is small compared with its length. The tube is closed at both ends and has a tap in the center. The left half of the tube is filled with gas A and the right half with gas B. The pressure of the gases should be equal in both parts of the tube. At time $t = 0$ the tap of the tube is opened. How does the concentration distribution develop along the tube with time? It is assumed that there is no cross-sectional concentration gradient.

Because the pressures in both parts of the tube are equal and remain so also after opening of the valve, it is sufficient to calculate one concentration (e.g. [A]) only. The partial pressure of B can then easily be determined via the total pressure.

Again, the tube is subdivided into n equally-spaced intervals, and then the differential equations for the different inner points are similar to those in the problem above. Only the differential equations for the first point x_1 and the last one x_{n+1} are different, because the concentrations at these points also vary with time. The differential equation system for $n = 10$ is the following if the concentration of A at the point x_i is written as c_i and $\frac{a}{h^2}$ as b_i

$$\frac{dc_1}{dt} = b \cdot (c_2 - c_1)$$

$$\frac{dc_2}{dt} = b \cdot (c_1 - 2 \cdot c_2 + c_3)$$

$$\frac{dc_3}{dt} = b \cdot (c_2 - 2 \cdot c_3 + c_4)$$

$$\frac{dc_4}{dt} = b \cdot (c_3 - 2 \cdot c_4 + c_5)$$

$$\frac{dc_5}{dt} = b \cdot (c_4 - 2 \cdot c_5 + c_6)$$

$$\frac{dc_6}{dt} = b \cdot (c_5 - 2 \cdot c_6 + c_7)$$

$$\frac{dc_7}{dt} = b \cdot (c_6 - 2 \cdot c_7 + c_8)$$

$$\frac{dc_8}{dt} = b \cdot (c_7 - 2 \cdot c_8 + c_9)$$

$$\frac{dc_9}{dt} = b \cdot (c_8 - 2 \cdot c_9 + c_{10})$$

$$\frac{dc_{10}}{dt} = b \cdot (c_9 - 2 \cdot c_{10} + c_{11})$$

$$\frac{dc_{11}}{dt} = b \cdot (c_{10} - c_{11})$$

The form of the differential equations for c_1 and c_{11} can be derived from Fick's law for the mass or concentration flow in diffusion. At x_1 and x_{11} there is diffusion only in one direction. The appropriate equations can be more easily derived from the principle of mass conservation. The total mass in the tube is constant; therefore the sum over all mass flows within the tube has to be zero

$$\frac{dc_1}{dt} + \frac{dc_2}{dt} + \frac{dc_3}{dt} + \frac{dc_4}{dt} + \frac{dc_5}{dt} + \frac{dc_6}{dt} + \frac{dc_7}{dt} + \frac{dc_8}{dt} + \frac{dc_9}{dt} + \frac{dc_{10}}{dt} + \frac{dc_{11}}{dt} = 0$$

Problem 151

A very long reaction tube is filled with a pure solvent. At both ends of this vessel there is — starting with time $t = 0$ — the solid substance A. A is assumed to have a low solubility in the solvent. Therefore it can be assumed that at both ends there is a saturated solution c_s at any time.

The concentration distribution along the tube can be calculated directly with program PDE. However, it is also assumed that the substance A reacts to product B with reaction order n, which makes this problem more complex. Now we do not have a pure diffusion problem, but a reaction diffusion problem. The modified partial differential equation has the following form

$$\frac{\partial c}{\partial t} = a \cdot \frac{\partial^2 c}{\partial x^2} - k \cdot c^n$$

This means that Fick's law has an additional reaction term.

Calculate the concentration distribution at different but fixed times considering the influence of the chemical reaction. Modify the system of differential equations (in our example there are 11) by adding the reaction term in the way shown for the fifth equation

$$\frac{dc_5}{dt} = b \cdot (c_4 - 2 \cdot c_5 + c_6) - k \cdot c_5^n$$

Problem 152

Let us consider now another reaction diffusion problem which is taken from the chemistry of the atmosphere under the influence of emission. This very simple model assumes that the two chemical substances A and B are emitted in a way so that they always have constant concentrations at the surface of the earth. The concentrations of both substances shall become zero at a certain height h_e above the surface. (This, for example, may be achieved by radiation.) Until now, this problem is only a diffusion problem with two different substances and therefore the dimension of the resulting ODE-system is doubled. But in addition a chemical reaction between A and B occurs which is described by the simple kinetics $k \cdot [A] \cdot [B]$.

What is the vertical concentration distribution of the two emitted substances? (This model assumes that there is no horizontal concentration gradient.) As an example the differential equations for the fifth point of subdivision are shown

$$\frac{d[A_5]}{dt} = b \cdot ([A_4] - 2 \cdot [A_5] + [A_6]) - k \cdot [A_5] \cdot [B_5]$$

$$\frac{d[B_5]}{dt} = b \cdot ([B_4] - 2 \cdot [B_5] + [B_6]) - k \cdot [A_5] \cdot [B_5]$$

The other differential equations have a similar form.

Problem 153

Radioactive waste which is homogeneously melted into glass pellets uniformly produces heat. The initial temperature distribution in the spheres at the time of production is $600\,°C$ (melting temperature of glass). After this the pellets are stored in a temperature bath having a temperature of $40\,°C$. This means that the temperature at the surface of a sphere is $40\,°C$ at all times.

In this case Fourier's law has to be extended by a heat production term. Because of the symmetry of a sphere, this three-dimensional problem in space can be transformed into a one-dimensional problem if the radial distribution is considered. To do this the radius is subdivided into n parts and for each of the shells an ordinary differential equation is established.

Calculate the radial temperature distribution at different times. What is the stationary temperature distribution after a very long time?

9.9 Stationary Solutions of Partial Differential Equations

Let us consider a homogeneous square metallic sheet, the four different sides of which are held at a constant, but for each side different, temperature. The sheet is assumed to be insulated in order to restrict the heat flow to the four sides.

The physical equation for such a problem is the second Fourier law in two spatial dimensions

$$\frac{\partial T(t, x, y)}{\partial t} = c \cdot \frac{\partial^2 T(t, x, y)}{\partial x^2} + c \cdot \frac{\partial^2 T(t, x, y)}{\partial y^2} \tag{1}$$

The temperature distribution is called stationary if there is no further change with time. This implies that the derivative of temperature with time becomes zero and the temperature $T(x, y)$ is no longer time dependent.

$$0 = c \cdot \frac{\partial^2 T(x, y)}{\partial x^2} + c \cdot \frac{\partial^2 T(x, y)}{\partial y^2} \tag{2}$$

If each side of the sheet is subdivided into $(n - 1)$ subintervals of the length h — in Fig. 9.2 they are divided into three intervals — the second derivative can again be approximated by the quotient of differences

$$\frac{\partial^2 T(x, y)}{\partial x^2} = \frac{T(x + h, y) - 2T(x, y) + T(x - h, y)}{h^2} \tag{3}$$

$$\frac{\partial^2 T(x, y)}{\partial y^2} = \frac{T(x, y + h) - 2T(x, y) + T(x, y - h)}{h^2} \tag{4}$$

With these approximations one obtains from equation (2)

$$T(x + h, y) + T(x - h, y) - 4T(x, y) + T(x, y + h) + T(x, y - h) = 0 \tag{5}$$

If Equation (5) is rewritten using indices, one obtains

$$T(x_{i+1}, y_j) + T(x_{i-1}, y_j) - 4T(x_i, y_j) + T(x_i, y_{j+1}) + T(x_i, y_{j-1}) = 0 \tag{6}$$

The equation applies to all inner points of the above square. For the points at the border, the temperature is fixed by the appropriate conditions for

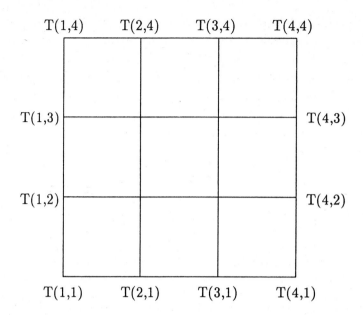

Fig. 9.2. Subdivision of a sheet to calculate the skipping temperature distribution

this side. The corner points are assumed to have the arithmetic averages of the temperatures of the adjacent sides.

If $(n - 1)$ intervals are constructed for each side then there are $(n - 2)^2$ linear difference equations to be solved.

The calculation should not be done with the algorithm of Gauss-Jordan. Because of the large size of the linear system and because of its special structure — it is diagonal dominant — an iterative procedure is faster and more efficient. In the literature on numerical analysis the method used here is called the Gauss-Seidel iteration. The method will be used without any further proof. It is most suitable if the matrix of the coefficients of the equation system is diagonal dominant. Equation (5) is transformed so that only $T(x, y)$ appears at the left hand side

$$T(x, y) = \frac{T(x + h, y) + T(x - h, y) + T(x, y + h) + T(x, y - h)}{4} \qquad (7)$$

Rewriting this equation using indices one obtains

$$T(x_i, y_j)_{\text{new}} = \frac{T(x_{i+1}, y_j) + T(x_{i-1}, y_j) + T(x_i, y_{j+1}) + T(x_i, y_{j-1})}{4} \qquad (8)$$

An arbitrary initial temperature distribution for the different points in the square is assumed, for example the average value of the temperatures of the four sides. Then for all i and j new $T(x_i, y_j)$ values are calculated according to the iteration formula (8). If the distribution was already the correct stationary one, then the new values are the same as the previous ones. However, if the new values are different, a new iteration cycle should be calculated using the new values. The iteration procedure is carried out until the differences for successive iterations become very small. This method is applied in the program STAT which is printed below.

```
0 REM "STAT    "          EBERT/EDERER        880624
1 REM ************************************************
2 REM *** Calculation of the stationary state of  ***
3 REM *** the 2nd Fourier law in 2 dimensions.    ***
4 REM *** Example: given is a square whose 4 sides ***
5 REM *** may be at different temperatures.       ***
9 REM ************************************************
100   DIM T(30,30),T0(30,30),S(4)
200   PRINT "Into how many subintervals shall a side"
250   PRINT "be subdivided "; : INPUT N
300   N=N+1
400   FOR I=1 TO 4
500     PRINT "Temperature of the";I;"'th side ";
510     INPUT S(I)
600   NEXT I
650   T5=(S(1)+S(2)+S(3)+S(4))/4
700   FOR I=2 TO (N-1) : T(I,1)=S(1) : NEXT I
720   FOR I=2 TO (N-1) : T(N,I)=S(2) : NEXT I
730   FOR I=2 TO (N-1) : T(I,N)=S(3) : NEXT I
740   FOR I=2 TO (N-1) : T(1,I)=S(4) : NEXT I
800   T(1,1)=(S(1)+S(4))/2
810   T(N,1)=(S(1)+S(2))/2
820   T(N,N)=(S(2)+S(3))/2
830   T(1,N)=(S(3)+S(4))/2
900   FOR I=2 TO (N-1) : FOR J=2 TO (N-1)
910       T(I,J)=T5 : NEXT J : NEXT I : GOSUB 10000
1000  REM ************************************************
1010  REM *** Iteration procedure for solving the  ***
1020  REM *** system of linear equations.          ***
1030  REM ************************************************
1050  E9=.1 : REM Accuracy
1060  E5=0
```

```
1100   FOR I=2 TO N-1
1200     FOR J=2 TO N-1
1300       T(I,J)=T0(I+1,J)+T0(I-1,J)+T0(I,J+1)
1310       T(I,J)=T(I,J)+T0(I,J-1)
1320       T(I,J)=T(I,J)*.25
1400       IF ABS(T(I,J)-T0(I,J))>E9 THEN E5=1
1600     NEXT J
1700   NEXT I
1750   REM GOSUB 5000 : REM intermediate results
1800   IF E5=0 THEN GOTO 2000 : REM output
1900   GOSUB 10000 : GOTO 1000
2000   GOSUB 5000 : END
2200   REM #########################################
2210   REM ### Main program ends here.          ###
2220   REM #########################################
5000   REM *****************************************
5002   REM *** Subroutine: OUTPUT                ***
5004   REM *** The temperature distribution is   ***
5006   REM *** written on the screen.            ***
5010   REM *****************************************
5100   PRINT
5200   FOR J=N TO 1 STEP -1
5300     PRINT
5400     FOR I=1 TO N
5500       PRINT TAB(2+(I-1)*5);INT(T(I,J)+.5);
5600   NEXT I : PRINT : NEXT J
5700   RETURN
10000  REM *****************************************
10002  REM *** Subroutine: MATRIX ID             ***
10003  REM *** Matrix T0 = Matrix T              ***
10010  REM *****************************************
10100  FOR I=1 TO N : FOR J=1 TO N
10200      T0(I,J)=T(I,J)
10300  NEXT J : NEXT I
12000  RETURN

RUN
Into how many subintervals shall a side
be subdivided ? 4
Temperature of the 1 'th side ? 0
Temperature of the 2 'th side ? 100
Temperature of the 3 'th side ? 0
Temperature of the 4 'th side ? 0
```

```
0     0     0     0     50

0     7     19    43    100

0     10    25    53    100

0     7     19    43    100

0     0     0     0     50
Ok

RUN
Into how many subintervals shall a side
be subdivided ? 12
Temperature of the 1 'th side ? 0
Temperature of the 2 'th side ? 100
Temperature of the 3 'th side ? 200
Temperature of the 4 'th side ? 300
```

```
250  200  200  200  200  200  200  200  200  200  200  200  150

300  247  224  212  204  198  194  190  185  180  170  150  100

300  263  237  219  206  196  187  180  172  163  150  130  100

300  269  243  222  205  191  180  170  161  150  137  120  100

300  270  243  220  201  185  172  161  150  139  128  115  100

300  269  241  215  194  177  162  150  139  130  120  110  100

300  266  235  207  184  165  150  138  128  120  113  106  100

300  261  225  194  170  150  135  123  115  109  104  102  100

300  252  210  176  150  130  116  106  99   95   94   96   100

300  237  187  150  124  106  93   85   80   78   81   88   100

300  209  150  113  90   75   65   59   57   57   63   76   100

300  150  91   63   48   39   34   31   30   31   37   53   100

150  0    0    0    0    0    0    0    0    0    0    0    50
Ok
```

In line 100 of program STAT two matrices are defined, each of them containing a temperature distribution; field T contains the new, and field TO the previous distribution. The number of intervals N is asked from the user by the PRINT- and INPUT-statements in line 200 and 250 of the program. In the next line N is increased by 1 representing the number of subdivision points for one side of the rectangle.

In the loop from line 400 to 600 the user has to enter in the temperatures of the four sides. In line 650 the average of these temperatures is calculated and assigned to T5.

In the four loops in lines 700, 720, 730 and 740 the boundary points are assigned to the temperatures of the respective sides. In the following four lines the temperatures of the corner points are calculated and stored in the variables T(1,1), T(N,1), T(1,N) and T(N,N). The double loop of line 900 sets the initial temperature for the inner points to the temperature average T5. The call of the subroutine GOSUB 10000 in the same line causes all elements of the matrix TO to be identical to the elements of the actual matrix T. With this statement the initial phase is finished and the iteration cycle can begin.

In line 1050 the required accuracy for the calculation of the temperature is fixed. In our case it is E9 = 0.1. In line 1060 a pointer variable E5 is set to zero, which indicates in a later IF-statement whether the desired accuracy is already attained.

The two nested loops from line 1100 to 1700 contain the iteration process. Here elements T(I,J) of the matrix are calculated from the appropriate elements of the matrix TO using the formula (8) above. In the same loop all elements T(I,J) are compared with TO(I,J). If two of these elements have a difference greater than the required accuracy E9, then the pointer variable E5 is set to 1. This means the variable E5 leaves only the loops with the value 0, if all elements of the matrices T and TO differ less than the accuracy E9. Line 1750 is an auxiliary line. If the REM at the beginning of the line is deleted, then intermediate results are printed on the screen. In line 1800 the pointer variable E5 is inquired. If E5 = 0 then the program jumps to line 2000. There the output routine is called and the program is terminated. If the accuracy isn't satisfactory the program jumps to line 1900. There the subroutine 10000 is called, which overwrites the old temperature distribution in TO with the new, better one in T. Afterwards the program jumps back to line 1000 and a new iteration is started.

The output subroutine from line 5000 to 5700 prints the temperature distribution in the form of a rectangle. This is realized by using the tabulator function. To get clearly arranged and short values, the numbers are rounded to integers.

Two different temperature distributions are shown in the examples after the program STAT.

Problem 154

Write a program similar to the program STAT to calculate the stationary temperature distribution of a cube. Restrict the number of subintervals to less than 10 for each side.

Try to make the program more general in order to calculate temperature distributions in rectangles and boxes.

Problem 155

Calculate the temperature distribution in a rectangular catalytic pellet. The temperature of the 6 surfaces is given. (Usually they are the same for every surface.) A chemical reaction within the catalytic pellet shall homogeneously and constantly produce heat.

To take this into consideration, a constant summand, which represents the heat production term, must be added to the iteration formula. In the hypothetical two-dimensional case you only have to change line 1300 of the program STAT:

```
1300 T(I,J)=(TO(I+1,J) + TO(I-1,J) + TO(I,J+1) + TO(I,J-1))*.25 + 10
```

10 Interpolation

Sometimes a great effort is required to calculate function values, for example if the y-value must be determined by a search routine, or if y is the solution of a differential equation. In such cases very often only a limited number of (x, y)-values are available. If, for additional calculations more (x, y)-points are needed, then approximate values between the original calculated points are used. This might be the case for plotting the graph of a certain function with a computer. Then frequently not only a few additional points are necessary but a rather fine grid for the plotting points is needed.

If the points are the results of measurements, the same arguments can be used. Each value which has to be measured costs time and money. Therefore the experimenter wishes to restrict the measurements to the number of points, which are necessary to give an impression of the course of the measurement. Again, for further calculations or for plotting procedures, additional interpolated points are required.

Interpolation routines are the methods for solving such problems. In this chapter we will present two procedures: the first one is the method of Lagrange, which is easier to understand, and the second one is the more frequently used Spline method. Spline originates from the methods used earlier by English ship builders in manufacturing boats.

10.1 The Interpolation Routine of Lagrange

The principle of this method is the idea of constructing a polynomial, which passes through n given points (x_i, y_i). If all the x_i-values are different, then a mathematical law states that this can be done by a polynomial of degree $(n - 1)$.

In this case, the polynomial due to the Lagrange formula can be written down directly

$$P(x) = \sum_{k=1}^{n} \left(\prod_{\substack{i=1 \\ i \neq k}}^{n} \frac{x - x_i}{x_k - x_i} \right) \cdot y_k$$

The proof of this rather complicated formula is relatively simple. The formula is an intricate presentation of a polynomial of degree $(n-1)$. This can be recognized best at the product sign, which runs over $(n-1)$ indices. If x in the polynomial $P(x)$ is substituted by x_m, then a sum is obtained which includes n products of the following form

$$\prod_{\substack{i=1 \\ i \neq k}}^{n} \frac{x_m - x_i}{x_k - x_i} = \begin{cases} 1 & \text{, if } m = k \\ 0 & \text{, if } m \neq k \end{cases}$$

The factors of the product include all i which are unequal to k. In the case $m \neq k$, the product is zero, because it contains the factor $(x_m - x_m)$, which is zero. In forming the product, only the factor with the index $i = k$ is omitted. Therefore if $m = k$, all factors of the respective product are of the form

$$\frac{x_m - x_i}{x_m - x_i} = 1$$

Therefore each factor and the whole product is equal to 1. $P(x_m)$ can be written

$$P(x_m) = 0 \cdot y_1 + 0 \cdot y_2 + \cdots + 1 \cdot y_m + 0 \cdot y_{m+1} + \cdots + 0 \cdot y_n = y_m$$

With this calculation it is shown that the polynomial has the result y_m if it is evaluated at $x = x_m$, and the validity of the Lagrange formula is proven. It remains to program this formula, which is performed in the following program LAGRANGE.

```
0 REM "LAGRANGE"          EBERT/EDERER            880711
1 REM ************************************************
2 REM *** Interpolation through n data points by   ***
3 REM *** Lagrange's method.                        ***
9 REM ************************************************
```

```
1000   DIM X(70),Y(70)
1100   INPUT "Number of data points ";N
1200   FOR I=1 TO N
1300     PRINT "x(";I;") , y(";I;")  ";
1400     INPUT X(I),Y(I)
1500   NEXT I
2000   PRINT
2020     PRINT "At which x-value shall an interpolation "
2040     PRINT "point be calculated ";
2100     INPUT X
5000     Y=0
6000     FOR K=1 TO N : P=1
6400       FOR I=1 TO N
6600         IF I=K THEN 7000
6800         P=P*(X-X(I))/(X(K)-X(I))
7000       NEXT I
7200       Y = Y+P*Y(K)
7400     NEXT K
8000     PRINT " x  = ";X
8100     PRINT " y  = ";Y
9000   GOTO 2000
9999   END


RUN
Number of data points ? 5
x( 1 ) , y( 1 )  ? 0,0
x( 2 ) , y( 2 )  ? 1,1
x( 3 ) , y( 3 )  ? 2,4
x( 4 ) , y( 4 )  ? 3,9
x( 5 ) , y( 5 )  ? 4,16
At which x-value shall an interpolation
point be calculated ? 2.5
 x  =  2.5
 y  =  6.25


At which x-value shall an interpolation
point be calculated ? 0.5
 x  =  .5
 y  =  .25


At which x-value shall an interpolation
point be calculated ? 1.1
 x  =  1.1
 y  =  1.21
Break in 2100
```

In line number 1000 a vector for the x-values and the y-values must be defined. With an INPUT-statement in the next line the number of pairs of values is requested. The input of the values occurs in the FOR-NEXT-loop from line 1200 to 1500. The x-value, for which an interpolated value for y is required is read in with the INPUT-statement in line number 2100.

From line number 5000 to 7400 the calculation of the Lagrange formula above is performed. The summing loop — the K-loop — includes lines 6000 to 7400. Within this loop, the loop for calculating the product, the I-loop, is embedded. Each time before the I-loop is entered, the variable P, which contains the product, is set to unity. In line number 6800 the actually calculated product is multiplied with the next factor and stored again in the variable P. In the Lagrange formula the factor is skipped if $i = k$; this is done in the program by an IF-statement in line 6600.

The variable Y for the sum is set to zero in line 5000 before the external summation loop is entered. The products calculated in the inner loop are added together in line number 7200 and stored in variable Y. In lines 8000 and 8100 the x-value and the respective interpolated y-value are printed on the screen.

In the example shown five pairs of figures are given that exactly follow a parabola. It can also be seen that the interpolated values fit exactly to the parabola.

Problem 156

Given are five pairs of values, with x-values between 0 and 2, which correspond exactly with an exponential function. Calculate interpolation values and compare them with those ones lying exactly on the exponential function.

Another similar problem is the interpolation of pairs of values which exactly fit a logarithmic function.

Problem 157

Try to interpolate assumed experimental values. To simulate this, you can calculate values which are given by an arbitrary function and then scatter the values by using a random number generator, e.g.:

```
Y(I) = X(I)*EXP(-X(I)*X(I))*(0.95 + 0.1*RND)
```

Try this procedure first with a little scattering and then increase the scattering.

Problem 158

If there are a lot of points and you still have to make interpolations, it doesn't make sense to construct the Lagrange polynomial for all points. It is usually sufficient to take only a few (e.g. 4) neighboring points for calculating an interpolated point.

Write an interpolation program which takes only a certain number of neighboring points for the interpolation routine. Naturally it is advisable to use points which surround the unknown point in a way that the latter one lies between them. Only for endpoints you have to be content with data points lying asymmetrically or completely on only one side of the point for which an interpolation is to be calculated.

The main task in treating this problem is to search for the appropriate neighboring points. The Lagrange interpolation is done the same way as in the printed program.

Often four points are enough for practical reasons, but for special problems it can be advantageous to consider more neighboring points.

Construct the new program as a structured program composed of blocks of subroutines.

Problem 159

If you wish to use the former program for experimental data which are more or less scattered, you can treat the data with a smoothing routine before entering the interpolation program. The degree of smoothing should be adjusted to the data because there are data which don't need to be smoothed and others which must be smoothed considerably. This can be realized by using a different number of neighboring points for smoothing, but it can also be done by smoothing the smoothed data again and repeating this several times.

Problem 160

One of the most unpleasant jobs in treating experimental data is the differentiation of measured and scattered data because very small variations in the data can lead to very high differences in the numerical derivatives. Therefore, very often the following procedure is applied: first the data are smoothed, then through these smoothed values an interpolation polynomial is drawn and then this polynomial is differentiated.

Write a program for differentiating experimental data. Apply the interpolation of Lagrange, then calculate two closely neighbored points and calculate the difference quotient.

Test this new program with simulated 'experimental' data, i.e.: calculate exact values with a mathematical function and scatter them by means of a random number generator.

10.2   Interpolation with Spline Functions

The spline interpolation is appropriately suited for getting a smooth interpolation curve. Therefore it is frequently used for graphical outputs. Again
n experimental points or calculated points are given; the spline interpolation puts a polynomial of order three through two neighboring — eventually
smoothed — points. It is obvious that a polynomial of order three is not
uniquely determined by only two points, because two points determine only
a polynomial of order one, which is a straight line.

This condition can be written as a formula for the ith polynomial $p_i(x)$,
which shall be applied between the points x_i and x_{i+1}

$$p_i(x_i) = y_i \tag{1}$$

$$p_i(x_{i+1}) = y_{i+1} \tag{2}$$

The transition from the polynomial p_{i-1}, which is used between the points
(x_{i-1}, y_{i-1}) and (x_i, y_i), to the polynomial p_i, which connects the points
(x_i, y_i) and (x_{i+1}, y_{i+1}), has to be smooth.

This condition is attained by the spline function considering the following mathematical conditions: Each two neighboring polynomials $p_{i-1}(x)$
and $p_i(x)$ must have the same value for the first and the second derivative
in their common point (x_i, y_i). This is expressed in formulas

$$\frac{dp_{i-1}}{dx}(x_i) = \frac{dp_i}{dx}(x_i) \tag{3}$$

$$\frac{d^2 p_{i-1}}{dx^2}(x_i) = \frac{d^2 p_i}{dx^2}(x_i) \tag{4}$$

For all inner polynomials we have now four conditions which are necessary
for determining polynomials of order three. These conditions are coupled
and therefore a system of coupled equations is obtained.

Both polynomials at the ends have only one neighbor. Therefore an
additional condition for these polynomials is needed. It is reasonable to
assume that outside the end points a linear extrapolation can be done and
then the last two needed conditions are obtained, which can be expressed
mathematically

$$\frac{d^2 p_1}{dx^2}(x_1) = 0 \quad \text{and} \quad \frac{d^2 p_{n-1}}{dx^2}(x_n) = 0 \tag{5}$$

With the following abbreviations

$$h_i = (x_{i+1} - x_i)$$

$$d_i = \frac{(y_{i+1} - y_i)}{h_i}$$

$$t = \frac{(x - x_i)}{h_i}$$

it is possible to write the ith polynomial $p_i(x)$ as the following equation

$$p_i(x) = ty_{i+1} + (1 - t)y_i + h_i(1 - t)t\Big((k_i - d_i)(1 - t) - (k_{i+1} - d_i)t\Big) \quad (6)$$

One can easily see that this polynomial is of order three with respect to t, and because t is only a linear transformation of x, the polynomial is also of order three with respect to x. The polynomials $p_i(x)$ contain the — until now — unknown parameters k_i. In the following it will be shown that the $p_i(x)$'s defined by equation (6) fulfil the conditions (1) – (5) for a spline function if a linear equation system for the k_i's is solved.

If you set $t = 0$, which means that $x = x_i$, then you will get from equation (6)

$$t = 0: \quad \Longrightarrow \quad x = x_i \quad \Longrightarrow \quad p_i(x_i) = y_i \quad (1')$$

This again is our former condition (1). If one puts $t = 1$, which means that $x = x_{i+1}$, then we can derive from equation (6)

$$t = 1: \quad \Longrightarrow \quad x = x_{i+1} \quad \Longrightarrow \quad p_i(x_{i+1}) = y_{i+1} \quad (2')$$

This is identical to condition (2), which is now fulfilled by the polynomial (6). The unknown parameters in (6), the k_i's, have to be determined with the help of equations (4) and (5). Equation (6) satisfies condition (3). This is realized if you calculate the derivative of (6) to x.

$$\frac{dp_i}{dx}(x) = d_i + (1 - 2t)\Big((k_i - d_i)(1 - t) - (k_{i+1} - d_i)t\Big)$$
$$+ (t - t^2)(2d_i - k_i - k_{i+1}) \quad (7)$$

If $t = 0$ and $t = 1$, then from (7) the following relations are obtained

$$t = 0: \quad \Longrightarrow \quad \frac{dp_i}{dx}(x_i) = k_i \tag{3'}$$

$$t = 1: \quad \Longrightarrow \quad \frac{dp_i}{dx}(x_{i+1}) = k_{i+1} \tag{3''}$$

These relations are exactly condition (3), which can be seen if i in equation (3'') is substituted by $i - 1$. If the second derivative of (6) is determined, then we obtain the equation

$$\frac{d^2 p_i}{dx^2}(x) = \frac{2}{h_i} \cdot (1 - 2t)(2d_i - k_i - k_{i+1})$$
$$- \frac{2}{h_i}\left((k_i - d_i)(1 - t) - (k_{i+1} - d_i)t\right) \tag{8}$$

If, according to this equation, the $\frac{d^2 p_{i-1}}{dx^2}(x_i)$-values and the $\frac{d^2 p_i}{dx^2}(x_i)$-values are calculated and if both expressions are set equal according to condition (4) the following equation is obtained, after some rearrangements

$$h_i k_{i-1} + 2(h_{i-1} + h_i)k_i + h_{i-1}k_{i+1} = 3h_{i-1}d_i + 3h_i d_{i-1} \tag{9}$$

This equation holds for all i from $i = 2$ until $i = (n - 1)$, which means that for n unknown k-values there are $(n - 2)$ linear equations. The two still missing linear equations are obtained from the two conditions (5) for the end points

$$2k_1 + k_2 = 3d_1$$

$$k_{n-1} + 2k_n = 3d_{n-1} \tag{10}$$

These n linear equations for the n unknown k's can be solved with a routine for solving a system of linear equations. After these k's are determined, all parameters in the spline polynomials (6) are known and therefore for every arbitrary x-value the respective y-value can be calculated. In doing so the polynomial $p_i(x)$, which belongs to the arbitrary x-value, must be looked for; first one has to search for the data points between which the x-value is situated. Then from x according to the formula above, the variable t is calculated and then the polynomial $p_i(x)$. This can be done using the program SPLINE printed below.

```
0 REM "SPLINE  "          EBERT/EDERER          880711
1 REM ***********************************************
2 REM *** Approximation of a function with n given ***
3 REM *** grid points using a SPLINE function.     ***
4 REM *** If the maximum of 25 grid points is sur- ***
5 REM *** passed the dimensions of line 1000 have  ***
6 REM *** to be changed.                           ***
9 REM ***********************************************
1000  DIM A(25,26),H(25),D(25),K(25),X(25),Y(25)
1100  INPUT "Number of data points ";N
1200  FOR I=1 TO N
1300    PRINT "x(";I;") ,  y(";I;")   ";
1400    INPUT X(I),Y(I) : NEXT I
1500  REM ***********************************************
1502  REM *** The datapoints are sorted in x.      ***
1510  REM ***********************************************
1550  FOR I=1 TO N-1 : H=X(I):H1=I
1570    FOR J=I+1 TO N
1600      IF H<X(J) THEN 1700
1650      H=X(J) : H1=J
1700    NEXT J
1750    X(H1)=X(I) : X(I)=H
1760    H=Y(H1) : Y(H1)=Y(I) : Y(I)=H
1800  NEXT I
2000  FOR I=1 TO N-1
2100    H(I)=X(I+1)-X(I)
2200    D(I)=(Y(I+1)-Y(I))/H(I)
2300  NEXT I
5000  A(1,1)=2 : A(N,N)=2
5100  A(1,2)=1 : A(N,N-1)=1
5200  A(1,N+1)=D(1)*3 : A(N,N+1)=D(N-1)*3
6000  FOR I=2 TO N-1
6500    A(I,N+1)=(D(I)*H(I-1)+D(I-1)*H(I))*3
7000    A(I,I-1)=H(I):A(I,I+1)=H(I-1)
7200    A(I,I)=2*(H(I)+H(I-1))
7500  NEXT I
8000  GOSUB 50000 : REM GJ
10000 INPUT "Which x-value ";X
11000   GOSUB 20000 : REM Which INTERVAL
12000   T=(X-X(J))/H(J)
12100   Y=(A(J,N+1)-D(J))*(1-T)-(A(J+1,N+1)-D(J))*T
12400   Y=T*Y(J+1)+(1-T)*Y(J)+H(J)*(1-T)*T*Y
13000   PRINT "Spline interpolation at position "
13100   PRINT " x = ";X;"  :  ";Y : PRINT
14000 GOTO 10000
```

```
15010 REM ###########################################
15020 REM ### Main program ends here.           ###
15030 REM ###########################################
20000 REM *******************************************
20002 REM *** Subroutine: INTERVAL              ***
20004 REM *** Searching for the interval for x. ***
20010 REM *******************************************
20020 IF X>X(1) THEN 20050
20030 J=1 : GOTO 22000
20050 IF X<X(N) THEN 20100
20060 J=N-1 : GOTO 22000
20100 FOR I=1 TO N
20200    IF X(I)<X THEN 21000
20500    J=I-1 : I=N
21000 NEXT I
22000 RETURN
50000 GOTO 50900
50001 REM *******************************************
50002 REM *** Subroutine: GJ                    ***
50004 REM *** The linear equations are solved.  ***
50010 REM *******************************************
50100 INPUT "Number of equations ";N
50200 PRINT "Input of the extended coefficient matrix"
50300 FOR I=1 TO N
50400    FOR J=1 TO N
50500       PRINT "A(";I;",";J;")="; : INPUT A(I,J)
50600    NEXT J
50700    PRINT "B(";I;")="; : INPUT A(I,N+1)
50800 NEXT I : PRINT
50900 FOR S=1 TO N
51000    FOR T=S TO N
51100       IF A(T,S)<>0 THEN 51300
51200    NEXT T
51210    PRINT "No solution is possible" : GOTO 63999
51300    GOSUB 53000 : REM EXCHANGE lines
51400    C=1/A(S,S)
51500    GOSUB 54000 : REM NORMALIZE the actual line
51600    FOR T=1 TO N
51700       IF T=S THEN 52000
51800       C=-A(T,S)
51900       GOSUB 55000 : REM make zeroes in the column
52000    NEXT T
52100 NEXT S
52200 RETURN
```

```
53000 REM ****************************************
53002 REM *** Subroutine: EXCHANGE            ***
53004 REM *** Lines S and T are exchanged.    ***
53010 REM ****************************************
53100   FOR J=1 TO N+1
53200     B=A(S,J) : A(S,J)=A(T,J) : A(T,J)=B
53300   NEXT J
53400 RETURN : REM end of subroutine EXCHANGE
54000 REM ****************************************
54002 REM *** Subroutine: NORMALIZE           ***
54004 REM *** Normalizing the actual line S.  ***
54010 REM ****************************************
54100   FOR J=1 TO N+1
54200     A(S,J)=C*A(S,J) : NEXT J
54400 RETURN : REM end of subroutine NORMALIZE
55000 REM ****************************************
55002 REM *** Subroutine: MAKEZERO            ***
55004 REM *** The elements of line S are multi- ***
55006 REM *** plied with C and subtracted from the ***
55008 REM *** elements of line T.             ***
55010 REM ****************************************
55100   FOR J=1 TO N+1
55200     A(T,J)=A(T,J)+C*A(S,J)
55300   NEXT J
55400 RETURN : REM end of subroutine MAKEZERO
56000 REM ****************************************
56002 REM *** Subroutine: OUTPUT              ***
56004 REM *** Output of the solution.         ***
56010 REM ****************************************
56100   FOR T=1 TO N
56200     PRINT "X(";T;")=";A(T,N+1)
56300   NEXT T
56400 RETURN : REM end of subroutine OUTPUT
63999 END


RUN
Number of data points ? 5
x( 1 ) ,  y( 1 )   ? 0,0.1
x( 2 ) ,  y( 2 )   ? 1,0.9
x( 3 ) ,  y( 3 )   ? 2.,3.8
x( 4 ) ,  y( 4 )   ? 3,9.2
x( 5 ) ,  y( 5 )   ? 4,16
Which x-value ? 3
Spline interpolation at position
 x = 3   :    9.2
```

```
Which x-value ? 2.5
Spline interpolation at position
  x = 2.5   :    6.23817

Which x-value ? 7
Spline interpolation at position
  x = 7    :    30.78573

Which x-value ? 0.25
Spline interpolation at position
  x = .25   :    .2041573

Which x-value ? 0
Spline interpolation at position
  x = 0    :    .1

Which x-value ? 4.2
Spline interpolation at position
  x = 4.2   :    17.40491

Break in 10000
```

This program can handle interpolations of up to 25 data points. Therefore in line 1000 the auxiliary vectors H, D, K and the vectors for the data points X and Y are dimensioned to 25. The resulting system of linear equations has then a maximum dimension of 25, the accompanying expanded matrix of coefficients is defined as A(25,26).

The second section of the program SPLINE, beginning with line 50000, consists of the program GJ for solving a system of linear equations. The Gauss-Jordan procedure, however, seems to be too expensive in computing time and computer memory for this sort of linear equations. In this system of equations only the diagonal and the two neighboring diagonals are unequal to zero. But because we don't want to introduce a new method for linear equations and because it is a good programming practice to use available and tested program blocks, and because GJ is a robust and satisfactory code, it makes sense to use the reliable but not so efficient Gauss-Jordan code. But in case the number of data points you have to consider for a spline interpolation is substantially greater, it is advisable to find a more suitable method in the literature on numerical mathematics for so-called tri-diagonal matrices.

In the program part from line 1100 through 1400, the n data points are read into the variables $X(I)$ and $Y(I)$. From line number 1500 through 1800 these points are sorted for ascending $X(I)$-values. This is done by a sorting algorithm which was discussed previously. This is useful for experimental points, which often are not given in an order of ascending x-values.

In the program loop from line 2000 to 2300 the auxiliary values $H(I)$ and $D(I)$, defined by equation (6), are evaluated. In the following lines from 5000 to 7500 the extended coefficient matrix is filled according to equations (9) and (10). In lines 5000, 5100 and 5200 the calculation of the first and the nth line of the matrix according to equation (10) is carried out. The programming loop from line 6000 to 7500 fills up the $(n-2)$ remaining lines of the matrix after calculating the elements with equation (9).

As soon as the matrix A is established, the GJ-subroutine for solving the system of linear equations is called up in line 8000. The solutions for the k-values are contained in the $(n+1)^{th}$ column of the matrix A, after the calculation in the subroutine is ended.

All parameters are now known and calculated. The user can put in x-values whose interpolated y-value shall be evaluated. This is done in line 10000.

After the input of an x-value, the subroutine 20000 is called. Therein a number J is determined which points to the interval $(X(J),X(J+1))$ where x is located. J also determines which polynomial has to be used for calculation. If the x-value is outside the limits of the data points, then the first or the last polynomial is used for the interpolation. This is decided with the help of the two IF-statements in lines 20020 and 20050. If the index J is known, then the auxiliary variable T (see the definition above) can be calculated. In lines 12100 and 12400 the interpolation value of y according to formula (6) is evaluated. Because this equation is rather extensive, this is done in two parts.

The x-value together with the calculated y-value is printed on the screen in lines 13000 and 13100. The program then jumps back to line 10000 to inquire for the next x-value.

In the example shown, 5 points are specified which approximately lie on a parabola. The spline values for these grid points are necessarily exact, the same is true for the interpolated values, as expected. Only the value which lies outside the 5 points has to be extrapolated and a value is obtained which is clearly different from the parabolic value. It is always a good rule to try to avoid extrapolations.

Problem 161

Modify the input part of the program SPLINE for reading in the points with a READ-statement and expand the program by a smoothing subroutine, which may be called if it is required.

Problem 162

Use program SPLINE and a program 'Smoothed Spline' to write a new program for solving numerical differentiation problems. You can use the smoothing algorithm which is shown in Chapter 8.1.

Problem 163

Rewrite the Gauss-Jordan subroutine for the special purpose of a band matrix where three elements along the main diagonal have numbers unequal to zero. An example (8×8) matrix with such a band structure is shown below.

$$\begin{pmatrix} a_{11} & a_{12} & 0 & 0 & 0 & 0 & 0 & 0 \\ a_{21} & a_{22} & a_{23} & 0 & 0 & 0 & 0 & 0 \\ 0 & a_{32} & a_{33} & a_{34} & 0 & 0 & 0 & 0 \\ 0 & 0 & a_{43} & a_{44} & a_{45} & 0 & 0 & 0 \\ 0 & 0 & 0 & a_{54} & a_{55} & a_{56} & 0 & 0 \\ 0 & 0 & 0 & 0 & a_{65} & a_{66} & a_{67} & 0 \\ 0 & 0 & 0 & 0 & 0 & a_{76} & a_{77} & a_{78} \\ 0 & 0 & 0 & 0 & 0 & 0 & a_{87} & a_{88} \end{pmatrix}$$

It is sufficient to define an expanded coefficient matrix of the form A(N,4), with three columns for the respective diagonals and one for the right hand side of the equation system. For $n = 8$ the new matrix filled with the elements a_{ij} and b_i is the following

$$\begin{pmatrix} 0 & a_{11} & a_{12} & b_1 \\ a_{21} & a_{22} & a_{23} & b_2 \\ a_{32} & a_{33} & a_{34} & b_3 \\ a_{43} & a_{44} & a_{45} & b_4 \\ a_{54} & a_{55} & a_{56} & b_5 \\ a_{65} & a_{66} & a_{67} & b_6 \\ a_{76} & a_{77} & a_{78} & b_7 \\ a_{87} & a_{88} & 0 & b_8 \end{pmatrix}$$

11　Nonlinear Systems

In earlier chapters a number of applications in chemistry and engineering sciences have been given for linear systems of equations and for linear regression. But more often problems and applications arise in these subjects which concern nonlinear systems of equations and nonlinear regression. The nonlinear case is the more general one, the linear case can be considered as a part of it. Linear problems can, of course, also be treated with suitable nonlinear methods — but this is, in most cases, too complicated and troublesome.

11.1　Nonlinear Systems of Equations

As an example, we consider the following system of nonlinear equations

$$f_1(x_1, x_2, x_3) = x_1^2 - x_2 x_1 - 3$$
$$f_2(x_1, x_2, x_3) = x_3 x_2 x_1 - 1.5 x_2^2$$
$$f_3(x_1, x_2, x_3) = x_1^3 + x_2^3 + x_3^3 - 36$$

First we are looking for a common zero-position of all three equations.

$$f_1(x_1, x_2, x_3) = f_2(x_1, x_2, x_3) = f_3(x_1, x_2, x_3) = 0$$

Treating only one equation it could be shown that all algebraic, transcendental and implicit equations can be transformed to the problem of finding the root of an equation. With a system of nonlinear equations this is done in a similar way with each equation. For example the following system of two equations

$$x_1^2 + x_2^2 = 2$$

$$\sin(x_1 \cdot x_2) = 0.5$$

can be transformed into

$$f_1(x_1, x_2) = x_1^2 + x_2^2 - 2$$

$$f_2(x_1, x_2) = \sin(x_1 \cdot x_2) - 0.5$$

This system of nonlinear equations can still be treated analytically and a common root is given by

$$x_1 = \sqrt{1 + \sqrt{1 - \frac{\pi^2}{36}}} = 1.3608692$$

$$x_2 = \frac{\pi}{6x_1} = 0.3847532$$

The first example of this chapter containing three unknowns cannot be solved analytically, however, by trial and error, one can get the following triple which is a solution of the equation system

$$x_1 = 3 \qquad x_2 = 2 \qquad x_3 = 1$$

What is the procedure for solving a system of nonlinear equations systematically? For answering this question we consider the method of solving the one-dimensional case with Newton's method.

The equation $f(x) = 0$ is solved iteratively by using a starting value x_0 for x and calculating an improved value x_1 according to the following iteration formula

$$x_1 = x_0 - \frac{f(x_0)}{f'(x_0)}$$

or more generally

$$x_{i+1} = x_i - \frac{f(x_i)}{f'(x_i)}$$

where we use the abbreviation

$$f'(x) = \frac{df(x)}{dx}$$

It shall be assumed that for the multidimensional case a completely analogous formula can be applied

$$\mathbf{X}_{i+1} = \mathbf{X}_i - \left(\mathbf{F}'(\mathbf{X}_i)\right)^{-1} \cdot \mathbf{F}(\mathbf{X}_i) \tag{1}$$

The bold symbols $\mathbf{X}$, $\mathbf{F}(\mathbf{X})$ and $\mathbf{F}'(\mathbf{X})$ have the following meanings, if for example three nonlinear equations with three unknowns are assumed; the generalization to dimension n is easily done.

$$\mathbf{X} = \begin{pmatrix} x_1 \\ x_2 \\ x_3 \end{pmatrix} \qquad \text{the vector of the 3 variables}$$

$$\mathbf{F}(\mathbf{X}) = \begin{pmatrix} f_1(x_1, x_2, x_3) \\ f_2(x_1, x_2, x_3) \\ f_3(x_1, x_2, x_3) \end{pmatrix} \quad \text{the vector of the 3 function values}$$

$$\mathbf{F}'(\mathbf{X}) = \begin{pmatrix} \dfrac{\partial f_1}{\partial x_1} & \dfrac{\partial f_1}{\partial x_2} & \dfrac{\partial f_1}{\partial x_3} \\ \dfrac{\partial f_2}{\partial x_1} & \dfrac{\partial f_2}{\partial x_2} & \dfrac{\partial f_2}{\partial x_3} \\ \dfrac{\partial f_3}{\partial x_1} & \dfrac{\partial f_3}{\partial x_2} & \dfrac{\partial f_3}{\partial x_3} \end{pmatrix} \qquad \text{the Jacobian}$$

The power -1 is a simple division in the one-dimensional case; applied to $\mathbf{F}'(\mathbf{X})$ it means the inversion of the Jacobian. The multiplication symbol in equation (1) stands for the multiplication of the inverted Jacobian with the vector of the function values. From this again a vector results. This vector, which is a correction vector, is subtracted from the estimation vector $\mathbf{X}_i$, previously used, and the new improved estimation vector $\mathbf{X}_{i+1}$ is obtained.

The procedure doesn't always lead to improved values for $\mathbf{X}$, that means that in the course of the iteration the correct solution vector $\mathbf{X}$ may not be obtained. But the better the initial estimation of the values is, the higher is the probability for a successful iteration procedure.

The main computational task for searching roots of a system of nonlinear equations is the calculation of the n^2 partial derivatives for obtaining the Jacobian and the calculation of the inverted Jacobian.

Note:

From a technical point of view it is better to avoid this cumbersome inversion of the matrix. If the iteration formula (1) is multiplied from the right hand side by the Jacobian $\mathbf{F}'(\mathbf{X})$ one obtains after some rearrangements

$$(\mathbf{X}_i - \mathbf{X}_{i+1}) = \left(\mathbf{F}'(\mathbf{X}_i)\right)^{-1} \cdot \mathbf{F}(\mathbf{X}_i)$$
$$\mathbf{F}'(\mathbf{X}_i) \cdot (\mathbf{X}_i - \mathbf{X}_{i+1}) = \mathbf{F}'(\mathbf{X}_i) \cdot \left(\mathbf{F}'(\mathbf{X}_i)\right)^{-1} \cdot \mathbf{F}(\mathbf{X}_i) \qquad (2)$$
$$\mathbf{F}'(\mathbf{X}_i) \cdot (\mathbf{X}_i - \mathbf{X}_{i+1}) = \mathbf{F}(\mathbf{X}_i)$$

This is a system of linear equations with the matrix $\mathbf{F}'(\mathbf{X})$ and the right hand side $\mathbf{F}(\mathbf{X})$. The solution vector of this equation system is the difference vector $(\mathbf{X}_i - \mathbf{X}_{i+1})$. From this $\mathbf{X}_{i+1}$ can easily be calculated because $\mathbf{X}_i$ is known.

The fastest calculation procedure for the Jacobian is obtained if the n^2 partial derivatives are calculated analytically and programmed respectively. But because this is very often cumbersome and erroneous, the Jacobian shall be calculated numerically using the difference quotient to make the program easier to use. This is done in the following program MNEWTON.

```
0 REM "MNEWTON "          EBERT/EDERER          880225
1 REM *************************************************
2 REM *** A multidimensional Newton method for    ***
3 REM *** calculating the roots of a system of    ***
4 REM *** coupled nonlinear equations.            ***
5 REM *** The functions have to be programmed in  ***
6 REM *** the subroutine FUNCTIONS line # 12000   ***
7 REM *** and the dimension n in line # 20100.    ***
9 REM *************************************************
100   DIM A(20,20),B(20,20),U(20,20)
110   DIM V(20,20),W(20,20)
200   DIM X(20),X9(20),F(20),F9(20)
210   DIM F0(20),H(20)
250   GOTO 20000 : REM jump to the MAIN program
300   REM *****************************************
305   REM *** Subroutine  INVERS              ***
310   REM *** Input Matrix    in A()          ***
315   REM *** Inverted Matrix in W()          ***
320   REM *****************************************
```

```
400    FOR I=1 TO  N
420      FOR J=1 TO  N
460        B(I,J)=0
480        V(I,J)=0
500        IF I<>J THEN GOTO 600
520        B(I,J)=1
540        V(I,J)=1
600      NEXT J
620    NEXT I
1000   FOR Z=1 TO N
1050     S=0
1100     REM ****************************************
1105     REM *** Searching for the pivot element.   ***
1110     REM ****************************************
1120     FOR I=Z TO N
1150       IF S>ABS(A(I,Z)) THEN GOTO 1300
1200       S=ABS(A(I,Z)) : T=I
1300     NEXT I
2000     REM ****************************************
2001     REM *** Lines Z and T are exchanged.     ***
2005     REM ****************************************
2050     FOR I=1 TO N
2100       S=A(Z,I) : A(Z,I)=A(T,I) : A(T,I)=S
2200     NEXT I
2300     IF ABS(A(Z,Z))>1E-30 THEN GOTO 2400
2350     PRINT "Inversion is not possible " : END
2400     V(Z,Z)=0 : V(T,T)=0
2450     V(Z,T)=1 : V(T,Z)=1
3000     REM ****************************************
3020     REM *** Gauss-Jordan elimination         ***
3040     REM ****************************************
3100     FOR I=1 TO N
3200       FOR J=1 TO N
3300         IF I=Z THEN GOTO 4000
3350         IF J=Z THEN GOTO 4500
3400         U(I,J)=A(I,J)-A(Z,J)*A(I,Z)/A(Z,Z)
3500         GOTO 5000
4000         IF I=J THEN GOTO 4350
4050         U(I,J)=-A(I,J)/A(Z,Z)
4100         GOTO 5000
4350         U(Z,Z)=1/A(Z,Z)
4400         GOTO 5000
4500         U(I,Z)=A(I,Z)/A(Z,Z)
5000       NEXT J : NEXT I
```

```
5100    REM ******************************************
5102    REM *** Multiplication  B = V * B        ***
5110    REM ******************************************
5200    FOR I=1 TO N
5250      FOR J=1 TO N
5300        W(I,J)=0
5350        FOR K=1 TO N
5400          W(I,J)=W(I,J)+V(I,K)*B(K,J)
5450    NEXT K : NEXT J : NEXT I
5500    FOR I=1 TO N : FOR J=1 TO N
5550        B(I,J)=W(I,J)
5600    NEXT J : NEXT I
6000    FOR I=1 TO N
6050      FOR J=1 TO N
6100        A(I,J)=U(I,J) : V(I,J)=0
6250        IF I=J THEN V(I,J)=1
6300    NEXT J : NEXT I
6500    NEXT Z
7000    REM ******************************************
7005    REM *** The result is obtained by multiply- ***
7010    REM *** ing matrix A with the permutation   ***
7020    REM *** matrix B.                           ***
7030    REM ******************************************
7100    FOR I=1 TO N
7200      FOR J=1 TO N
7300        W(I,J)=0
7400        FOR K=1 TO N
7500          W(I,J)=W(I,J)+A(I,K)*B(K,J)
7600    NEXT K : NEXT J : NEXT I
7700    RETURN : REM end of the subroutine INVERS
12000 REM ******************************************
12010 REM *** Subroutine  FUNCTIONS               ***
12020 REM *** The functions, whose roots must     ***
12030 REM *** be calculated, are programmed here.  ***
12060 REM ******************************************
12100 F(1)=X(1)*X(1)-X(2)*X(1)-3
12200 F(2)=X(3)*X(2)*X(1)-X(2)*X(2)*1.5
12300 F(3)=X(1)^3+X(2)^3+X(3)^3-36
12999 RETURN
15000 REM ******************************************
15010 REM *** Subroutine  PARTIAL                 ***
15020 REM *** The partial derivatives at the po-   ***
15030 REM *** sition X9() are calculated and       ***
15035 REM *** stored in matrix A().                ***
15040 REM ******************************************
```

```
15100 FOR I=1 TO N : X(I)=X9(I) : NEXT I
15200 FOR I=1 TO N : DI=ABS(X(I)/100000!)+1E-09
15210   X(I)=X(I)+DI : GOSUB 12000
15240   FOR J=1 TO N : A(J,I)=F(J)
15260 NEXT J : X(I)=X9(I) : NEXT I
15300 FOR I=1 TO N : DI=ABS(X(I)/100000!)+1E-09
15310   X(I)=X(I)-DI : GOSUB 12000
15340   FOR J=1 TO N : A(J,I)=(A(J,I)-F(J))/2/DI
15360 NEXT J : X(I)=X9(I) : NEXT I
15999 RETURN
19000 REM ******************************************
19010 REM *** Subroutine  OUTPUT              ***
19020 REM ******************************************
19100 PRINT
19120 FOR I=1 TO N
19140   PRINT "x(";I;")=";X9(I);
19160   PRINT TAB(25);"  f(";I;")=";F9(I)
19180 NEXT I : RETURN
20000 REM ******************************************
20010 REM *** MAIN PROGRAM                    ***
20020 REM ******************************************
20100 N=3 : REM dimension of the equation system
20200 PRINT "Input of the estimated ";
20220 PRINT "x-values of the roots" : PRINT
20300 FOR I=1 TO N : PRINT "x(";I;")=";
20320   INPUT X(I) : X9(I)=X(I) : NEXT I
20400 GOSUB 15000 : REM PARTIAL derivatives
20410   FOR I=1 TO N: X(I)=X9(I): NEXT I : GOSUB 12000
20420   FOR I=1 TO N: F9(I)=F(I): NEXT I
20450   GOSUB 19000 : REM Output
20500   GOSUB 300 : REM matrix inversion
20600   FOR I=1 TO N : H(I)=0
20700     FOR J=1 TO N : H(I)=H(I)+W(I,J)*F9(J)
20720   NEXT J: NEXT I
20900   GOSUB 25000 : REM Optimization factor OF
21000   FOR I=1 TO N : X9(I)=X9(I)-OF*H(I) : NEXT I
21100 GOTO 20400
21500 REM ##########################################
21510 REM ### MAIN PROGRAM ends here          ###
21520 REM ##########################################
25000 REM ******************************************
25001 REM *** Subroutine  OPTIMUM             ***
25002 REM *** The optimization factor OF is deter- ***
25003 REM *** mined here.                     ***
25005 REM ******************************************
```

```
25010 DF=.6 : OF=1 : FF=OF
25020 FOR I=1 TO N : X(I)=X9(I)-FF*H(I) : NEXT I
25040 S5=0 : GOSUB 12000
25060 FOR I=1 TO N : S5=S5+F(I)*F(I) : NEXT I
25100 FOR J=0 TO 3 : DF=DF*(-.5)^J
25120    FF=OF+DF:S7=S5
25140    FOR I=1 TO N : X(I)=X9(I)-FF*H(I) : NEXT I
25160    S6=0 : GOSUB 12000
25170    FOR I=1 TO N : S6=S6+F(I)*F(I) : NEXT I
25180    IF S6<S5 THEN S5=S6 : OF=FF
25200    IF S6<.9*S7 THEN GOTO 25120
25300 NEXT J
26990 RETURN
60000 END


RUN
Input of the estimated x-values of the roots

x( 1 )=? 1
x( 2 )=? 1
x( 3 )=? 1


x( 1 )= 1                   f( 1 )=-3
x( 2 )= 1                   f( 2 )=-.5
x( 3 )= 1                   f( 3 )=-33

x( 1 )= 3.09125             f( 1 )= 1.018333E-02
x( 2 )= 2.117475            f( 2 )= 1.82303
x( 3 )= 1.305995            f( 3 )= 5.261101

x( 1 )= 3.000239            f( 1 )=-2.579689E-03
x( 2 )= 2.001179            f( 2 )=-1.101208E-02
x( 3 )= .9986754            f( 3 )= 1.663971E-02

x( 1 )= 2.999998            f( 1 )=-4.768372E-07
x( 2 )= 1.999998            f( 2 )=-1.478195E-05
x( 3 )= .9999971            f( 3 )=-7.629395E-05

x( 1 )= 3                   f( 1 )= 0
x( 2 )= 2                   f( 2 )= 0
x( 3 )= 1                   f( 3 )= 0
Break in 5110
```

```
RUN
Input of the estimated x-values of the roots

x( 1 )=? 4
x( 2 )=? 5
x( 3 )=? 6


x( 1 )= 4                      f( 1 )=-7
x( 2 )= 5                      f( 2 )= 82.5
x( 3 )= 6                      f( 3 )= 369


x( 1 )= 1.20705                f( 1 )=-.3891592
x( 2 )=-.9559436               f( 2 )=-5.791017
x( 3 )= 3.830821               f( 3 )= 21.10307


x( 1 )= 1.645583               f( 1 )=-.2150462
x( 2 )=-4.679752E-02           f( 2 )=-.2532366
x( 3 )= 3.245736               f( 3 )= 2.649246


x( 1 )= 1.736455               f( 1 )= 1.040292E-02
x( 2 )= 2.805717E-03           f( 2 )= 1.525556E-02
x( 3 )= 3.133696               f( 3 )= 8.945465E-03


x( 1 )= 1.732001               f( 1 )=-1.604557E-03
x( 2 )= 8.257674E-04           f( 2 )= 4.482521E-03
x( 3 )= 3.134843               f( 3 )= 2.540589E-03


x( 1 )= 1.732399               f( 1 )= 1.015186E-03
x( 2 )= 1.09892E-04            f( 2 )= 5.967459E-04
x( 3 )= 3.134648               f( 3 )= 3.814697E-04


x( 1 )= 1.73215                f( 1 )= 1.759529E-04
x( 2 )= 9.683471E-05           f( 2 )= 5.257799E-04
x( 3 )= 3.134722               f( 3 )= 3.28064E-04


x( 1 )= 1.732099               f( 1 )= 1.683235E-04
x( 2 )=-5.491165E-08           f( 2 )=-2.981514E-07
x( 3 )= 3.134726               f( 3 )= 0


x( 1 )= 1.732051               f( 1 )=-4.768372E-07
x( 2 )=-1.296741E-12           f( 2 )=-7.040692E-12
x( 3 )= 3.134741               f( 3 )= 0


x( 1 )= 1.732051               f( 1 )= 0
x( 2 )= 1.084202E-19           f( 2 )= 5.886709E-19
x( 3 )= 3.134741               f( 3 )= 3.814697E-06
```

```
x( 1 )= 1.732051        f( 1 )= 0
x( 2 )= 0               f( 2 )= 0
x( 3 )= 3.134741        f( 3 )=-3.814697E-06

x( 1 )= 1.732051        f( 1 )= 0
x( 2 )= 0               f( 2 )= 0
x( 3 )= 3.134741        f( 3 )= 3.814697E-06
Break in 4050
```

After the dimensions of different matrices and vectors have been defined in the lines 100 to 210 the program jumps to line 20000. There the main program starts. From lines 300 to 7700 there is as a subroutine, the inversion of a matrix, a program discussed earlier. The matrix to be inverted is put into the field A of dimension 20 and the inverted matrix, calculated by this subroutine, is returned in the field W.

After some REM-statements the main program actually starts in line 20100 by assigning the dimension of the system of equations. This is one of the lines which has to be changed by the user if another system of nonlinear equations is to be solved. From line 20200 to 20320 the program is asking the user for the initial values of the N different x-values. These values are read into the vector variables X(I) by an INPUT-statement and are then additionally stored in the variables X9(I). The variables X9(I) are used for saving the 'up until now' best X-values until one entire iteration loop is finished. In the next line 20400 the subroutine for the calculation of the Jacobian is called by the statement GOSUB 15000. Immediately afterwards in lines 20410 and 20420 the function values for the 'up until now' best X-values, which are stored in the variables X9(I), are calculated and assigned to the field variables F9(I). In the next line 20450 the output subroutine, which begins at line 19000, is called. Here the actual optimum X-values X9(I) together with the N function values F9(I) — which should all become zero — are printed out. Continuing the course of the main program in line 20500 the subroutine for the matrix inversion is called. After the completion of this subroutine the inverted matrix is stored in the field variable W. Now all the necessary values are available to calculate the iteration formula (1). The multiplication of the inverted Jacobian W with the vector of the function values F9 is done in the nested loops from line 20600 to 20720. The result is then contained in the vector H. These values H(I) are the correction values to improve the variables X9(I) using formula (1) to obtain new and hopefully better values X9(I). This is done in the loop at line 21000. Differently to equation (1), the correction is done by multiplying with an optimizing factor OF. This factor has been determined in the preceding line by calling up the subroutine 25000. This optimizing factor very often will be close to 1, but especially with poor initial guesses, it may deviate from unity considerably, sometimes it may even have a negative value. In line 21100 the iteration cycle is finished and the program jumps back to line 20400 to start a new iteration loop.

There is no termination criterion included in the main program, the user must stop the program as soon as he regards the obtained results as sufficiently accurate. The function values should be as close as possible to zero. If this is achieved, then the **X**-values usually change only negligibly with further iterations.

The subroutine for the calculation of the functions starts at line 12000 and ends with the RETURN-statement in line 12999. In the three lines 12100, 12200 and 12300 the three functions of the first example in this chapter are programmed. Of course, these lines have to be changed if the solutions of other equation systems are required. But the new system of equations has to be programmed in the same form as shown before; the function values have to be assigned to the indexed variables F(I) and the function formulas must be programmed by using the independent field variables X(I).

The subroutine for the calculation of the Jacobian is contained in the section from line 15000 through 15999. The first thing to do is to fill the variables X(I) with the 'up until now' best values for the **X**'s. In line 15200 an I-loop starts for the values X(I), where the actual X(I) is increased by the 10^{-5}-fold of its absolute value. This is programmed in the second statement of line 15200 and in the first statement of line 15210. The addition of the small quantity 10^{-9} is necessary if X(I) approaches 0. After X(I) has been increased by the small quantity DI, then the calculation of the functions is called by GOSUB 12000. These values of the functions are stored in the variables A(J,I) in the next two program lines. After NEXT J the X(I) is reset to the value of X9(I) and the I-loop is continued. In the lines from 15300 to 15360 the same calculations are done with an X(I) diminished by DI. The partial derivatives $\frac{\partial f_j}{\partial x_i}$ are approximated in line 15340 using the following equation are then assigned to the matrix element A(J,I).

$$\frac{\partial f_j}{\partial x_i} \approx \frac{f_j(x_1, \ldots, x_i + DI, \ldots, x_n) - f_j(x_1, \ldots, x_i - DI, \ldots, x_n)}{2 \cdot DI}$$

Before the loop variable I is incremented at the end of the I-loop, the old value X(I) is restored to the previous optimum value by the statement X(I)=X9(I).

The subroutine for the determination of the optimizing factor OF starts with calculation of X(I) by correcting the X9(I) with the theoretical optimizing factor 1 (line 25020). Then the function values are calculated and their squared sum — which shall become zero — is formed and stored in variable S5 (line 25060). In the following J-loop (lines 25100 to 25300) which runs from 0 to 3, the correction factor is consecutively changed by the quantity $(-0.5)^J$. After the correction with this new optimizing factor, the function values and their square sum are again calculated in lines 25140 to 25170. If the squared sum is now smaller, then this correction factor becomes — maybe temporarily — the new optimizing factor (line 25180). If the improvement is better than 10% then the same variation of the correction factor is repeated (line 25200). Otherwise, by increasing the loop counter J, a new variation with the opposite sign and the half of the previous variation quantity is attempted.

As in the example shown above, it can be seen that poor initial guesses may lead to poor intermediate X-values, but they will finally converge rather quickly towards the correct results. But, occasionally, using very poor initial values, it may happen that this procedure diverges. The function values are then steadily growing instead of decreasing and finally an overflow error (the quantities become too big) stops the program. If this or similar things happen, you should try again and restart the program with new and hopefully better initial guesses for the solution.

Problem 164

Solve the following system of coupled nonlinear equations

$$x^2 + y^2 = x^3 + y^3$$
$$xy + \log(xy) + \log 2 = \log(x + y) + 1$$

If identical initial guesses for x und y are used in this example, the program will always stop with an error message. This happens even if the starting values chosen are very close to the solution values. But if the x-and y-values are only slightly different then these difficulties are avoided.

Problem 165

Solve the following system of coupled nonlinear equations

$$x + y + z = yz$$

$$xyz = z^2 - z$$

$$x^2 + y^2 + z^2 = 7y$$

Problem 166

Given is the following chemical reaction system:

$$A \xrightarrow{k_1} \text{product 1} \quad \text{(thermal reaction)}$$

$$A \xrightarrow{k_2} \text{product 2} \quad \text{(photochemical reaction)}$$

From this reaction mechanism the following reaction rate equation can be derived

$$\frac{d[A]}{dt} = -(k_1 + k_2) \cdot [A] = -\left(k_{0,1} \cdot e^{-\frac{E_a}{RT}} + k_2\right) \cdot [A]$$

$k_{0,1}$ is assumed to be 10^{12} s^{-1}.

At the temperature $T = 300$ K the half time for the decay of A is 0.5985 s; at $T = 350$ K it is 0.14 s. Use the program MNEWTON to calculate the activation energy E_a for the thermal reaction and the photochemical reaction rate constant k_2.

For help: $$t_{\frac{1}{2}} = \frac{\log 2}{k_1 + k_2}$$

Problem 167

The van der Waals equation is written as

$$\left(p + \frac{a}{V^2}\right) \cdot (V - b) = nRT$$

The van der Waals constants a and b should be determined from measurements on ethane at two different volumes and at two pressures. Use the program MNEWTON for this calculation.

If all physical constants are calculated and measured in the dimension system which uses liters, atmospheres, moles and Kelvin, then the gas constant R is given as

$$R = 0.08206 \left[\frac{l \cdot atm}{mol \cdot K}\right]$$

The experimental values for 1 mole of ethane at 400 K are:

p = 3.2486 atm   for   V = 10 l
p = 29.577 atm   for   V = 1 l

If more than two data pairs for the volumes and the pressures have been measured, then the nonlinear equation system is over-determined. The determination of the parameters a and b is then a problem of nonlinear regression which will be treated later.

Problem 168

Given is the consecutive chemical reaction:

$$A \xrightarrow{k_1} B \xrightarrow{k_2} C$$

The expression for the reaction rate of [B] follows the equation

$$r = \frac{d[B]}{dt} = k_1 \cdot [A] - k_2 \cdot [B]$$

$$= k_{0,1} \cdot e^{-\frac{E_1}{RT}} \cdot [A] - k_{0,2} \cdot e^{-\frac{E_2}{RT}} \cdot [B]$$

The reaction rate r has been determined at four different temperatures and concentrations. The dimensions are not given, it is assumed that they are consistent.

[A]	[B]	T	r
1	1	300	+0.001966
1	2	310	−0.0975
2	1	320	+0.2435
1	1	330	−0.06219

Use the program **MNEWTON** to calculate the four Arrhenius parameters $k_{0,1}$, E_1, $k_{0,2}$ and E_2. The correct results are:

$k_{0,1} = 2 \cdot 10^{12}$ $E_1/R = 9500$
$k_{0,2} = 1 \cdot 10^{13}$ $E_2/R = 10000$

Knowing these results, you can test how far the initial guesses can be off to still obtain the correct answers.

Problem 169

Write a new program for the multidimensional Newton method. Don't use the matrix inversion but rather the linear equation method which was described in a note immediately before the listing of program **MNEWTON**.

Use the other subroutines of program **MNEWTON** unchanged and substitute the matrix inversion subroutine by a Gauss-Jordan subroutine with partial pivoting for solving a system of linear equations. In addition to this you must change the iteration procedure in the main program to be able to use the new iteration procedure of formula (2) from the beginning of this chapter.

Improve the program by introducing a termination criterion into the main program.

Problem 169a

Another problem which can be solved with the Newton method is the minimization or maximization of a multidimensional function, that means the finding of a minimum or maximum of a function with more than one independent variable. Usually the method proposed cannot distinguish between minima and maxima, you will find extrema of the function. Given is the multidimensional function

$$f(x_1, \ldots, x_n)$$

A necessary condition for an extremum is that all partial derivatives of $f(x_1, \ldots, x_n)$ have to be zero

$$\frac{\partial f(x_1, \ldots, x_n)}{\partial x_1} = \cdots = \frac{\partial f(x_1, \ldots, x_i, \ldots, x_n)}{\partial x_i} = \cdots = \frac{\partial f(x_1, \ldots, x_n)}{\partial x_n} = 0$$

These n — usually nonlinear — equations with n independent variables whose values are unknown must be solved.

Solve the following problem with the program **MNEWTON** and compare the solution with the easily obtainable analytical solution. Find the minimum of the following function

$$f(x, y) = x^2 \cdot y^2 + (x + 1)^2 \cdot (y + 1)^2$$

The partial derivatives have to be zero

$$\frac{\partial f(x, y)}{\partial x} = 2xy^2 + 2(x + 1)(y + 1)^2 = 0$$

$$\frac{\partial f(x, y)}{\partial y} = 2x^2 y + 2(x + 1)^2(y + 1) = 0$$

The solution is found for $x = -0.5$ and $y = -0.5$.

Find values for x and y which are extrema for the following function

$$f(x, y) = e^{-(x^2 + y^2)} \cdot \sin(xy)$$

If the partial derivatives are set to zero you will get the two equations

$$-2x \cdot \sin(xy) + y \cdot \cos(xy) = 0$$
$$-2y \cdot \sin(xy) + x \cdot \cos(xy) = 0$$

One obvious solution is $(x = 0, y = 0)$, another one is $x = 0.6809$ and $y = -0.6809$, but there are more.

Problem 169b

We want to find the extrema of a function $f(x, y)$ but with an additional constraint for the independent variables x and y which is given as $g(x, y) = 0$. The method of Lagrange's multiplier gives an easy strategy to solve such problems. A new function $F(x, y, \lambda)$ is built from $f(x, y)$ and $g(x, y)$.

$$F(x, y, \lambda) = f(x, y) + \lambda \cdot g(x, y)$$

Then the partial derivatives of F with respect to x, y and λ are calculated and set to zero.

$$\frac{\partial F}{\partial x} = \frac{\partial f(x, y)}{\partial x} + \lambda \cdot \frac{\partial g(x, y)}{\partial x} = 0$$

$$\frac{\partial F}{\partial y} = \frac{\partial f(x, y)}{\partial y} + \lambda \cdot \frac{\partial g(x, y)}{\partial y} = 0$$

$$\frac{\partial F}{\partial \lambda} = g(x, y) = 0$$

This set of three equations has to be solved and the solutions for x and y are the extrema with the constraint $g(x, y) = 0$. Calculate with this method the maximum of

$$\text{function:} \quad f(x, y) = e^{-(x^2 + y^2)}$$
$$\text{constraint:} \quad g(x, y) = x - y + 2$$

The absolute maximum of $f(x, y)$ is at $(x = 0, y = 0)$ but this solution doesn't fit the constraint $g(x, y)$. Following Langrange's multiplier method we get

$$F(x, y, \lambda) = e^{-(x^2 + y^2)} + \lambda(x - y + 2)$$

$$\frac{\partial F}{\partial x} = -2x \cdot e^{-(x^2 + y^2)} + \lambda$$

$$\frac{\partial F}{\partial y} = -2y \cdot e^{-(x^2 + y^2)} - \lambda$$

$$\frac{\partial F}{\partial \lambda} = x - y + 2$$

Now solve the system of three equations which are obtained by setting the partial derivatives to zero. The solution is $x = -1$ and $y = 1$.

Lagrange's method can be generalized to a function $f(x_1, \ldots, x_n)$ with n independent variables with m different constraints $g_i(x_1, \ldots, x_n)$. The function F is created

$$F(x_1, \ldots, x_n, \lambda_1, \ldots, \lambda_m) = f(x_1, \ldots, x_n) + \sum_{i=1}^{m} \lambda_i \cdot g_i(x_1, \ldots, x_n)$$

From the n partial derivatives of F with respect to the x's and the m partial derivatives with respect to the λ's a set of $n + m$ equations is obtained and this has to be solved.

11.2   Nonlinear Regression

An arbitrary theoretical relation is given between the two measured quantities x and y and n physical parameters $k_1, k_2, \ldots, k_n$.

$$y = f(x, k_1, k_2, \ldots, k_n)$$

In addition there are m data pairs (x_i, y_i) which are experimentally determined and may have small errors. From the theoretical relation and the measurements, the physical parameter shall be determined with the condition that the sum of the squared errors becomes a minimum.

Example: Given is the adsorption isotherm for two energetically different sites on a surface

$$\theta = \frac{k_1 p}{1 + k_1 p} \cdot k_2 + \frac{k_3 p}{1 + k_3 p} \cdot (1 - k_2)$$

θ :   Fraction of surface covered by adsorbed molecules
p :   Pressure in the gas phase
k_1, k_2, k_3 :   Adsorption parameters

The parameters k_1, k_2, k_3 of this isotherm will be determined from a series of measurements of θ_i and p_i. The problem is formulated as the minimization of S in the following equation

$$S = \sum_{i=1}^{m} (y_{i,\text{theor}} - y_{i,\text{exp}})^2$$

If the experimental points $(y_{i,\text{exp}}, x_{i,\text{exp}})$ are abbreviated as (y_i, x_i) the equation can be reformulated as

$$S = \sum_{i=1}^{m} \Big(f(x_i, k_1, k_2, \ldots, k_n) - y_i \Big)^2$$

To obtain the minimum of the sum of the squared errors, all of the n partial derivatives of S with respect to the parameters k_j have to be zero.

$$\frac{\partial S}{\partial k_1} = \frac{\partial S}{\partial k_2} = \ldots = \frac{\partial S}{\partial k_n} = 0$$

By this differential procedure we get n equations. The partial derivative with respect to k_j has the following form

$$\frac{\partial S}{\partial k_j} = 2 \cdot \sum_{i=1}^{m} \Big(f(x_i, k_1, \ldots, k_j, \ldots, k_n) - y_i \Big) \cdot \frac{\partial f(x_i, k_1, \ldots, k_n)}{\partial k_j} = 0$$

For the determination of the unknown parameters $k_1, k_2, \ldots, k_n$, a system of n coupled nonlinear equations must be solved.

To solve this problem the program MNEWTON may be used as the core program for carrying out the nonlinear regression.

To make the program more comfortable, the partial derivatives $\frac{\partial S}{\partial k_j}$ shall be calculated numerically. Then there is no need for the user to program the rather complicated formulas above. This is done at the same position in the program as in the original program, where the system of nonlinear equations was contained.

For this procedure the complete program NL-REGR is listed below:

```
0 REM "NL-REGR "          EBERT/EDERER              881212
1 REM ************************************************
2 REM *** Nonlinear regression by a multi-        ***
3 REM *** dimensional Newton method. The function ***
4 REM *** of regression has to be programmed in   ***
5 REM *** line # 50000, the number of parameters  ***
6 REM *** in line # 20100 and the data input      ***
7 REM *** section starts in line # 60000.         ***
9 REM ************************************************
```

```
100    DIM A(20,20),B(20,20),U(20,20)
110    DIM V(20,20),W(20,20)
200    DIM X(20),X9(20),F(20),F9(20)
210    DIM F0(20),H(20),K(20)
220    DIM XX(100),YY(100)
250    GOTO 20000  :  REM jump to the MAIN program
300    REM *******************************************
305    REM *** Subroutine: INVERS               ***
310    REM *** Input Matrix    in A()           ***
315    REM *** Inverted Matrix in W()           ***
320    REM *******************************************
400    FOR I=1 TO  N
420      FOR J=1 TO  N
460        B(I,J)=0
480        V(I,J)=0
500        IF I<>J THEN GOTO 600
520        B(I,J)=1
540        V(I,J)=1
600      NEXT J
620    NEXT I
1000   FOR Z=1 TO N
1050     S=0
1100     REM ***************************************
1105     REM *** Searching for the pivot element.  ***
1110     REM ***************************************
1120     FOR I=Z TO N
1150       IF S>ABS(A(I,Z)) THEN GOTO 1300
1200       S=ABS(A(I,Z)) : T=I
1300     NEXT I
2000     REM ***************************************
2001     REM *** Lines Z and T will be exchanged.  ***
2005     REM ***************************************
2050     FOR I=1 TO N
2100       S=A(Z,I) : A(Z,I)=A(T,I) : A(T,I)=S
2200     NEXT I
2300     IF ABS(A(Z,Z))>1E-30 THEN GOTO 2400
2350     PRINT "No inversion is possible " : END
2400     V(Z,Z)=0 : V(T,T)=0
2450     V(Z,T)=1 : V(T,Z)=1
3000     REM ***************************************
3020     REM *** Gauss-Jordan elimination         ***
3040     REM ***************************************
3100     FOR I=1 TO N
3200       FOR J=1 TO N
3300         IF I=Z THEN GOTO 4000
3350         IF J=Z THEN GOTO 4500
```

```
3400        U(I,J)=A(I,J)-A(Z,J)*A(I,Z)/A(Z,Z)
3500        GOTO 5000
4000        IF I=J THEN GOTO 4350
4050        U(I,J)=-A(I,J)/A(Z,Z)
4100        GOTO 5000
4350        U(Z,Z)=1/A(Z,Z)
4400        GOTO 5000
4500    U(I,Z)=A(I,Z)/A(Z,Z)
5000    NEXT J : NEXT I
5100    REM ****************************************
5102    REM *** Multiplication  B = V * B         ***
5110    REM ****************************************
5200    FOR I=1 TO N
5250      FOR J=1 TO N
5300        W(I,J)=0
5350        FOR K=1 TO N
5400          W(I,J)=W(I,J)+V(I,K)*B(K,J)
5450    NEXT K : NEXT J : NEXT I
5500    FOR I=1 TO N : FOR J=1 TO N
5550        B(I,J)=W(I,J)
5600    NEXT J : NEXT I
6000    FOR I=1 TO N
6050      FOR J=1 TO N
6100        A(I,J)=U(I,J) : V(I,J)=0
6250        IF I=J THEN V(I,J)=1
6300    NEXT J : NEXT I
6500  NEXT Z
7000  REM ****************************************
7005  REM *** The result is obtained by multiply-  ***
7010  REM *** ing matrix A with the permutation    ***
7020  REM *** matrix B.                            ***
7030  REM ****************************************
7100  FOR I=1 TO N
7200    FOR J=1 TO N
7300      W(I,J)=0
7400      FOR K=1 TO N
7500        W(I,J)=W(I,J)+A(I,K)*B(K,J)
7600  NEXT K : NEXT J : NEXT I
7700  RETURN : REM end of INVERS
12000 REM ****************************************
12010 REM *** Subroutine: FUNCTIONS              ***
12020 REM *** The functions, whose roots have to  ***
12030 REM *** be calculated, are programmed here. ***
12040 REM ***    dS/dK(J)                        ***
12060 REM ****************************************
```

```
12100 FOR J=1 TO N : K(J)=X(J) : NEXT J
12200 GOSUB 30000 : F0=SS
12300 FOR J=1 TO N : DJ=ABS(X(J)/10000!)+1E-08
12400   K(J)=K(J)+DJ : GOSUB 30000
12410   F(J)=SS : K(J)=K(J)-2*DJ : GOSUB 30000
12500   F(J)=(F(J)-SS)/2/DJ
12600   K(J)=X(J) : NEXT J
12999 RETURN
15000 REM ******************************************
15010 REM *** Subroutine: PARTIAL             ***
15020 REM *** The partial derivatives at the po-  ***
15030 REM *** sition X9() are calculated and     ***
15035 REM *** stored in matrix A().            ***
15040 REM ***    d (dS/dK(J) ) / dK(I)         ***
15050 REM ******************************************
15100 FOR I=1 TO N : X(I)=X9(I) : NEXT I
15120 GOSUB 12000 : REM FUNCTIONS calling
15160 FOR I=1 TO N : F9(I)=F(I) : NEXT I
15200 FOR I3=1 TO N : DI=ABS(X(I3)/10000!)+1E-08
15210   X(I3)=X(I3)+DI : GOSUB 12000
15240   FOR J3=1 TO N : A(J3,I3)=F(J3)
15260   NEXT J3 : X(I3)=X9(I3) : NEXT I3
15300 FOR I3=1 TO N : DI=ABS(X(I3)/10000!)+1E-08
15310   X(I3)=X(I3)-DI : GOSUB 12000
15340   FOR J3=1 TO N : A(J3,I3)=(A(J3,I3)-F(J3))/2/DI
15360   NEXT J3 : X(I3)=X9(I3) : NEXT I3
15999 RETURN
19000 REM ******************************************
19010 REM *** Subroutine: OUTPUT              ***
19020 REM ******************************************
19100 PRINT
19120 FOR I=1 TO N
19140   PRINT "k(";I;") =";X9(I);"+/-";
19150   PRINT SQR(ABS(W(I,I)*SS/(NP-N)))
19180 NEXT I
19185 PRINT "Sum of squared errors = ";SS
19190 RETURN
20000 REM ******************************************
20010 REM *** MAIN program                   ***
20020 REM ******************************************
20100 N=2 : REM number of parameters
20130 FF=1 : SM=1E+30
20150 GOSUB 25000
20200 PRINT "Input of the estimated parameters"
20220 PRINT
```

```
20300 FOR I=1 TO N : PRINT"k(";I;") = ";
20320   INPUT X(I) : X9(I)=X(I) : NEXT I
20400 GOSUB 15000 : REM Partial derivatives
20500   GOSUB 300 : REM matrix inversion
20600   FOR I=1 TO N : H(I)=0
20700     FOR J=1 TO N : H(I)=H(I)+W(I,J)*F9(J)
20720   NEXT J : NEXT I
20750   GOSUB 19000 : REM Output
20800   GOSUB 35000 : REM Optimization factor
21000   FOR I=1 TO N : X9(I)=X9(I)-FF*H(I) : NEXT I
21100 GOTO 20400
21500 REM ##########################################
21510 REM ### MAIN program ends here.           ###
21520 REM ##########################################
25000 REM ******************************************
25010 REM *** Subroutine: READ_DATA            ***
25080 REM ******************************************
25100 READ NP : REM number of points
25200 FOR I=1 TO NP : READ XX(I),YY(I) : NEXT I
25400 RETURN
30000 REM ******************************************
30010 REM *** Subroutine: SQUARED_SUM          ***
30020 REM *** The sum of the squared deviations  ***
30030 REM *** is evaluated in this section.      ***
30100 REM ******************************************
30200 SS=0
30300 FOR I=1 TO NP
30350   X=XX(I) : GOSUB 50000 : Y=Y-YY(I)
30400   SS=SS+Y*Y : NEXT I
30500 RETURN
35000 REM ******************************************
35001 REM *** Subroutine: OPTIMUM              ***
35002 REM *** The optimization factor FF is deter- ***
35003 REM *** mined here.                        ***
35005 REM ******************************************
35200 FF=.6 : FM=0 : DF=.5
35310 GOSUB 35600
35320 IF SS<SM THEN SM=SS: FM=FF: FF=FF+DF: GOTO 35310
35330 FF=FM-DF
35335 GOSUB 35600
35340 IF SS<SM THEN SM=SS: FM=FF: FF=FF-DF: GOTO 35335
35360 FOR J=1 TO 7
35380   DF=DF/2
35390   FF=FM+DF : GOSUB 35600
35400   IF SS<SM THEN SM=SS : FM=FF : GOTO 35430
35410   FF=FM-DF : GOSUB 35600
```

```
35420    IF SS<SM THEN SM=SS : FM=FF
35430 NEXT J
35450 FF=FM
35500 RETURN
35600 FOR I=1 TO N : K(I)=X9(I)-FF*H(I) : NEXT I
35620 GOSUB 30000
35650 RETURN
50000 REM *********************************************
50010 REM *** Subroutine: FUNCTION_OF_REGRESSION   ***
50100 REM *********************************************
50200 Y= K(1)*EXP(-K(2)*X)+K(1)*K(2)
50999 RETURN
60000 REM #########################################
60010 REM ### Data section:                     ###
60020 REM ### The first value is the number of  ###
60030 REM ### data points, followed by the data ###
60040 REM ### points as pairs (X,Y).            ###
60050 REM #########################################
61000 DATA  6          : REM number of data points
62010 DATA  0,2
62020 DATA  1,1.368
62030 DATA  2,1.1353
62040 DATA  3,1.04979
62050 DATA  0.5,1.6065
62060 DATA  1.5,1.2231
63999 END


RUN
Input of the estimated parameters

k( 1 ) = ? 3
k( 2 ) = ? 2

k( 1 ) = 3 +/- .1766976
k( 2 ) = 2 +/- .1636607
Sum of squared errors =   177.5771

k( 1 ) = 1.089905 +/- 1.972289E-02
k( 2 ) = .8377619 +/- 3.329766E-02
Sum of squared errors =   3.521497E-03

k( 1 ) = 1.000212 +/- 1.176791E-03
k( 2 ) = .9985304 +/- 2.178653E-03
Sum of squared errors =   5.956755E-06
```

```
k( 1 ) = 1.000117 +/- 2.073943E-04
k( 2 ) = .9997891 +/- 3.847024E-04
Sum of squared errors =  1.637333E-07

k( 1 ) = 1.000044 +/- 2.053504E-04
k( 2 ) = .9999226 +/- 3.810273E-04
Sum of squared errors =  1.600127E-07

k( 1 ) = 1.000044 +/- 2.048676E-04
k( 2 ) = .9999241 +/- 3.801784E-04
Sum of squared errors =  1.599464E-07

^C
Break in 5400
```

The program contains an additional subroutine from line 30000 to 30500 in which the sum of the squared errors is calculated and stored in variable SS. The experimental data pairs (x_i, y_i) are read in the field variables XX(I) and YY(I). For calculating the theoretical y-values, the regression function $f(x, k_1, k_2, ..., k_n)$ is called in line 30350 with the statement GOSUB 50000. In the programmed example this is

$$Y = k_1 \cdot e^{-k_2 x} + k_1 k_2$$

The function has to be programmed using the variables Y, X and K(1), K(2) etc. and is written into the program as follows:

```
50200 Y=K(1)*EXP(-K(2)*X)+K(1)*K(2)
```

The partial derivatives $\frac{\partial S}{\partial k_j}$ of the sum S of squared errors with respect to k_j are approximately calculated in the subroutine from line 12000 to 12999 using the difference quotient. To do this, the sum of the squared errors is calculated using the actual parameters k_j and is stored in the variable FO (line 12200). Then a parameter k_j is increased by a small amount DJ and again the sum of the squared errors is calculated (line 12300 and 12400). After storing the error sum in F(J) a new sum of squared errors is calculated with the again changed parameter k_j, but this time it is decreased by the small amount DJ (line 12410). The difference of the two sums, divided by 2*DJ, is an approximation of the partial derivative $\frac{\partial S}{\partial k_j}$ and is stored in the variable F(J) (line 12500). In the first line of this subroutine (line 12100) the assignment K(J)=X(J) is programmed. The reason for doing this is based on the fact that the program MNEWTON uses the X(J) as the unknown parameters and MNEWTON is used unchanged as a part of the new program for nonlinear regression. In nonlinear regression however, the K(J) are used for the unknown parameters, which shall be determined.

The main program starting in line 20000 is substantially unchanged. Only the unknowns, for which the user has to provide an initial guess in the INPUT-statement, are now called K(J) and no longer X(J). The output subroutine 19000 is called in line 20750. There the actual best parameters k_j, their standard deviations and the sum of the squared errors are printed out.

In line 20800 the subroutine 35000 for the determination of the optimum correction factor FF is called. The calculation is carried out in a very similar way as in the program MNEWTON. The factor FF is searched for which the sum of the squared errors is a minimum. This search is accomplished in the J-loop from line 35360 to 35430 by halving the search interval for each pass of the loop (bisection method). Because this is done 7 times in our program, the resulting value of FF is rather accurate. However, this costs a lot of calculation time and may not be necessary for many problems.

At the beginning of the main program in line 20150, a new subroutine 25000 is called which reads the measured data pairs into the variables XX(I) and YY(I). These quantities are written as DATA-statements starting in line 60000. The first value is the number of data pairs, the following numbers are the (x_i, y_i) data pairs.

Summarizing, the changes the user has to carry out for solving his own regression problem with the program NL-REGR are as follows:

a) All DATA-statements, starting from line 60000, have to be replaced by a new number of data pairs and by the new experimental data.

b) Line 20100 : N=2 has to be substituted by the new number of parameters k_j which should be determined by the program.

c) The function for which the regression shall be carried out has to be programmed in line 50200.

d) In addition, the subroutine 35000 for the calculation of the optimizing factor FF may be changed to increase the calculation speed.

In the example listed with the program, it can be seen that the sum of the squared errors is decreasing by one order of magnitude at each iteration pass. This cannot always be expected for real problems, because the given example is constructed in a way that the assumed measurements lie exactly on the theoretical curve. But because experimental measurements always have errors, the minimum of the sum of the squared errors frequently does not approach zero. The partial derivatives $\frac{\partial S}{\partial k_j}$ however, must converge to zero if the regression run should be considered as successful.

Problem 170

Given is a chemical reaction of first order; the Arrhenius equation describes the temperature dependence of the reaction rate constant. The following experimental data for the reaction rate constants and the temperatures were experimentally obtained:

$k \ [\mathrm{s}^{-1}]$	$4 \cdot 10^{-8}$	$7 \cdot 10^{-7}$	$8 \cdot 10^{-6}$	$7 \cdot 10^{-5}$	$5 \cdot 10^{-4}$	$3 \cdot 10^{-3}$	$2 \cdot 10^{-2}$
$T \ [\mathrm{K}]$	320	340	360	380	400	420	440

Determine the preexponential factor k_0 and the activation energy E_a by linear regression of the relation $\log k$ against $1/T$. Compare the result with the result of the nonlinear regression of the original Arrhenius equation using the program NL-REGR. For the nonlinear regression rather good initial guesses are needed; a promising way is to start with the results of the linear regression. If you generate your own examples, then you can see that the results of the nonlinear and the linear regression agree more, the better the experimental values coincide with the Arrhenius equation. Otherwise stated: the less scattering of the experimental values the better the result of the regression procedure. The differences of the results of both methods result from the different sums of squared errors, which shall be minimized.

$$S_{\text{nonlinear}} = \sum_{i=1}^{m} \left(k_i - k_0 \cdot e^{-\frac{E_a}{RT_i}} \right)^2$$

$$S_{\text{linear}} = \sum_{i=1}^{m} \left(\log k_i - \log k_0 + \frac{E_a}{RT_i} \right)^2$$

Problem 171

Given is the adsorption isotherm shown at the beginning of Chapter 11.2 and the following 10 measured data pairs:

θ	0.05	0.09	0.15	0.23	0.34	0.48	0.64	0.77	0.86	0.93
p	2	4	8	15	30	60	130	250	500	1000

Calculate the parameters k_1, k_2 and k_3 using the program **NL-REGR**. Use an optimizing factor for this problem which is considerably less than one.

Very often small experimental values for y have the same relative error as large ones. The calculation procedure however minimizes the sum of the squared error, and small values are only considered by their absolute value and not by the relative one. For this reason it is valuable for practical applications to be able to weight the experimental data and find the weighted least squares solution of S.

$$S = \sum_{i=1}^{m} w_i \cdot (y_{i,\text{theor}} - y_{i,\text{exp}})^2$$

Then for every experimental point three quantities have to be read into the program, the third number of this triple is the weight of this particular data point. The weight factor can be read for instance in line 25200 into an additional variable **WW(I)**, but it must not be forgotten to provide a dimension statement for this variable at the beginning of the program. Calculating the sum of the squared error in line 30400, the squares have to be multiplied with the respective factor **WW(I)**.

Without proof, another optimizing factor **FG** shall be given; add the following statements as line 15900:

```
15900 FOR I3=1 TO N : A(I3,I3)=A(I3,I3)+FG : NEXT I3
```

This optimizing factor should be greater than zero. It is also possible to use for each **I3** a particular optimizing factor **FG(I3)**. This procedure is usually called the Gauss-Newton method in the literature. If the optimizing factor **FG** becomes zero, then we arrive at the old Newton method. The Gauss-Newton method generally converges slower but it is safer and more robust.

It may happen that a mathematical minimum for the least squares sum was found with parameter values which have no physical meaning. Negative parameters can be avoided if instead of K(J) the absolute value ABS(K(J)) is used. Other values without a physical meaning can be prevented using so-called 'punishment' functions. As an example, the parameter k_2 in the last problem stands for the sizes of different surface fractions and can only have values between zero and one. To enforce this, the subroutine for the calculation of the error sum asks for the value of k_2. If k_2 is greater than the allowed value, then we can add to the sum of the squared errors for example the 4th power of k_2. This constraint brings the parameter back into the allowed region.

Problem 172

If the user has no idea or no hint about the numerical values of the parameters, the described method very often does not lead to reasonable results. Then we have to use other primary searching strategies. A rather slow method, which is simple to program, is the search of the parameter values in given intervals. Random points within these intervals are determined and the error sum is calculated and compared with the previous best results. If a better solution, with a smaller sum of squared errors, is found, then this solution is stored and used as the up until now best parameter value. This Monte-Carlo method should be used only if all other methods fail.

Usually a better method is to fix all parameters with the exception of the first one. Then the minimum of the error sum is searched varying only this first parameter. This can be done with the one-dimensional Newton method or with the more robust bisection method. If the minimum is found, the corresponding value of the first parameter is fixed and then the second parameter is varied to find a better minimum. This procedure is continued through all the parameters. If this procedure is finished then a new search varying the first parameter can be started again and so on. Write a program and test this method.

If there arise greater difficulties in solving a problem with nonlinear regression, as for instance very long computing times or the occurence of convergency problems (sometimes it is impossible to determine a special parameter unambiguously, which may not be realized from the problem itself), then consult your colleagues from the numerical mathematics department.

Problem 173

Extend the program **NL-REGR** to calculate a nonlinear regression for a function which is the numerical solution of a differential equation.

For this the subroutine beginning with line 50000 has to be substituted by a subroutine for solving the respective differential equation.

Problem 174

Expand the program for nonlinear regression to carry out multiple nonlinear regression. This means that the given theoretical relation consists not only of a relation between parameters k_j and pairs of measurements (x, y)

$$y = f(x, k_1, k_2, ..., k_n)$$

but it is also a relation between parameters and tupels of measurements.

$$y = f(x_1, x_2, ..., x_m, k_1, k_2, ..., k_n)$$

Given are the measured data tupels $(y, x_1, x_2, ..., x_m)$. Determine the parameters $k_1, k_2, ..., k_n$, for which the sum of the squared errors is a minimum.

Problem 175

Write a program for nonlinear regression in which not only the parameters of one relation but the parameters of a family of theoretical relations can be determined using a family of experimental data. The different relations may have one or more of the parameters in common.

If an acceleration of the nonlinear regression procedure is required, or if the same regression function is used frequently, then it is recommended to use the analytical derivation for the partial derivatives of the sum of squared errors.

$$S = \sum_{i=1}^{m} \left(y_i - f(\vec{k}, x_i) \right)^2 \tag{1}$$

The vector of the parameters $(k_1, k_2, ..., k_n)$ is abbreviated by $\vec{k}$. The partial derivatives with respect to k_j, which shall become zero, are

$$\frac{\partial S}{\partial k_j} = -2 \cdot \sum_{i=1}^{m} \left(y_i - f(\vec{k}, x_i) \right) \cdot \frac{\partial f(\vec{k}, x_i)}{\partial k_j} \tag{2}$$

Rearranging and differentiating again give the following equation for the second partial derivatives

$$\frac{\partial^2 S}{\partial k_j \partial k_l} = -2 \cdot \sum_{i=1}^{m} \left(y_i - f(\vec{k}, x_i) \right) \cdot \frac{\partial^2 f(\vec{k}, x_i)}{\partial k_j \partial k_l}$$
$$+ 2 \cdot \sum_{i=1}^{m} \frac{\partial f(\vec{k}, x_i)}{\partial k_j} \cdot \frac{\partial f(\vec{k}, x_i)}{\partial k_l} \tag{3}$$

The following program `NL-REGRA` uses these analytical derivatives (2) and (3). The numerical algorithm is still the same as in the program `NL-REGR`.

```
0 REM "NL-REGRA"         EBERT/EDERER         880120
1 REM **************************************************
2 REM *** Nonlinear regression using a multi-     ***
3 REM *** dimensional Newton method with analyti- ***
4 REM *** cal derivatives. The function of regres- ***
5 REM *** sion and the derivatives are programmed  ***
6 REM *** in line # 50000 and # 51000. The data   ***
7 REM *** points start at line # 60000, the number ***
8 REM *** of parameters is contained in line 20100.***
9 REM **************************************************
100    DIM A(20,20),B(20,20),U(20,20)
110    DIM V(20,20),W(20,20)
200    DIM X(20),X9(20),F(20),F9(20)
205    DIM F1(20),F2(20,20)
210    DIM F0(20),H(20),K(20)
220    DIM XX(100),YY(100)
250    GOTO 20000 : REM Jump to the MAIN program
300    REM **************************************************
305    REM *** Subroutine: INVERS                 ***
310    REM *** Input Matrix   in A()              ***
315    REM *** Inverted Matrix in W()             ***
320    REM **************************************************
400    FOR I=1 TO  N
420      FOR J=1 TO  N
460        B(I,J)=0
480        V(I,J)=0
500        IF I<>J THEN GOTO 600
520        B(I,J)=1
540        V(I,J)=1
600      NEXT J
620    NEXT I
1000   FOR Z=1 TO N
1050     S=0
1100     REM ****************************************
1105     REM *** Searching for the pivot element.  ***
1110     REM ****************************************
1120     FOR I=Z TO N
1150       IF S>ABS(A(I,Z)) THEN GOTO 1300
1200       S=ABS(A(I,Z)) : T=I
1300     NEXT I
2000     REM ****************************************
2001     REM *** Lines Z and T will be exchanged.  ***
2005     REM ****************************************
2050     FOR I=1 TO N
2100       S=A(Z,I) : A(Z,I)=A(T,I) : A(T,I)=S
2200     NEXT I
```

```
2300      IF ABS(A(Z,Z))>1E-30 THEN GOTO 2400
2350      PRINT "No inversion is possible " : END
2400      V(Z,Z)=0 : V(T,T)=0
2450      V(Z,T)=1 : V(T,Z)=1
3000      REM *****************************************
3020      REM *** Gauss-Jordan elimination       ***
3040      REM *****************************************
3100      FOR I=1 TO N
3200        FOR J=1 TO N
3300          IF I=Z THEN GOTO 4000
3350          IF J=Z THEN GOTO 4500
3400          U(I,J)=A(I,J)-A(Z,J)*A(I,Z)/A(Z,Z)
3500          GOTO 5000
4000          IF I=J THEN GOTO 4350
4050          U(I,J)=-A(I,J)/A(Z,Z)
4100          GOTO 5000
4350          U(Z,Z)=1/A(Z,Z)
4400          GOTO 5000
4500          U(I,Z)=A(I,Z)/A(Z,Z)
5000      NEXT J : NEXT I
5100      REM *****************************************
5102      REM *** Multiplication  B = V * B        ***
5110      REM *****************************************
5200      FOR I=1 TO N
5250        FOR J=1 TO N
5300          W(I,J)=0
5350          FOR K=1 TO N
5400            W(I,J)=W(I,J)+V(I,K)*B(K,J)
5450      NEXT K : NEXT J : NEXT I
5500      FOR I=1 TO N : FOR J=1 TO N
5550          B(I,J)=W(I,J)
5600      NEXT J : NEXT I
6000      FOR I=1 TO N
6050        FOR J=1 TO N
6100          A(I,J)=U(I,J) : V(I,J)=0
6250          IF I=J THEN V(I,J)=1
6300      NEXT J : NEXT I
6500      NEXT Z
7000      REM *****************************************
7005      REM *** The result is obtained by multiply- ***
7010      REM *** ing matrix A with the permutation   ***
7020      REM *** matrix B.                           ***
7030      REM *****************************************
7100      FOR I=1 TO N
7200        FOR J=1 TO N
7300          W(I,J)=0
```

```
7400      FOR K=1 TO N
7500        W(I,J)=W(I,J)+A(I,K)*B(K,J)
7600  NEXT K : NEXT J : NEXT I
7700  RETURN : REM end of INVERS
16000 REM *******************************************
16010 REM *** Subroutine: PARTIAL                 ***
16020 REM *** The partial derivatives at the po-  ***
16030 REM *** sition X9() are calculated and      ***
16035 REM *** stored in matrix A().               ***
16050 REM *******************************************
16100 SS=0 : FOR I=1 TO N : F9(I)=0 : K(I)=X9(I)
16110   FOR J=1 TO N : A(I,J)=0 : NEXT J : NEXT I
16150 FOR IP=1 TO NP
16160   X=XX(IP) : GOSUB 50000
16180   DI=YY(IP)-Y : SS=SS+DI*DI
16200   FOR I3=1 TO N
16220     F9(I3)=F9(I3)-DI*F1(I3)
16240     FOR J3=I3 TO N
16260     A(I3,J3)=A(I3,J3)-DI*F2(I3,J3)+F1(I3)*F1(J3)
16280     NEXT J3
16300   NEXT I3
16350 NEXT IP
16400 FOR I3=1 TO N
16410   FOR J3=I3 TO N
16420     A(J3,I3)=A(I3,J3)
16430   NEXT J3
16440 NEXT I3
16999 RETURN
19000 REM *******************************************
19010 REM *** Subroutine: OUTPUT                  ***
19020 REM *******************************************
19100 PRINT
19120 FOR I=1 TO N
19140   PRINT "k(";I;") =";X9(I);"+/-";
19150   PRINT SQR(ABS(W(I,I)*SS/(NP-N)))
19180 NEXT I
19185 PRINT "Sum of squared errors = ";SS
19190 RETURN
20000 REM *******************************************
20010 REM *** MAIN program                        ***
20020 REM *******************************************
20100 N=3 : REM number of parameters
20130 FF=1 : SM=1E+30
20150 GOSUB 25000
20200 PRINT "Input of the estimated parameters"
20220 PRINT
```

```
20300 FOR I=1 TO N : PRINT"k(";I;") = ";
20320   INPUT X(I) : X9(I)=X(I) : NEXT I
20400 GOSUB 16000 : REM Partial derivatives
20500   GOSUB 300 : REM matrix inversion
20600   FOR I=1 TO N : H(I)=0
20700     FOR J=1 TO N : H(I)=H(I)+W(I,J)*F9(J)
20720   NEXT J : NEXT I
20750   GOSUB 19000 : REM Output
20800   GOSUB 35000 : REM Optimization factor
21000   FOR I=1 TO N : X9(I)=X9(I)-FF*H(I) : NEXT I
21100 GOTO 20400
21500 REM ##########################################
21510 REM ### MAIN program ends here            ###
21520 REM ##########################################
25000 REM ******************************************
25010 REM *** Subroutine: READ_DATA             ***
25080 REM ******************************************
25100 READ NP : REM number of points
25200 FOR I=1 TO NP : READ XX(I),YY(I) : NEXT I
25400 RETURN
30000 REM ******************************************
30010 REM *** Subroutine: SQUARED_SUM           ***
30020 REM *** The sum of the squared deviations ***
30030 REM *** is evaluated in this section.     ***
30100 REM ******************************************
30200 SS=0
30300 FOR I=1 TO NP
30350   X=XX(I) : GOSUB 51000 : Y=Y-YY(I)
30400   SS=SS+Y*Y : NEXT I
30500 RETURN
35000 REM ******************************************
35001 REM *** Subroutine: OPTIMUM               ***
35002 REM *** The optimization factor FF is deter- ***
35003 REM *** mined here.                       ***
35005 REM ******************************************
35200 FF=.6 : FM=0 : DF=.5
35310 GOSUB 35600
35320 IF SS<SM THEN SM=SS: FM=FF: FF=FF+DF: GOTO 35310
35330 FF=FM-DF
35335 GOSUB 35600
35340 IF SS<SM THEN SM=SS: FM=FF: FF=FF-DF: GOTO 35335
35360 FOR J=1 TO 7
35380   DF=DF/2
35390   FF=FM+DF : GOSUB 35600
35400   IF SS<SM THEN SM=SS : FM=FF : GOTO 35430
35410   FF=FM-DF : GOSUB 35600
```

```
35420   IF SS<SM THEN SM=SS : FM=FF
35430 NEXT J
35450 FF=FM
35500 RETURN
35600 FOR I=1 TO N : K(I)=X9(I)-FF*H(I) : NEXT I
35620 GOSUB 30000
35650 RETURN
50000 REM ********************************************
50010 REM *** Subroutine: FUNCTIONS              ***
50020 REM *** The function of regression and the  ***
50030 REM *** first and second derivatives to the ***
50040 REM *** parameter k's are calculated here.  ***
50050 REM *** In Y the function itself is stored,  ***
50060 REM *** in F1() the first derivatives and   ***
50070 REM *** in F2() the second derivatives.     ***
50100 REM ********************************************
50200 REM   Y=K(1)+K(2)*EXP(-K(3)*X)
50210 F1(1)=1 : F2(1,1)=0 : F2(1,2)=0 : F2(1,3)=0
50220 F1(2)=EXP(-K(3)*X) : F2(2,2)=0: F2(2,3)=-X*F1(2)
50230 F1(3)=K(2)*F2(2,3) : F2(3,3)=-X*F1(3)
50240 Y=K(1)+K(2)*F1(2)
50999 RETURN
51000 REM ********************************************
51010 REM *** Subroutine: FUNCTION OF REGRESSION  ***
51100 REM ********************************************
51200 Y= K(1)+K(2)*EXP(-K(3)*X)
51999 RETURN
60000 REM ########################################
60010 REM ### Data section:                    ###
60020 REM ### The first datum is the number of  ###
60030 REM ### data points, followed by the values ###
60040 REM ### of the data points themselves.   ###
60050 REM ########################################
61000 DATA  64        : REM number of data points
62010 DATA  .000 , 1.000000
62020 DATA  .005 ,  .854415
62030 DATA  .010 ,  .734266
62040 DATA  .015 ,  .627988
62050 DATA  .020 ,  .535742
62060 DATA  .025 ,  .459174
62070 DATA  .030 ,  .395416
62080 DATA  .035 ,  .337545
62090 DATA  .040 ,  .288678
62100 DATA  .045 ,  .249475
62110 DATA  .050 ,  .213886
62120 DATA  .055 ,  .185657
```

```
62130 DATA  .060 ,  .157447
62140 DATA  .065 ,  .135157
62150 DATA  .070 ,  .118451
62160 DATA  .075 ,  .099275
62170 DATA  .080 ,  .085432
62180 DATA  .085 ,  .074251
62190 DATA  .090 ,  .063035
62200 DATA  .095 ,  .054768
62210 DATA  .100 ,  .047787
62220 DATA  .105 ,  .039426
62230 DATA  .110 ,  .034518
62240 DATA  .115 ,  .030223
62250 DATA  .120 ,  .026509
62260 DATA  .125 ,  .023732
62270 DATA  .130 ,  .019806
62280 DATA  .135 ,  .014693
62290 DATA  .140 ,  .011894
62300 DATA  .145 ,  .009542
62310 DATA  .150 ,  .006593
62320 DATA  .155 ,  .005535
62330 DATA  .160 ,  .005540
62340 DATA  .165 ,  .007713
62350 DATA  .170 ,  .003393
62360 DATA  .175 ,  .004965
62370 DATA  .180 ,  .001981
62380 DATA  .185 ,  .003614
62390 DATA  .190 ,  .000934
62400 DATA  .195 ,  .0000215
62410 DATA  .200 ,  .0002637
62420 DATA  .205 ,  .0013670
62430 DATA  .210 ,  .0011571
62440 DATA  .215 , -.0006458
62450 DATA  .220 ,  .0019994
62460 DATA  .225 , -.0035763
62470 DATA  .230 , -.0026426
62480 DATA  .235 , -.0023520
62490 DATA  .240 , -.0026480
62500 DATA  .245 , -.0054493
62510 DATA  .250 , -.0026830
62520 DATA  .255 , -.0028068
62530 DATA  .260 , -.0018326
62540 DATA  .265 , -.0008719
62550 DATA  .270 , -.0006889
62560 DATA  .275 , -.0008531
62570 DATA  .600 , -.0052879
62580 DATA  .605 , -.0058611
```

```
62590 DATA   .610 , -.0047900
62600 DATA   .615 , -.0052341
62610 DATA   .620 , -.0022847
62620 DATA   .625 , -.0024500
62630 DATA   .630 , -.0030974
62640 DATA   .635 , -.0
63999 END

RUN

Input of the estimated parameters

k( 1 ) = ? 1
k( 2 ) = ? 1
k( 3 ) = ? 10

k( 1 ) = 1 +/- .1560369
k( 2 ) = 1 +/- .400806
k( 3 ) = 10 +/- 1.681519
Sum of squared errors =  92.1335

k( 1 ) =-.1958427 +/- 4.274544E-02
k( 2 ) = 1.087971 +/- 7.949695E-02
k( 3 ) = 10.90902 +/- 2.067494
Sum of squared errors =  1.167863

k( 1 ) =-3.187487E-02 +/- 1.522411E-02
k( 2 ) = .9628192 +/- 5.497994E-02
k( 3 ) = 16.53135 +/- 1.276722
Sum of squared errors =   .4637885

k( 1 ) =-3.727321E-02 +/- 7.118935E-03
k( 2 ) = .8726169 +/- 2.367817E-02
k( 3 ) = 21.63466 +/- 1.06516
Sum of squared errors =  9.136758E-02

k( 1 ) = 1.208596E-03 +/- 1.433948E-03
k( 2 ) = .9788397 +/- 5.952308E-03
k( 3 ) = 29.02453 +/- .2939613
Sum of squared errors =   .0048159

k( 1 ) =-2.425226E-03 +/- 3.487824E-04
k( 2 ) = .9982698 +/- 1.454179E-03
k( 3 ) = 30.50817 +/- 7.867524E-02
Sum of squared errors =  2.798721E-04
^C
Break in 51000
```

The function for which the regression shall be carried out is programmed in subroutine 51000. This function, together with its first and second partial derivatives with respect to the parameters k_j, is programmed in subroutine 50000. The subroutines 12000 and 15000 of program NL-REGR are omitted in this program. They are substituted by the new subroutine starting in line 16000. In this part of the program the first derivatives of the sum of squared errors are calculated according to equation (2) and stored in the field F9. The second derivatives are calculated according to equation (3) and stored in the matrix A. The rest of this program section, the main program, the matrix inversion program and the search of the optimizing factor remain unchanged.

The experimental data pairs, which are used for the demonstration of this program, are measured values from a light scattering experiment; they are the autocorrelation values of a polymer solution. If these experimental data are evaluated with the theoretical relation which is programmed in line 51200, then one obtains directly from the parameter K(3) the diffusion coefficient of the polymer sample if K(3) is multiplied with a calibration factor.

Problem 176

Use the data and the regression function of the program NL-REGRA and process these with the program NL-REGR. Compare the results and the needed computing time for both program runs.

A modified method in which only the first derivatives of the regression function are used, is listed below. The method is based on equation (3), where the second derivatives of $f(\vec{k}, x)$ have been neglected.

```
0 REM "NL-MULT "        EBERT/EDERER        881231
1 REM ************************************************
2 REM *** Nonlinear regression using a multi-    ***
3 REM *** dimensional Newton method with analyti- ***
4 REM *** cal derivatives. The function of regres- ***
5 REM *** sion in line # 50000, the data points  ***
6 REM *** in line 60000, the number of parameters ***
7 REM *** in 20100 and the derivatives in subrou- ***
8 REM *** tine DERIV at 15000 are user defined.   ***
9 REM ************************************************
```

```
100    DIM A(20,20),B(20,20),U(20,20)
110    DIM V(20,20),W(20,20)
200    DIM X(20),X9(20),F(20),F9(20)
210    DIM F0(20),H(20),K(20)
220    DIM XX(100),YY(100)
250    GOTO 20000 : REM jump to the MAIN program
300    REM ****************************************
305    REM *** Subroutine: INVERS              ***
310    REM *** Input Matrix    in A()          ***
315    REM *** Inverted Matrix in W()          ***
320    REM ****************************************
400    FOR I=1 TO  N
420      FOR J=1 TO  N
460        B(I,J)=0
480        V(I,J)=0
500        IF I<>J THEN GOTO 600
520        B(I,J)=1
540        V(I,J)=1
600      NEXT J
620    NEXT I
1000   FOR Z=1 TO N
1050     S=0
1100     REM ****************************************
1105     REM *** Searching for the pivot element.  ***
1110     REM ****************************************
1120     FOR I=Z TO N
1150       IF S>ABS(A(I,Z)) THEN GOTO 1300
1200       S=ABS(A(I,Z)) : T=I
1300     NEXT I
2000     REM ****************************************
2001     REM *** Lines Z and T will be exchanged.  ***
2005     REM ****************************************
2050     FOR I=1 TO N
2100       S=A(Z,I) : A(Z,I)=A(T,I) : A(T,I)=S
2200     NEXT I
2300     IF ABS(A(Z,Z))>1E-30 THEN GOTO 2400
2350     PRINT "No inversion is possible " : END
2400     V(Z,Z)=0 : V(T,T)=0
2450     V(Z,T)=1 : V(T,Z)=1
3000     REM ****************************************
3020     REM *** Gauss-Jordan elimination        ***
3040     REM ****************************************
3100     FOR I=1 TO N
3200       FOR J=1 TO N
3300         IF I=Z THEN GOTO 4000
3350         IF J=Z THEN GOTO 4500
```

```
3400        U(I,J)=A(I,J)-A(Z,J)*A(I,Z)/A(Z,Z)
3500        GOTO 5000
4000        IF I=J THEN GOTO 4350
4050        U(I,J)=-A(I,J)/A(Z,Z)
4100        GOTO 5000
4350        U(Z,Z)=1/A(Z,Z)
4400        GOTO 5000
4500        U(I,Z)=A(I,Z)/A(Z,Z)
5000    NEXT J : NEXT I
5100    REM ****************************************
5102    REM *** Multiplication  B = V * B        ***
5110    REM ****************************************
5200    FOR I=1 TO N
5250      FOR J=1 TO N
5300        W(I,J)=0
5350        FOR K=1 TO N
5400          W(I,J)=W(I,J)+V(I,K)*B(K,J)
5450    NEXT K : NEXT J : NEXT I
5500    FOR I=1 TO N : FOR J=1 TO N
5550        B(I,J)=W(I,J)
5600    NEXT J : NEXT I
6000    FOR I=1 TO N
6050      FOR J=1 TO N
6100        A(I,J)=U(I,J) : V(I,J)=0
6250        IF I=J THEN V(I,J)=1
6300    NEXT J : NEXT I
6500    NEXT Z
7000    REM ****************************************
7005    REM *** The result is obtained by multiply- ***
7010    REM *** ing matrix A with the permutation    ***
7020    REM *** matrix B.                            ***
7030    REM ****************************************
7100    FOR I=1 TO N
7200      FOR J=1 TO N
7300        W(I,J)=0
7400        FOR K=1 TO N
7500          W(I,J)=W(I,J)+A(I,K)*B(K,J)
7600    NEXT K : NEXT J : NEXT I
7700    RETURN : REM end of INVERS
12000 REM ****************************************
12010 REM *** Subroutine: NORMAL                   ***
12020 REM *** The matrix A() for the normal        ***
12030 REM *** equations is filled here.            ***
12060 REM ****************************************
12100 FOR I=1 TO N : FOR J=1 TO N : A(I,J)=0 : NEXT J
12150   K(I)=X9(I) : F9(I)=0 : NEXT I
```

```
12200 FOR IP=1 TO NP : X=XX(IP) : GOSUB 50000 : YI=Y
12300   GOSUB 15000 : REM derivatives
12400   FOR I3=1 TO N
12410     FOR J3=I3 TO N
12440       A(I3,J3)=A(I3,J3)+F(I3)*F(J3)
12460     NEXT J3
12500     F9(I3)=F9(I3)-(YY(IP)-YI)*F(I3)
12550   NEXT I3
12600 NEXT IP
12700 FOR I=1 TO N : FOR J=I TO N
12720   A(J,I)=A(I,J) : NEXT J : NEXT I
12999 RETURN
15000 REM ******************************************
15010 REM *** Subroutine: DERIV                  ***
15020 REM *** The derivatives of the regr.function ***
15030 REM *** to the parameter K(J) at the posi-  ***
15040 REM *** tion x is calculated here. This may ***
15050 REM *** be done analytically or numerically ***
15060 REM *** by activating the GOTO at 15105.    ***
15070 REM ******************************************
15100 FOR I=1 TO N : K(I)=X9(I) : NEXT I
15105 REM goto 15300 : REM jump to the numerical sol.
15110 F(1)=1
15120 F(2)=EXP(-K(3)*X)
15160 F(3)=-K(2)*X*F(2)
15200 RETURN
15300 FOR I=1 TO N
15310   DI=ABS(X9(I)/10000!)+1E-08
15340   K(I)=X9(I)+DI : GOSUB 50000:F(I)= Y
15360   K(I)=X9(I)-DI : GOSUB 50000:F(I)=(F(I)-Y)/2/DI
15400   K(I)=X9(I) : NEXT I
15999 RETURN
19000 REM ******************************************
19010 REM *** Subroutine: OUTPUT                 ***
19020 REM ******************************************
19100 PRINT
19120 FOR I=1 TO N
19140   PRINT "k(";I;") =";X9(I);"+/-";
19150   PRINT SQR(ABS(W(I,I)*SS/(NP-N)))
19180 NEXT I
19185 PRINT "Sum of squared errors = ";SS
19190 RETURN
20000 REM ******************************************
20010 REM *** MAIN program                       ***
20020 REM ******************************************
```

```
20100 N=3 : REM number of parameters
20130 FF=1 : SM=1E+30
20150 GOSUB 25000
20200 PRINT "Input of the estimated parameters"
20220 PRINT
20300 FOR I=1 TO N : PRINT"k(";I;") = ";
20320   INPUT X(I) : X9(I)=X(I) : NEXT I
20400 GOSUB 12000 : REM Matrix of the normal equations
20410   FOR I=1 TO N : FO(I)=A(I,I) : NEXT I
20420   FOR I=1 TO N : K(I)=X9(I) : NEXT I
20430   GOSUB 30000 : SM=SS
20450   GOSUB 19000 : REM Output
20500   GOSUB 300 : REM matrix inversion
20600   FOR I=1 TO N : H(I)=0
20700     FOR J=1 TO N : H(I)=H(I)+W(I,J)*F9(J)
20720   NEXT J : NEXT I
20800   GOSUB 35000 : REM Optimization factor
21000   FOR I=1 TO N : X9(I)=X9(I)-FF*H(I) : NEXT I
21010   FOR I=1 TO N : H(I)=F9(I) : NEXT I
21020   GOSUB 35000
21030   FOR I=1 TO N : X9(I)=X9(I)-FF*H(I) : NEXT I
21100 GOTO 20400
21500 REM ##########################################
21510 REM ### MAIN program ends here            ###
21520 REM ##########################################
25000 REM ******************************************
25010 REM *** Subroutine: READ DATA             ***
25080 REM ******************************************
25100 READ NP : REM number of points
25200 FOR I=1 TO NP : READ XX(I),YY(I) : NEXT I
25400 RETURN
30000 REM ******************************************
30010 REM *** Subroutine: SQUARED SUM           ***
30020 REM *** The sum of the squared deviations ***
30030 REM *** is evaluated in this section.     ***
30100 REM ******************************************
30200 SS=0
30300 FOR I=1 TO NP
30350   X=XX(I) : GOSUB 50000 : Y=Y-YY(I)
30400   SS=SS+Y*Y : NEXT I
30500 RETURN
35000 REM ******************************************
35001 REM *** Subroutine: OPTIMUM               ***
35002 REM *** The optimization factor FF is deter- ***
35003 REM *** mined here.                       ***
35005 REM ******************************************
```

```
35200 FF=.6 : FM=0 : DF=.5
35310 GOSUB 35600
35320 IF SS<SM THEN SM=SS: FM=FF: FF=FF+DF: GOTO 35310
35330 FF=FM-DF
35335 GOSUB 35600
35340 IF SS<SM THEN SM=SS: FM=FF: FF=FF-DF: GOTO 35335
35360 FOR J=1 TO 7
35380   DF=DF/2
35390   FF=FM+DF : GOSUB 35600
35400   IF SS<SM THEN SM=SS : FM=FF : GOTO 35430
35410   FF=FM-DF : GOSUB 35600
35420   IF SS<SM THEN SM=SS : FM=FF
35430 NEXT J
35450 FF=FM
35500 RETURN
35600 FOR I=1 TO N : K(I)=X9(I)-FF*H(I) : NEXT I
35620 GOSUB 30000
35650 RETURN
50000 REM ********************************************
50010 REM *** Subroutine: FUNCTION OF REGRESSION   ***
50100 REM ********************************************
50200 Y=K(1)+K(2)*EXP(-K(3)*X)
50999 RETURN
60000 REM ##########################################
60010 REM ### Data section:                     ###
60020 REM ### The first datum is the number of  ###
60030 REM ### data points, followed by the values ###
60040 REM ### of the data points themselves.    ###
60050 REM ##########################################
61000 DATA  64        : REM number of data points
62010 DATA  .000 , 1.000000
62020 DATA  .005 ,  .854415
62030 DATA  .010 ,  .734266
62040 DATA  .015 ,  .627988
62050 DATA  .020 ,  .535742
62060 DATA  .025 ,  .459174
62070 DATA  .030 ,  .395416
62080 DATA  .035 ,  .337545
62090 DATA  .040 ,  .288678
62100 DATA  .045 ,  .249475
62110 DATA  .050 ,  .213886
62120 DATA  .055 ,  .185657
62130 DATA  .060 ,  .157447
62140 DATA  .065 ,  .135157
62150 DATA  .070 ,  .118451
62160 DATA  .075 ,  .099275
```

```
62170 DATA   .080 ,   .085432
62180 DATA   .085 ,   .074251
62190 DATA   .090 ,   .063035
62200 DATA   .095 ,   .054768
62210 DATA   .100 ,   .047787
62220 DATA   .105 ,   .039426
62230 DATA   .110 ,   .034518
62240 DATA   .115 ,   .030223
62250 DATA   .120 ,   .026509
62260 DATA   .125 ,   .023732
62270 DATA   .130 ,   .019806
62280 DATA   .135 ,   .014693
62290 DATA   .140 ,   .011894
62300 DATA   .145 ,   .009542
62310 DATA   .150 ,   .006593
62320 DATA   .155 ,   .005535
62330 DATA   .160 ,   .005540
62340 DATA   .165 ,   .007713
62350 DATA   .170 ,   .003393
62360 DATA   .175 ,   .004965
62370 DATA   .180 ,   .001981
62380 DATA   .185 ,   .003614
62390 DATA   .190 ,   .000934
62400 DATA   .195 ,   .0000215
62410 DATA   .200 ,   .0002637
62420 DATA   .205 ,   .0013670
62430 DATA   .210 ,   .0011571
62440 DATA   .215 , -.0006458
62450 DATA   .220 ,   .0019994
62460 DATA   .225 , -.0035763
62470 DATA   .230 , -.0026426
62480 DATA   .235 , -.0023520
62490 DATA   .240 , -.0026480
62500 DATA   .245 , -.0054493
62510 DATA   .250 , -.0026830
62520 DATA   .255 , -.0028068
62530 DATA   .260 , -.0018326
62540 DATA   .265 , -.0008719
62550 DATA   .270 , -.0006889
62560 DATA   .275 , -.0008531
62570 DATA   .600 , -.0052879
62580 DATA   .605 , -.0058611
62590 DATA   .610 , -.0047900
62600 DATA   .615 , -.0052341
62610 DATA   .620 , -.0022847
62620 DATA   .625 , -.0024500
```

```
62630 DATA   .630 , -.0030974
62640 DATA   .635 , -.0
63999 END

RUN
Input of the estimated parameters

k( 1 ) = ? 1
k( 2 ) = ? 1
k( 3 ) = ? 10

k( 1 ) = 1 +/- 0
k( 2 ) = 1 +/- 0
k( 3 ) = 10 +/- 0
Sum of squared errors =   92.1335

k( 1 ) =-2.022886E-02 +/- 1.873578E-02
k( 2 ) = .7265076 +/- 2.847408E-02
k( 3 ) = 18.14177 +/- .6582113
Sum of squared errors =  .227631

k( 1 ) = 5.576546E-04 +/- 1.722552E-03
k( 2 ) = 1.009808 +/- 4.584228E-03
k( 3 ) = 29.88757 +/- .2284332
Sum of squared errors =   4.163651E-03

k( 1 ) =-1.779021E-03 +/- 3.386148E-04
k( 2 ) = .9977629 +/- 1.390104E-03
k( 3 ) = 30.60199 +/- .0727403
Sum of squared errors =   2.608841E-04

k( 1 ) =-1.752929E-03 +/- 3.356516E-04
k( 2 ) = .9981278 +/- 1.40101E-03
k( 3 ) = 30.62243 +/- 7.572292E-02
Sum of squared errors =   2.604783E-04
^C
Break in 30350
```

The analytical equations of the first derivatives are programmed in subroutine starting in line 15000. The second part of this subroutine is a dormant program section for the numerical calculation of the derivatives. It is activated if the REM-statement in front of the GOTO-statement of line 15105 is removed. In the subroutine 12000, the matrix A and the vector F9 are filled according to the modified normal equations.

This program is extended by an additional procedure to reduce the sum of squared errors. In lines 20800 and 21000 the usual Newton iteration step is made after the determination of the optimizing factor. In the following three lines from 21010 to 21030 it is tried to reduce the error sum further by applying the method of the steepest descent. The vector of the first derivatives F9 is used as the correction vector. The method of the steepest descent alone is rarely successful, but the combination of different methods sometimes accelerates the convergence of the program substantially.

Problem 177

Compare the results and the required computing times, if the program is running with Newton's method or the steepest descent alone and with the combination of both in NL-MULT.

Problem 178

Extend the output of the parameters $\vec{k}$ by the output of the covariance matrix. The covariance matrix $\mathbf{KO}$ is obtained from the inverted matrix $\mathbf{W}$ by multiplying from the left and from the right with a diagonal matrix $\mathbf{D}$, which has the following diagonal elements:

$$D(I, I) = \sqrt{\frac{1}{W(I, I)}}$$

$$D(I, J) = 0 \qquad \text{if} \quad I \neq J$$

Covariance matrix: $\qquad \mathbf{KO} = \mathbf{D} \cdot \mathbf{W} \cdot \mathbf{D}$

The non-diagonal elements of the covariance matrix $\mathbf{KO}$ are a measure for the dependences of the different parameters on each other. If the matrix element $KO(I, J)$ is zero, then the parameters k_i and k_j are independent from each other; if the absolute value equals $KO(I, J) = 1$ then the two parameters k_i and k_j are completely dependent on each other.

There is another method for nonlinear regression which uses a completely different algorithm. It is called the simplex method[1] and it requires only the evaluation of the function. No derivatives have to be computed. The method may be the best one if only bad estimations of the parameters $(k_1, k_2, \ldots, k_n)$

---

[1] J.A. Nelder, R. Mead, *Computer Journal*, **7**, 308 (1965)

are available. The simplex method is robust but it is not very efficient because the convergence is usually slower than with the Newton method. The algorithm is easy to program and can be described from a geometrical point of view.

A simplex in n dimensions is a geometrical figure which consists of $n + 1$ vertices or points. This means that in two dimensions a simplex is a triangle and in three dimensions a tetrahedron. Such a simplex has to be constructed for the unknown parameters $(k_1, k_2, \ldots, k_n)$. We abbreviate a set of parameters $(k_1, k_2, \ldots, k_n)$ as $\vec{k}$. Then $n+1$ simplex points $\vec{k}$ are chosen to build a non-degenerate simplex, which means that the simplex should have a finite n-dimensional volume. The n vector directions $\vec{k}_2 - \vec{k}_1$, $\vec{k}_3 - \vec{k}_1$, ..., $\vec{k}_{n+1} - \vec{k}_1$ should be linearly independent and span the n-dimensional vector space. For $n = 2$ the three points should form a real triangle and they should not lie on a straight line. If the initial guess for the parameters $(k_1, k_2, \ldots, k_n)$ is given as

$$
\vec{k}_1 = \begin{pmatrix} k_1 \\ k_2 \\ \vdots \\ k_i \\ \vdots \\ k_n \end{pmatrix}
$$

then the other vertices $\vec{k}_{i+1}$ can be constructed by

$$
\vec{k}_{i+1} = \vec{k}_1 + \lambda_i \cdot \vec{e}_i = \begin{pmatrix} k_1 \\ k_2 \\ \vdots \\ k_i \\ \vdots \\ k_n \end{pmatrix} + \begin{pmatrix} 0 \\ 0 \\ \vdots \\ \lambda_i \\ \vdots \\ 0 \end{pmatrix} = \begin{pmatrix} k_1 \\ k_2 \\ \vdots \\ k_i + \lambda_i \\ \vdots \\ k_n \end{pmatrix}
$$

This has to be done for all $i = 1, 2, \ldots, n$ to get a set of points which form a non-degenerate simplex. λ_i is a constant which should be chosen by the user and which should be within the problem's characteristic length scale.

The sum of the squared errors has to be calculated for each point of the initial simplex. The point with the least squared sum is called the low point, the point with the largest squared sum is called the high point. After sorting, $\vec{k}_1$ is the low point and $\vec{k}_{n+1}$ is the high point. Now four different movements or changes of the simplex are made:

1) Reflection away from the high point
    First a mean point $\vec{k}_m$ of all vertices without the high point is calculated and the high point is reflected on the mean point to get a new point $\vec{k}_{\text{reflect}}$.

$$\vec{k}_m = \frac{1}{n} \sum_{i=1}^{n} \vec{k}_i$$

$$\vec{k}_{\text{reflect}} = \vec{k}_m + (\vec{k}_m - \vec{k}_{n+1}) = 2\vec{k}_m - \vec{k}_{n+1}$$

2) Double reflection away from the high point
    The high point is reflected as in 1) and in addition is expanded in the same direction to get the new point $\vec{k}_{\text{dreflect}}$.

$$\vec{k}_{\text{dreflect}} = \vec{k}_m + 2 \cdot (\vec{k}_m - \vec{k}_{n+1}) = 3\vec{k}_m - 2\vec{k}_{n+1}$$

3) Contraction of the high point to the mean point
    The new point $\vec{k}_{\text{contr}}$ is the middle between the high point $\vec{k}_{n+1}$ and the mean point $\vec{k}_m$.

$$\vec{k}_{\text{contr}} = \frac{1}{2} \cdot (\vec{k}_m + \vec{k}_{n+1})$$

4) Shrinking of all points towards the low point
    The low point $\vec{k}_1$ remains unchanged but all others move towards the low point and at the half of the distance the new points $\vec{k}_{i,\text{new}}$ are established.

$$\vec{k}_{i,\text{new}} = \frac{1}{2} \cdot (\vec{k}_i + \vec{k}_1) \qquad \text{for all } i = 2, 3, \ldots, n+1$$

The points $\vec{k}_{\text{reflect}}$, $\vec{k}_{\text{dreflect}}$ and $\vec{k}_{\text{contr}}$ are constructed and the squared error sum for each of these points is calculated. The best of these sums is determined and if it is lower than the error sum of the high point, then the high point is substituted by the best of the new points and a new simplex is obtained. If all the points $\vec{k}_{\text{reflect}}$, $\vec{k}_{\text{dreflect}}$ and $\vec{k}_{\text{contr}}$ have error sums higher than the high point, then item 4) is used to shrink the simplex towards the low point.

An iteration of this procedure will always converge to a minimum (perhaps a local one) of the regression sum. The following program SIMPLEX handles as an example the same problem as in the preceding program NL-MULT.

```
0 REM "SIMPLEX "          EBERT/EDERER          880808
1 REM *************************************************
2 REM *** Nonlinear regression using the simplex   ***
3 REM *** method which is due to Nelder and Mead.  ***
4 REM *** The method needs no derivatives.         ***
5 REM ***                                          ***
6 REM *** Regression Function      # 15000 ff      ***
7 REM *** Number of Parameters     # 20100         ***
8 REM *** Data Points              # 60000 ff      ***
9 REM *************************************************
100   DIM XX(100),YY(100),K(20),HE(20)
120   DIM SIMP(21,20),LSQ(21),OP(20)
140   DIM MEAN(20),REFL(20),DREFL(20),CONTR(20)
400   GOTO 20000 : REM  start the MAIN program
15000 REM *************************************************
15010 REM *** Subroutine: FUNCTION OF REGRESSION   ***
15100 REM *************************************************
15200 Y=K(1)+K(2)*EXP(-K(3)*X)
15999 RETURN
20000 REM *************************************************
20010 REM *** MAIN program                         ***
20020 REM *************************************************
20100 N=3 : REM number of parameters
20110 N1 = N + 1
20150 GOSUB 25000
20200 PRINT "Input of the estimated parameters"
20210 PRINT "These parameters must not be zero"
20220 PRINT
20300 FOR I=1 TO N : PRINT"k(";I;") = ";
20320   INPUT K(I) : OP(I)=K(I) : NEXT I
20350 GOSUB 22000 : REM initial simplex is generated
20380 GOSUB 23000 : REM sorting of the simplex
21100 REM GOSUB 26000 : REM output of the whole simplex
21110 GOSUB 24000 : REM output of the best point
21120 GOSUB 27000 : REM mean vector without high point
21200 GOSUB 32000 : GOSUB 33000 : GOSUB 34000
21210 LSQH=LSQ(N1)
21220 IF (LREFL>LSQ(N1))  THEN GOTO 21280
21230    FOR I=1 TO N : SIMP(N1,I)=REFL(I) : NEXT I
21240    LSQ(N1)= LREFL
```

```
21280 IF (LDREFL>LSQ(N1)) THEN GOTO 21340
21290    FOR I=1 TO N : SIMP(N1,I)=DREFL(I) : NEXT I
21300    LSQ(N1)= LDREFL
21340 IF (LCONTR>LSQ(N1)) THEN GOTO 21400
21350    FOR I=1 TO N : SIMP(N1,I)=CONTR(I) : NEXT I
21360    LSQ(N1)= LCONTR
21400 IF (LSQ(N1)<LSQH) THEN GOTO 21500
21420 GOSUB 35000 : REM Contraction towards best point
21500 REM
21600 GOTO 20380
21890 STOP
21900 REM #########################################
21910 REM ### MAIN program ends here          ###
21920 REM #########################################
22000 REM *****************************************
22010 REM *** Subroutine: Initial Simplex      ***
22020 REM *****************************************
22040 FOR I=1 TO N
22050    SIMP(1,I)=K(I)
22060 NEXT I
22070 GOSUB 30000
22080 LSQ(1) = SS
22100 FOR J=1 TO N
22120    K(J) = 1.2*K(J)
22140    GOSUB 30000
22160    LSQ(J+1) = SS
22200    FOR I=1 TO N
22220      SIMP(J+1,I) = K(I)
22240    NEXT I
22260    K(J) = OP(J)
22300 NEXT J
22900 RETURN
23000 REM *****************************************
23010 REM *** Subroutine: Sorting the Simplex  ***
23020 REM *****************************************
23100 FOR J=1 TO N
23120    LSQMIN=LSQ(J) : IMIN = J
23140    FOR I=J TO N1
23160      IF LSQMIN < LSQ(I) THEN GOTO 23240
23180      LSQMIN=LSQ(I) : IMIN = I
23240    NEXT I
23300    FOR I=1 TO N
23320      HE(I)=SIMP(J,I)
23340      SIMP(J,I)=SIMP(IMIN,I)
23360      SIMP(IMIN,I)=HE(I)
23380    NEXT I
```

```
23400   LSQH=LSQ(J) : LSQ(J)=LSQ(IMIN) : LSQ(IMIN)=LSQH
23500 NEXT J
23700 RETURN
24000 REM ********************************************
24010 REM *** Subroutine: OUTPUT                  ***
24020 REM ********************************************
24100 PRINT
24120 FOR I=1 TO N
24140   PRINT ,"k(";I;") = ",SIMP(1,I);TAB(45);"(";SIMP(N1,I);")"
24180 NEXT I
24185 PRINT "Sum of squared errors = ",LSQ(1);
24186 PRINT TAB(45);"(";LSQ(N1);")"
24200 RETURN
25000 REM ********************************************
25010 REM *** Subroutine: READ DATA               ***
25080 REM ********************************************
25100 READ NP : REM number of points
25200 FOR I=1 TO NP : READ XX(I),YY(I) : NEXT I
25400 RETURN
26000 REM ********************************************
26010 REM *** Subroutine: Printing all Simplex    ***
26020 REM ********************************************
26100 FOR I=1 TO N1
26120   FOR J=1 TO N
26140     PRINT SIMP(I,J);
26160   NEXT J
26180   PRINT LSQ(I)
26200 NEXT I
26500 RETURN
27000 REM ********************************************
27010 REM *** Subroutine: Mean without high point ***
27020 REM ********************************************
27100 FOR I=1 TO N : MEAN(I)=0 : NEXT I
27120 FOR J=1 TO N
27130   FOR I=1 TO N
27140     MEAN(J)=MEAN(J)+SIMP(I,J)
27160   NEXT I
27180   MEAN(J)=MEAN(J)/N
27200 NEXT J
27290 RETURN
30000 REM ********************************************
30010 REM *** Subroutine: SQUARED SUM             ***
30020 REM *** The sum of the squared deviations   ***
30030 REM *** is evaluated in this section.       ***
30100 REM ********************************************
```

```
30200 SS=0
30300 FOR I=1 TO NP
30350   X=XX(I) : GOSUB 15000 : Y=Y-YY(I)
30400   SS=SS+Y*Y : NEXT I
30500 RETURN
32000 REM ********************************************
32010 REM *** Subroutine: Reflection point        ***
32020 REM ********************************************
32100 FOR I=1 TO N
32120   REFL(I)=2*MEAN(I)-SIMP(N1,I)
32140   K(I)=REFL(I)
32160 NEXT I
32180 GOSUB 30000
32200 LREFL=SS
32500 RETURN
33000 REM ********************************************
33010 REM *** Subroutine: Double Reflection Point ***
33020 REM ********************************************
33100 FOR I=1 TO N
33120   DREFL(I)=3*MEAN(I)-2*SIMP(N1,I)
33140   K(I)=DREFL(I)
33160 NEXT I
33180 GOSUB 30000
33200 LDREFL=SS
33500 RETURN
34000 REM ********************************************
34010 REM *** Subroutine: Contraction Point       ***
34020 REM ********************************************
34100 FOR I=1 TO N
34120   CONTR(I)=(MEAN(I)+SIMP(N1,I))*.5
34140   K(I)=CONTR(I)
34160 NEXT I
34180 GOSUB 30000
34200 LCONTR=SS
34500 RETURN
35000 REM ********************************************
35010 REM *** Subroutine: Contraction towards the ***
35020 REM ***                best point          ***
35030 REM ********************************************
35100 FOR IC=2 TO N1
35120   FOR J=1 TO N
35140     SIMP(IC,J)=.5*(SIMP(1,J)+SIMP(IC,J))
35160     K(J)=SIMP(IC,J)
35180   NEXT J
35200   GOSUB 30000 : LSQ(IC)=SS
35220 NEXT IC
35900 RETURN
```

```
60000 REM ############################################
60010 REM ### Data section:                        ###
60020 REM ### The first datum is the number of     ###
60030 REM ### data points, followed by the values  ###
60040 REM ### of the data points themselves.       ###
60050 REM ############################################
61000 DATA  64          : REM number of data points
62010 DATA  .000 , 1.000000
62020 DATA  .005 ,  .854415
62030 DATA  .010 ,  .734266
62040 DATA  .015 ,  .627988
62050 DATA  .020 ,  .535742
62060 DATA  .025 ,  .459174
62070 DATA  .030 ,  .395416
62080 DATA  .035 ,  .337545
62090 DATA  .040 ,  .288678
62100 DATA  .045 ,  .249475
62110 DATA  .050 ,  .213886
62120 DATA  .055 ,  .185657
62130 DATA  .060 ,  .157447
62140 DATA  .065 ,  .135157
62150 DATA  .070 ,  .118451
62160 DATA  .075 ,  .099275
62170 DATA  .080 ,  .085432
62180 DATA  .085 ,  .074251
62190 DATA  .090 ,  .063035
62200 DATA  .095 ,  .054768
62210 DATA  .100 ,  .047787
62220 DATA  .105 ,  .039426
62230 DATA  .110 ,  .034518
62240 DATA  .115 ,  .030223
62250 DATA  .120 ,  .026509
62260 DATA  .125 ,  .023732
62270 DATA  .130 ,  .019806
62280 DATA  .135 ,  .014693
62290 DATA  .140 ,  .011894
62300 DATA  .145 ,  .009542
62310 DATA  .150 ,  .006593
62320 DATA  .155 ,  .005535
62330 DATA  .160 ,  .005540
62340 DATA  .165 ,  .007713
62350 DATA  .170 ,  .003393
62360 DATA  .175 ,  .004965
62370 DATA  .180 ,  .001981
62380 DATA  .185 ,  .003614
62390 DATA  .190 ,  .000934
```

```
62400 DATA   .195 ,   .0000215
62410 DATA   .200 ,   .0002637
62420 DATA   .205 ,   .0013670
62430 DATA   .210 ,   .0011571
62440 DATA   .215 , -.0006458
62450 DATA   .220 ,   .0019994
62460 DATA   .225 , -.0035763
62470 DATA   .230 , -.0026426
62480 DATA   .235 , -.0023520
62490 DATA   .240 , -.0026480
62500 DATA   .245 , -.0054493
62510 DATA   .250 , -.0026830
62520 DATA   .255 , -.0028068
62530 DATA   .260 , -.0018326
62540 DATA   .265 , -.0008719
62550 DATA   .270 , -.0006889
62560 DATA   .275 , -.0008531
62570 DATA   .600 , -.0052879
62580 DATA   .605 , -.0058611
62590 DATA   .610 , -.0047900
62600 DATA   .615 , -.0052341
62610 DATA   .620 , -.0022847
62620 DATA   .625 , -.0024500
62630 DATA   .630 , -.0030974
62640 DATA   .635 , -.0
63999 END
```

Input of the estimated parameters
These parameters must not be zero

```
k( 1 ) = ? 10
k( 2 ) = ? 10
k( 3 ) = ? 1
```

k(1) =	10	(12)
k(2) =	10	(10)
k(3) =	1.2	(1)
Sum of squared errors =	20719.17	(26248.35)
k(1) =	6	(10)
k(2) =	12	(12)
k(3) =	1.2	(1)
Sum of squared errors =	15651.56	(25431.6)
k(1) =	6	(10)
k(2) =	8	(10)
k(3) =	1.4	(1)
Sum of squared errors =	9497.307	(21328.81)

```
                 k( 1 ) =          2             ( 10 )
                 k( 2 ) =         10             ( 10 )
                 k( 3 ) =          1.800001      ( 1.2 )
Sum of squared errors =        5619.572         ( 20719.17 )


                 k( 1 ) =         -6             ( 6 )
                 k( 2 ) =         10             ( 12 )
                 k( 3 ) =          2.000001      ( 1.2 )
Sum of squared errors =         279.1642        ( 15651.56 )


                 k( 1 ) =         -4.666667      ( 6 )
                 k( 2 ) =          6.666666      ( 8 )
                 k( 3 ) =          2.266668      ( 1.4 )
Sum of squared errors =         105.4541        ( 9497.307 )



                  . . . . .       . . . . .      . . . . . . .



                 k( 1 ) =         -.2273843      (-.3224511 )
                 k( 2 ) =          .7858639      ( .7740815 )
                 k( 3 ) =         5.931441       ( 4.233292 )
Sum of squared errors =        1.240389         ( 1.57306 )


                 k( 1 ) =         -.2273843      (-.2630174 )
                 k( 2 ) =          .7858639      ( .7495368 )
                 k( 3 ) =         5.931441       ( 4.247579 )
Sum of squared errors =        1.240389         ( 1.54343 )


                 k( 1 ) =         -.1691527      (-.1710384 )
                 k( 2 ) =          .8967436      ( .8231341 )
                 k( 3 ) =         8.087646       ( 6.560191 )
Sum of squared errors =        1.174159         ( 1.395626 )


                 k( 1 ) =         -.1691527      (-.2498325 )
                 k( 2 ) =          .8967436      ( .8604226 )
                 k( 3 ) =         8.087646       ( 6.011205 )
Sum of squared errors =        1.174159         ( 1.36011 )


                 k( 1 ) =         -.1567467      (-.2598746 )
                 k( 2 ) =          .8339815      ( .8722193 )
                 k( 3 ) =         8.790012       ( 6.793336 )
Sum of squared errors =         .8730438        ( 1.254521 )



                  . . . .         . . . .        . . . .
```

```
                    k( 1 ) =        -2.390061E-02    (-1.61313E-03 )
                    k( 2 ) =         .8319288        ( .7720496 )
                    k( 3 ) =         24.56122        ( 19.2899 )
Sum of squared errors =             9.160028E-02    ( .1952779 )


                    k( 1 ) =        -3.103259E-02    (-3.481808E-02 )
                    k( 2 ) =         .9206961        ( .7854691 )
                    k( 3 ) =         24.3131         ( 18.52007 )
Sum of squared errors =             4.795649E-02    ( .1778263 )


                    k( 1 ) =         4.696029E-03    ( 9.750121E-03 )
                    k( 2 ) =         .9974279        ( .9217208 )
                    k( 3 ) =         28.94494        ( 22.3232 )
Sum of squared errors =             1.285707E-02    ( .1587476 )


                    k( 1 ) =         4.696029E-03    (-2.390061E-02 )
                    k( 2 ) =         .9974279        ( .8319288 )
                    k( 3 ) =         28.94494        ( 24.56122 )
Sum of squared errors =             1.285707E-02    ( 9.160028E-02 )


                    k( 1 ) =         4.696029E-03    (-3.497802E-03 )
                    k( 2 ) =         .9974279        ( .9192026 )
                    k( 3 ) =         28.94494        ( 24.13148 )
Sum of squared errors =             1.285707E-02    ( 5.214024E-02 )



                    . . . . .         . . . .          . . . .


                    k( 1 ) =        -1.750963E-03    (-1.752954E-03 )
                    k( 2 ) =         .9981389        ( .9981434 )
                    k( 3 ) =         30.62295        ( 30.62263 )
Sum of squared errors =             2.604785E-04    ( 2.60479E-04 )


                    k( 1 ) =        -1.750963E-03    (-1.752634E-03 )
                    k( 2 ) =         .9981389        ( .9981406 )
                    k( 3 ) =         30.62295        ( 30.62271 )
Sum of squared errors =             2.604785E-04    ( 2.604788E-04 )
^C
Break in 15200
```

The dimensions of the fields in the lines 100, 120 and 140 are given for a maximum of 100 data points and 20 unknown parameters. If you want to use more data or more parameters just change these dimension statements.

The regression function $f(x, k_1, k_2, \ldots, k_n)$ is programmed in the subroutine beginning at line 15000. The data, the subroutine at line 25000 for reading the data, and the subroutine 30000 for calculating the sum of the squared errors are identical to program NL-MULT.

The main program starts at line 20000. In the example the number of parameters is three; this is written in line 20100. The subroutine 25000 for reading the data is called in line 20150. From line 20200 to 20320 the user is asked for his guess of all the parameters k_i. Of course this guess should be as good as possible but for the simplex procedure this initial guess isn't as important as for the Newton methods. Now subroutine 22000 is called where the initial simplex is generated. This is done in a special way and may be adapted by the user to be suitable for the problem under consideration. $\vec{k}_1 = (k_1, k_2, \ldots, k_n)$ is given by the user and none of the parameters can be zero. The additional n vertices are calculated by multiplying the numbers k_i by 1.2 to get the point $\vec{k}_{i+1}$.

$$
\vec{k}_{i+1} = \begin{pmatrix} k_1 \\ k_2 \\ \vdots \\ 1.2k_i \\ \vdots \\ k_n \end{pmatrix} \qquad \text{for all } i = 1, 2, \ldots, n
$$

For each of the points of the simplex (variable SIMP(I,J)) the error sum is evaluated by calling subroutine 30000. The values are stored in the variables LSQ(I). The main program continues to sort the simplex. The sorting criterion is the squared error sum. This is done by calling subroutine 23000. The low point is now stored in SIMP(1,*) and the high point in SIMP(N+1,*). Next, the subroutine 24000 is called to write the parameters of the best and the worst point, including their squared error sums, on the screen. The subroutine 27000 which is called in line 21120 calculates the mean point of all simplex points excluding only the high point. In the next line 21200 three subroutines are called to determine the reflection point, the double reflection point, the contraction point and the appropriate error sums. From line 21220 to 21360 the best of these points is searched, and if it has a lower error sum than the high point then the high point is substituted by the best of the new points. The IF-statement in line 21400 decides whether a better point than the high point has been found or not. If it was possible to substitute the high point then the main program jumps back to line 20380 to repeat the optimization steps. If no better point could be found, then the subroutine 35000 is called where the contraction towards the low point is performed to construct a new simplex. Then the program jumps back to repeat the iteration.

There is no termination criterion in the program. You have to inspect the results on the screen and stop the program with the <CNTR>-<BREAK> keys. If the reader wants to improve the program by a termination criterion he/she has to provide the program with a tolerance value ϵ_i for each parameter k_i and a tolerance level ϵ_{lss} for the least squares sum. The program should be stopped if from one iteration to the next all changes remain within these given tolerances. But such a criterion may cause trouble because it may happen that the simplex is moving through a very small valley and the program may be fooled to terminate without having obtained the real minimum. Therefore, it is often a good idea to restart the simplex procedure with another (very) different initial simplex.

The example shown gives the same results in parameter values and least squares sum as the programs NL-MULT and NL-REGRA.

Problem 178a

With the simplex method it is easy to add restrictions such as inequality restrictions to the parameters. Examples for this are

$$k_2 > 0$$

$$k_4 + k_7 < 10$$

$$0 < k_3 < 1$$

$$-1 < k_1 + k_5 < 1$$

$$10 < \frac{1}{k_6} + \frac{1}{k_8 + 1} < 100$$

You can impose such restrictions by 'punishment functions' or by direct programming of these restrictions in the program section for the construction of the new simplex.

Write a program where only positive parameter values are allowed ($k_i > 0$) and test it with the data and the regression function of the program SIMPLEX.

Problem 178b

The simplex method is not only appropriate for least squares problems. It is also useful for finding the minimum of arbritrary functions and for solving systems of nonlinear equations.

Modify the program **SIMPLEX** in order to solve the nonlinear equations given as examples in the program **MNEWTON**.

Problem 178c

How does **SIMPLEX** work in dimension $n = 1$? Try to find the two roots of the equation

$$\sin x = x^2$$

and compare the results with the solutions given in Chapter 6 which treats equations.

12   Non-Numerical Data Processing

Although an exact distinction between numerical and non-numerical data handling is hardly possible, the following rule of thumb may be helpful: In programs of numerical data processing the number crunching is the main part of the program; in the non-numerical case the following things are done with the data: sorting, searching, storing, restoring, deleting, copying, reading, printing, and so on.

Examples for the application of non-numerical data processing are: booking of airline tickets, lending books from a library, searching in the Chemical Abstracts file, searching for crimes and persons by the police, making time schedules for schools and universities, and computer games.

In this chapter, three examples from the field of non-numerical data processing shall be treated. The first is the establishment of a system of differential equations for the calculation of reaction kinetics from a given reaction mechanism which consists of elementary chemical reactions. In the second example we will try to program a computer game, and in the third one a simulation program for geometrical patterns is written.

These examples should encourage you to get acquainted with additional BASIC-statements not used in this book, but which are important or useful for programming non-numerical problems (for example: LEN, VAL, STR$, ASC, CHR$, LEFT$, MID$, RIGHT$ ...).

In the following examples, however, we are trying to avoid too many new statements and we will try to restrict ourselves mainly to those already used in this book.

12.1   Generation of an ODE-System From a Reaction Mechanism

If a chemist has to construct an appropriate system of differential equations (which is often called the time law) from a reaction mechanism consisting of elementary reactions (ER), then this is usually considered a routine and rather easy task. If the reaction mechanism is large this task is by no means more difficult but it may become more cumbersome and time consuming and may call for a lot more writing and typing work. It may sometimes be very hard to do this relatively easy job without making any errors. At least at this point it becomes obvious that this sort of work is best done by a computer. A program that converts a chemical reaction mechanism into a system of ordinary differential equations is often called a 'chemical compiler'.

Given is the following rather small reaction mechanism consisting of 9 elementary reactions, which describes the thermal decomposition of ethane. The dot which is usually used to designate a free radical is omitted here.

(1)	C_2H_6			$\longrightarrow$	CH_3	$+$	CH_3
(2)	C_2H_6	$+$	CH_3	$\longrightarrow$	C_2H_5	$+$	CH_4
(3)	C_2H_5			$\longrightarrow$	C_2H_4	$+$	H
(4)	C_2H_6	$+$	H	$\longrightarrow$	C_2H_5	$+$	H_2
(5)	CH_3	$+$	C_2H_5	$\longrightarrow$	C_3H_8		
(6)	CH_3	$+$	H	$\longrightarrow$	CH_4		
(7)	C_2H_5	$+$	H	$\longrightarrow$	C_2H_4	$+$	H_2
(8)	C_2H_5	$+$	C_2H_5	$\longrightarrow$	C_2H_4	$+$	C_2H_6
(9)	H	$+$	H	$\longrightarrow$	H_2		

A reaction rate expression which is the product of the reaction rate constant and the respective concentrations of the reactants belongs to each of these elementary reactions.

$$r_1 = k_1 \cdot [C_2H_6]$$
$$r_2 = k_2 \cdot [C_2H_6] \cdot [CH_3]$$
$$r_3 = k_3 \cdot [C_2H_5]$$
$$r_4 = k_4 \cdot [C_2H_6] \cdot [H]$$
$$r_5 = k_5 \cdot [C_2H_5] \cdot [CH_3]$$

$$r_6 = k_6 \cdot [CH_3] \cdot [H]$$
$$r_7 = k_7 \cdot [C_2H_5] \cdot [H]$$
$$r_8 = k_8 \cdot [C_2H_5] \cdot [C_2H_5]$$
$$r_9 = k_9 \cdot [H] \cdot [H]$$

Because the mechanism consists of unimolecular and bimolecular reactions only, the expressions for r_i contain a maximum of two concentrations in the products. The reactive impact of three molecules can be described in a sequence of three elementary reactions.

Instead of	A + A + A	$\longrightarrow$	Products
write:	A + A	$\longrightarrow$	B
	B	$\longrightarrow$	A + A this is a fast reaction
	B + A	$\longrightarrow$	Products

If the second reaction, the decay of the intermediate complex B, is very fast, then it can be shown that a time law of third order (impact of three particles) is obtained if a quasi steady state assumption for B is made.

The close agreement of these two formulations can also be shown if the third order reaction is integrated analytically and compared with the physical analogue evaluated from the three reactions. The latter calculation should be done numerically using a program like SYS-RKN.

If the differential equation for the time dependence of a substance must be constructed, then all the elementary reactions are checked for occurrence of this substance. If the substance is present on the left hand side (the reactants) of the elementary reaction, then the reaction rate r_i of this ER is written with a negative sign in the differential equation. If the substance is found on the right hand side (the products) then r_i is taken with a positive sign. If a species is found twice in an elementary reaction — like H in ER (9) in our example — then r_i must also appear twice in the ODE. For our example the system of differential equations as follows

(1) $d[C_2H_6]/dt = -r_1 - r_2 - r_4 + r_8$

(2) $d[CH_3]/dt = -r_2 - r_5 - r_6 + r_1 + r_1$

(3) $d[C_2H_5]/dt = -r_3 - r_5 - r_7 - r_8 - r_8 + r_2 + r_4$

(4) $d[CH_4]/dt = +r_2 + r_6$

(5) $d[C_2H_4]/dt = +r_3 + r_7 + r_8$

(6) $d[H]/dt = -r_4 - r_6 - r_7 - r_9 - r_9 + r_3$

(7) $d[H_2]/dt \quad = \quad +r_4 + r_7 + r_9$

(8) $d[C_3H_8]/dt \quad = \quad +r_5$

A computer program to execute this procedure is listed as the following program ER-KIN:

```
0 REM "ER-KIN  "           EBERT/EDERER           881111
1 REM ************************************************
2 REM *** Construction of the differential equa-   ***
3 REM *** tions for the chemical kinetics from a   ***
4 REM *** reaction mechanism consisting of elemen- ***
5 REM *** tary reactions.                          ***
9 REM ************************************************
200    DIM ER$(40),N$(20),L$(40,2),L(40),R$(40,4),R(40)
1000   DATA "C2H6               ==>  CH3     + CH3"
1010   DATA "C2H6   + CH3    ==>  C2H5    + CH4"
1020   DATA "C2H5               ==>  C2H4    + H"
1030   DATA "C2H6   + H      ==>  C2H5    + H2"
1040   DATA "CH3    + C2H5   ==>  C3H8 "
1050   DATA "CH3    + H      ==>  CH4"
1060   DATA "C2H5   + H      ==>  H2      + C2H4"
1070   DATA "C2H5   + C2H5   ==>  C2H4    + C2H6"
1080   DATA "H      + H      ==>  H2 "
1999   DATA "END"
2000   N=1 : RESTORE
2020   READ ER$(N)
2030   IF ER$(N)="END" THEN GOTO 2100
2050   GOSUB 10000 : REM DECOMPOSITION
2090   N=N+1 : GOTO 2020
2100   N=N-1
2200   GOSUB 11000 : REM SUBSTANCES
2300   GOSUB 12000 : REM REACTION_RATES
2400   GOSUB 13000 : REM ODE's
2800   REM ######################################
2810   REM ### MAIN program ends here         ###
2820   REM ######################################
9999   END
10000 REM ************************************************
10001 REM *** Subroutine: DECOMPOSITION          ***
10002 REM *** The elementary reactions are decom- ***
10003 REM *** posed to substances belonging either ***
10004 REM *** to the left hand side or to the     ***
10005 REM *** right hand side of the ER.          ***
10006 REM ************************************************
```

```
10020 H$=""
10030 B=LEN(ER$(N))
10033 PRINT N,ER$(N)
10040 FOR I=1 TO B
10060    IF MID$(ER$(N),I,1)=" " THEN GOTO 10090
10080    H$=H$+MID$(ER$(N),I,1)
10090 NEXT I
10100 B=LEN(H$)
10130 Z1=1 : B1=1 : I=1
10160 IF MID$(H$,I,1)="+" THEN GOTO 10220
10170 IF MID$(H$,I,1)="=" THEN GOTO 10250
10180 I=I+1 : GOTO 10160
10220 L$(N,Z1)=MID$(H$,B1,I-B1)
10230 I=I+1 : Z1=2 : B1=I : GOTO 10160
10250 L$(N,Z1)=MID$(H$,B1,I-B1)
10260 L(N)=Z1
10280 I=I+3 : Z1=1 : B1=I
10290 IF I=B THEN 10400
10300 IF MID$(H$,I,1)="+" THEN GOTO 10320
10310 I=I+1 : GOTO 10290
10320 R$(N,Z1)=MID$(H$,B1,I-B1)
10330 I=I+1 : Z1=Z1+1 : B1=I : GOTO 10290
10400 R$(N,Z1)=MID$(H$,B1,I-B1+1)
10410 R(N)=Z1
10500 RETURN
11000 REM ****************************************
11001 REM *** Subroutine: SUBSTANCES          ***
11002 REM ****************************************
11010 PRINT
11020 S=1 : N$(1)=L$(1,1)
11030 FOR I=1 TO N
11040    FOR J=1 TO L(I)
11070       S1=0
11080       FOR K=1 TO S
11100          IF L$(I,J)<>N$(K) THEN GOTO 11140
11120          S1=1
11140       NEXT K
11150       IF S1=1 THEN GOTO 11170
11160       S=S+1 : N$(S)=L$(I,J)
11170    NEXT J
11240    FOR J=1 TO R(I)
11270       S1=0
11280       FOR K=1 TO S
11300          IF R$(I,J)<>N$(K) THEN GOTO 11340
11320          S1=1
11340       NEXT K
```

```
11350     IF S1=1 THEN GOTO 11370
11360     S=S+1 : N$(S)=R$(I,J)
11370   NEXT J
11400 NEXT I
11500 FOR I=1 TO S: PRINT "Y(";I;") = ";N$(I): NEXT I
11510 PRINT
11900 RETURN
12000 REM ********************************************
12001 REM *** Subroutine: REACTION_RATES          ***
12002 REM ********************************************
12030 FOR I=1 TO N
12040   PRINT "R(";I;") = K(";I;") * ";
12050   FOR J=1 TO L(I)
12060     FOR K=1 TO S
12070       IF N$(K)<>L$(I,J) THEN GOTO 12100
12080       PRINT "Y(";K;")";
12090       IF J<>L(I) THEN PRINT " * ";
12100     NEXT K
12120   NEXT J : PRINT : NEXT I
12200 PRINT : RETURN
13000 REM ********************************************
13001 REM *** Subroutine: ODE                     ***
13002 REM ********************************************
13040 FOR I=1 TO S
13050   PRINT "dY(";I;")/dt = ";
13080   FOR J=1 TO N
13090     FOR K=1 TO L(J)
13100       IF N$(I)=L$(J,K) THEN PRINT "-R(";J;")";
13120     NEXT K : NEXT J
13180   FOR J=1 TO N
13190     FOR K=1 TO R(J)
13200       IF N$(I)=R$(J,K) THEN PRINT "+R(";J;")";
13220     NEXT K : NEXT J
13240 PRINT : NEXT I
13300 PRINT : RETURN


RUN
1         C2H6                ==>  CH3    + CH3
2         C2H6  + CH3         ==>  C2H5   + CH4
3         C2H5                ==>  C2H4   + H
4         C2H6  + H           ==>  C2H5   + H2
5         CH3   + C2H5        ==>  C3H8
6         CH3   + H           ==>  CH4
7         C2H5  + H           ==>  H2     + C2H4
8         C2H5  + C2H5        ==>  C2H4   + C2H6
9         H     + H           ==>  H2
```

```
Y( 1 )  =  C2H6
Y( 2 )  =  CH3
Y( 3 )  =  C2H5
Y( 4 )  =  CH4
Y( 5 )  =  C2H4
Y( 6 )  =  H
Y( 7 )  =  H2
Y( 8 )  =  C3H8

R( 1 )  =  K( 1 ) * Y( 1 )
R( 2 )  =  K( 2 ) * Y( 1 ) * Y( 2 )
R( 3 )  =  K( 3 ) * Y( 3 )
R( 4 )  =  K( 4 ) * Y( 1 ) * Y( 6 )
R( 5 )  =  K( 5 ) * Y( 2 ) * Y( 3 )
R( 6 )  =  K( 6 ) * Y( 2 ) * Y( 6 )
R( 7 )  =  K( 7 ) * Y( 3 ) * Y( 6 )
R( 8 )  =  K( 8 ) * Y( 3 ) * Y( 3 )
R( 9 )  =  K( 9 ) * Y( 6 ) * Y( 6 )

dY( 1 )/dt  =  -R( 1 )-R( 2 )-R( 4 )+R( 8 )
dY( 2 )/dt  =  -R( 2 )-R( 5 )-R( 6 )+R( 1 )+R( 1 )
dY( 3 )/dt  =  -R( 3 )-R( 5 )-R( 7 )-R( 8 )-R( 8 )+R( 2 )+R( 4 )
dY( 4 )/dt  =  +R( 2 )+R( 6 )
dY( 5 )/dt  =  +R( 3 )+R( 7 )+R( 8 )
dY( 6 )/dt  =  -R( 4 )-R( 6 )-R( 7 )-R( 9 )-R( 9 )+R( 3 )
dY( 7 )/dt  =  +R( 4 )+R( 7 )+R( 9 )
dY( 8 )/dt  =  +R( 5 )

Ok
```

The program can handle systems of up to 40 elementary reactions and 20 substances. If the number of ER's or species is higher, then the dimensions in line 200 have to be changed.

Lines 1000 to 1999 contain the chemical reaction system. This must be written into the program by the user. Each elementary reaction is written in a DATA-statement within double quotes. The reaction arrow is symbolized by ==>. If a product or reactant appears twice in an elementary reaction, it is written down in two terms and not in the usual way with a figures 2 preceding. The separator between the species is the plus sign (+). Therefore this sign mustn't occur in a symbol of a chemical substance. A particular substance must always have the same symbol within a mechanism. If you are using CH4 and METHANE in the same mechanism, the program will interprete this as two different substances. Blanks have no effect within a mechanism, therefore for the sake of clearness you can write as many blanks as you wish. The reaction system must end (line 1999) with "END" in the last DATA-statement. These restrictions are necessary to keep the analysis and the program from being unnecessarily big. It can be stated as a general rule: The less restrictions given for the user, the more extensive the program has to be.

In lines 2000 to 2100 the elementary reactions are read into the string field ER$(I) and the number of ER's is counted in N. The input is terminated if the word END appears. Then the program jumps to line 2100, where N is decreased by one, because the last reading of END is, obviously, no elementary reaction.

After each reading of an ER the program calls up the subroutine 10000 in line 2050. There the elementary reaction is decomposed into the substances of the reactant and the product side. In the first part of this subroutine from line 10020 to 10100 the ER is printed on the screen and the blanks are removed from the ER. To do this the variable H$, which will store the ER without blanks, is set equal to the null string (line 10020). In the following line 10030 the number of characters of the string in the variable ER$(N) is determined by the statement B=LEN(ER$(N)) and then stored in the number variable B.

Length of a String

$$\boxed{\text{LEN(B\$)}}$$

Syntax: z a = LEN(B$)

z   :   line number
a   :   number variable
B$:   string variable

LEN(B$) results in an integer number which is the length of the string stored in the variable B$.

From line 10040 to 10090 an I-loop runs from 1 to B. In this loop each character is transferred from ER$(N) to H$ if the character is not a blank. The ith character of the text in variable ER$(N) can be determined by the expression MID$(ER$(N),I,1).

Part of a String

$$\boxed{\text{MID\$(A\$,P,L)}}$$

Syntax: z D\$ = MID\$(A\$,p,l)

z : line number
A\$: string variable from which a part, a substring, shall be taken
p : position of the beginning of this substring
l : length of this substring
D\$: string variable containing this substring

If the ith character is a blank (IF-statement in line 10060) then the program jumps to the NEXT I in line 10090, without transferring this blank to the variable H\$. The combining of strings is done using the plus (+) sign.

As soon as the removal of the blanks has ended, then the length of the remaining ER is determined in line 10100 and stored in B.

In the second part of this subroutine the number of substances on the right and on the left side of the nth ER is determined and stored in the variables L(N) and R(N) respectively. The substances contained in this ER are identified and stored in the variables L\$(N,I) and R\$(N,I). From line 10160 to 10180 the string in H\$ is investigated if the characters '+' or '=' (first character of the reaction arrow) are present. In case a '+' is found, the program branches to line 10220, where the substring from the variable H\$ beginning with the first character and ending with the character immediately before the '+' is stored as the first substance of the left hand side of the nth ER in the variable L\$(N,1).

After marking the position of the '+' character, the program jumps back to line 10160. Then the '=' character must be found which identifies the end of the left hand side of the reaction equation. If the '=' character is found, then in line 10250 the substring beginning with the mark and ending with the character immediately before the '=' character is stored as the symbol of a substance in the variable L\$(N,2) or L\$(N,1) depending on whether it is the first or the second species of the reactants. In variable L(N) the number of reactants of the nth elementary reaction is stored.

The same analysis is done from line 10280 to 10410 with the right hand side of the ER. In R\$(N,1), R\$(N,2), ... the symbols of the substances of the right side of the ER are stored and in R(N) the number of these products. After this is done for all ER's, the decomposition of the ER's in reactants and products is completed. It should be noted that in line 10033 the ER's are printed on the screen unchanged together with their identification number N.

After all ER's are read in and decomposed, the main program at line 2200 calls up the subroutine 11000. There the different substances are identified and printed. To begin this task in line 11020, the first substance of the first elementary reaction is stored as the first substance and the identification number S of the different substances found is attributed the value 1. An I-loop from line 11030 to 11400 runs over all n elementary reactions. This I-loop contains two J-loops (11040–11170 and 11240–11370). These J-loops run over all reactants L(I) and all products R(I) of the ith elementary reaction. Within each of these J-loops a further loop, the K-loop, is nested. Before starting a K-loop the pointer S1 is set to zero. In both K-loops all of the different substances identified at this point (the number is stored in variable S) are compared with the jth educt and the jth product of ER number i. If the comparison fails, then the pointer remains unchanged. If two substances are identical, which means that it is an already identified substance symbol, the pointer S1 is set to one. If after the end of the K-loop the pointer is still zero, this particular substance didn't occur until now. The number S of the substances is increased by one and is stored together with the name of the substance in the variable N$(S).

At the end of this subroutine (line 11500), the names of the substances, together with their identification numbers, are printed in the following form:

    Y(I) = name of the substance

This assignment of the name of a species to a field variable is done in order to form a system of ordinary differential equations with indexed variables rather than with the substance names. The substance names are substituted by the respective Y(I), representing the variable for the concentration of the species I. This results in an ODE-system which can be further processed in other computer programs, for example in one of the programs from the chapter on differential equations.

The main program now jumps into the subroutine 12000 where the expression for the reaction rates for each ER is determined. This subroutine consists of an I-loop running over all ER's (and therefore all reaction rates). This contains a J-loop for all reactants. Again a K-loop identifying the species number of the reactant L$(I,J) is nested in the J-loop. This number

K is printed on the screen as Y(K) in line 12080. If there is a second reactant (line 12090) then an asterisk '*' is added. Each output in this subroutine is started in line 12040 by writing on the screen:

```
R(I)  =  K(I) *
```

If all reaction rates are printed out, then the main program calls up the subroutine 13000 where the differential equations are established. This subroutine consists of an I-loop over all S substances (line 13040–13240). Therein are nested two J-loops (lines 13080–13120 and 13180–13220) for searching through all reactants and all products. If a reactant is found that is identical to the ith substance, then -R(J) is printed (line 13100). If it is a product then +R(J) is printed (line 13200). After finishing with this subroutine the main program is terminated. The computer example given is the same which we analyzed by hand at the beginning of this chapter.

Problem 179

Extend the program ER-KIN so that the expressions for the reaction rates are substituted into the differential equations. For calculation purposes this may be disadvantageous but this usual form of the differential equations may be helpful and easier to interpret by the chemist.

Problem 180

Analyze the following reaction mechanism of the oxygen–hydrogen reaction using the program ER-KIN.[1]

$$OH + H_2 \longleftrightarrow H_2O + H$$
$$H + O_2 \longleftrightarrow OH + O$$
$$O + H_2 \longleftrightarrow OH + H$$
$$OH + OH \longleftrightarrow H_2O + O$$

$$H + H \longrightarrow H_2$$
$$H + OH \longrightarrow H_2O$$
$$H + O_2 \longrightarrow HO_2$$

---

[1] J. Warnatz, *Chemistry of Stationary and Non-Stationary Combustion*, p.177 in: K.H. Ebert, P. Deuflhard, W. Jäger, *Modelling of Chemical Reaction Systems, Springer Series in Chemical Physics*, **18**, Springer Berlin Heidelberg New York (1981)

$$H \quad + \quad HO_2 \quad \longrightarrow \quad OH \quad + \quad OH$$
$$H \quad + \quad HO_2 \quad \longrightarrow \quad H_2 \quad + \quad O_2$$
$$OH \quad + \quad HO_2 \quad \longrightarrow \quad H_2O \quad + \quad O_2$$
$$O \quad + \quad HO_2 \quad \longrightarrow \quad OH \quad + \quad O_2$$

This reaction mechanism is a set of 15 elementary reactions because the first four equations consist of forward and backward reactions.

Problem 181

Extend program ER-KIN to check each elementary reaction for its stoichiometric correctness. This can be accomplished by reading in molecular formulas for each substance of the mechanism. The stoichiometric correctness for an ER is proved if the same number of atoms of each element is found on the right and left side of the ER.

12.2   A Computer Game

Computer games are found everywhere nowadays. Most of them are games in which the manual skill of a human player is tested (many of them are shooting games). Some other games belong to the field of artificial intelligence — like chess, checkers, go or reversi etc.

For writing programs for shooting games the most important element is a fast graphic system for presenting the movement of different objects more or less continuously.

Programming intelligent games requires a profound theoretical knowledge of these games and, of course, some programming experience in artificial intelligence.

For another little group of games the computer is used as a tool for one person games, such as Solitaire or the Rubik's cube.

It is common to nearly all computer games that they become boring after a certain time. The reason for this is the inability of computer programs to learn during playing and the absence of the dynamics of a human playing partner.

The computer game which we here are concerned with may be an exception. It is more pleasant to play against a computer than against a partner. The reason for this is relatively simple: it is in fact a two person game, but only one person is challenged by the game, the other person is only needed to give information to the intellectual player. For the intellectual player the game may be interesting, but the other player has only the boring job of giving information. This part of the game can as well be done by a specialized 'idiot'. A computer and a computer program are excellently suited for such an unpleasant task. The game we want to program (the part of presenting information) is known as Mastermind.

Let's introduce quickly the rules of the game: from the first 10 characters a combination of 5 letters is chosen. A character can occur more than once. The intellectual player tries to find out this unknown combination of characters. The number of guesses should be as small as possible. After each guess the information partner tells how many positions are now correct and how many characters are additionally correct but are in the wrong position. The maximum number of allowed guesses should be 15.

```
0 REM "MM        "        EBERT/EDERER        880504
1 REM ************************************************
2 REM *** Mastermind - a computer game          ***
9 REM ************************************************
10    RANDOMIZE (TIMER/3) MOD 32767
50    DIM A$(10),Z$(10),B$(10),H$(10),K$(10)
100   PRINT " M a s t e r m i n d"
200   PRINT: PRINT "You must guess a combination of"
300   PRINT "5 letters out of :   a b c d e f g h i j"
400   PRINT "The program tells you the number of "
500   PRINT "correct positions and the number of "
600   PRINT "additionally correct letters."
700   PRINT : PRINT "You have 15 guesses!" : PRINT
800   PRINT "Enter the 5 chosen letters"
900   PRINT "to the computer with the RETURN key."
1000  PRINT : PRINT "Using a letter k-z cancels the "
1100  PRINT "input and you can choose a new one."
2000  REM ********************************************
2002  REM *** Selection of the letter combination. ***
2006  REM ********************************************
2010  Z$(1)="a"  : Z$(2)="b"  : Z$(3)="c"
2012  Z$(4)="d"  : Z$(5)="e"
2015  Z$(6)="f"  : Z$(7)="g"  : Z$(8)="h"
2017  Z$(9)="i"  : Z$(10)="j"
2020  FOR I=1 TO 5 : X=INT(RND*10+1)
2040    A$(I)=Z$(X) : NEXT I
2900  PRINT : PRINT
2930  PRINT "Now your guess of a combination": PRINT
```

```
3000    REM ******************************************
3002    REM *** Letters from the player           ***
3003    REM ******************************************
3005    FOR M=1 TO 15
3010      FOR I=1 TO 5 : B$(I)=A$(I) : NEXT I
3020      PRINT M;TAB(5);
3030      FOR I=1 TO 5
3040        G$="" : G$=INKEY$ : IF G$="" THEN GOTO 3040
3070        IF G$>"j" THEN GOTO 3200
3072        IF G$<"a" THEN GOTO 3200
3080        K$(I)=G$ : PRINT K$(I);"   "; : H$(I)=K$(I)
3100      NEXT I
3110      G$="" : G$=INKEY$ : IF G$="" THEN GOTO 3110
3112      IF ASC(G$)=13 THEN GOTO 4000
3114      IF G$>"j" THEN GOTO 3200
3118      IF G$<"a" THEN GOTO 3200
3120      GOTO 3110
3150      GOTO 4000
3200      LOCATE CSRLIN-1,1
3210      PRINT : PRINT "                         ";
3220      LOCATE CSRLIN-1,1
3240      PRINT : GOTO 3020
4000    REM ******************************************
4002    REM *** Analysis of the input letters      ***
4003    REM ******************************************
4100      P=0 : REM analysis of the positions
4120      FOR I=1 TO 5 : IF A$(I)><K$(I) THEN GOTO 4160
4140        P=P+1 : B$(I)="0" : H$(I)="1"
4160      NEXT I
4200      PRINT "     Pos  = ";P;
4500      REM analysis of additionally correct letters
4540      B=0
4560      FOR I=1 TO 5 : FOR J=1 TO 5
4580        IF H$(I)<>B$(J) THEN GOTO 4700
4600        B=B+1 : H$(I)="1" : B$(J)="0"
4700      NEXT J : NEXT I
4800      PRINT "    Letters =  ";B
5000      IF P=5 THEN GOTO 9000
8000    NEXT M
8100    PRINT : PRINT "Sorry - you didn't guess "
8110    PRINT " The correct combination is "
8140    PRINT TAB(5);
8150    FOR I=1 TO 5 : PRINT A$(I);"   "; : NEXT I
8999    STOP
9000    PRINT : PRINT "My congratulations"
9999    END
```

```
RUN

 M a s t e r m i n d

You must guess a combination of
5 letters out of :   a b c d e f g h i j
The program tells you the number of
correct positions and the number of
additionally correct letters.

You have 15 guesses!

Enter the 5 chosen letters
to the computer with the RETURN key.

Using a letter k-z cancels the
input and you can choose a new one.


Now your guess of a combination

    1  a  a  b  b  c       Pos  =  0     Letters  =  1
    2  c  d  d  e  e       Pos  =  0     Letters  =  2
    3  f  f  g  g  h       Pos  =  0     Letters  =  2
    4  h  i  i  j  j       Pos  =  0     Letters  =  0
    5  g  g  a  d  d       Pos  =  3     Letters  =  1
    6  g  g  e  d  a       Pos  =  5     Letters  =  0

My congratulations
Ok
```

The initial PRINT-lines contain a short narrative for the game which is printed on the screen.

In the program part from line 2000 to 2040 the computer chooses a random combination of five characters. This target combination should be found out by the competitor. (Of course, it is unfair to change the program in a way to produce some more or less hidden information — like the numbers of blanks in the instruction text — in the combination of letters.)

Because the built-in random generator only supplies random numbers, there has to be some sort of relation between random numbers and random characters. In lines 2010 through 2017 this relation is defined as a table.

The integer numbers from 1 to 10 are attributed the characters from 'A' to 'J' using the assignments Z$(1)="A", ..., Z$(10)="J".

In the following I-loop five integer numbers between 1 and 10 are calculated using the RND-function. The character which is related to the random number X, is stored in the field variables A$(I) by the statement:

    A$(I) = Z$(X)

In the variables A$(1), A$(2), A$(3), A$(4) and A$(5), the target combination of the characters are stored after the end of the loop.

An M-loop which counts the number of guesses runs from line 3005 to 8000. This number M is printed on the screen in line 3020.

In the program from line 3020 to 3240 the player has to type in his guess of character combinations. This is stored both in the field variable K$(I) and in H$(I). Because exactly 5 characters are read from the keyboard the I-loop from line 3030 to 3100 runs from 1 to 5. The input of the letters is programmed with the INKEY$-command which reads the typed characters from the keyboard buffer which may be filled during program execution. This INKEY$-function is programmed as the second statement of line 3040 (G$=INKEY$). If no key has been pressed, then there is nothing in the keyboard buffer. The following IF-condition is affirmed and the program jumps back to the beginning of line 3040. The program remains in this infinite loop as long as the input buffer is empty, that means, as long as no key has been pressed.

If there is a character from 'A' to 'J' in the input buffer, then the next two IF-interrogations in lines 3070 and 3072 are false and the character is stored as well in K$(I) as in H$(I) and is printed on the screen in line 3080.

If the character in the input buffer is greater than 'J' (that means 'K', 'L', ... etc.) or less than 'A' (that means a special character or a digit), then the program jumps to line 3200. There the whole input line is cleared on the screen and the program jumps back to line 3020. Then a new guess of the player is started.

The input of a wrong character can also be used for deleting and restarting an actual guess. This may be important if there is a typing error or if a new idea comes to mind while typing. If five allowed characters are read in, then the I-loop is terminated in line 3100.

The character combination may still be corrected because in line 3110 the program waits for a further input. If an allowed letter from 'A'–'J' is typed in, then nothing happens and the program jumps back by IF-statements (line 3114, 3118 and 3120) to the input loop in line 3110. If another character is typed in then the program jumps to line 3200 where the whole input line is erased and the user can restart a new character combination.

Only if the user presses the <ENTER>- or <RETURN>-key, is the input of his guess finished and the program jumps to line 4000 where the analysis of the character combination is done. How can one test for the <ENTER>-key? By using the function ASC(G$) from the first character stored in G$, an integer number is determined which is characteristic for this character. This code which is a relation between an integer number and a character, is probably printed as a table in your computer manual.

Code of a Character

ASC

Syntax: z A = ASC(B$)

A   : integer variable
B$: string variable

B$ stands for a string variable or a string expression. ASC can be considered as a function which has as the argument the first character of a string expression and as a result an integer number according to a code table. The code used is the decimal representation of the ASCII-code.

The decimal number 13 is the ASCII-code for the <RETURN>-key. Therefore line 3112 is written as:

```
3112  IF ASC(G$)=13 THEN GOTO 4000
```

If the <RETURN>-key is pressed then this IF-statement is confirmed and the program jumps to line 4000 to begin the analysis.

Before the analysis of a guess is described, it should be mentioned that in line 3010 each time before the input of a guess takes place, the correct character combination in A$(I) is also stored in B$(I). This is necessary because B$(I) is changed during the analysis.

The number of correct positions is determined in the program from line 4100 to 4200. To do this, the counter variable P for the correct positions is set to zero in line 4100, and each letter of the correct letter combination A$(I) is compared with the respective guess combination K$(I). If an equality is found, the position counter is increased by one and the respective letters in the correct combination and the guessed combination are destroyed by B$(I)="0" and H$(I)="1". This destruction in the position analysis is necessary to avoid counting these letters again in the following analysis of correct letters on a false position.

If all five letters are compared, then the number of correct positions is printed in line 4200.

The analysis of the additional correct letters is made in the program from line 4540 to 4800. Within the double loop each letter of the above modified correct combination is compared with each letter of the modified guessed combination. If two identical letters are found then the correct-letter counter which was set to zero at the beginning, is increased by one and the respective letters are destroyed in the same manner as above to avoid double counting. After the end of the double loop the number of additional correct letters is printed on the screen in line 4800.

In line 5000 it is asked if 5 positions are correct. If this is the case — the correct combination has been found — then the program jumps to the end of the program at line 9000, where congratulations to the player are printed on the screen.

Otherwise the program runs to **NEXT M** at line 8000 and the M-loop will run through again. At the beginning of this loop in line 3010 the combination B$(I) which were destroyed in the previous analysis is renewed by the statement B$(I)=A$(I).

If the M-loop is terminated regularly after 15 runs, that means after 15 guesses the correct combination of letters has not been found by the player. Then after printing a short regret (line 8100), the correct combination of letters is printed (line 8150) and the program stops in line 8999.

Problem 182

Write the program **MM** using **INPUT**-statements instead of the **INKEY$**.

Problem 183

Write a new or modified program **MASTERMIND**, where the computer may or may not tell lies once during the first 5 guesses of the player. The player should not know during which analysis the computer is lying.

Problem 184

Write a computer-based learning program for the following simple tutorial chemical subject: molecular formulas of alkanes, alkenes and alkines. More than one double and triple bond shall be allowed to in addition cyclic structures.

Computer based learning means: the computer program asks the student different questions from a certain subject. If the answer is correct the computer praises the student, if the answer is wrong the program will explain the problem and then ask (other) questions again.

Problem 185

Write a computer program for the simulation of the movements of Rubik's cube. For this you first have to think about a reasonable notation for the different rotations of the cube. The next problem is the presentation of the result. You may do it by using a letter code for the different colors; on the other hand you can restrict the output to changes only.

Generalize this program to 4 × 4 × 4 and 5 × 5 × 5 cubes.

12.3   The Game of Life

This simulation method for the generation of geometric patterns was popularized by Scientific American.[2] The simulation is done on an area which is divided into rectangular cells. An example is checkered paper. In the original model the area is considered as unlimited. But in order to follow the generation of patterns on the screen, we will consider the limited area of the screen assume the left margin is connected with the right one and also the upper line of the screen with the lower one. This modification is like playing the game of life on the surface of a torus.

At the beginning an initial population is assumed, that means a number of cells is considered as occupied. The occupied fields will be marked by a star on the screen, the empty cells result in a blank. From the initial population the following generation is calculated using the following rules:

a)   Occupied cells will survive only if they have two or three occupied neighboring cells. Neighboring cells are cells with a common line but also cells with a common corner.

---

[2] M. Gardner, *Mathematical Games, Scientific American*, **10**, 120 (1970)

b) Occupied cells are dying and will result in empty cells if there are more than three neighbors (overcrowded) or if there is only one or no neighbor at all (isolation death).

c) A birth occurs at an empty cell if this cell has exactly three occupied neighboring cells.

With these very simple laws geometrical structures are generated which are of a surprising diversity in their dynamics. There are, for example, populations which bloom up at the beginning and later die out completely. There are others which develop stable islands of constant population. There can be oscillating figures or stable figures, which move in different directions within the playing area.

Simulation games of this sort are not only made for entertainment but they can be used — of course with other rules — for solving scientific problems, for example for the simulation of physical-chemical transition points or for simulating the spreading of cancerous cells in a tissue. In the following program the Game of Life covers an area of 20×20 cells. The area is the surface of a torus which is rolled up for inspection on the screen. Therefore the first and 20th line are neighbors as are the first and 20th column.

```
0 REM "GOL     "          EBERT/EDERER          880211
1 REM ************************************************
2 REM ***           Game of Life                ***
3 REM *** Scientific American Oct 1970 and Oct 1983***
4 REM *** The simulation is done on a torus, which ***
5 REM *** means that the upper and lower end of    ***
6 REM *** the screen and the left and right hand   ***
7 REM *** margin are connected.                    ***
9 REM ************************************************
50    DIM X(21,21),Y(21,21),L(17),M$(20)
60    PRINT "You have three different possibilities to "
62    PRINT "choose the initial population"
65    PRINT "1 - free programmed population"
70    PRINT "2 - initial population with a mask"
75    PRINT "3 - random initial population"
80    INPUT "Enter your selection ";Q
100   IF Q=1 THEN GOSUB 1000
105   IF Q=2 THEN GOSUB 5000
110   IF Q=3 THEN GOSUB 2500
120   GOSUB 2000 : REM table of surviving and dying
130   CLS : REM Clear screen
140   GOSUB 3000 : REM Output
```

```
160    GOSUB 4000 : REM a new generation
200    GOTO 140
999    STOP
1000   REM *******************************************
1001   REM *** Subroutine: INITIAL_POPULATION      ***
1005   REM *******************************************
1020   FOR I=1 TO 20
1030     FOR J=1 TO 20
1040       X(I,J)=0
1060     NEXT J
1080   NEXT I
1100   X(8,10)=1 : X(8,11)=1
1120   X(9,9)=1  : X(9,10)=1
1140   X(10,10)=1
1890   G=1
1900   RETURN
2000   REM *******************************************
2001   REM *** Subroutine: TABLE                   ***
2003   REM *** Table of survival and dying.        ***
2005   REM *******************************************
2020   FOR I=1 TO 17 : L(I)=0 : NEXT I
2040   L(3)=1
2060   L(11)=1
2080   L(12)=1
2200   RETURN
2500   REM *******************************************
2501   REM *** Subroutine: RANDOM_POPULATION       ***
2502   REM *******************************************
2510   PRINT "Enter population density between 0 and 1"
2530   RANDOMIZE (TIMER/3) MOD 32767
2540   INPUT P
2620   FOR I=1 TO 20
2630     FOR J=1 TO 20
2640       X(I,J)=0
2650       IF RND(5)<P THEN X(I,J)=1
2660     NEXT J
2680   NEXT I
2940   G=1
2960   RETURN
3000   REM *******************************************
3001   REM *** Subroutine: OUTPUT                  ***
3002   REM *******************************************
3020   LOCATE 1,1 : REM cursor home
3040   FOR I=1 TO 20
3060     PRINT : PRINT"                    ";
3080     FOR J=1 TO 20
```

```
3100      IF X(I,J)=0 THEN PRINT "   ";
3120      IF X(I,J)=1 THEN PRINT " *";
3140    NEXT J
3160  NEXT I : PRINT
3200  FOR I=1 TO 70 : PRINT " "; : NEXT I
3250  LOCATE CSRLIN  ,1 : PRINT "                    ";G;
3260  PRINT " .       generation";
3300  RETURN
4000  REM *****************************************
4001  REM *** Subroutine: NEW_GENERATION        ***
4002  REM *****************************************
4020  FOR I=1 TO 20 : FOR J=1 TO 20
4030      Y(I,J)=X(I,J) : NEXT J : NEXT I
4040  FOR I=1 TO 20 : Y(I,0)=Y(I,20)
4050    Y(I,21)=Y(I,1) : NEXT I
4060  FOR J=0 TO 21 : Y(0,J)=Y(20,J)
4070    Y(21,J)=Y(1,J) : NEXT J
4100  FOR I=1 TO 20 : FOR J=1 TO 20
4120      H=9*Y(I,J)+Y(I-1,J-1)+Y(I-1,J)+Y(I-1,J+1)
4130      H=H+Y(I,J-1)+Y(I,J+1)
4140      H=H+Y(I+1,J-1)+Y(I+1,J)+Y(I+1,J+1)
4160      X(I,J)=L(H)
4200  NEXT J : NEXT I
4300  G=G+1
4990  RETURN
5000  REM *****************************************
5001  REM *** Subroutine: MASK                  ***
5002  REM *** Initial population with a mask    ***
5005  REM *****************************************
5101  M$(1)  ="                    "
5102  M$(2)  ="                    "
5103  M$(3)  ="                    "
5104  M$(4)  ="                    "
5105  M$(5)  ="          **        "
5106  M$(6)  ="           **       "
5107  M$(7)  ="                    "
5108  M$(8)  ="        ****        "
5109  M$(9)  ="     ** *     *      "
5110  M$(10) ="     ** **    *      "
5111  M$(11) ="        * * * **     "
5112  M$(12) ="       * *   * **    "
5113  M$(13) ="         ****        "
5114  M$(14) ="                    "
5115  M$(15) ="          **        "
5116  M$(16) ="          **        "
5117  M$(17) ="                    "
```

```
5118  M$(18) ="                    "
5119  M$(19) ="                     "
5120  M$(20) ="                     "
5200  FOR I=1 TO 20 : FOR J=1 TO 20
5220      X(I,J)=0
5240      IF MID$(M$(I),J,1)="*" THEN X(I,J)=1
5260  NEXT J : NEXT I
5280  G=1
5300  RETURN
```

RUN
You have three different possibilities to
choose the initial population
1 - free programmed population
2 - initial population with a mask
3 - random initial population
Enter your selection ? 2

```
                  * *                           * *
                  * *                           * *

              * * * *                       * * * *
    * *     *           *           * *     *       *     *
    * *   * *           *           * *   * *             *
          *     *   *   * *               *   *       *   * *
          *     *   *   * *               *         *   * *
          * * * *                         * * * *

          * *                             * *
          * *                             * *

    1  .     generation        2  .     generation

                  * *                           * *
                  * *                           * *

              * * * *                       * * * *
    * *     *     *   *             * *     *             *
    * *     *   *       *           * *     *       *   *
          *       * *   * *               *       * *   * *
          *           *   * *             *   *       *   * *
          * * * *                         * * * *

          * *                             * *
          * *                             * *

    3  .     generation        4  .     generation
```

```
             * *
             * *

          * * * *
   * *    *          *
   * *    * *        *
          *     *    *    * *
          *  *       *    * *
          * * * *

          * *
          * *

   5   .      generation
```

RUN
You have three different possibilities to
choose the initial population
1 - free programmed population
2 - initial population with a mask
3 - random initial population
Enter your selection ? 1

```
             * *
          * *
             *

   1   .         generation
```

```
          * * *
          *
          * *

   2   .         generation
```

```
                       *
*  *  *  *             *
*     *  *        *     *                         *
         *        *                             *  *
*                 *  *  *                          *
      *  *           *  *
   *  *  *           *  *
   *  *     *        *  *
         *  *  *     *  *
         *  *  *  *
               *  *
```

                25   .   generation

```
*  *  *             *
*  *  *  *       *  *                    *
*  *     *          *                 *
*  *  *  *  *                       *  *
*  *  *     *                       *  *
      *  *  *     *  *                  *
      *  *        *
         *  *  *

         *  *     *  *
```

                27   .   generation

```
   *           *     *
      *  *           *                      *
      *  *  *     *  *  *                  *  *
         *  *                           *  *
         *        *                       *
            *  *                             *
                  *
         *  *  *
         *  *
```

                29   .   generation

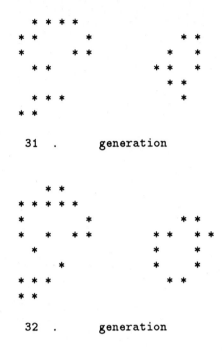

```
    * * * *
  *  *        *              * *
  *         * *            *    *
    * *                * *    *
                          * *
    * * *                  *
  * *
```

31 .   generation

```
      * *
  * * * * *
  *          *              * *
  *   *   * *        * *    * *
    *                  *      *
      *                *      *
  * * *                  * *
  * *
```

32 .   generation

Break in 4100

The program starts by assigning dimensions to the fields X and Y and the vectors M\$ and L. The dimension used is 21×21. For simplifying the calculations, line 1 is made equal to line 21 and column 1 is made equal to column 21. The actual population is stored in the matrix Y. The new population is generated in the matrix X. The vector L serves as the so-called surviving and dying table. The initial population is stored in the vector M\$.

During the next program lines the user is asked how he wants to generate the initial population. If the initial population is chosen arbitrarily then the question in line 80 has to be answered by 1, which causes the program to jump to the subroutine 1000. The whole field X is made empty, which means that each element of the matrix X is set to 0. Then the user has to program by the statements X(I,J)=1 which cells are occupied and the respective elements are filled by a 1. Before this subroutine is terminated the generation counter is set to one.

If the generation of the initial distribution shall be done with a given pattern (the INPUT of line 80 has to be answered with 2) the subroutine 5000 is called. In this subroutine a pattern of 20 simple string assignments is specified. Because all 20 strings (this is the text between quotation marks) have the length 20, the pattern on the screen is the same as the pattern obtained by listing this program. From line 5200 to 5260 the matrix X is filled with 1's or 0's depending on the occurrence of a star at the respective site in the pattern (line 5240).

If the initial population is generated by random numbers, the question in line 80 has to be answered with a 3 and then the subroutine starting in line 2500 is called. There, the user is asked for the population density. In the double loop between line 2620 and 2680, a random number is generated for all 400 cells. If the respective random number is smaller than the specified population density then the cell is filled with a 1, otherwise it is filled with a zero. At the end of the last two subroutines the generation counter G is set to one.

After the initial population is generated in one of the three given ways, the program jumps from line 120 to the subroutine starting with line 2000 where the surviving, birth and dying table is generated by the given rules. The number of occupied neighbors is counted for each cell and if the cell itself is occupied then a 9 is added. This method characterizes the state of a cell unambiguously by a reference number between 0 and 17. In the surviving and dying table every reference number is associated by a one, if the respective state of the cell results in an occupied cell in the next generation. According to the rules of our game only the reference numbers L(3) (birth) and L(11) and L(12) (survival) will get a 1. For all other states the cell will either die or remain empty. Using this table technique the rules of the game can be changed easily by changing or expanding the assignments between lines 2020 and 2200 in the program.

After jumping back to the main program, line 130 causes the screen to be cleared and in the next line the subroutine 3000 is called up for the output of the next generation on the screen. This subroutine poses the cursor in the upper left corner and prints 20 lines with 20 characters each. If there is a zero in X(I,J) then a blank is printed at the Jth column of the Ith line (line 3100); an asterisk is printed if X(I,J)=1 (line 3120). To move the whole pattern to the middle of the screen the statements in line 3060 are used to start each printing line with 20 blanks. The number of the generation is printed in line 3250.

Immediately afterwards the main program directs the control to the subroutine 4000 in which the new generation is calculated. To start with, the matrix Y is made equal to the matrix X (line 4020), column 0 is made equal to column 20 and the column 21 is made equal to column 1 (line 4040). In line 4060 the same procedure is done with the first and the last line of the pattern matrix. The following double loop from line 4100 through 4200 determines the state of each cell by counting the number of living neighbors and storing the result in the variable H. The number 9 is added to H if the cell itself occupied. With the statement X(I,J)=L(H) a one or a zero is read from the table L (depending on the value of H) and assigned to the cell X(I,J) as the new state. After the new generation has been established in this way the last statement before jumping back increases the generation counter by one.

The main program now jumps back (this is an infinite loop) to line 140 and starts the determination of the next generation.

Problem 186

Work with different initial populations and try to find interesting figures like 'sliders' or 'blinkers'.

Problem 187

Change the rules of the game in GOL. This can be done in a deterministic way by changing the survival and birth table L. Also change the rules into stochastic rules, which means that for each state of the cell you have to specify a probability for being an occupied cell in the next generation.

Problem 188

Such geometric pattern simulation programs may be used for the simulation of real problems like extraction (making coffee), structure of polymers, the conductivity of alloys, the effectivity of a telephone network or the spreading of forest fires or infectious diseases.

Write a program for the simulation of the two-dimensional Ising model for ferromagnetism. Each cell can be in one of two spin states. In this model only four neighbors are considered which have a common horizontal or vertical line. If a cell has two neighbors with spin up and two with spin down then the cell may have one of the two spins next generation with the same probability. If there are more neighbors with one sort of spins, then the cell is considered to have a higher probability to occupy this spin. Additionally the probabilities are temperature dependent. For example, assume that three identical neighbors result in a probability expression of $0.5+0.25/(1+a*T)$ for occupying the same spin as the majority; if there are four identical neighbors this probability may be $0.5+0.5/(1+a*T)$. The constant a is a characteristic of the material and is related to the Curie temperature of the material considered.

Of course you cannot expect that this simple two-dimensional model supplies a sharp Curie point for magnetizing and demagnetizing as it is in reality. But this model shows clearly the more statistical distribution of the spins at higher temperatures and large regions of identical spins at lower temperatures. In addition you can see that at a temperature of 0 K the antiferromagnetic state is as stable as the ferromagnetic state.

You can add to the model the influence of an external magnetic field by assuming an asymmetric probability.

Problem 189

Write a program for the generation of certain DNA-molecules as an example of a self-organizing system. Use the following simple model: choose 100 arbitrary DNA-chains of length 12 by generating random sequences of the letters A, T, C, G. Choose another sequence of DNA and declare it as the ideal one, which is thought to have the best chances for reproduction and survival. Make a reduplication step of 100 DNA-chains to 200 DNA-chains. But the reproduction should be done with an error rate (call it mutations) of 1%. From the 200 DNA-chains 100 shall die in the next step. The chances of survival will be the better the more similar a DNA-sequence is to the ideal sequence. This can be done by reducing the dying probability by a factor of 2 for each agreement in a letter with the ideal sequence. This reproduction and dying procedure shall be repeated very frequently. After a great number of generations almost all DNA-chains will have the ideal structure, though the probability for generation of the ideal chain by chance is only $1/16777216$. What happens if the mutation rate in the reproduction step is too high or too low?

Problem 189a

Write a program to simulate a special case of the 'percolation' problem.

Given is an $n \times n$ matrix filled with A's and B's. This should be done by a random process where the probability for filling a certain matrix element with A is given by p_A. In order to fill the matrix completely the probability for B has to be $p_B = 1 - p_A$. This may be considered as a very simple model for an alloy.

What is the minimum probability for p_A to get structures in the matrix where all (or nearly all) A's are connected?

A very similar problem is given by a quadratic grid where neighboring grid points are connected by electric resistors. The resistances may be either very high (∞) or very low (0) and their distribution is again given by a random process. Determine the minimum probability for the low resistances p_0 in order to have an electric conductivity from the left side of the grid to the right side. This may be considered as a very simple model for superconductivity.

13 Computer Graphics

The results of measurements or calculations don't reveal the whole information immediately to the observer when the results are given in tables.
To make this more evident, consider the following experiment. A person
is given a long table consisting of two columns of numbers, which are the
sines and cosines of a list of independent values. If the person is asked for
the relation between the two columns hardly anyone will guess the correct
answer immediately. But if the values of the two columns are plotted in an
(x, y)-diagram everyone will recognize the resulting circle at once.

It is often useful to present results in a graphical form. In some of the
problems it was recommended to set up plots or diagrams manually. In this
chapter we will show some methods to do this with the computer.

To make graphics with a computer it is necessary to have suitable hardware. Either a screen is needed which has plotting ability, or a connection to
a plot terminal or to a paper plotter. Low resolution pictures can be made
with almost every screen; the resolution is restricted to the number of lines
and the number of characters in a line. The BASIC-commands for plotting
may be very different depending on your hardware specification. But the
programs of this chapter use only statements available for every computer.
This refers to the semantics, the syntax may be different. We are using here
the BASICA statements for an IBM AT or compatible computers.

To make it easier for the user to translate the plot commands of the
statements used, their semantics and their syntax are shown. But we will
discuss only the most important graphic statements and the descriptions of
statements are restricted to the most important parameters. For most of
the statements there is a great variety of expansions and parameter combinations for color, linetype, etc. To use these, consult your computer and
language handbook.

Changing to the Graphics Mode

> SCREEN m

Syntax: z SCREEN m

z   :   line number
m   :   mode (integer number)

The variable m stands for mode; it may have four different values:

mode=0 is the text mode,

mode=1 gives medium resolution graphics with 320×200 pixels and four different colors, e.g. black, white, blue and red,

mode=2 results in high resolution graphics with 640×200 pixels and 2 colors (black and white),

mode=9 is used for the high resolution EGA mode which has 640×350 pixels and different 16 colors.

World Coordinates are Defined

> WINDOW (X1,Y1)-(X2,Y2)

Syntax: z WINDOW (x_1, y_1)-(x_2, y_2)

z   :   line number
x, y :   variables, constants or expressions

The screen coordinates without the WINDOW-command are 0 to 319 (or 639 for mode=2) in X and 0 to 199 in Y if mode=1 is used. If the figure you want to display doesn't fit into these coordinates you have to use the WINDOW-command. The pairs (x_1, y_1) and (x_2, y_2) represent the coordinates of the opposite diagonal corners. Only points or lines which lie between these two points can be plotted. The connection of the WINDOW-statement to the whole screen is cancelled if a VIEW-statement has been used. The WINDOW-statement refers now to the part of the screen which is defined by the VIEW-statement.

A Part of the Screen is Defined as Graphics Display

> VIEW (X1,Y1)-(X2,Y2)

Syntax: z VIEW (x_1, y_1)-(x_2, y_2)

z   :   line number

x, y :   variables, constants or expressions

This command is needed if not the entire screen is used for the graphic display. The physical coordinates of the VIEW-command define the part of the screen to be reserved for the display graphics. Using VIEW as it is given above — together with the WINDOW-statement — reduces the whole contents of the figure in the window to the part of the screen defined by VIEW. If you are using VIEW as: VIEW SCREEN (x_1, y_1)-(x_2, y_2), then this extension SCREEN causes the statement to show only the defined part of the screen and only the part of the defined window, the figure is not reduced.

Coloring an Individual Point (x, y)

> PSET (X,Y),C

Syntax: z PSET (x, y),c

z   :   line number

x, y :   variables, constants or expressions

c   :   color attribute value

The point defined by the coordinates (x, y) is given the color C. For medium resolution there are four color attribute values (0–3), for high resolution there are only two values (0 and 1).

Drawing a Line From (x_1, y_1) to (x_2, y_2)

> LINE (X1,Y1)-(X2,Y2)

Syntax: z LINE (x_1, y_1)-(x_2, y_2)

z   :   line number

x, y :   variables, constants or expressions

This statement draws a line from the point (x_1, y_1) to the point (x_2, y_2) in the coordinates defined by WINDOW. Because this statement is used very often in the examples shown, the syntax for the extensions is given too:

```
LINE (X1,Y1)-(X2,Y2), Color, B, Style
```

The Color may be an integer 0, 1, 2 or 3 for the color of the line. The option B draws not a line but a box with the two points in the opposite corners. If the option BF is used the box is filled with the specified color. The parameter Style is used for drawing lines other than solid lines. Style stands for a two byte variable or constant which specifies by its bit pattern the way in which the line is to be drawn. A dotted line is obtained for example if &HAAAA is used, because the hexadecimals AAAA are written in bits as 1010101010101010. The first point in this statement (x_1, y_1) may be omitted. Then a line is drawn from the last point used in a former plot statement to (x_2, y_2).

<table>
<tr><td>

Drawing a Circle Around the Point (x, y) with Radius r

</td><td>

CIRCLE (X,Y),R

</td></tr>
</table>

Syntax: z CIRCLE $(x, y), r$

z   : line number
x, y, r : variables, constants or expressions

The center of the circle has the coordinates (x, y) and the circle's radius is r, both in the coordinates defined by WINDOW. The radius r refers to the coordinates in the x-direction. Additional parameters may be used for different colors, for an aspect ratio to draw ellipses, or for drawing a sector of the circle or the ellipse only.

13.1   Graphical Games

A short program shall now be presented which serves as an introduction to plotting. With this program, arbitrary but nice figures will be plotted on the screen. The program with the name PLAYING is listed below:

```
0 REM "PLAYING "            EBERT/EDERER            880104
1 REM ************************************************
2 REM *** Graphical random figures with sines     ***
3 REM *** cosines and random numbers.             ***
9 REM ************************************************
40      RANDOMIZE (TIMER/3) MOD 32767
100     SCREEN 2 : CLS : KEY OFF
200     WINDOW (-3,-3)-(3,3)
260     A4=RND : A3=RND
270     B4=RND : B3=RND
280     C1=RND*2
300     FOR T=0 TO 12.6 STEP .05
310       D=T+C1
400       X=2*COS(A4*T)-COS(A3*2*T)
410       X2=2*COS(A4*D)-COS(A3*2*D)
500       Y=2*SIN(B4*T)-SIN(B3*2*T)
510       Y2=2*SIN(B4*D)-SIN(B3*2*D)
590       LINE (X2,Y2)-(X,Y)
700     NEXT T
800     INPUT A$
900     SCREEN 0 : WIDTH 80
999     END
```

In the four examples shown in Fig. 13.1 randomly generated figures are presented. The program starts after the usual comment block by initializing the random number generator in line 40. In line 100 the statement SCREEN 2 switches to the graphic screen with high resolution. It makes sense to use the statements CLS (for clear screen) and KEY OFF to remove the key menu at the bottom of the screen. The statement WINDOW in line 200 defines the mathematical coordinates within which lines will be drawn. The points which will be plotted are expected to be between -3 and $+3$ as well for X as for Y.

The plot to be generated is a parameterized function for x and y. The x-values are a cosine function with a parameter t (lines 400 and 410), the y-values are a sine function with the same parameter t (lines 500 and 510). Four additional parameters are needed for the frequency and the offset. They are generated in lines 260 and 270 with random numbers.

With each run through the parameter loop (T-loop) two points (two x- and two y-values) are generated. The second point is formed from the first one by simply increasing the parameter T by C1. C1 is again a random number between 0 and 2 which is calculated in line 280. The LINE-statement in line 590 draws a solid line from point (X2,Y2) to (X,Y) on the screen. The small increase of the parameter T produces a line which is not too far from the previous one. The resulting figures may be considered as beautiful and esthetic plots.

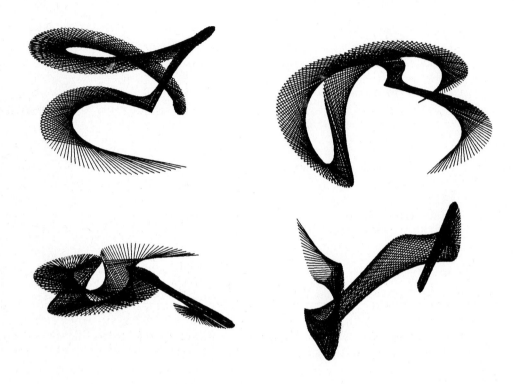

Fig. 13.1. Different examples of the output of the program PLAYING

After the end of the plotting the computer waits for an arbitrary INPUT (line 800) and then restores the text mode with the statement SCREEN 0. In addition the command WIDTH 80 sets the number of characters for a line to 80.

Problem 190

Write a program to plot the graphs of the parameterized functions used in Problem 36. Use similar functions to test your program.

Problem 191

Expand the program BISECT with a plotting subroutine. Plot the function, whose roots are searched together with the *x*-axis. Now you can see the approximate locations of the roots before you put in the interval for the bisection search.

Problem 192

Extend the program HEAT-CON with a plotting section, a criterion for terminating the calculation and a loop for the location variable X. Draw plots for the temperature distribution calculated for the rod at different times.

Expand other programs of this book with plot routines. This is useful if the graphical output gives faster and clearer information about the results than the numerical output.

13.2 Linear Regression with Graphics

The following program is based on the program for linear regression REGLINW and is extended by a plot subroutine:

```
0 REM "REGLINP "         EBERT/EDERER         880217
1 REM *************************************************
2 REM *** Regression to a straight line with cal-  ***
3 REM *** culation of errors for the parameters.   ***
9 REM *************************************************
1000   DIM X(100),Y(100),W(100)
1010   A$="                            "
1100   INPUT "How many data points (y,x) ";N
1120   FOR I=1 TO N : W(I)=1 : NEXT I : PRINT
1140   PRINT "Do you want to weight your data points ";
1150   INPUT B$
1200   FOR I=1 TO N
1300     PRINT "  y(";I;") = ";
1310     INPUT Y(I)  : LOCATE CSRLIN-1,1
1330     PRINT A$;A$ : LOCATE CSRLIN-1,1
1350     PRINT "  y(";I;") =";Y(I);
1400     PRINT "  x(";I;") = ";
1410     INPUT X(I)  : LOCATE CSRLIN-1,1
1450     IF B$<>"yes" THEN GOTO 1615
1490     PRINT A$;A$ : LOCATE CSRLIN-1,1
1500     PRINT "  y(";I;") =";Y(I);
1600     PRINT "  x(";I;") =";X(I);
```

```
1610    PRINT "  w(";I;") = ";
1613    INPUT W(I)  : LOCATE CSRLIN-1,1
1615    PRINT A$;A$ : LOCATE CSRLIN-1,1
1620    PRINT "  y(";I;") =";Y(I);
1630    PRINT "  x(";I;") =";X(I);
1640    PRINT "  w(";I;") =";W(I)
1700  NEXT I
2000  S1=0
2100  S2=0 : S3=0 : S4=0 : S5=0
2200  FOR I=1 TO N
2250    S1=S1+W(I)
2300    S2=S2+X(I)*W(I)
2400    S3=S3+Y(I)*W(I)
2500    S4=S4+X(I)*X(I)*W(I)
2600    S5=S5+Y(I)*X(I)*W(I)
2700  NEXT I
3000  D1=S1*S4-S2*S2
3100  D2=S3*S4-S5*S2
3200  D3=S1*S5-S2*S3
4000  A=D2/D1
4100  B=D3/D1
5000  PRINT "The regression line for the data is"
5100  PRINT " y  = ";A;"  +  ";B;"* x"
6000  S=0
6100  FOR I=1 TO N
6200    S=S+W(I)*(Y(I)-(A+B*X(I)))^2
6300  NEXT I
6500  PRINT : PRINT "Sum of squared deviations =";S
6600  D=SQR(S/(N-2))
6605  PRINT : PRINT "Standard deviation  =";D
6610  DA=D*SQR(S4/D1) : DB=D*SQR(S1/D1)
6620  PRINT : PRINT "Number of data points =";N : PRINT
6630  PRINT "a = ";A;" +/- ";DA : PRINT
6640  PRINT "b = ";B;" +/- ";DB : PRINT
7000  PRINT : PRINT "Do you want to compare the input "
7100  PRINT "values with the regression values ";
7200  INPUT A$
7300  IF A$="yes" THEN GOTO 8000
7500  GOTO 8400
8000  PRINT
8050  PRINT " x           y(input)   y(regression) "
8100  FOR I=1 TO N
8200    PRINT X(I);TAB(12);Y(I);TAB(23);A+B*X(I)
8300  NEXT I
8400  PRINT : PRINT "Do you want to interpolate data"
8500  PRINT "using the regression line ";
```

```
8600   INPUT A$
8700   IF A$<>"yes" THEN GOTO 9800
9000   PRINT "x-value"; : INPUT X
9050   PRINT A$;A$; : LOCATE CSRLIN-1,1
9100   PRINT "y(";X;") = ";A+B*X
9200   GOTO 8400
9800   GOSUB 20000 : REM PLOT
9850   G$=INKEY$ : IF G$="" THEN GOTO 9850
9860   SCREEN 0
9999   END
20000 REM ******************************************
20001 REM *** Subroutine: PLOT                  ***
20002 REM ******************************************
20010 X1=X(1) : X9=X(1)
20020 Y1=Y(1) : Y9=Y(1)
20100 FOR I=1 TO N
20110    IF X(I)<X1 THEN X1=X(I)
20120    IF X(I)>X9 THEN X9=X(I)
20130    IF Y(I)<Y1 THEN Y1=Y(I)
20140    IF Y(I)>Y9 THEN Y9=Y(I)
20150 NEXT I
20160 DX=(X9-X1)*.1 : DY=(Y9-Y1)*.1
20170 X1=X1-DX : X9=X9+DX
20180 Y1=Y1-DY : Y9=Y9+DY
20200 SCREEN 2 : KEY OFF : CLS
20220 WINDOW (X1,Y1)-(X9,Y9)
20230 LINE (X1,Y1)-(X9,Y9),,B
20300 FOR I=1 TO N
20310    CIRCLE (X(I),Y(I)),DX/10
20340 NEXT I
20500 Y2=A+B*X1 : Y8=A+B*X9
20503 M=N+1
20505 IF B=0 THEN : X2=1E+30 : X8=1E+30 : GOTO 20520
20510 X2=(Y1-A)/B : X8=(Y9-A)/B
20520 IF Y2>=Y1 AND Y2<=Y9 THEN:X(M)=X1:Y(M)=Y2:M=M+1
20530 IF Y8>=Y1 AND Y8<=Y9 THEN:X(M)=X9:Y(M)=Y8:M=M+1
20540 IF X2>=X1 AND X2<=X9 THEN:X(M)=X2:Y(M)=Y1:M=M+1
20550 IF X8>=X1 AND X8<=X9 THEN:X(M)=X8:Y(M)=Y9:M=M+1
20600 LINE (X(N+1),Y(N+1))-(X(N+2),Y(N+2))
20900 RETURN


RUN
How many data points (y,x) ? 10

Do you want to weight your data points ? yes
```

```
y( 1 ) = 10    x( 1 ) = 9    w( 1 ) = 1
y( 2 ) = 12    x( 2 ) = 8    w( 2 ) = 2
y( 3 ) = 13    x( 3 ) = 7    w( 3 ) = 1
y( 4 ) = 15    x( 4 ) = 5    w( 4 ) = 1
y( 5 ) = 14    x( 5 ) = 6    w( 5 ) = 1
y( 6 ) = 17    x( 6 ) = 3    w( 6 ) = 2
y( 7 ) = 18    x( 7 ) = 1    w( 7 ) = 1
y( 8 ) = 19    x( 8 ) = 2    w( 8 ) = 1
y( 9 ) = 20    x( 9 ) = 0    w( 9 ) = 2
y( 10 ) = 22   x( 10 ) =-2   w( 10 ) = 1
The regression line for the data is
y  =  20.02703  +  -1.027027 * x

Sum of squared deviations = 2.810811

Standard deviation  = .592749

Number of data points = 10

a =  20.02703  +/-  .2466671

b = -1.027027  +/-  4.781287E-02

Do you want to compare the input
values with the regression values ? yes
 x           y(input)   y(regression)
 9           10         10.78378
 8           12         11.81081
 7           13         12.83784
 5           15         14.89189
 6           14         13.86487
 3           17         16.94595
 1           18         19
 2           19         17.97297
 0           20         20.02703
-2           22         22.08108

Do you want to interpolate data
using the regression line ? no
Ok
```

The only statements which have been added to the main program of REGLINW are lines 9800–9860. In line 9800 the plot subroutine is called after all calculations for regression are finished. Line 9850 is a waiting loop; the program loops in this line until a key is pressed. Then the graphic screen disappears and the text screen is reinstalled in line 9860. The example given shows both a numerical output and a graphical output (Fig. 13.2) of the regression problem. The graphical output facilitates the interpretation of the results and it is easier to detect input errors.

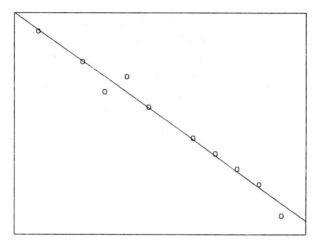

Fig. 13.2. Plot of the input data points and of the regression line generated by the program REGLINP

The plot subroutine in line 20000 starts with a search for the maximum and minimum x- and y-values. The extremes are stored in the variables X1, X9, Y1 and Y9. At the beginning they are set equal to X(1) and Y(1) respectively. Then in the loop from 20100 to 20150 every X- and Y-value is compared with the minimum and maximum and they are corrected if new extreme values are found. The following three lines serve to expand slightly the found region for the coordinates. This is done in order to get a small offset from the axes for the smallest and largest points. The x-region and the y-region are multiplied with 0.1 and this small amount is subtracted from the smallest values and added to the largest ones.

In line 20200 the high resolution screen is activated, the screen is cleared and the key menu is switched off. In line 20220 the WINDOW-statement defines the coordinates for the plot. Line 20230 draws a frame for the plotting window. In the loop from 20300 to 20340 the experimental points are plotted on the screen using small circles.

The rest of the plot subroutine calculates and draws the regression line. A problem that arises is to start and end the straight line at the borders of the coordinate window. First the y-values of the regression line (Y2 and Y8) are determined which are the straight line values of the minimum (X1) and maximum (X9) of x. In the same manner the x-values (X2 and X8) are calculated which result in the minimum and maximum y-values. If the slope of the regression line is zero then these x-values are set to the very big value of 10^{30}. With these x- and y-values we have obtained four points of intersection of the regression line with the prolonged border lines of the window. In addition it must be determined which of these points of intersection are within the window. The IF-interrogations of the four lines 20520 to 20550 are used for this task. In these IF-statements two conditions are connected with the logical relation AND. Besides the AND, three other logical relations OR, XOR and NOT are of interest. To be able to use these logical operations one has to know how they are connecting conditions within an IF-statement. If a condition is satisfied, the condition is called 'true' and is symbolized by the number 1. If the contrary is obtained the condition is called 'false' and is symbolized by the number 0. The respective tables for the logical operations are called truth tables:

Table 13.1. Truth tables for the boolean operators AND, OR, XOR and NOT

AND	1	0
1	1	0
0	0	0

OR	1	0
1	1	1
0	1	0

XOR	1	0
1	0	1
0	1	0

NOT 1	results in 0
NOT 0	results in 1

If expressions are calculated which include arithmetic expressions, comparison operators, and logical operators, then the hierarchy of the operations is: first the arithmetic operations, then the comparison operations and at last the logical operations. But it is strongly recommended to use parentheses to eliminate doubts and to make the expressions clear and easy to understand.

With the help of IF-statements the points of intersection with the window are determined and stored in the X- and Y-vectors as the elements number (N+1) and (N+2). Therefore M is set to N+1 in the line 20503. The line 20520 considers the left border of the plotting window. If the intersection point of the regression line (X1,Y2) lies on this left border, i.e. if Y2 is smaller than Y9 and greater than Y1, then (X1,Y2) is one point for which we are searching. It is stored in X(M) and Y(M) and M is increased by one. In the same way the other three borders of the window are analyzed and the second point of intersection is found. The plot statement in the last line of the subroutine (line 20600) finally draws this line on the screen.

Problem 193

Improve the program REGLINP above to become more useful for applications. Add a grid of dotted lines within the window at certain x- and y-values. Make an additional expansion by printing tick marks, values and a label on the axis.

Problem 194

Expand one of the programs for nonlinear regression with a plotting subroutine. Write the program so that after every iteration the improved regression function is plotted together with its input points.

13.3 Differential Equations Using Graphs

The concentration of four substances as a function of the reaction time shall be plotted. The reaction is governed by the following reaction kinetics:

$$A \; + \; B \; \longrightarrow \; C \qquad \text{(reaction 1)}$$
$$C \; \longrightarrow \; D \qquad \text{(reaction 2)}$$

For the numerical solution of the kinetics of these consecutive reactions we use the Runge-Kutta method for the solution of a system of ordinary differential equations (program SYS-RKN). To simplify the input and output all quantities are considered to be without physical dimensions. The system of differential equations which results from the kinetics above is programmed as subroutine 10000 in SYS-RKN. This is the core of the following program KINETIC which includes a plotting section.

```
0 REM "KINETIC "          EBERT/EDERER          880206
1 REM ************************************************
2 REM *** Simulation of the kinetics of the      ***
3 REM *** following reaction mechanism:          ***
4 REM ***     A   +   B   ==>   C       (g1)     ***
5 REM ***     C           ==>   D       (g2)     ***
9 REM ************************************************
100    DIM Y(20),Y1(20),D(20),Y9(20),YA(20)
120    DIM K1(20),K2(20),K3(20),K4(20)
200    N=4 : REM N is the dimension of the ODE-system
310    G1=1       : REM Reaction rate constant 1
320    G2=1       : REM Reaction rate constant 2
410    Y1(1)=1.5 : REM Initial concentration of A
420    Y1(2)=1   : REM Initial concentration of B
430    Y1(3)=0   : REM Initial concentration of C
440    Y1(4)=0   : REM Initial concentration of D
600    CM=1.5     : REM Maximum of concentrations
700    REM x-values  start, end, step size
710    XA=0 : XE=5 : XD=.1
715    GOSUB 2000 : REM INIT_PLOT
720    FOR I=1 TO N : Y(I)=Y1(I) : YA(I)=Y1(I) : NEXT I
740    N1=1
1000   FOR A=XA TO XE STEP XD
1020     E=A+XD
1040     H=(E-A)/N1 : GOSUB 5000 : REM ODE
1050     FOR I=1 TO N : YA(I)=Y(I) : NEXT I
1060     N1=N1*2 : H=(E-A)/N1 : GOSUB 5000
1080     FOR I=1 TO N
1090       IF ABS(Y(I))+ABS(YA(I))<.000001 THEN 1120
1100       YG=ABS(YA(I)-Y(I))/(ABS(YA(I))+ABS(Y(I)))
1110       IF YG >.01 THEN 1050
1120     NEXT I
1200     GOSUB 15000 : REM OUTPUT
1220     FOR I=1 TO N : Y1(I)=Y(I) : NEXT I
1240     N1=INT(N1/4) : IF N1=0 THEN N1=1
1260   NEXT A
1300   A$="" : A$=INKEY$ : IF A$="" THEN GOTO 1300
1400   SCREEN 0
1899   END
1900   REM ##########################################
1910   REM ### Main program ends here.          ###
1920   REM ##########################################
2000   REM ************************************************
2001   REM *** Subroutine:   INIT_PLOT              ***
2002   REM ************************************************
2010   KEY OFF : SCREEN 2
```

```
2020   WINDOW (XA,0) - (XE,CM)
2030   LINE   (XA,0) - (XE,CM) ,,B
2200   RETURN
5000   REM ******************************************
5002   REM *** Subroutine: ODE                    ***
5004   REM *** The solution of the ODE is approxi- ***
5006   REM *** mated by using N1 subintervals.     ***
5010   REM ******************************************
5100   FOR I=1 TO N : Y(I)=Y1(I) : NEXT I
5200   FOR I9=0 TO N1-1 : X=A+I9*H
5300     X9=X : FOR I=1 TO N : Y9(I)=Y(I) : NEXT I
5320     GOSUB 10000
5340     FOR I=1 TO N : K1(I)=H*D(I) : NEXT I: X9=X+H/2
5400     FOR I=1 TO N : Y9(I)=Y(I)+K1(I)/2 : NEXT I
5420     GOSUB 10000
5440     FOR I=1 TO N : K2(I)=H*D(I) : NEXT I: X9=X+H/2
5500     FOR I=1 TO N : Y9(I)=Y(I)+K2(I)/2 : NEXT I
5520     GOSUB 10000
5540     FOR I=1 TO N : K3(I)=H*D(I) : NEXT I : X9=X+H
5600     FOR I=1 TO N : Y9(I)=Y(I)+K3(I) : NEXT I
5620     GOSUB 10000
5640     FOR I=1 TO N : K4(I)=H*D(I) : NEXT I
5700     FOR I=1 TO N
5720       Y(I)=Y(I)+(K1(I)+2*K2(I)+2*K3(I)+K4(I))/6
5740     NEXT I : X=E
5800   NEXT I9 : RETURN
10000  REM ******************************************
10002  REM *** Subroutine: RHS                    ***
10004  REM *** The right hand sides of the ODE's   ***
10006  REM *** are calculated here as              ***
10008  REM *** D(I) = f(X9,Y9(1),Y9(2),..Y9(N))    ***
10010  REM ******************************************
10100  D(1)=-G1*Y9(1)*Y9(2)
10200  D(2)=D(1)
10300  D(3)=G1*Y9(1)*Y9(2)-G2*Y9(3)
10400  D(4)=G2*Y9(3)
12000  RETURN
15000  REM ******************************************
15001  REM *** Subroutine: OUTPUT                 ***
15002  REM ******************************************
15040  FOR I=1 TO N
15050    LINE (A,Y1(I)) - (E,Y(I))
15080  NEXT I
15500  RETURN
```

The example shows only the plot output (Fig. 13.2). We can see the decrease of the substances A and B, the increase of the final product D and the intermediate species C, whose concentration is running through a maximum. From line 5000 to 12000 the program KINETIC is equivalent to program SYS-RKN. But a few lines have been added to the main program, subroutines for the graphical output and for initializing the graphics.

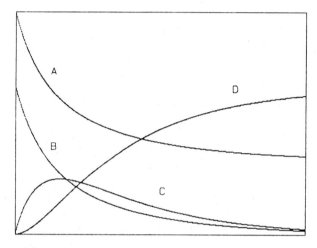

Fig. 13.3. Plot of the concentrations of the substances A, B, C and D versus reaction time. The y-axis for the concentrations is scaled from 0 to 1.5. The x-axis shows the time for 0 to 5.

In line 100 the vector YA(20) is added. The dimension of the system of differential equations is N=4 (line 200). In the following lines the reaction rate constants are set to 1 (G1 and G2). The initial concentrations of A and B are 1.5 and 1.0 and these values are stored in variables Y1(1) and Y1(2), whereas the initial concentrations of substances C and D are 0 and are stored in variables Y1(3) and Y1(4). The reaction time shall be between 0 and 5; for each time interval of 0.1 the concentration of the substances shall be calculated. This information for the time scale is programmed in line 710 using the variables XA, XE and XD. The variable CM is set to the maximum concentration for constructing the plotting window. The next statement (line 715) is the call of the plot initializing subroutine. Here the high resolution graphic screen is initialized, the plot window is defined and a frame is plotted. Line 720 stores the initial concentrations in the vectors Y and YA. To be able to understand the following A-loop it should be remembered that subroutine 5000 solves the ODE-system for the initial values, which are stored in the vector Y1. Solution with GOSUB 5000 means, that new Y-values are calculated with N1 Runge-Kutta steps and with the assumption that the variables A and E define the interval of the independent variable X (which we interpret as reaction time) over which the integration of the ODE-system must be done.

The total time interval from XA to XE, for the concentrations we are interested in, is calculated and divided into subintervals of length XD (there are 50 subintervals in the example). These subintervals are treated in the A-loop from line 1000 to 1260. Line 1020 defines the end of the subinterval considered and the loop variable A defines the beginning of the actual subinterval. After the determination of the step length H for the ODE-subroutine this subroutine is called up by the statement GOSUB 5000. The result of the calculated concentrations is contained in vector Y and kept in field YA by the statements of line 1050. Then the same calculation is done with a halved step size and twice the number of steps (N1=N1*2). In the loop from line 1080 to 1120 it is determined if the accuracy of the calculation is adequate. In this example the accuracy is considered as sufficient if for every element (for every concentration) of the Y-vector, the absolute value of the difference of two consecutive calculations divided by the absolute value of the sum of these calculations is smaller than 0.01 or if the concentrations are smaller than 10^{-6}. If these accuracy requirements are not achieved, the program jumps back to line 1050 where the latest results are kept in YA and a new calculation with smaller step size is started.

If the accuracy requirement is fulfilled, then the program jumps to the graphical output subroutine 15000. The calculated concentrations Y for the final time E of the actual subinterval are the initial concentrations for the next subinterval (line 1220). The number of integration steps N1 is reduced by a factor of 4, but it remains at least 1; in line 1260 a run through the A-loop ends and the program jumps back to line 1000 to start the calculation for the next subinterval. After the calculation of all subintervals, the program remains in a waiting loop (line 1300). After hitting any key the graphic screen disappears and the text mode is restored.

The plot subroutine 15000 draws the concentrations of all four substances with the statement LINE (A,Y1(I)) - (E,Y(I)). This means that the initial and final concentrations of a subinterval are connected by a straight line. An inspection of the concentration curves shows that they are made of short linear pieces. If smoother curves for the concentrations are required, the variable XD in line 710 has to be decreased.

Problem 195

Try to calculate the predator-prey dynamics of Problem 137. Make a graphical presentation of the population densities with time. Another way to present the results graphically is to plot the concentration of the preys against the concentration of the predators. Oscillations in the time region result in this representation in spiral figures. A stable oscillation can be identified in this way by a closed curve which is called a limit cycle.

Problem 196

Solve the reaction model OREGONATOR for a chemical oscillation (Problem 139) and present the results graphically.

A collision of two particles having a potential energy $V(r)$ — r is the distance between the two particles — shall be calculated and plotted on the screen.[1] To reduce the number of differential equations, the impact is considered from the common center of mass and polar coordinates (r, ϕ) are used (ϕ is the angle). The Hamiltonian operator H, which describes this problem, is given by

$$H = \frac{1}{2}\mu(\dot{r}^2 + r^2\dot{\phi}^2) + V(r)$$

The dotted symbols $\dot{r}$ or $\dot{\phi}$ are used as abbreviations for the time derivatives.

$$\dot{r} = \frac{dr}{dt} \qquad \dot{\phi} = \frac{d\phi}{dt}$$

The kinetic energy T is

$$T = \frac{1}{2}\mu(\dot{r}^2 + r^2\dot{\phi}^2)$$

---

[1] I.W.M. Smith, *Kinetics and Dynamics of Elementary Gas Reactions*, p.61ff, Butterworths, London, Boston (1980)

The generalized momenta p_r and p_ϕ are derived as

$$p_r = \frac{\partial T}{\partial \dot{r}} = \mu \dot{r}$$

$$p_\phi = \frac{\partial T}{\partial \dot{\phi}} = \mu r^2 \dot{\phi}$$

Using these momenta, the Hamiltonian equation can be rewritten

$$H = \frac{1}{2\mu}\left((\mu \dot{r})^2 + (\mu r^2 \dot{\phi})^2 \cdot \frac{1}{r^2}\right) + V(r)$$

$$H = \frac{1}{2\mu}\left((p_r^2 + p_\phi^2 \cdot \frac{1}{r^2}\right) + V(r)$$

From this final Hamiltonian operator four differential equations can be derived which describe the dynamics of the particle collision by calculating the partial derivatives to the momenta p_r and p_ϕ and the position coordinates r and ϕ

$$\frac{\partial H}{\partial \phi} = -\dot{p}_\phi = -\frac{dp_\phi}{dt} = 0$$

$$\frac{\partial H}{\partial r} = -\dot{p}_r = -\frac{dp_r}{dt} = -\frac{p_\phi^2}{\mu r^3} + \frac{\partial V(r)}{\partial r}$$

$$\frac{\partial H}{\partial p_\phi} = \dot{\phi} = \frac{d\phi}{dt} = \frac{p_\phi}{\mu r^2}$$

$$\frac{\partial H}{\partial p_r} = \dot{r} = \frac{dr}{dt} = \frac{p_r}{\mu}$$

To use the Runge-Kutta procedure the physical quantities are mapped to variables in the following way

$$\phi \longrightarrow \text{Y9(1)}$$

$$r \longrightarrow \text{Y9(2)}$$

$$p_r \longrightarrow \text{Y9(3)}$$

The differential equation for the angular momentum (the time derivative) is equal to zero, which means that the angular momentum is constant.

$$p_\phi = \mu \cdot b \cdot w$$

In this formula μ is the reduced mass and w is the relative velocity of the two particles. The impact parameter b can be interpreted as the distance of the closest approach of the two particles along their hypothetical undisturbed trajectory.

If, as a potential function $V(r)$, the Lennard-Jones potential for argon is taken, then one obtains the following expressions for the potential energy and its derivatives with respect to r

$$V(r) = 4\epsilon\left[\left(\frac{\sigma}{r}\right)^{12} - \left(\frac{\sigma}{r}\right)^{6}\right]$$

$$\frac{\partial V(r)}{\partial r} = \frac{4\epsilon}{\sigma}\left[-12\left(\frac{\sigma}{r}\right)^{13} + 6\left(\frac{\sigma}{r}\right)^{7}\right]$$

The parameters σ and ϵ/k for argon can be found in the table for Lennard-Jones potentials in Chapter 5.7. The Boltzmann constant k is expressed in molecular quantities:

$$k = 1.3807 \quad \left[\frac{\text{g} \cdot \text{\AA}^2}{\text{s}^2 \cdot \text{K}}\right]$$

Using this we obtain the formula which is programmed in line 160 in the following program IMPACT.

```
0 REM "IMPACT  "         EBERT/EDERER          881228
1 REM *************************************************
2 REM *** A model calculation for a collision of   ***
3 REM *** two particles is made. According to      ***
4 REM *** I. W. M. Smith, Kinetics and Dynamics    ***
5 REM *** Butterworth 1989, pages 62ff. The impact ***
6 REM *** parameter B is assigned in line 250.     ***
9 REM *************************************************
100    DIM Y(20),Y1(20),D(20),Y9(20)
120    DIM K1(20),K2(20),K3(20),K4(20)
150    REM dV/dr of the potential of argon
160    DEF FNV(R)=(-12*(3.405/R)^13+6*(3.405/R)^7)*194.3
200    N=3 : REM N is the dimension of the ODE-system
210    MY=20/6.022E+23
220    W=1.245E+12 : REM initial velocity in A/sec
250    B=8!: REM impact parameter in angstroem
```

```
400    Y1(2)=16 : REM initial radius in angstroem
450    Y1(1)=ATN(B/SQR(Y1(2)^2-B^2)) : REM initial angle
470    Y1(3)=-MY*W*COS(Y1(1)) : REM initial impact(r)
700    REM input "x-values begin,end ";a,e
710    A=0 : E=1E-10
1000   N1=1000
1100   H=(E-A)/N1
2000   REM initializing the graphics
2010   KEY OFF
2020   SCREEN 2
2050   PP=20
2060   WINDOW (-PP,-PP)-(PP,PP)
2080   LINE   (-PP,-PP)-(PP,PP),,B
2100   CIRCLE (0,0),PP/20
4200   GOSUB 5000
4220   A$="" : A$=INKEY$ : IF A$="" THEN GOTO 4220
4240   SCREEN 0
4300   STOP
4400   END
4900   REM ###########################################
4910   REM ### Main program ends here.            ###
4920   REM ###########################################
5000   REM *******************************************
5002   REM *** Subroutine: ODE                     ***
5004   REM *** The solution of the ODE is approxi- ***
5006   REM *** mated by using N1 subintervals.     ***
5010   REM *******************************************
5100   FOR I=1 TO N : Y(I)=Y1(I) : NEXT I
5110   PI=0 : GOSUB 15000
5200   FOR I9=0 TO N1-1 : X=A+I9*H
5300     X9=X : FOR I=1 TO N : Y9(I)=Y(I) : NEXT I
5320     GOSUB 10000
5340     FOR I=1 TO N : K1(I)=H*D(I) : NEXT I: X9=X+H/2
5400     FOR I=1 TO N : Y9(I)=Y(I)+K1(I)/2 : NEXT I
5420     GOSUB 10000
5440     FOR I=1 TO N : K2(I)=H*D(I) : NEXT I: X9=X+H/2
5500     FOR I=1 TO N : Y9(I)=Y(I)+K2(I)/2 : NEXT I
5520     GOSUB 10000
5540     FOR I=1 TO N : K3(I)=H*D(I) : NEXT I : X9=X+H
5600     FOR I=1 TO N : Y9(I)=Y(I)+K3(I) : NEXT I
5620     GOSUB 10000
5640     FOR I=1 TO N : K4(I)=H*D(I) : NEXT I
5700     FOR I=1 TO N
5720       Y(I)=Y(I)+(K1(I)+2*K2(I)+2*K3(I)+K4(I))/6
5740     NEXT I
5760     GOSUB 15000 : REM OUTPUT
5800   NEXT I9 : RETURN
```

```
10000 REM ******************************************
10002 REM *** Subroutine: RHS                   ***
10004 REM *** The right hand sides of the ODE's ***
10006 REM *** are calculated here as            ***
10008 REM *** D(I) = f(X9,Y9(1),Y9(2),..Y9(N))  ***
10010 REM ******************************************
10100 D(1)=B*W/Y9(2)/Y9(2)
10200 D(2)=Y9(3)/MY
10300 D(3)=MY*B*B*W/(Y9(2)^3)*W-FNV(Y9(2))
12000 RETURN
15000 REM ******************************************
15001 REM *** Subroutine: OUTPUT                ***
15002 REM ******************************************
15020 REM PRINT X,Y(1),Y(2),Y(3)
15030 PX=Y(2)*SIN(Y(1)):PY=Y(2)*COS(Y(1))
15040 LINE -(PX,PY),PI
15060 PI=1
15500 RETURN
```

Here again the program SYS-RKN is used for numerical calculation of the system of differential equations for the collision problem. These are programmed in lines 10100 to 10300. The dimension of the system of ODE's, the reduced mass, the initial velocity, the impact parameter, the initial angle, and the initial momentum are assigned or calculated in the program from lines 200 to 470. The initial and end values of the independent variable X are set to 0 and 10^{-10} in line 710. This variable X corresponds to the time coordinate in the underlying physical problem.

For calculating the values of the graphical output, another way was chosen in this program as compared to the last program KINETIC. The ODE-subroutine 5000 is called up after setting N1=1000, which means 1000 subintervals are chosen. It was tested previously that this subdivision is fine enough to obtain results for the ODE-system which are correct to the first three decimal places. After the calculation of each subinterval, the plot subroutine 15000 is called up from line 5760. The variables A, E and N1 are chosen to give a far bigger time interval than necessary to integrate for the generation of the plots.

The program section from line 2000 to 2100 initializes the graphics, defines the plotting window, plots the frame and draws a small circle in the middle of the screen. This circle has a radius of $\frac{1}{20}$ of the x-axis and represents the center of mass of the collision system.

In the plot output subroutine the polar coordinates are converted into cartesian coordinates (line 15030) which are plotted in the following statement
(line 15040). In this statement a parameter variable PI is used. This parameter stands for the color of the line to be drawn. PI=0 is applied for the
first call from line 5110. This causes a line to be drawn in the background
color which means it is invisible. This is necessary in order to omit a line on
the screen to the initial point of the trajectories. Examples in which only
the impact parameter b was changed are shown in Fig. 13.4. You can see the
small scattering if the particles are passing at great distances ($b = 14$ Å) and
a strong repulsion if the particles undergo a nearly central impact ($b = 1$ Å).
The right figure shows trajectories with impact parameters from 7.4 to 8.4 Å.
This is the region for the impact parameter where strong interactions happen
between the two particles which sometimes lead to two or three rotations of
one atom around the other one. This is sometimes called a van der Waals
molecule.

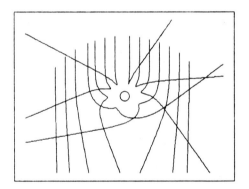
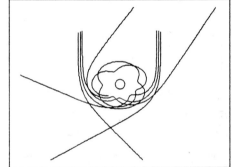

Fig. 13.4. Plot of relative trajectories for an impact of two argon atoms. The circle
in the middle of screen is the common center of mass. The left figure
shows the trajectories for impact parameters b of −13, −11, −9, −7, −5,
−3, −1, 2, 4, 6, 8, 10, 12, 14 Å and the right figure for parameters b of
−8.2, −7.8, −7.4, 7.6, 8, 8.4 Å.

Problem 197

Use the physical parameters for the potential energy between helium atoms and
compare the trajectories for argon with the trajectories for helium.

Problem 198

Use the program IMPACT to produce a series of figures of trajectories with fixed impact parameter b and varying initial velocities w. Expand the plot subroutine to include the relative velocities in the graphical output. One way to do this is to mark the trajectories after every i^{th} subinterval. The closer the marks are on the trajectory, the slower is the relative velocity.

In the following example, which is very similar to the previous one, no effort is made to keep the number of differential equations small. This results in a little more computing time but you don't have to use tricks and problem analysis to transform the physical problem for computations.

Again a collision between two argon atoms is considered whose intermolecular forces are described by a Lennard-Jones potential

$$V(r^2) = 4\epsilon \left(\left(\frac{\sigma^2}{r^2} \right)^6 - \left(\frac{\sigma^2}{r^2} \right)^3 \right) = V(x_1, x_2, y_1, y_2)$$

This time the problem shall be calculated in cartesian coordinates. It is sufficient to consider only a plane, because a biatomic collision always remains in the same plane. This reduces the number of differential equations from 12 to 8.

The Hamiltonian for this system is

$$H = T + V(x_1, x_2, y_1, y_2) = T + V\left(\sqrt{(x_1 - x_2)^2 + (y_1 - y_2)^2} \right)$$

The kinetical energy T is given by

$$T = \frac{1}{2} m_1 (\dot{x}_1^2 + \dot{y}_1^2) + \frac{1}{2} m_2 (\dot{x}_2^2 + \dot{y}_2^2)$$

The derivation of the generalized momenta results in the usual momentum definitions

$$p_{x_1} = \frac{\partial T}{\partial \dot{x}_1} = m_1 \dot{x}_1$$

$$p_{y_1} = \frac{\partial T}{\partial \dot{y}_1} = m_1 \dot{y}_1$$

$$p_{x_2} = \frac{\partial T}{\partial \dot{x}_2} = m_2 \dot{x}_2$$

$$p_{y_2} = \frac{\partial T}{\partial \dot{y}_2} = m_2 \dot{y}_2$$

Now the Hamiltonian equation can be rewritten to have only location and momentum variables

$$H = \frac{1}{2m_1}\left(p_{x_1}^2 + p_{y_1}^2\right) + \frac{1}{2m_2}\left(p_{x_2}^2 + p_{y_2}^2\right) + V(x_1, x_2, y_1, y_2)$$

If the partial derivatives of H with respect to all variables are calculated, then the eight differential equations which describe the physical problem are constructed

$$\frac{\partial H}{\partial p_{x_1}} = \dot{x}_1 = \frac{p_{x_1}}{m_1} \qquad D(1)$$

$$\frac{\partial H}{\partial p_{y_1}} = \dot{y}_1 = \frac{p_{y_1}}{m_1} \qquad D(2)$$

$$\frac{\partial H}{\partial p_{x_2}} = \dot{x}_2 = \frac{p_{x_2}}{m_2} \qquad D(3)$$

$$\frac{\partial H}{\partial p_{y_2}} = \dot{y}_2 = \frac{p_{y_2}}{m_2} \qquad D(4)$$

$$\frac{\partial H}{\partial x_1} = -\dot{p}_{x_1} = \frac{\partial V}{\partial x_1} \qquad D(5)$$

$$\frac{\partial H}{\partial y_1} = -\dot{p}_{y_1} = \frac{\partial V}{\partial y_1} \qquad D(6)$$

$$\frac{\partial H}{\partial x_2} = -\dot{p}_{x_2} = \frac{\partial V}{\partial x_2} \qquad D(7)$$

$$\frac{\partial H}{\partial y_2} = -\dot{p}_{y_2} = \frac{\partial V}{\partial y_2} \qquad D(8)$$

In the last column of this table of formulas the variables D(I) are shown that are used in the following program IMPACT2 to calculate the right hand side of the ODE's starting in line 10000.

```
0 REM "IMPACT2 "          EBERT/EDERER          880202
1 REM ************************************************
2 REM *** A model calculation for a collision of  ***
3 REM *** two particles is made. According to     ***
4 REM *** I. W. M. Smith, Kinetics and Dynamics   ***
5 REM *** Butterworth 1980, pages 62 ff with      ***
7 REM *** 'brute force' method.                    ***
9 REM ************************************************
100    DIM Y(20),Y1(20),D(20),Y9(20),YA(20)
120    DIM K1(20),K2(20),K3(20),K4(20)
150    REM FNV(r) is the potential for argon
160    REM FNV(r)=((3.405/r)^12-(3.405/r)^6)*479.2
200    N=8 : REM N is the dimension of the ODE-system
210    M1=20/6.022E+23 : M2=M1
400    REM Initial values in Angstroem , g , sec
410    Y1(1)=-15          : REM x-coordinate 1. particle
420    Y1(2)=-3           : REM y-coordinate 1. particle
430    Y1(3)=15           : REM x-coordinate 2. particle
440    Y1(4)=3            : REM y-coordinate 2. particle
450    Y1(5)=1.7E+12*M1   : REM impact x  of 1. particle
460    Y1(6)=0*M1         : REM impact y  of 1. particle
470    Y1(7)=-1.7E+12*M2  : REM impact x  of 2. particle
480    Y1(8)=0*M2         : REM impact y  of 2. particle
700    REM x start, end, step size
710    XA=0 : XE=1E-10 : XD=3E-13
715    GOSUB 2000 : REM Initializing the graphics
720    FOR I=1 TO N
725      Y(I)=Y1(I) : YA(I)=Y1(I)
730    NEXT I : GOSUB 15000
740    N1=1
1000   FOR A=XA TO XE STEP XD
1020     E=A+XD
1040     H=(E-A)/N1 : GOSUB 5000
1050     FOR I=1 TO N : YA(I)=Y(I) : NEXT I
1060     N1=N1*2 : H=(E-A)/N1 : GOSUB 5000
1080     FOR I=1 TO N
1090       IF ABS(Y(I))+ABS(YA(I))<.000001 THEN 1120
1100       YG=ABS(YA(I)-Y(I))/(ABS(YA(I))+ABS(Y(I)))
1110       IF YG >.01 THEN 1050
1120     NEXT I
1200     GOSUB 15000 : REM OUTPUT
1220     FOR I=1 TO N : Y1(I)=Y(I) : NEXT I
1240     N1=INT(N1/4) : IF N1=0 THEN N1=1
1260   NEXT A
1300   A$="" : A$=INKEY$ : IF A$="" THEN GOTO 1300
1400   SCREEN 0
1899   END
```

```
1900  REM ##########################################
1910  REM ### Main program ends here.          ###
1920  REM ##########################################
2000  REM ******************************************
2001  REM *** Subroutine:   INIT_PLOT          ***
2002  REM ******************************************
2010  KEY OFF
2020  SCREEN 2
2030  VIEW (80,0)-(559,199)
2050  PP=16
2060  WINDOW (-PP,-PP)-(PP,PP)
2080  LINE   (-PP,-PP)-(PP,PP),,B
2200  RETURN
5000  REM ******************************************
5002  REM *** Subroutine: ODE                  ***
5004  REM *** The solution of the ODE is approxi- ***
5006  REM *** mated by using N1 subintervals.     ***
5010  REM ******************************************
5100  FOR I=1 TO N : Y(I)=Y1(I) : NEXT I
5200  FOR I9=0 TO N1-1 : X=A+I9*H
5300    X9=X : FOR I=1 TO N : Y9(I)=Y(I) : NEXT I
5320    GOSUB 10000
5340    FOR I=1 TO N : K1(I)=H*D(I) : NEXT I: X9=X+H/2
5400    FOR I=1 TO N : Y9(I)=Y(I)+K1(I)/2 : NEXT I
5420    GOSUB 10000
5440    FOR I=1 TO N : K2(I)=H*D(I) : NEXT I: X9=X+H/2
5500    FOR I=1 TO N : Y9(I)=Y(I)+K2(I)/2 : NEXT I
5520    GOSUB 10000
5540    FOR I=1 TO N : K3(I)=H*D(I) : NEXT I : X9=X+H
5600    FOR I=1 TO N : Y9(I)=Y(I)+K3(I) : NEXT I
5620    GOSUB 10000
5640    FOR I=1 TO N : K4(I)=H*D(I) : NEXT I
5700    FOR I=1 TO N
5720      Y(I)=Y(I)+(K1(I)+2*K2(I)+2*K3(I)+K4(I))/6
5740    NEXT I : X=E
5800  NEXT I9 : RETURN
10000 REM ******************************************
10002 REM *** Subroutine: RHS                  ***
10004 REM *** The right hand sides of the ODE's   ***
10006 REM *** are calculated here as              ***
10008 REM *** D(I) = f(X9,Y9(1),Y9(2),..Y9(N))    ***
10010 REM ******************************************
10100 D(1)=Y9(5)/M1
10200 D(2)=Y9(6)/M1
10300 D(3)=Y9(7)/M2
10400 D(4)=Y9(8)/M2
```

```
10500 RX=Y9(1)-Y9(3):RY=Y9(2)-Y9(4)
10560 R2=RX*RX+RY*RY
10600 VR=-342.4*((11.6/R2)^4-2*(11.6/R2)^7)
11500 D(5)=RX*VR
11600 D(6)=RY*VR
11700 D(7)=-D(5)
11800 D(8)=-D(6)
12000 RETURN
15000 REM *******************************************
15001 REM *** Subroutine: OUTPUT               ***
15002 REM *******************************************
15030 LINE (Y(1),Y(2)) -(Y(1),Y(2))
15040 LINE (Y(3),Y(4)) -(Y(3),Y(4))
15080 A$="" : A$=INKEY$ : IF A$="q" THEN GOTO 1400
15500 RETURN
```

The derivation of the potential energy V to the coordinate variables x_1, x_2, y_1, y_2 is done by differentiation with respect to r^2 first

$$\frac{\partial V(r^2)}{\partial (r^2)} = -12\frac{\epsilon}{\sigma^2} \cdot \left(2\left(\frac{\sigma^2}{r^2}\right)^7 - \left(\frac{\sigma^2}{r^2}\right)^4 \right)$$

Using the molecular data for argon one obtains

$$\frac{\partial V(r^2)}{\partial (r^2)} = 171.2 \left(\left(\frac{11.6}{r^2}\right)^4 - 2\left(\frac{11.6}{r^2}\right)^7 \right)$$

For r^2 and its partial derivatives with respect to the coordinates we obtain the following expressions

$$r^2 = (x_1 - x_2)^2 + (y_1 - y_2)^2$$

$$\frac{\partial r^2}{\partial x_1} = 2(x_1 - x_2)$$

$$\frac{\partial r^2}{\partial x_2} = -2(x_1 - x_2)$$

$$\frac{\partial r^2}{\partial y_1} = 2(y_1 - y_2)$$

$$\frac{\partial r^2}{\partial y_2} = -2(y_1 - y_2)$$

Using the chain rule, the partial derivatives of V with respect to the coordination variables are finally obtained

$$\frac{\partial V}{\partial x_i} = \frac{\partial V(r^2)}{\partial r^2} \cdot \frac{\partial r^2}{\partial x_i} \qquad\qquad \frac{\partial V}{\partial y_i} = \frac{\partial V(r^2)}{\partial r^2} \cdot \frac{\partial r^2}{\partial y_i}$$

The actual physical parameters and the initial values for the coordinates and the momenta are programmed in the lines through 480. The remaining main program and the loop from line 1000 to line 1260 are nearly the same as in program KINETIC. Only in line 700 other values are used for the time scale. From line 715 the plot initializing subroutine 2000 is called where the window and the part of the screen for plotting are defined.

In subroutine 15000 for the graphical output, points — not lines — are plotted at the coordinates for the two particles. This is done with the LINE-command. Another way to do the same thing would be with the command PSET. In this way trajectories are obtained where the distance of the points is a measure for the particle velocities.

In the following examples trajectories for different initial conditions are shown. Table 13.2 presents these initial conditions.

Table 13.2. Table of the initial conditions used for the plots shown in Fig. 13.5. Y1(1),Y1(2) and Y1(3),Y1(4) are the (x, y)-coordinates of the first and the second particle in Å, Y1(5),Y1(6) and Y1(7),Y1(8) are the (x, y)-impacts of the first and the second particle in gÅs^{-1}, and M1,M2 are the molar masses of the particles.

Variable	Initial Values of the Examples					
	1	2	3	4	5	6
Y1(1)	−15	−15	−15	−15	−15	−15
Y1(2)	−3	−3	−3	−3	−3	−3
Y1(3)	15	15	15	15	15	15
Y1(4)	3	3	3	3	3	3
Y1(5)[$\cdot 10^{-12}$]	1.7	2	1.97	1.7	1.71	1.5
Y1(6)	0	0	0	0	0	0
Y1(7)[$\cdot 10^{-12}$]	−1.7	−2	−1.97	−2.7	−2.71	−2.5
Y1(8)	0	0	0	0	0	0
M1	20	40	20	40	40	40
M2	20	10	20	10	10	10

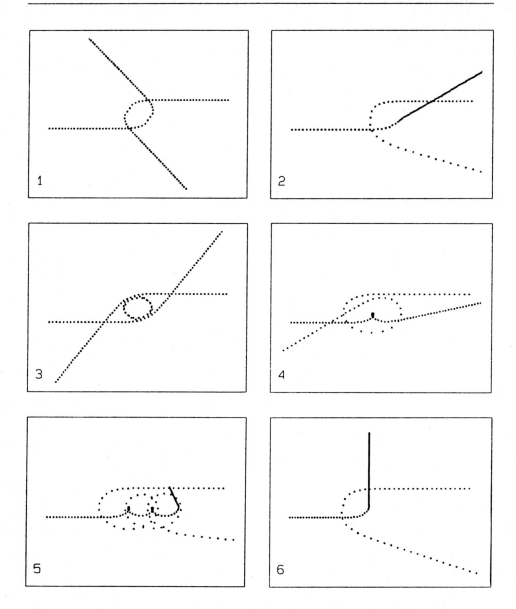

Fig. 13.5. Trajectories of two atoms using different initial conditions which are given in Table 13.2 and using the Lennard-Jones potential of argon

Although the potential for argon atoms is used, the mass of the atoms is also varied to demonstrate the resulting trajectories for different initial conditions. Sometimes rather complicated collisions can be obtained, which may be combined with a drastical change in the velocities. Fig. 13.6 show simpler collisions and scattering behavior.

Table 13.3. Table of the initial conditions used for the plots shown in Fig. 13.6. Y1(1),Y1(2) and Y1(3),Y1(4) are the (x,y)-coordinates of the first and the second particle in Å, Y1(5),Y1(6) and Y1(7),Y1(8) are the (x,y)-impacts of the first and the second particle in gÅs^{-1}, and M1,M2 are the molar masses of the particles.

Variable	Initial Values of the Examples					
	1	2	3	4	5	6
Y1(1)	−15	−15	−15	−15	−15	−15
Y1(2)	−3.3	−2.7	−3	−1.3	−0.5	−2
Y1(3)	15	15	15	15	15	15
Y1(4)	3.3	2.7	3	1.3	0.5	2
Y1(5)[·10^{-12}]	1.7	1.7	1.7	1.7	1.7	1.7
Y1(6)	0	0	0	0	0	0
Y1(7)[·10^{-12}]	−1.7	−1.7	−1.7	−1.7	−1.7	−1.7
Y1(8)	0	0	0	0	0	0
M1	20	20	20	20	20	20
M2	20	20	20	20	20	20

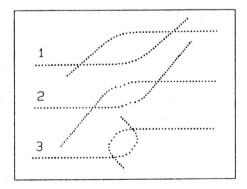
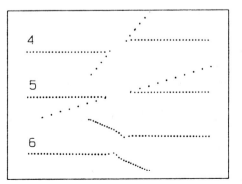

Fig. 13.6. Trajectories of two atoms with different initial conditions (Table 13.3) yielding simpler collision behavior

Problem 199

Write a simulation program for a molecular beam experiment using the elastic collision of two atoms. For doing so use fixed initial momenta in x and fixed initial x-positions. The momenta in y-direction shall be zero. Fix the y-coordinate of one particle and vary the y-position of the other one in small steps. How does the scattering angle change with the position of y? After this is done in the real physical coordinate system (which is called laboratory system) recalculate the same in the center of mass system.

13.4   3-D-Graphics

There are a large number of problems and applications in which the result is dependent not only on one variable but on two or more. The description of a quantity z, which is a function $z = f(x, y)$ of two independent variables x and y, results in a three-dimensional model. The projection of a three-dimensional model into a plane is called a 3-D-plot. This shall be demonstrated in the following example. The 'cowboy hat' of Fig. 13.7 which is slightly deformed is a 3-D-graph of the function

$$z = 3 \cos \sqrt{x^2 + y^2}$$

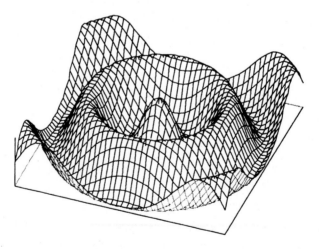

Fig. 13.7. 3-D-plot of the 'cowboy hat function' $z = 3 \cos \sqrt{x^2 + y^2}$ with $\rho = 70°$ and $\tau = 50°$

The projection is characterized by two angles: the rotation angle ρ about the z-axis, which rotates the (x, y)-plane, and the rotation angle τ about the x-axis. If the angle τ is zero, then the viewer looks at the model from above and only the rectangular grid in the (x, y)-plane is seen. The three-dimensional impression of this graphic is caused by the grid which connects neighboring (x, y)-points, by not drawing the lines which are hidden by the front parts of the object, and by the different line types, which are used if a part of the object is seen from below (a better 3-D-impression is obtained using different colors). The method to generate such a 3-D-graph is shown in the following program PLOT3D.

```
0 REM "PLOT3D  "          EBERT/EDERER          880115
1 REM ********************************************
2 REM *** A general 3-D-plot routine          ***
9 REM ********************************************
20      DIM T(3),IM(20),IZ(20)
30      DIM D(3,3),D1(3,3),D2(3,3)
40      DIM X(50),Y(50),Z(50,50)
50      DIM M1(251),M9(251)
60      PI=ATN(1)*4
200     DEF FNY(X)=(YA+SL*(X-XA))
1000    REM ********************************************
1002    REM *** Rotation and viewing matrix         ***
1004    REM ********************************************
1020    RH=30 : TA=30 : REM in degrees between 0 and 90
1040    D1(1,1)=COS(RH/180*PI) : D1(2,2)= D1(1,1)
1060    D1(2,1)=SIN(RH/180*PI) : D1(1,2)=-D1(2,1)
1080    D1(3,3)=1 : D1(1,3)=0 : D1(2,3)=0
1090    D1(3,1)=0 : D1(3,2)=0
1140    D2(2,2)= COS(TA/180*PI) : D2(3,3)= D2(2,2)
1160    D2(3,2)=-SIN(TA/180*PI) : D2(2,3)=-D2(3,2)
1180    D2(1,1)=1 : D2(1,2)=0 : D2(1,3)=0
1190    D2(2,1)=0 : D2(3,1)=0
1200    FOR I=1 TO 3 : FOR J=1 TO 3
1210       D(I,J)=0 : FOR K=1 TO 3
1220         D(I,J)=D(I,J)+D2(I,K)*D1(K,J) : NEXT K
1230    NEXT J : NEXT I
1400    REM ********************************************
1402    REM *** Calculation of X(), Y() and Z()     ***
1404    REM ********************************************
1410    NX=0 : Z5=1E+30 : Z6=-1E+30
1430    FOR X0=-10 TO 10 STEP .5 : NX=NX+1 : NY=0
1450      FOR Y0=-10 TO 10 STEP .5 : NY=NY+1
```

```
1460      X(NX)=X0 : Y(NY)=Y0
1470      Z(NX,NY)=3*COS(SQR(X0*X0+Y0*Y0))
1480      IF Z(NX,NY)<Z5 THEN Z5=Z(NX,NY)
1490      IF Z(NX,NY)>Z6 THEN Z6=Z(NX,NY)
1500   NEXT Y0 : NEXT X0
1600   REM *********************************************
1602   REM *** Determination of Xmin and Xmax after ***
1604   REM *** transformation in variables X1 and X9***
1606   REM *********************************************
1607   T(1)=X(1) : T(2)=Y(1) : T(3)=Z5
1608   GOSUB 8000 : X1=XN : X9=XN : Y5=YN
1610   T(1)=X(1) : T(2)=Y(NY) : GOSUB 8000
1620   IF XN<X1 THEN X1=XN
1625   IF XN>X9 THEN X9=XN
1630   T(1)=X(NX) : T(2)=Y(NY) : T(3)=Z6
1635   GOSUB 8000 : Y6=YN
1640   IF XN<X1 THEN X1=XN
1650   IF XN>X9 THEN X9=XN
1660   T(1)=X(NX) : T(2)=Y(1) : GOSUB 8000
1670   IF XN<X1 THEN X1=XN
1680   IF XN>X9 THEN X9=XN
1690   DX=250/(X9-X1)
1700   FOR I=1 TO 251
1710      M1(I)= 1E+30
1720      M9(I)=-1E+30
1730   NEXT I
2000   REM *********************************************
2002   REM *** Start of the plot routine            ***
2006   REM *********************************************
2010   KEY OFF : SCREEN 2
2020   SM=ABS(Y6-Y5)/250*.2
2030   LT0=&HFFFF : LT1=&HAAAA : LT9=&H0
2040   LT=LT0
2100   WINDOW (0,Y5) - (252,Y6)
2150   T(1)=X(1) : T(2)=Y(1) : T(3)=Z(1,1)
2180   GOSUB 8200 : REM transformations
2200   LINE -(XN,YN),,,LT9
2220   XA=XN : YA=YN
2240   FOR I=2 TO NY
2260      T(1)=X(1) : T(2)=Y(I) : T(3)=Z(1,I)
2280      GOSUB 8200 : GOSUB 9000
2290   NEXT I
2310   T(1)=X(1) : T(2)=Y(1) : T(3)=Z(1,1)
2320   GOSUB 8200 : REM transformations
2340   LINE -(XN,YN),,,LT9 : XA=XN : YA=YN
2360   FOR I=2 TO NX
```

```
2380    T(1)=X(I) : T(2)=Y(1) : T(3)=Z(I,1)
2390    GOSUB 8200 : GOSUB 9000
2400  NEXT I
2500  FOR I=2 TO NY
2510    J=I : K=2
2520    T(1)=X(1) : T(2)=Y(I) : T(3)=Z(1,I)
2525    GOSUB 8200 : LINE -(XN,YN),,,LT9
2530    XA=XN : YA=YN
2550    T(1)=X(K) : T(2)=Y(J) : T(3)=Z(K,J)
2560    GOSUB 8200 : GOSUB 9000
2580    J=J-1 : IF J<1 THEN GOTO 2700
2600    T(1)=X(K) : T(2)=Y(J) : T(3)=Z(K,J)
2610    GOSUB 8200 : GOSUB 9000
2620    K=K+1 : IF K>NX THEN GOTO 2700
2640    GOTO 2550
2700  NEXT I
2800  FOR I=2 TO NX-1
2810    J=NY : K=I+1
2820    T(1)=X(I) : T(2)=Y(J) : T(3)=Z(I,NY)
2825    GOSUB 8200 : LINE -(XN,YN),,,LT9
2830    XA=XN : YA=YN
2850    T(1)=X(K) : T(2)=Y(J) : T(3)=Z(K,J)
2860    GOSUB 8200 : GOSUB 9000
2880    J=J-1 : IF J<1 THEN GOTO 3000
2900    T(1)=X(K) : T(2)=Y(J) : T(3)=Z(K,J)
2910    GOSUB 8200 : GOSUB 9000
2920    K=K+1 : IF K>NX THEN GOTO 3000
2940    GOTO 2850
3000  NEXT I
3100  T(1)=X(1) : T(2)=Y(NY) : T(3)=Z6
3110  GOSUB 8200 : LINE -(XN,YN),,,LT9
3120  T(1)=X(1) : T(2)=Y(NY) : T(3)=Z5
3130  GOSUB 8200 : LINE -(XN,YN)
3160  T(2)=Y(1)  : GOSUB 8200 : LINE -(XN,YN)
3180  T(1)=X(NX) : GOSUB 8200 : LINE -(XN,YN)
3190  XA=XN : YA=YN
3200  T(2)=Y(NY) : GOSUB 8200 : GOSUB 9000
3220  T(1)=X(1)  : GOSUB 8200 : GOSUB 9000
3990  A$="" : A$=INKEY$ : IF A$="" THEN GOTO 3990
3995  SCREEN 0 : KEY ON
3999  END
4900  REM #############################################
4910  REM ### Main program ends here.          ###
4920  REM #############################################
8000  REM *********************************************
8001  REM *** Subroutine: TRANSFORMATION_1        ***
8002  REM *********************************************
```

```
8060  XN=0 : FOR J9=1 TO 3 : XN=XN+D(1,J9)*T(J9) : NEXT J9
8080  YN=0 : FOR J9=1 TO 3 : YN=YN+D(2,J9)*T(J9) : NEXT J9
8100  RETURN
8200  REM ********************************************
8201  REM *** Subroutine: TRANSFORMATION_2        ***
8202  REM ********************************************
8220  XN=0 : FOR J9=1 TO 3 : XN=XN+D(1,J9)*T(J9) : NEXT J9
8240  YN=0 : FOR J9=1 TO 3 : YN=YN+D(2,J9)*T(J9) : NEXT J9
8260  XN=INT((XN-X1)*DX+1.5)
8300  RETURN
9000  REM ********************************************
9001  REM *** Subroutine: PLOT                    ***
9002  REM ********************************************
9030  IF XA=XN THEN GOTO 9800
9040  SL=(YN-YA)/(XN-XA)
9060  IM(1)=XA : S1=SGN(XN-XA) : IL=2 : YK=YA
9100  IZ(1)=0 : IF YK>M1(XA) THEN GOTO 9140
9120  IZ(1)=-1 : M1(XA)=YK
9140  IF YK<M9(XA) THEN GOTO 9200
9160  IZ(1)=1 : M9(XA)=YK
9200  FOR R=XA+S1 TO XN STEP S1
9220    YK=FNY(R) : IZ(IL)=0
9240    IF R=XN THEN YK=YN
9260    IF YK>M1(R) THEN GOTO 9300
9280    IZ(IL)=-1 : YL=M1(R) : M1(R)=YK
9300    IF YK<M9(R) THEN GOTO 9340
9320    IZ(IL)=1 : YL=M9(R) : M9(R)=YK
9340    IF IZ(IL)=IZ(IL-1) THEN GOTO 9380
9350    IM(IL)=R
9355    KO=S1*(FNY(R-S1)-YL)/(FNY(R-S1)-YK)
9360    IF ABS(SL)>SM AND ABS(KO)<1 THEN IM(IL)=R-S1+KO
9370    IL=IL+1
9380  NEXT R : IM(IL)=XN
9500  IF IL>2 THEN GOTO 9600
9520  IF IZ(2)=-1 THEN LT=LT1
9530  IF IZ(2)= 0 THEN LT=LT9
9540  LINE -(XN,YN),,,LT : XA=XN : YA=YN
9560  LT=LT0 : GOTO 9990
9600  FOR R=1 TO IL-2
9620    IF IZ(R)=-1 THEN LT=LT1
9630    IF IZ(R)= 0 THEN LT=LT9
9640    LINE -(IM(R+1),FNY(IM(R+1))),,,LT
9650    LT=LT0
9660  NEXT R
9680  IF IZ(IL-1)=-1 THEN LT=LT1
9690  IF IZ(IL-1)= 0 THEN LT=LT9
```

```
9700   LINE -(XN,YN),,,LT
9710   LT=LTO : XA=XN : YA=YN
9720   GOTO 9990
9800   REM xa=xn
9820   IZ(1)=0 : IF YA>M1(XA) THEN GOTO 9860
9840   IZ(1)=-1 : M1(XA)=YA
9860   IF YA<M9(XA) THEN GOTO 9900
9880   IZ(1)=1 : M9(XA)=YA
9900   IZ(2)=0 : IF YN>M1(XA) THEN GOTO 9920
9910   IZ(2)=-1 : YK=M1(XA) : M1(XA)=YN
9920   IF YN<M9(XA) THEN GOTO 9940
9930   IZ(2)=1 : YK=M9(XA) : M9(XA)=YN
9940   IF IZ(1)=IZ(2) THEN GOTO 9970
9945   IF IZ(1)=-1 THEN LT=LT1
9946   IF IZ(1)= 0 THEN LT=LT9
9950   LINE -(XA,YK),,,LT : LT=LTO
9955   IF IZ(2)=-1 THEN LT=LT1
9956   IF IZ(2)= 0 THEN LT=LT9
9960   LINE -(XA,YN),,,LT : LT=LTO : YA=YN
9965   GOTO 9990
9970   IF IZ(1)=-1 THEN LT=LT1
9975   IF IZ(1)= 0 THEN LT=LT9
9980   LINE -(XN,YN),,,LT : YA=YN : LT=LTO
9990   RETURN
```

The rectangular (x, y)-area, for which the function $z = f(x, y)$ shall be plotted, is divided in the x- and y-direction into 40 subintervals. The z-values are calculated for all mesh points of the grid (maximum of 2500). The dimensions of the vectors X and Y and the matrix Z are defined in line 40. Matrix Z holds the function values of $z = f(x, y)$ for all combinations of 41 x-values and 41 y-values. The calculation of the function values is carried out in the program section from line 1400 to 1500. The X-, Y- and Z-values are stored and the minimum (Z5) and maximum (Z6) of Z is determined. The x-value X(I) and y-value Y(J) belong to the function value Z(I,J).

The projection of every triple (X(I),Y(J),Z(I,J)) into a plane is done by a simple matrix multiplication. The common rotation matrix D is constructed from the matrix D1 for rotation about the z-axis and D2 for rotation about the x-axis (D=D2*D1).

The transformation matrix D is calculated at the front part of the program up to line 1230. The projection is performed in the subroutine starting with line 8000. The results of the coordinates in the projection plane are stored in the variables XN and YN. This is done using an auxiliary vector T, which holds the x-, y- and z-value of one point; T is multiplied with the matrix D. The z-value after the projection is of no interest, therefore it is not calculated.

$$D2 = \begin{pmatrix} \cos\rho & -\sin\rho & 0 \\ \sin\rho & \cos\rho & 0 \\ 0 & 0 & 1 \end{pmatrix} \qquad D1 = \begin{pmatrix} 1 & 0 & 0 \\ 0 & \cos\tau & -\sin\tau \\ 0 & \sin\tau & \cos\tau \end{pmatrix}$$

$$D = D2 * D1 = \begin{pmatrix} \cos\rho & -\sin\rho\cos\tau & \sin\rho\sin\tau \\ \sin\rho & \cos\rho\cos\tau & -\cos\rho\sin\tau \\ 0 & \sin\tau & \cos\tau \end{pmatrix}$$

The minimum and maximum of x and y after the transformation are searched (X1, X9, Y5 and Y6) in the program section from line 1600 to 1690. The scaling factor DX is calculated by dividing 250 by the length of the x-region. Then the two vectors M1(251) and M9(251) are filled with extremely large numbers in the loop starting in line 1700. These two vectors serve to hold for each x-position the part of the screen which is already filled with graphical output. This is necessary to prohibit the plotting of hidden lines.

Table 13.4. Table of the zigzag pattern; showing the plotting sequence for the x-, y- and z-values

X(1)	Y(10)	Z(1,10)	dark
X(2)	Y(10)	Z(2,10)	bright
X(2)	Y(9)	Z(2, 9)	bright
X(3)	Y(9)	Z(3, 9)	bright
X(3)	Y(8)	Z(3, 8)	bright
X(4)	Y(8)	Z(4, 8)	bright
. . .	. . .	. . .	
X(8)	Y(4)	Z(8, 4)	bright
X(8)	Y(3)	Z(8, 3)	bright
X(9)	Y(3)	Z(9, 3)	bright
X(9)	Y(2)	Z(9, 2)	bright
X(10)	Y(2)	Z(10, 2)	bright
X(10)	Y(1)	Z(10, 1)	bright

The process of drawing is best followed on the screen. Plotting should start with the fixed first x along the y-axis (lines 2220–2290) and then continue with fixed the first y along the x-axis (lines 2310–2400). Plotting is ensued in each case with a call of subroutine 9000. The plotted net is generated in the program section from line 2500 to 3000 beginning with the lowest corner up to the topmost corner. The points are drawn in a zigzag sequence. This method can be understood more clearly if the loop from line 2500 to 3000 is analyzed, for example with I=10, by looking at the x-, y- and z-values (Table 13.4), which are transformed and then plotted. The variables XA and YA in this part of the program are used to store the actual position of the plotting pointer; XN and YN are generated by the transformation matrix and mark the new position to which a line has to be drawn.

The analysis of the plot subroutine 9000 is left to the reader. Plot examples are shown in Fig. 13.8 and 13.9.

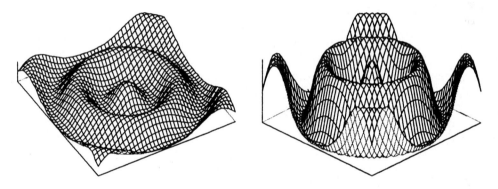

Fig. 13.8. Plot of the 'cowboy hat' with $\rho = 30°$, $\tau = 30°$ and $\rho = 45°$, $\tau = 70°$

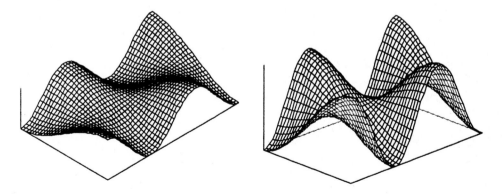

Fig. 13.9. Plot of the 'wavy function' $z = \cos x \cdot \cos \frac{y}{2}$ with $\rho = 45°$, $\tau = 60°$ and
$\rho = 45°$, $\tau = 80°$

13.5   Contour Plots

Another method for presenting three-dimensional functions and objects in two dimensions is a contour plot, where for the variable z several contour level curves are drawn in the (x, y)-plane. The program CONTOUR which generates such diagrams is given below:

```
0 REM "CONTOUR "          EBERT/EDERER          880117
1 REM *************************************************
2 REM *** The function z=f(x,y) shall be plotted   ***
3 REM *** as a CONTOUR plot .                       ***
9 REM *************************************************
40    DIM H(20),X(50),XP(4),Y(50),YP(4),Z(50,50),LT(3)
50    LT(0)=&HFFFF : LT(1)=&HF0F0
60    LT(2)=&HCCCC : LT(3)=&HAAAA
100   NH=8 : REM number of different contour levels
1400  REM *******************************************
1401  REM *** Calculation of the function values.   ***
1402  REM *** The triples are stored in the         ***
1403  REM *** vectors X(), Y() and Z().             ***
1405  REM *******************************************
1410  NX=0 : Z5=1E+30 : Z6=-1E+30
1430  FOR X0=-10 TO 10 STEP .5 : NX=NX+1 : NY=0
1450    FOR Y0=-10 TO 10 STEP .5 : NY=NY+1
1460      X(NX)=X0 : Y(NY)=Y0
1470      Z(NX,NY)=3*COS(SQR(X0*X0+Y0*Y0))
1480      IF Z(NX,NY)<Z5 THEN Z5=Z(NX,NY)
1490      IF Z(NX,NY)>Z6 THEN Z6=Z(NX,NY)
1500  NEXT Y0 : NEXT X0
1600  FOR I=1 TO NH
1610    H(I)=Z5+(I-.75)*.95*(Z6-Z5)/(NH-1)
1620  NEXT I
2000  REM *******************************************
2001  REM *** Start of the plot routine             ***
2005  REM *******************************************
2010  KEY OFF : SCREEN 2
2030  VIEW (80,0)-(559,199)
2100  WINDOW (X(1),Y(1)) - (X(NX),Y(NY))
2200  LINE (X(1),Y(1)) - (X(NX),Y(NY)),,B
2500  FOR IH=1 TO NH : HV=H(IH)
2510    LTY=LT(IH MOD 4)
2520    FOR IX=2 TO NX
2530      FOR IY=2 TO NY
2540        IP=0
```

```
2550        Z1=HV-Z(IX-1,IY-1) : Z2=HV-Z(IX-1,IY)
2560        Z3=HV-Z(IX,IY)     : Z4=HV-Z(IX,IY-1)
2700        IF SGN(Z1)=SGN(Z2) THEN 2800
2720        IP=IP+1 : XP(IP)=X(IX-1)
2740        YP(IP)=Y(IY-1)+(Y(IY)-Y(IY-1))*Z1/(Z1-Z2)
2800        IF SGN(Z2)=SGN(Z3) THEN 2900
2820        IP=IP+1 : YP(IP)=Y(IY)
2840        XP(IP)=X(IX-1)+(X(IX)-X(IX-1))*Z2/(Z2-Z3)
2900        IF SGN(Z3)=SGN(Z4) THEN 3000
2920        IP=IP+1 : XP(IP)=X(IX)
2940        YP(IP)=Y(IY-1)+(Y(IY)-Y(IY-1))*Z4/(Z4-Z3)
3000        IF SGN(Z4)=SGN(Z1) THEN 3100
3020        IP=IP+1 : YP(IP)=Y(IY-1)
3040        XP(IP)=X(IX-1)+(X(IX)-X(IX-1))*Z1/(Z1-Z4)
3100        IF IP<>2 THEN GOTO 3200
3120        LINE (XP(1),YP(1))-(XP(2),YP(2)),,,LTY
3200 NEXT IY : NEXT IX : NEXT IH
7900 A$="" : A$=INKEY$ : IF A$="" THEN GOTO 7900
7910 SCREEN 0 : KEY ON
7990 END
```

After some definitions, the statement in line 100 fixes the number of contour levels; it is set to 8 in the printed program example. In the following program section the grid for the (x, y)-plane is defined and the $z = f(x, y)$ values are calculated and stored in the field variables X, Y and Z. The minimum (Z5) and maximum (Z6) for the z-values are determined. From these extremes, the contour levels are calculated in lines 1600–1620 and stored in the variables H(I). Sometimes it is better to define the contour levels manually because the automatic equidistant levels may not be adequate for the 3-D-object presentation.

In the following section through lines 2010–2200 the plot screen is initialized and a frame is plotted. The grid itself is not plotted, but sometimes it may be of advantage to print the grid too. The rest of the program is the outer loop (IH-loop) for the different contour levels. For each mesh of the grid it is analyzed whether the contour level line HV=H(IH) is running through it. This is done in the inner loops (IX-loop and IY-loop) from line 2520 to 3200. For this examination, the contour level value HV is subtracted from all function values z at each corner of the mesh and these differences are stored in the variables Z1, Z2, Z3 and Z4 (line 2550 and 2560). If the signs of all four values are the same, then the actual contour level doesn't cross the mesh in question and the program continues with the next mesh. But if different signs are found, then the intersection points with the border lines of the cell are calculated with linear interpolation and stored. If the meshes are small enough, it can be assumed that a contour level line is crossing it either twices or not at all. The two intersection points found are connected by a straight line, which is the last statement (line 3120) within the loops. The final parameter in this **LINE**-command defines the line type. It seems to be useful to plot a different line type, or better a different line color, for every contour level. We have used only four different line types. It remains to the reader to improve this; of course this depends on the possibilities your computer system offers.

The figures shown in Fig. 13.10 are the contour plots of the same functions we used as the examples for the program **PLOT3D**.

$$z = 3 \cos \sqrt{x^2 + y^2}$$

$$z = \cos x \cdot \cos \frac{y}{2}$$

Fig. 13.10. Contour plots of the 'cowboy hat function' and of the 'wavy function'

As an example for the application of such plots, the contour plot for the potential energy surface (Fig. 13.11) of the linear triatomic molecule (H—C—H) was chosen. The combined Morse potential shown here is taken from the literature.[2] The program and the contour plot of the potential energy surface are shown below.

---

[2] R.F. Nalewajski, R.E. Wyatt, *Chem. Phys.*, **81**, 357 (1983)

$$E = D \cdot \left(e^{-2c(x-r_{eq})} - 2e^{-c(x-r_{eq})} + e^{-2c(y-r_{eq})} - 2e^{-c(y-r_{eq})} \right)$$

E	:	Energy
x	:	Length of the first H—C bond
y	:	Length of the second C—H bond
r_{eq}	:	Equilibrium distance of the C—H bond
D	:	Dissociation energy ($= 0.1339$)
c	:	1.03613

```
0 REM "MORSE    "          EBERT/EDERER        880203
1 REM ********************************************
2 REM *** The Morse potential of the molecule    ***
3 REM *** H-C-H shall be plotted with the program ***
4 REM *** CONTOUR.                               ***
5 REM *** Lit.: R. F. Nalewajski, R. E. Wyatt    ***
6 REM *** Chem.Phys. 81, 357 (1983)              ***
9 REM ********************************************
40    DIM H(40),X(50),XP(4),Y(50),YP(4),Z(50,50),LT(3)
50    LT(0)=&HFFFF : LT(1)=&HF0F0
60    LT(2)=&HCCCC : LT(3)=&HAAAA
100   NH=30 : REM number of different contour levels
1400  REM ********************************************
1401  REM *** Calculation of the function values.    ***
1402  REM *** The triples are stored in the          ***
1403  REM *** vectors X(), Y() and Z().              ***
1405  REM ********************************************
1410  NX=0 : Z5=1E+30 : Z6=-1E+30
1430  FOR X0=1 TO 5 STEP .1 : NX=NX+1 : NY=0
1450    FOR Y0=1 TO 5 STEP .1 : NY=NY+1
1460      X(NX)=X0 : Y(NY)=Y0
1470      ZX=EXP(-2.07226*(X0-2.1163))
1471      ZX=ZX-2*EXP(-1.03613*(X0-2.1163))
1472      ZY=EXP(-2.07226*(Y0-2.1163))
1473      ZY=ZY-2*EXP(-1.03613*(Y0-2.1163))
1475      Z(NX,NY)=(ZX+ZY)*.1339
1480      IF Z(NX,NY)<Z5 THEN Z5=Z(NX,NY)
1490      IF Z(NX,NY)>Z6 THEN Z6=Z(NX,NY)
1500  NEXT Y0 : NEXT X0
1600  FOR I=1 TO NH
1610    H(I)=Z5+(I-.75)*.95*(Z6-Z5)/(NH-1)
1620  NEXT I
2000  REM ********************************************
2001  REM *** Start of the plot routine              ***
2005  REM ********************************************
```

```
2010  KEY OFF : SCREEN 2
2030  VIEW (80,0)-(559,199)
2100  WINDOW (X(1),Y(1)) - (X(NX),Y(NY))
2200  LINE (X(1),Y(1)) - (X(NX),Y(NY)),,B
2500  FOR IH=1 TO NH : HV=H(IH)
2510    LTY=LT(IH MOD 4)
2520    FOR IX=2 TO NX
2530      FOR IY=2 TO NY
2540        IP=0
2550        Z1=HV-Z(IX-1,IY-1) : Z2=HV-Z(IX-1,IY)
2560        Z3=HV-Z(IX,IY)     : Z4=HV-Z(IX,IY-1)
2700        IF SGN(Z1)=SGN(Z2) THEN 2800
2720        IP=IP+1 : XP(IP)=X(IX-1)
2740        YP(IP)=Y(IY-1)+(Y(IY)-Y(IY-1))*Z1/(Z1-Z2)
2800        IF SGN(Z2)=SGN(Z3) THEN 2900
2820        IP=IP+1 : YP(IP)=Y(IY)
2840        XP(IP)=X(IX-1)+(X(IX)-X(IX-1))*Z2/(Z2-Z3)
2900        IF SGN(Z3)=SGN(Z4) THEN 3000
2920        IP=IP+1 : XP(IP)=X(IX)
2940        YP(IP)=Y(IY-1)+(Y(IY)-Y(IY-1))*Z4/(Z4-Z3)
3000        IF SGN(Z4)=SGN(Z1) THEN 3100
3020        IP=IP+1 : YP(IP)=Y(IY-1)
3040        XP(IP)=X(IX-1)+(X(IX)-X(IX-1))*Z1/(Z1-Z4)
3100        IF IP<>2 THEN GOTO 3200
3120        LINE (XP(1),YP(1))-(XP(2),YP(2)),,,LTY
3200  NEXT IY : NEXT IX : NEXT IH
7900  A$="" : A$=INKEY$ : IF A$="" THEN GOTO 7900
7910  SCREEN 0 : KEY ON
7990  END
```

Fig. 13.11. Contour plot of the linear H—C—H molecule

In this plot the two C—H distances are chosen as axes. Therefore the result
is a symmetrical plot. At very small distances the potential energy surface
is very steep and high, at large distances the plain of a totally dissociated
molecule is obtained. Between these two extremes a minimum of the potential energy surface can be found (smallest closed curve) which is the most
stable configuration of the molecule. If the potential energy surface is cut
at one big C—H distance, then the Morse potential for a diatomic species
is obtained.

In the following program TRAJECT, this potential energy surface is used
to calculate several trajectories for the molecule motions.

```
0 REM "TRAJECT "         EBERT/EDERER         880206
1 REM **********************************************
2 REM *** Trajectories of colinear H-C-H movements ***
3 REM *** using the Morse potential of the program ***
4 REM *** MORSE .                                ***
5 REM *** Lit.: R. F. Nalewajski, R. E. Wyatt    ***
6 REM *** Chem.Phys. 81, 357 (1983)              ***
9 REM **********************************************
100    DIM Y(20),Y1(20),D(20),Y9(20),YA(20)
120    DIM K1(20),K2(20),K3(20),K4(20)
150    DEF FNV(R)=-4*(EXP(-4*(R-2.1))-EXP(-2*(R-2.1)))
200    N=6 : REM n the dimension of the ODE-system
210    MA=1 : MB=12 : MC=1
410    Y1(1)=-1.8 : REM x-coordinate of particle a
420    Y1(2)= 0! : REM x-coordinate of particle b
430    Y1(3)= 5! : REM x-coordinate of particle c
440    Y1(4)= 0! : REM x-impact     of particle a
450    Y1(5)= 0! : REM x-impact     of particle b
460    Y1(6)=-.05 : REM x-impact     of particle c
700    REM INPUT "x-value start,end,step_size";xa,xe,xd
710    XA=0 : XE=10000 : XD=.1
715    GOSUB 2000 : REM Initializing the graphics
720    FOR I=1 TO N
725       Y(I)=Y1(I) : YA(I)=Y1(I)
730    NEXT I : GOSUB 15000
740    N1=1
1000   FOR A=XA TO XE STEP XD
1020      E=A+XD
1040      H=(E-A)/N1 : GOSUB 5000
1050      FOR I=1 TO N : YA(I)=Y(I) : NEXT I
1060      N1=N1*2 : H=(E-A)/N1 : GOSUB 5000
1080      FOR I=1 TO N
```

```
1090      IF ABS(Y(I))+ABS(YA(I))<.000001 THEN 1120
1100      YG=ABS(YA(I)-Y(I))/(ABS(YA(I))+ABS(Y(I)))
1110      IF YG >.01 THEN 1050
1120    NEXT I
1200    GOSUB 15000 : REM OUTPUT
1220    FOR I=1 TO N : Y1(I)=Y(I) : NEXT I
1240    N1=INT(N1/4) : IF N1=0 THEN N1=1
1260  NEXT A
1300  A$="" : A$=INKEY$ : IF A$="" THEN GOTO 1300
1400  SCREEN 0 : KEY ON
1899  END
1900  REM #########################################
1910  REM ### Main program ends here.          ###
1920  REM #########################################
2000  REM *****************************************
2001  REM *** Subroutine:   INIT_PLOT          ***
2002  REM *****************************************
2010  KEY OFF : SCREEN 2
2020  FJ=&H0
2030  VIEW (80,0)-(559,199)
2050  PP=16
2060  WINDOW (1,1)-(5,5)
2080  LINE   (1,1)-(5,5),,B
2200  RETURN
5000  REM *****************************************
5002  REM *** Subroutine: ODE                  ***
5004  REM *** The solution of the ODE is approxi- ***
5006  REM *** mated by using N1 subintervals.     ***
5010  REM *****************************************
5100  FOR I=1 TO N : Y(I)=Y1(I) : NEXT I
5200  FOR I9=0 TO N1-1 : X=A+I9*H
5300    X9=X : FOR I=1 TO N : Y9(I)=Y(I) : NEXT I
5320    GOSUB 10000
5340    FOR I=1 TO N : K1(I)=H*D(I) : NEXT I: X9=X+H/2
5400    FOR I=1 TO N : Y9(I)=Y(I)+K1(I)/2 : NEXT I
5420    GOSUB 10000
5440    FOR I=1 TO N : K2(I)=H*D(I) : NEXT I: X9=X+H/2
5500    FOR I=1 TO N : Y9(I)=Y(I)+K2(I)/2 : NEXT I
5520    GOSUB 10000
5540    FOR I=1 TO N : K3(I)=H*D(I) : NEXT I : X9=X+H
5600    FOR I=1 TO N : Y9(I)=Y(I)+K3(I) : NEXT I
5620    GOSUB 10000
5640    FOR I=1 TO N : K4(I)=H*D(I) : NEXT I
5700    FOR I=1 TO N
5720      Y(I)=Y(I)+(K1(I)+2*K2(I)+2*K3(I)+K4(I))/6
5740    NEXT I : X=E
5800  NEXT I9 : RETURN
```

```
10000 REM ********************************************
10002 REM *** Subroutine: RHS                     ***
10004 REM *** The right hand sides of the ODE's   ***
10006 REM *** are calculated here as              ***
10008 REM *** D(I) = f(X9,Y9(1),Y9(2),..Y9(N))    ***
10010 REM ********************************************
10100 D(1)=Y9(4)/MA
10200 D(2)=Y9(5)/MB
10300 D(3)=Y9(6)/MC
10500 RX=ABS(Y9(1)-Y9(2)) : RY=ABS(Y9(2)-Y9(3))
10560 V1=FNV(RX) : V2=FNV(RY)
10600 D(4)=-(Y9(1)-Y9(2))*V1/RX
10800 D(6)=(Y9(2)-Y9(3))*V2/RY
11500 D(5)=-D(4)-D(6)
12000 RETURN
15000 REM ********************************************
15001 REM *** Subroutine: OUTPUT                  ***
15002 REM ********************************************
15020 RX=ABS(Y(1)-Y(2)) : RY=ABS(Y(2)-Y(3))
15030 LINE -(RX,RY),,,FJ : FJ=&HFFFF
15080 A$="" : A$=INKEY$ : IF A$="q" THEN GOTO 1400
15500 RETURN
```

Because only colinear motions of the molecule are considered, it is sufficient to consider only one coordinate (e.g. the x-coordinate) of the three atoms. The six differential equations are constructed similar to the method shown for the program IMPACT2. The time was normalized to units of 10^{-13} seconds and the mass unit was rescaled with the factor $6.022 \cdot 10^{23}$. With these rescalings we obtain more common numbers. For the trajectories shown in the plot examples of Fig. 13.12 we used the initial conditions of Table 13.5.

The potential energy surfaces on which the last two program examples are based are Morse potentials, which can be described by the following formulas

$$V(r_i) = D_i \cdot \left(e^{-2c_i(r_i - g_i)} - 2e^{-c_i(r_i - g_i)} \right)$$

$$\frac{\partial V}{\partial r_i} = -2c_i D_i \cdot \left(e^{-2c_i(r_i - g_i)} - e^{-c_i(r_i - g_i)} \right)$$

$$V(A\text{---}B\text{---}C) = V(A\text{---}B) + V(B\text{---}C)$$
$$= V(r_{AB}) + V(r_{BC})$$

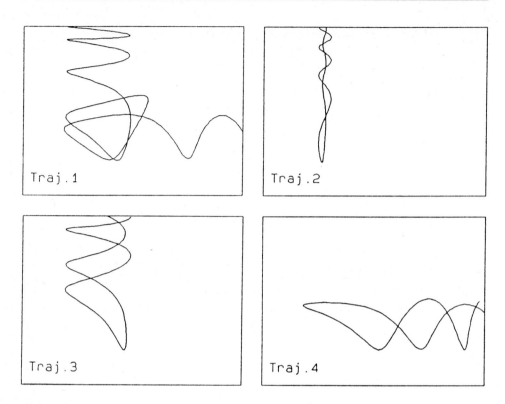

Fig. 13.12. Several plots of trajectories of the H—C—H molecule

Table 13.5. Table of the initial conditions for four different trajectories which are shown as plots in Fig. 13.12. Y(1), Y(2), Y(3) are the x-coordinates of the atoms in the H–C–H molecule, Y(4), Y(5), Y(6) are the x-impacts of the atoms in the H–C–H molecule.

	Traj. 1	Traj. 2	Traj. 3	Traj. 4
Y1(1)	−1.8	−2	−3	−3
Y1(2)	0	0	0	2
Y1(3)	5	5	5	5
Y1(4)	0	0	0	0.2
Y1(5)	0	0	0	0
Y1(6)	−0.05	−0.2	−0.2	−0.2

Problem 199a

Construct a new program which combines the two programs **MORSE** and **TRAJECT** in order to get a trajectory within its contour plot of the Morse potential.

14 Data Acquisition and Data Handling

This chapter is an introduction to the acquisition of laboratory data using a computer. The path from the measurement of the physical quantity to the storage of a number (which corresponds to this physical quantity) in the computer is shown in a diagram (examples are given in parentheses):

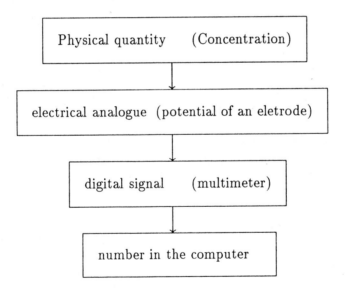

We will treat here only the part that is represented by the last arrow; the conversion of the digital signal into a number in the computer. The hardware for this connection are the interfaces. A very common interface is the IEEE-interface or bus. With this bus many different instruments can be connected to the computer through only one IEEE-port. The other port which is more common for personal computers is the RS232-C-interface, but a separate interface is needed for each instrument connected to the computer. Both interfaces mentioned can be used for data input and data output. But we don't want to describe technical details; this information is given in the literature on computer equipment.

Additionally in this chapter we will present some programs which may be useful for data handling: among them are programs for quick sorting and spline algorithms for smoothing experimental data.

14.1   Data Acquisition

A simple chemical experiment, the Belousov-Zhabotinski reaction, easily carried out in a beaker, is used to demonstrate data recording. The BZ reaction is the best known reaction for a chemical oscillation. The redox potential of the reaction mixture is recorded with a platinum and a calomel electrode. The potential values are read into the computer and plotted on the screen.

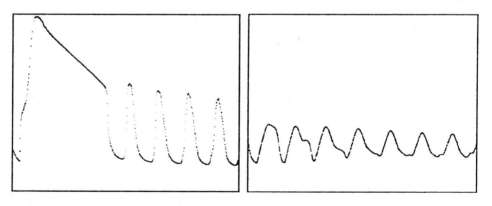

Fig. 14.1. The redox potential of the BZ reaction as a function of reaction time

The experimental conditions are as follows:

50 ml	malonic acid	(1.2 m)
50 ml	cerium-III-sulfate	(0.004 m)
50 ml	potassium bromate	(0.35 m)
50 ml	sulfuric acid	(15%)

In a beaker, the sulfuric acid is poured into the mixture of the other three components 20 seconds after the data acquisition has started. The result of this experiment is shown in Fig. 14.1. The x-axis corresponds to a time of 600 seconds. The y-axis is scaled to the potential interval from 1.65 V to 1.95 V. The left figure shows the start of the reaction, which stays a rather long time at a high potential level and then slowly changes to an oscillating system. The right figure shows the system after a long time; most of the initial substances are now consumed and the amplitudes of the oscillations are decreasing. The following program LABDATA1 was used for data acquisition and graphical output for this experiment:

```
0 REM "LABDATA1"        EBERT/EDERER        880211
1 REM *********************************************
2 REM *** Reading data into the computer using a   ***
3 REM *** digital multimeter DVM M2110 METRAWATT.  ***
4 REM *** The RS232-C interface for asynchronous   ***
5 REM *** data transfer is used.(with plot routine)***
9 REM *********************************************
100    PRINT "What is the time interval between two "
110    PRINT "data to be read into the computer"
115    PRINT "The time has to be given in seconds ";
120    INPUT DT : IF DT<.5 THEN GOTO 120
130    PRINT "What is the overall measurement time ";
135    INPUT TM
200    KEY OFF : SCREEN 2
210    Y1=16500:Y9=19500
220    WINDOW (0,Y1) - (TM,Y9)
230    LINE   (0,Y1) - (TM,Y9) ,,B
290    TI=TIMER
300    TO=TI+1 : IT=0
400    TV=TO+DT*IT
490      TI=TIMER
500      IF TI<TV THEN 490
550      T=TI : GOSUB 50000
570      REM PRINT (T-TO),A
580      PSET(T-TO,A)
600    IT=IT+1 : IF (T-TO)< TM THEN 400
700    A$="" : A$=INKEY$ : IF A$="" THEN GOTO 700
720    SCREEN 0 : KEY ON
999    STOP
1900   REM ##########################################
1910   REM ### Main program ends here.            ###
1920   REM ##########################################
```

```
50000 REM *********************************************
50001 REM *** Subroutine for taking a datum from  ***
50002 REM *** the digital multimeter.             ***
50003 REM *********************************************
50010 OPEN "com1:4800,o,7,2,rs,cs,ds,cd" AS #5
50015 PRINT #5,CHR$(5);
50020 A$=INPUT$(9,#5)
50040 A=VAL(A$)
50050 CLOSE 5
50060 RETURN
```

In line 120 the user is asked in which time interval DT he wants to take data. Because the digital multimeter used can only get a data value every 0.5 seconds, the input of shorter intervals is rejected. Then the maximum for the duration TM of the measurement is requested in line 135. In line 200 the graphics environment is initialized. In the variables Y1 and Y9 (line 210) the minimum and maximum values for the measured voltage are assigned and with that the screen window is defined and the frame is drawn (lines 220 and 230). In line 300 the measurement is started with a retardation of one second. To achieve this, the actual computer time, which can be obtained from the variable TIMER, is increased by 1 and stored in variable T0, the starting time. The counter DT for the data numbers is set to zero. The time TV for the next data input is calculated in line 400, that means, the time interval DT*IT is added to the starting time T0.

The program waits in the loop (lines 490 and 500) until the actual computer time is greater than the time for the measurement TV. The computer time is now stored in variable T and a datum is taken by calling the subroutine GOSUB 50000 which fills the variable A with the experimental value. In line 580 the datum A is plotted against the measurement time (T-T0) in the form of singular points on the screen. The data counter is increased by one, and the program jumps back to line 400 if the overall measurement time is not yet reached.

The subroutine 50000 for retrieving the data values starts in line 50010 with the statement:

```
50010 OPEN "com1:4800,o,7,2,rs,cs,ds,cd" AS #5
```

For details of the OPEN-statement consult your computer and BASIC-manual. This statement causes channel #5 to be opened for data transfer using the serial interface named COM1. The data flow rate is 4800 bits per second.

One byte consists of 7 data bits, the parity is odd and two stop bits are used.

With the statement PRINT #5 of line 50015 the multimeter, which is connected to channel #5, gets the command to hold the digital output until it has been read completely. The next statement in line 50020 is the read statement for taking the data:

```
50020 A$=INPUT$(9,#5)
```

This INPUT$-function reads 9 bytes from the channel #5. The result is stored in the string variable A$. The function VAL which is used in the next statement converts numbers, which are written as a string, into number variables. The last statement of this short subroutine closes the channel #5.

With the following example the data of an elution curve of a liquid chromatograph can be read into a computer. The experiment is done to analyze a polymer sample; the molecular weight distribution is calculated from the elution data using a known calibration curve.

```
0 REM "GPCDATA "          EBERT/EDERER          880211
1 REM ***********************************************
2 REM *** Taking data into the computer using a    ***
3 REM *** digital multimeter.                      ***
4 REM *** The RS232-C interface for asynchronous   ***
5 REM *** data transfer is used.                   ***
6 REM *** GPC (Gel Permeation Chromatography) of a ***
7 REM *** polymer; the elution curve is recorded   ***
8 REM *** and the molecular weight distribution    ***
9 REM *** is calculated.                           ***
10 REM***********************************************
50    DIM X(250),M(250),DM(250),XP(250)
100   PRINT "What is the time interval between two "
110   PRINT "data to be read into the computer"
115   PRINT "The time has to be given in seconds ";
120   INPUT DT : IF DT<.5 THEN GOTO 120
130   PRINT "What is the overall measurement time ";
135   INPUT TM
138   REM DT=5 : TM=1250
200   KEY OFF : SCREEN 2
210   Y1=-1000 : Y9=10000
220   WINDOW (0,Y1) - (TM,Y9)
230   LINE   (0,Y1) - (TM,Y9) ,,B
290   TI=TIMER
300   TO=TI+1 : IT=0
```

```
400     TV=T0+DT*IT
490       TI=TIMER
500       IF TI<TV THEN 490
550       T=TI : GOSUB 50000
570       REM PRINT (T-T0),A
580       PSET(T-T0,A)
590       X(IT)=A
600     IT=IT+1 : IF (T-T0)< TM THEN 400
650     GOSUB 12000 : REM Baseline correction
700     GOSUB 10000 : REM MWD calculation and plot
800     A$="" : A$=INKEY$ : IF A$="" THEN GOTO 800
820     SCREEN 0 : KEY ON
999     STOP
1900    REM ##########################################
1910    REM ### Main program ends here.          ###
1920    REM ##########################################
10000 REM ******************************************
10001 REM *** Calculation of the molecular weight  ***
10002 REM *** distribution and plot .              ***
10003 REM ******************************************
10100 DEF FNM(T)=EXP(13-.005*T)
10200 MM=0
10240 FOR I=0 TO 250
10260   M(I)=FNM(I*5) : NEXT I
10300 FOR I=1 TO 250
10320   D=M(I-1)-M(I)
10340   DM(I)=(X(I)+X(I-1))/(2*D)
10360   IF DM(I)>MM THEN MM=DM(I)
10380 NEXT I
10500 WINDOW (M(250),0) - (M(0),MM*1.01)
10510 X1=0 : X2=0 : X3=0 : X4=0
10515 LINE (XP(1),DM(1)) - (XP(1),DM(1))
10520 FOR I=1 TO 250
10540   XP(I)=(M(I)+M(I-1))/2
10560   S1=S1+DM(I)/XP(I)
10580   S2=S2+DM(I)
10600   S3=S3+DM(I)*XP(I)
10620   S4=S4+DM(I)*XP(I)*XP(I)
10680   LINE -(XP(I),DM(I))
10690   REM PRINT I,XP(I),DM(I)
10700 NEXT I
10740 REM PRINT "Number average    = ";S2/S1
10760 REM PRINT "Weight average    = ";S3/S2
10780 REM PRINT "Z -     average   = ";S4/S3
10800 FOR I=1 TO 250
10820   I1=I : IF DM(I)>MM/50 THEN GOTO 10900
10840 NEXT I
```

```
10900 FOR I=250 TO 1 STEP -1
10920   I9=I : IF DM(I)>MM/50 THEN GOTO 10980
10940 NEXT I
10980 WINDOW (M(I9),0) - (M(I1),MM*1.01)
10990 LINE (XP(1),DM(1)) - (XP(1),DM(1))
11000 FOR I=I1 TO I9
11020   LINE -(XP(I),DM(I))
11040 NEXT I
11990 RETURN
12000 REM ******************************************
12001 REM *** Baseline correction               ***
12002 REM ******************************************
12020 X0=X(0) : X9=X(250)
12040 FOR I=0 TO 250
12060   REM PRINT I,X(I);"    ";
12100   X(I)=X(I)-(X0+(X9-X0)/250*I)
12120   IF X(I)<0 THEN X(I)=0
12130   REM PRINT X(I)
12140 NEXT I
12300 RETURN
50000 REM ******************************************
50001 REM *** Subroutine for taking a data from  ***
50002 REM *** the digital multimeter.            ***
50003 REM ******************************************
50010 OPEN "com1:4800,o,7,2,rs,cs,ds,cd" AS #5
50015 PRINT #5,CHR$(5);
50020 A$=INPUT$(9,#5)
50040 A=VAL(A$)
50050 CLOSE 5
50060 RETURN
```

The subroutine for the external reading of data, beginning with line 50000, is the same as in the last example. In the main program there are only minor changes. In line 210 the interval for the expected voltage values is declared to $-1000\,\text{mV}$ to $10000\,\text{mV}$. In addition the recorded data are not only plotted but also stored (line 590) in the vector variable X(IT). The main program ends by the call of two subroutines. In the first subroutine the elution curve is corrected for the baseline and in the second subroutine the elution curve is converted into the molecular weight distribution. The measurement is programmed for 1250 seconds. A datum is read every 5 seconds; this results in 251 data points during the whole measurement time.

In subroutine 12000 concerning the baseline correction, it is assumed that the first value X(0) and the last value X(250) are representatives for the baseline. A straight line is laid through these points and the corresponding values of this straight line are subtracted from the measured values (line 12100). Corrected values, which would be less than zero, are set to zero in line 12120. The method for the baseline correction used here is very simple and can be used only if the data are only slightly scattered and if the baseline drift is almost linear and small.

The subroutine for the calculation of the molecular weight distribution from the elution curve starts with the definition of the calibration curve. This formula defines a relation between the molecular weight M and the elution time T, which starts at the time of injection of the polymer sample into the chromatographic system. In the loop from line 10240 to 10260 for each data read, the respective molecular weight is calculated and in the following loop (line 10300–10380) the conversion to the molecular weight distribution is carried out. This transformation has to be performed under the condition of constant area segments below the curves. (The area below the curves — or better the integral — is proportional to the amount of the substance, which must not change by the transformation.) The equidistant points in the time scale of the elution curve are shifted to smaller and smaller distances at lower molecular weight. This is caused by the exponential form of the calibration curve. Because the area between two measured points has to remain constant, the mean value of two neighboring points is divided by the distance in the molecular weight scale. The maximum value of the molecular weight distribution is determined and stored in the same loop. The molecular weight distribution is plotted in the loop from line 10520 to 10700 and shown in Fig. 14.2. In the same loop the moments S1, S2, S3 and S4 of this distribution are calculated and from these the different average values of the molecular weight distribution are determined in the following lines. The rest of this subroutine serves to plot the molecular weight distribution in an expanded form. To obtain this, the values at the beginning and the end of the distribution, which are less than $\frac{1}{50}$ of the maximum value, are suppressed.

Problem 200

Complete the programs for data recording in the plot section by adding a legend and titles for the axis.

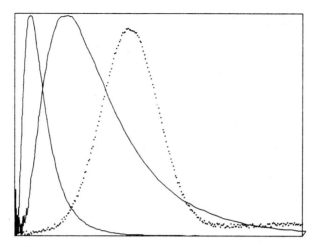

Fig. 14.2. Elution curve (dotted line) and molecular weight distribution (solid lines) of a polymer sample. The complete molecular weight distribution and the expanded version is shown.

Problem 201

Write a program for reading two different data sets from experiments; for example the redox potential and the temperature during a chemical reaction. Write the program more generally to be able to measure the different data (temperature and potential) with different sample rates. In the experimental set-up a second A–D converter, a thermoresistor or thermocouple, and a second interface are needed.

14.2   Quick Sorting of Data

In handling experimental data there is often a demand for sorting data. The programs SORTNUM and SORTTEXT (sometimes called the bubble sort method) in earlier chapters contain rather poor algorithms because they have computing times which are proportional to n^2, if n is the number of data points to be sorted. Therefore for large sets of data quicker algorithms are required.

Three different methods of quick sorting shall be presented here. They are called shell-sort, quick-sort, and address-sort. The best algorithms require only roughly several times $n \cdot \log n$ computer operations.

The first method to be treated is shell sort. The program QSORT is listed below:

```
0 REM "QSORT    "          EBERT/EDERER          880510
1 REM ************************************************
2 REM *** Sorting of numerical values. The data   ***
3 REM *** input can be done in arbitrary sequence. ***
4 REM *** The output follows in ascending order.  ***
6 REM *** Fast version.                           ***
9 REM ************************************************
100   RANDOMIZE INT(TIMER/3)
1000  DIM X(256),Y(256)
1100  INPUT "Number of data points ";N
1200  FOR I =1 TO N
1300    X(I)=RND(5):Y(I)=X(I)
1400  NEXT I
1410  T=TIMER
1500  D=INT(LOG(N)/LOG(2)) : D=2^D-1
2000  FOR I=1 TO N-D
3000    FOR J=I TO 1 STEP -D
3100      IF Y(J)<=Y(J+D) THEN 4100
3200      H=Y(J) : Y(J)=Y(J+D) : Y(J+D)=H
3400    NEXT J
4100  NEXT I
4200  D=INT(D/2)
4300  IF D>0 THEN GOTO 2000
4400  T=(TIMER-T)
5000  PRINT : PRINT "Unsorted data"
5100  FOR I=1 TO N : PRINT X(I); : NEXT I : PRINT
5200  PRINT : PRINT "  Sorted data"
5300  FOR I=1 TO N : PRINT Y(I); : NEXT I : PRINT
6000  PRINT "Sorting time:";T;"sec for";N;"data"
9999  END
```

The idea behind this method is to sort the data first into many blocks and then to reduce the number of blocks until only one block remains in which all data are contained. Thus the number of exchange and rearrangement operations (line 3200) is reduced considerably. The rearrangements within the sorting method are the most computing-time consuming operations.

In lines 1200 to 1400, random numbers are generated and stored in both vector variables X and Y. After the data in Y are sorted, they are printed together with the unsorted ones; this starts in line 5000. The lines 1410 and

4400 are needed for the determination of the computing time. The sorting itself is done in the program section from line 1500 to line 4300. The number of blocks which is used to start the sorting, is calculated in line 1500 and stored in variable D. D is defined as the greatest integer number for which $(2^D - 1)$ is smaller than the number N of data to be sorted.

In the two nested I- and J-loops (lines 2000 to 4100) the data, whose indices have a distance of D, are compared and exchanged if they have the wrong sequence. In line 4200 D is halved and the integer of D is formed. Then the sorting is done again with this smaller D. This is done until D is 1. With D=1 this method is very similar to the simple bubble sort, but at this time most of the data are already in the correct order. Therefore only a few further data rearrangements have to be made. To make this more evident: If the number of data is N=100, then for D the following series is used: 63, 31, 15, 7, 3 and 1.

One of the fastest sorting methods is quick-sort. A program implementation of this algorithm is given in the following QQSORT2:

```
0 REM "QQSORT2 "        EBERT/EDERER        880121
1 REM ****************************************
2 REM *** Sorting of numerical values. The data    ***
3 REM *** input can be done in arbitrary sequence. ***
4 REM *** The output is in ascending order.        ***
6 REM *** Very fast version with many data points. ***
9 REM ****************************************
100    RANDOMIZE TIMER/3
1000   DIM X(2000),Y(2000)
1010   DIM L(50),R(50)
1100   INPUT "Number of data points ";N
1200   FOR I =1 TO N
1300     X(I)=RND : Y(I)=X(I)
1400   NEXT I
1410   T=TIMER
2000   L(1)=1 : R(1)=N : SZ=1
3000   L1=L(SZ) : R1=R(SZ) : D=Y(R1)
3020   IF R1-L1< 5 THEN 3400
3040   M1=INT((L1+R1)/2)
3060   IF D>=Y(L1) THEN 3200
3080   IF D>=Y(M1) THEN 3400
3100   IF Y(L1)<Y(M1) THEN 3160
3120   D=Y(M1) : Y(M1)=Y(R1) : Y(R1)=D : GOTO 3400
3160   D=Y(L1) : Y(L1)=Y(R1) : Y(R1)=D : GOTO 3400
3200   IF D<=Y(M1) THEN 3400
```

```
3220   IF Y(L1)>Y(M1) THEN 3260
3240   D=Y(M1) : Y(M1)=Y(R1) : Y(R1)=D : GOTO 3400
3260   D=Y(L1) : Y(L1)=Y(R1) : Y(R1)=D : GOTO 3400
3400   REM The mean value has been determined
3500   L1=L1-1
3520   L1=L1+1 : IF R1<=L1 THEN 4000
3540   IF Y(L1)<D THEN 3520
3560   R1=R1-1 : IF R1<=L1 THEN 4000
3600   IF D<=Y(R1) THEN 3560
3700   YH=Y(R1) : Y(R1)=Y(L1) : Y(L1)=YH : GOTO 3520
4000   R1=R(SZ) : YH=Y(L1) : Y(L1)=Y(R1) : Y(R1)=YH
4060   IF (L1-L(SZ))<(R(SZ)-L1) THEN 4100
4080   L(SZ+1)=L(SZ): R(SZ+1)=L1-1: L(SZ)=L1+1: GOTO 4200
4100   L(SZ+1)=L1+1 : R(SZ+1)=R(SZ) : R(SZ)=L1-1
4200   SZ=SZ+1
4300   IF L(SZ)<R(SZ) THEN 3000
4340   SZ=SZ-1
4360   IF SZ>0 THEN 4300
4500   T=(TIMER-T)
5000   PRINT : PRINT "Unsorted data"
5100   FOR I=1 TO N : PRINT X(I); : NEXT I : PRINT
5200   PRINT : PRINT "  Sorted data"
5300   FOR I=1 TO N : PRINT Y(I); : NEXT I : PRINT
6000   PRINT "Sorting time:";T;"sec for";N;"data"
9999   END
```

In this program again random numbers are generated (lines 1200–1400)
and beginning with line 5000 the sorted and the unsorted data are printed
on the screen. The numbers are first sorted in two groups, one for the
smaller numbers and one for the bigger ones. Every group is split into two
further groups and again partitioned into smaller and larger numbers. This
is repeated until all groups consist of only one number. The sorting of one
group into two subgroups is done with an arbitrary number from the total
group as a separator. If the group has more than 4 numbers, then 3 numbers
are selected arbitrarily and the middle number is selected and stored in D
(lines 3000 to 3260). The main group is split into a subgroup of numbers
smaller than D and a subgroup of numbers greater than or equal to D (lines
3520 to 4000).

This method is demonstrated in the following example. Given are the following unsorted integer numbers:

$$4 \quad 3 \quad 7_L \quad 8 \quad 9 \quad 1 \quad 0 \quad 2_R \quad 5_S$$

Now the first number (4), the last one (5) and the number in the middle (9) are chosen. The middle number of these three (the median of the three numbers) is 5, which becomes the separator (S) for all numbers. This number 5 is put at the right end (in the above example it is already there). From the left side a number is looked for which is greater than 5 (L), and from the right side one which is less than or equal to 5 (R). The two numbers found (in our example 7 and 2) are exchanged. The new series reads:

$$4 \quad 3 \quad 2 \quad 8_L \quad 9 \quad 1 \quad 0_R \quad 7 \quad 5_S$$

This is repeated until both search sequences meet. The following series is obtained:

$$4 \quad 3 \quad 2 \quad 0 \quad 9_L \quad 1_R \quad 8 \quad 7 \quad 5_S$$

$$4 \quad 3 \quad 2 \quad 0 \quad 1 \quad 9 \quad 8 \quad 7 \quad 5_S$$

Now the separator number at the right end is put into the position where the two search sequences have met:

$$4 \ 3 \ 2 \ 0 \ 1 \quad 5 \quad 9 \ 8 \ 7$$

The number 5 is already in the correct position and both subgroups (4 3 2 0 1) and (9 8 7) are treated in the same way until the subgroups consist of one number only.

Problem 202

Modify the program QQSORT2 to print intermediate results. In this way you can supervise the whole sorting procedure. Write a modified program QQSORT2 for ascending and descending sorting.

A very special and fast sorting method can be used if the numbers to be sorted are selected from a limited set of different numbers. In the following program ADDRSORT, arbitrary integer numbers in the interval between 1 and 1000 are sorted. This special method is called address sorting and is listed below:

```
0 REM "ADDRSORT"        EBERT/EDERER        880125
1 REM ***************************************************
2 REM *** Sorting of n integer numbers. The data   ***
3 REM *** input can be done in arbitrary sequence. ***
4 REM *** The output follows in ascending order.   ***
6 REM *** Address sorting of integer numbers bet-  ***
7 REM *** ween 1 and m.                            ***
8 REM *** Maximum for n and m is 1000              ***
9 REM ***************************************************
100    RANDOMIZE TIMER/3
1000   DIM X(1001),A(1001),Y(1001)
1050   M=1000
1100   INPUT "Number of data points ";N
1120   K=0
1140   FOR I=1 TO M : A(I)=0 : NEXT I
1200   FOR I=1 TO N
1300     X(I)=INT(M*RND)+1
1400   NEXT I
1410   T=TIMER
2000   FOR I=1 TO N : A(X(I))=A(X(I))+1 : NEXT I
3000   FOR I=1 TO M : IF A(I)=0 THEN 4000
3200     FOR J=1 TO A(I) : K=K+1 : Y(K)=I : NEXT J
4000   NEXT I
4500   T=(TIMER-T)
5000   PRINT : PRINT "Unsorted data"
5100   FOR I=1 TO N : PRINT X(I); : NEXT I : PRINT
5200   PRINT : PRINT " Sorted data"
5300   FOR I=1 TO N : PRINT Y(I); : NEXT I : PRINT
6000   PRINT "Sorting time:";T;"sec for";N;"data"
9999   END
```

In lines 1200 to 1400 random integer numbers are generated. In the program section beginning with line 5000 the sorted and unsorted numbers are printed. In line 2000 the address vector A is calculated, which has been set to zero in line 1140. The number stored in X(I) is used to increase the respective address counter variable A(X(I)) by one. The counting vector A is therefore the frequency distribution of the unsorted numbers.

To obtain the sorted vector Y, Y is filled with the address number I as often as stored in the counting variable A(I).

Table 14.1 gives the approximate computing time (in seconds) needed to sort N random numbers using the different sorting programs and an IBM AT with BASICA.

Table 14.1. Computing time [s] needed to sort N numbers with different programs

N	SORTNUM	QSORT	QQSORT2	ADDRSORT
10	0.3	0.3	0.4	2.1
20	0.8	0.7	0.8	2.2
40	2.2	1.8	1.8	2.4
80	7.8	4.5	3.8	2.8
125	18	7.8	6.4	3.1
250	69	18.3	13.9	4.1
500	267	47	31	4.1
1000	1051	103	67	9.9
2000	4184	257	143	16.9
4000	16747	595	310	29.8

The bubble-sort needs more than one hour computing time for 2000 values, while the quick-sort method for the same data needs only a little more than two minutes. The address-sort is the fastest method in this comparison. But if the set of allowed numbers is increased, then the computing time and the memory needed may increase considerably.

14.3   Splines with Regression

In this section on data acquisition and data handling two methods shall be presented which are useful for data smoothing, data interpolation, differentiation of data and the graphical display of the smoothed data. These methods are very important if noisy data must be handled. The basic idea is to interpolate the noisy experimental data with a spline function on which a user-defined degree of smoothness is imposed. (For the pure spline function see Chapter 10.1).

The first method, which is explained in more detail in the book of Lawson and Hanson, is a spline through artificial points.[1] The pure spline function passes exactly through all experimental points and there remains no degree of freedom to smooth it. We now drop this restriction in a way to get a cubic spline function in n subintervals or n artificial points (n has to be less than the number of experimental points) and in addition we let

---

[1] C.L. Lawson, R.J. Hanson, *Solving Least Squares Problems*, Prentice Hall, Englewood Cliffs, New Jersey (1974), p.222ff

the spline function pass through the experimental points under the condition that the sum of the deviation squares becomes a minimum. The subintervals in this method have to be equidistant; the corresponding program REGSPL1 is listed below:

```
0 REM "REGSPL1 "          EBERT/EDERER          881207
1 REM ************************************************
2 REM *** A regression - spline function is       ***
3 REM *** calculated. Equidistant grid points     ***
4 REM *** are needed.                             ***
5 REM *** The algorithm is taken from:            ***
6 REM *** Lawson/Hanson: Solving Least Squares     ***
7 REM *** Problems. Prentice Hall 1974.           ***
9 REM ************************************************
100    DIM X(250),Y(250),Z(250)
110    DIM A(252,5),D(252,7)
120    DIM B(250),C(250)
200    DEF FNP(T)=.25*T*T*T
210    DEF FNQ(T)=1-.75*(1+T)*(1-T)*(1-T)
300    INPUT "How many grid points ";N
310    IF N<2 THEN N=2
320    GOSUB 10000 : REM Reading data in X() and Y()
330    GOSUB 11000 : REM Sorting data ascending in X()
360    IF N>=M-2 THEN N=M-2
400    B(1)=X(1) : B(N)=X(M)
410    H=(B(N)-B(1))/(N-1)
420    FOR I=2 TO N-1
430       B(I)=B(1)+H*(I-1)
440    NEXT I
460    GOSUB 12000 : REM Filling of matrix A
480    GOSUB 13000 : REM Filling of band matrices
481                  REM D=At*A und Z=At*Y
500    GOSUB 14000 : REM Solving equation D*C=Z
550    GOSUB 20000 : REM Calculating spline
600    PRINT : PRINT "For plotting press any key"
620    A$="" : A$=INKEY$ : IF A$="" THEN 620
650    GOSUB 16000 : REM Plotting
9900   A$="" : A$=INKEY$ : IF A$="" THEN 9900
9910   SCREEN 0 : KEY ON
9999   STOP
10000 REM ******************************************
10001 REM *** Reading data in X() and Y()         ***
10002 REM ******************************************
10010 READ M : REM Number of data points
```

```
10020 FOR I=1 TO M : READ X(I),Y(I) : NEXT I
10030 RETURN
11000 REM *********************************************
11001 REM ***   Data are sorted ascending in x.   ***
11002 REM ***   The minima and maxima in x and y   ***
11003 REM ***   are determined.                    ***
11005 REM *********************************************
11020 X0=X(1) : Y0=Y(1) : XM=X0 : YM=Y0
11040 FOR I=2 TO M
11060   IF X0>X(I) THEN X0=X(I)
11080   IF XM<X(I) THEN XM=X(I)
11100   IF Y0>Y(I) THEN Y0=Y(I)
11120   IF YM<Y(I) THEN YM=Y(I)
11140 NEXT I
11200 D=INT(LOG(M)/LOG(2)):D=2^D-1
11220 FOR I=1 TO M-D
11240   FOR J=I TO 1 STEP -D
11260     IF X(J)<=X(J+D) THEN 11300
11270     HY=Y(J) : Y(J)=Y(J+D) : Y(J+D)=HY
11280     HX=X(J) : X(J)=X(J+D) : X(J+D)=HX
11290   NEXT J
11300 NEXT I
11320 D=INT(D/2) : IF D>0 THEN GOTO 11220
11900 RETURN
12000 REM *********************************************
12001 REM ***   Filling of the matrix A           ***
12002 REM *********************************************
12010 IB=1 : REM Index for grid points
12020 IX=1 : REM Index for data points
12050 IF X(IX)>B(IB+1) THEN GOTO 12150
12060 T=(X(IX)-B(IB))/H
12070 A(IX,1)=FNP(1-T)
12080 A(IX,2)=FNQ(1-T)
12090 A(IX,3)=FNQ(T)
12100 A(IX,4)=FNP(T)
12110 A(IX,5)=IB
12120 IF IX=M THEN GOTO 12200
12130 IX=IX+1 : GOTO 12050
12150 IB=IB+1 : GOTO 12050
12200 RETURN
13000 REM *********************************************
13001 REM *** Filling of the band matrices        ***
13002 REM *** D=At*A and Z=At*Y                   ***
13003 REM *********************************************
13060 FOR I=1 TO N+2
13080   FOR J=1 TO 7 : D(I,J)=0
13100   NEXT J : Z(I)=0 : NEXT I
```

```
13140 FOR K=1 TO M
13160   FOR I=1 TO 4 : FOR J=1 TO I
13200       I0=I+A(K,5)-1 : J0=J+A(K,5)-1
13220       D(I0,J0-I0+4)=D(I0,J0-I0+4)+A(K,I)*A(K,J)
13240       NEXT J : Z(I0)=Z(I0)+A(K,I)*Y(K)
13260   NEXT I : NEXT K
13300 FOR I=1 TO N+2 : FOR J=1 TO 3
13320     IF D(I,J)=0 THEN GOTO 13400
13340     D(J+I-4,8-J)=D(I,J)
13400 NEXT J : NEXT I
13900 RETURN
14000 REM ******************************************
14001 REM *** Solving equation system D*C=Z     ***
14002 REM ******************************************
14040 FOR I=1 TO N+2 : D4=D(I,4) : FOR J=1 TO 7
14060     D(I,J)=D(I,J)/D4 : NEXT J : Z(I)=Z(I)/D4
14080   J=I+3 : IF J>(N+2) THEN J=N+2
14100   IF I=N+2 THEN GOTO 14400
14140   FOR K=I+1 TO J
14150     L3=D(K,4 -(K-I))
14160     FOR L=1 TO 4
14180       L1=L+3 : L2=L+3-(K-I)
14200       D(K,L2)=D(K,L2)-L3*D(I,L1)
14220     NEXT L : Z(K)=Z(K)-L3*Z(I)
14240   NEXT K
14400 NEXT I
14500 FOR I=1 TO N+5 : C(I)= 0 : NEXT I
14520 FOR I=N+2 TO 1 STEP -1
14540 C(I)=Z(I)-D(I,5)*C(I+1)-D(I,6)*C(I+2)-D(I,7)*C(I+3)
14560 NEXT I
14900 RETURN
16000 REM ******************************************
16001 REM *** Plotting the data points          ***
16002 REM ******************************************
16020 KEY OFF : SCREEN 2
16030 VIEW (80,0)-(559,199)
16040 DI=(XM-X0)/20 : X0=X0-DI : XM=XM+DI
16060 DI=(YM-Y0)/20 : Y0=Y0-DI : YM=YM+DI
16080 WINDOW (X0,Y0) - (XM,YM)
16120 FOR I=1 TO M
16140   CIRCLE (X(I),Y(I)),(XM-X0)/100
16180 NEXT I
16200 LINE    (X0,Y0) - (XM,YM),,B
16270 FOR I=1 TO N
16280   LINE (B(I),Y0) - (B(I),YM),,,&HAAAA
16300 NEXT I
```

```
16400 X=X(1) : GOSUB 22000 : LINE (X,Y) - (X,Y)
16420 FOR X=X(1) TO X(M) STEP (X(M)-X(1))/100
16440    GOSUB 22000 : LINE -(X,Y)
16460 NEXT X
16900 RETURN
20000 REM *********************************************
20001 REM *** Calculating the spline function     ***
20002 REM *********************************************
20020 FOR I=1 TO M
20030   X=X(I)
20040   GOSUB 21000 : REM determining spline interval j
20060   T=(X-B(J))/H
20080   Y=C(J)*FNP(1-T)+C(J+1)*FNQ(1-T)+C(J+2)*FNQ(T)
20090   Y=Y+C(J+3)*FNP(T)
20100   PRINT I,X(I),Y(I),Y
20200 NEXT I
20990 RETURN
21000 REM *********************************************
21001 REM *** Determining spline interval j       ***
21002 REM *********************************************
21010 J=0
21020 IF X<=B(1) THEN : J=1 : RETURN
21040 IF X>=B(N) THEN : J=N-1 : RETURN
21060 J=J+1
21080 IF X>B(J) THEN GOTO 21060
21100 J=J-1
21200 RETURN
22000 REM *********************************************
22001 REM ***  Evaluating the spline function     ***
22002 REM *********************************************
22040 GOSUB 21000 : REM determining spline interval j
22060 T=(X-B(J))/H
22070 Y=C(J)*FNP(1-T)+C(J+1)*FNQ(1-T)+C(J+2)*FNQ(T)
22080 Y=Y+C(J+3)*FNP(T)
22900 RETURN
60000 DATA 20
60010 DATA   0,   0.5
60020 DATA   1,   1
60030 DATA   2,   3.5
60040 DATA   3,   4.2
60050 DATA   4,   5.5
60060 DATA   5,   8.2
60070 DATA   6,   8.5
60080 DATA   7,  10.2
60085 DATA   8,  10.9
60090 DATA   9,  11.5
60100 DATA  10,  11.6
```

```
60110 DATA   11,    11.5
60120 DATA   12,    10.6
60130 DATA   13,     9.5
60140 DATA   14,     7.2
60150 DATA   15,     5.5
60160 DATA   16,     3.2
60170 DATA   17,     2.1
60180 DATA   18,     0.9
60190 DATA   19,     0.0
63999 END

RUN
How many grid points ? 7
```

1	0	.5	.3808096
2	1	1	1.47392
3	2	3.5	2.847342
4	3	4.2	4.387517
5	4	5.5	5.982124
6	5	8.2	7.527075
7	6	8.5	8.921636
8	7	10.2	10.06914
9	8	10.9	10.92006
10	9	11.5	11.45527
11	10	11.6	11.65401
12	11	11.5	11.44627
13	12	10.6	10.71273
14	13	9.5	9.335271
15	14	7.2	7.393847
16	15	5.5	5.275477
17	16	3.2	3.393112
18	17	2.1	1.996261
19	18	.9	.9337989
20	19	0	-5.666673E-03

```
For plotting press any key
Break in 9999
```

The example in Fig. 14.3 shows 20 noisy experimental points which are connected by spline functions of different smoothness. This is obtained by using 12, 7, 5 or 3 grid points, that means 11, 6, 4 or 2 subintervals for the spline functions. The fewer grid points used, the smoother is the resulting function, but the more inaccurate are the interpolated values.

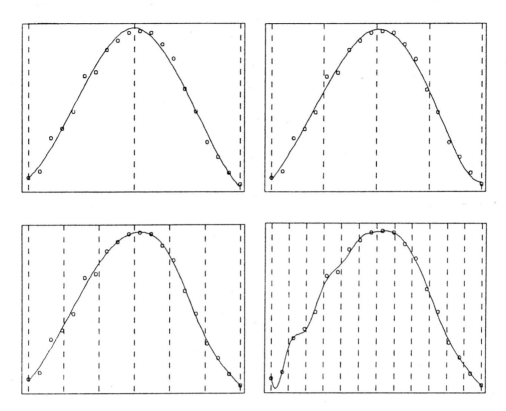

Fig. 14.3. Regression with spline functions of different smoothness on scattered data points

This method is based on the essential mathematical idea that it is possible to combine every spline function with n grid points from $n + 2$ independent basis spline functions in a linear way. All the basis spline functions can be expressed by the cubic polynomials $p(t)$ and $q(t)$ which are programmed in lines 200 and 210.

$$p(t) = 0.25\, t^2$$

$$q(t) = 1 - 0.75(1 + t)(1 - t)^2$$

The linear regression determines the coefficients for the linear combination of these basis spline functions; in this way the problem is reduced to an ordinary linear least squares problem.

In the first program section after the definition of P(T) and Q(T), the data are read in and sorted for ascending x-values with a quick sorting algorithm. The number of grid points is interrogated with an `INPUT` from the user. The values of the grid points are calculated and stored in the vector `B`. The matrix `A`, which is filled by calling from line 460 the subroutine 12000, is relatively big. But only the elements in the diagonal and in the four subdiagonals contain numbers unequal to zero. Therefore, to save computer memory, the matrix is treated as a band matrix `A(252,5)` where only the appropriate diagonal elements are stored. But the calculations have to be done with the mathematical rules for the whole matrix. In the following subroutine beginning with line 13000, the matrix `D` and the vector `Z` are calculated using the formulas:

$$D = A^T \cdot A$$

$$Z = A^T \cdot Y$$

A^T is the transposed of matrix A, D is a matrix with seven diagonals filled. This calculation corresponds to the setting up of a linear equation system for the general linear regression in Chapter 8.3.

The linear equation system is solved in the subroutine beginning with line 14000, that means the coefficients of the linear combination are calculated. To do this, the Gauss-Jordan method (Chapter 8.2) is used. But it must be remembered that not the whole matrix is stored but only the relevant band of elements. After the coefficients `C(I)` are calculated, the spline problem is solved. Within the subroutine 20000 the spline values for the x-values of the experimental points are calculated and compared with the experimental ones. The plot subroutine 16000 plots the experimental values, the grid and the least squares regression function on the graphical screen.

Problem 203

Use the spline smoothing method to integrate experimental data of your choice.

Problem 204

Expand the program `REGSPL1` to be able to use weighted data points.

Problem 205

Expand the program REGSPL1 with a subroutine which calculates the derivative function of the smoothed spline function. The basis functions P(T) and Q(T) are differentiated with respect to X and these derivatives are defined as new functions, for example:

```
201 DEFFN P1(T) = 0.75*T*T/H
211 DEFFN Q1(T) = (1.5*T+0.75-2.25*T*T)/H
```

In line 20080 the *y*-values of the spline function are calculated. In the same way the values of the derivatives can be evaluated:

```
20081 Y1=-C(J)*FNP1(1-T)-C(J+1)*FNQ1(1-T)+C(J+2)*FNQ1(T)
20082 Y1=Y1+C(J+3)*FNP1(T)
```

The numerical and graphical output of the derivative function should be programmed too. In the same way the second derivative of the spline function can be calculated. It should be remembered that by definition the second derivative is still continuous but it is not any more differentiable at the grid points. The second derivative is a polygon from one grid point to the next. To obtain higher smooth derivatives it is better to determine the first derivative with the method of this problem and then use the spline with regression to smooth the spline points again. This procedure can be continued to higher derivatives, but it is dangerous to trust these derivatives too much. It is well known that it is easy to integrate experimentally measured data points, but it is rather difficult to obtain the correct derivatives. You can examine this by using data of a known mathematical function, scattering the values a little bit with a random number generator and then determining the higher derivatives with this spline smoothing method.

The second method (program REGSPL2) can also be used to calculate a smoothed spline function. But the smoothness is now measured via the curvature, that is, the second derivatives of the resulting spline function. Every *x*-value is now used as a grid point for the spline. If no other condition exists then the least squares spline runs exactly through the data points (x_i, y_i). However, a spline function should be found where a linear combination of the sum of squared errors plus the integral over the second derivatives is a minimum. If this linear combination is weighted with 100% in the error squares (parameter PP in line 200), then the exact spline is obtained. On the contrary, if the integral of the squares of the second derivatives is weighted with 100%, then a straight line, that is a linear regression, is obtained. This may be considered as the maximum in smoothness. The program is listed in the following:

```
0 REM "REGSPL2 "        EBERT/EDERER        880327
1 REM ************************************************
2 REM *** Spline function with regression.        ***
3 REM *** The least square is determined for the  ***
4 REM *** sum of the deviations and the squares   ***
5 REM *** of the second derivative (curvature)    ***
6 REM *** of the spline function.                 ***
7 REM *** Smoothing factor PP for the spline      ***
8 REM *** function is given in line # 200.        ***
9 REM ************************************************
100    DIM X(100),Y(100),Z(300)
110    DIM D(300,10),A(100),B(100),C(100)
120    DIM H(100),IH(100),R(300)
200    PP=.5 : QQ=1-PP
220    PP=2*PP : QQ=QQ/3
320    GOSUB 10000 : REM Reading data in X() and Y()
330    GOSUB 11000 : REM Sorting data ascending in X()
350    GOSUB 11500 : REM Calculating h(i)
480    GOSUB 13000 : REM Filling of bandmatrix  D
500    GOSUB 14000 : REM Solving equation D*Z=R
520    GOSUB 12000 : REM Calculation of m, la, a, b, c
550    GOSUB 20000 : REM Calculating spline
600    PRINT : PRINT "For plotting press any key"
620    A$="" : A$=INKEY$ : IF A$="" THEN 620
650    GOSUB 16000 : REM Plotting
9900   A$="" : A$=INKEY$ : IF A$="" THEN 9900
9910   SCREEN 0 : KEY ON
9999   STOP
10000 REM ************************************************
10001 REM *** Reading data in X() and Y()           ***
10002 REM ************************************************
10010 READ N : N=N-1 : REM Number of data points
10020 FOR I=0 TO N : READ X(I),Y(I) : NEXT I
10030 RETURN
11000 REM ************************************************
11001 REM ***  Data are sorted ascending in x.     ***
11002 REM ***  The minima and maxima in x and y    ***
11003 REM ***  are determined.                     ***
11005 REM ************************************************
11020 XO=X(0) : YO=Y(0) : XM=XO : YM=YO
11040 FOR I=1 TO N
11060    IF XO>X(I) THEN XO=X(I)
11080    IF XM<X(I) THEN XM=X(I)
11100    IF YO>Y(I) THEN YO=Y(I)
11120    IF YM<Y(I) THEN YM=Y(I)
11140 NEXT I
```

```
11200 D=INT(LOG(N+1)/LOG(2)):D=2^D-1
11220 FOR I=0 TO N-D
11240   FOR J=I TO 0 STEP -D
11260     IF X(J)<=X(J+D) THEN 11300
11270     HY=Y(J) : Y(J)=Y(J+D) : Y(J+D)=HY
11280     HX=X(J) : X(J)=X(J+D) : X(J+D)=HX
11290   NEXT J
11300 NEXT I
11320 D=INT(D/2) : IF D>0 THEN GOTO 11220
11400 RETURN
11500 REM ********************************************
11501 REM ***  h(i) is calculated             ***
11502 REM ********************************************
11510 FOR I=1 TO N
11520   H(I)=X(I)-X(I-1) : IH(I)=1/H(I)
11530 NEXT I
11540 RETURN
12000 REM ********************************************
12001 REM ***  Calculation of m, la, a, b, c    ***
12002 REM ********************************************
12010 A(0)=0 : A(N)=0
12020 FOR I=1 TO N-1 : A(I)=Z(3*I-1) : NEXT I
12040 REM m in ih
12060 IH(0)=Z(0) : IH(N)=Z(M)
12080 FOR I=1 TO N-1 : IH(I)=Z(3*I-2) : NEXT I
12100 REM la in d(*,0)
12120 FOR I=1 TO N-1 : D(I,0)=Z(3*I) : NEXT I
12200 S1=0 : FOR I=0 TO N : S1=S1+(Y(I)-IH(I))^2 : NEXT I
12220 PRINT : PRINT "sum of squared deviations =";S1
12240 S2=0 : FOR I=0 TO N-1
12250 S2=S2+H(I+1)*(A(I)*A(I)+A(I)*A(I+1)+A(I+1)*A(I+1))
12260 NEXT I : S2=S2/3
12280 PRINT "Integral on the square of the "
12290 PRINT "2nd derivatives = ";S2 : PRINT
12300 FOR I=0 TO N-1
12320 C(I)=IH(I)-A(I)*H(I+1)*H(I+1)/6
12340 B(I)=(IH(I+1)-IH(I))/H(I+1)-H(I+1)/6*(A(I+1)-A(I))
12360 NEXT I
12900 RETURN
13000 REM ********************************************
13001 REM *** Filling band matrix D             ***
13002 REM ********************************************
13020 M=3*N-2
13030 FOR I=0 TO M : FOR J=0 TO 10
13040     D(I,J)=0 : NEXT J : R(I)=0
13050 NEXT I
```

```
13080 D(0,5)=PP : D(0,8)=-IH(1) : R(0)=PP*Y(0)
13100 D(1,5)=PP : D(1,7)=IH(1)+IH(2)
13110 D(1,10)=-IH(2) : R(1)=PP*Y(1)
13120 D(M-3,4)=-IH(N-1) : D(M-3,5)=PP
13130 D(M-3,7)=IH(N)+IH(N-1) : R(M-3)=PP*Y(N-1)
13140 D(M,4)=-IH(N) : D(M,5)=PP : R(M)=PP*Y(N)
13160 FOR I=2 TO N-2 : J=3*I-2
13180   D(J,5)=PP : R(J)=PP*Y(I)
13200   D(J,4)=-IH(I) : D(J,7)=IH(I)+IH(I+1)
13210   D(J,10)=-IH(I+1)
13220 NEXT I
13260 D(2,5)=2*QQ*(H(1)+H(2)) : D(2,6)=(H(1)+H(2))/3
13280 D(2,8)=QQ*H(2) : D(2,9)=H(2)/6
13300 D(M-2,2)=QQ*H(N-1) : D(M-2,3)=H(N-1)/6
13320 D(M-2,5)=2*QQ*(H(N-1)+H(N))
13330 D(M-2,6)=(H(N-1)+H(N))/3
13340 FOR I=2 TO N-2 : J=3*I-1
13360   D(J,2)=QQ*H(I) : D(J,3)=H(I)/6
13370   D(J,5)=2*QQ*(H(I)+H(I+1))
13380   D(J,6)=(H(I)+H(I+1))/3
13390   D(J,8)=QQ*H(I+1) : D(J,9)=H(I+1)/6
13400 NEXT I
13540 D(3,2)=-IH(1) : D(3,3)=IH(1)+IH(2)
13550 D(3,4)=(H(1)+H(2))/3
13560 D(3,6)=-IH(2) : D(3,7)=H(2)/6
13580 D(M-1,0)=-IH(N-1) : D(M-1,1)=H(N-1)/6
13590 D(M-1,6)=-IH(N)
13600 D(M-1,3)=IH(N)+IH(N-1): D(M-1,4)=(H(N-1)+H(N))/3
13640 FOR I=2 TO N-2 : J=3*I
13660   D(J,0)=-IH(I) : D(J,1)=H(I)/6
13670   D(J,3)=IH(I)+IH(I+1)
13680   D(J,4)=(H(I)+H(I+1))/3
13690   D(J,6)=-IH(I+1):D(J,7)=H(I+1)/6
13700 NEXT I
13900 RETURN
14000 REM ******************************************
14001 REM *** Solving the equation system D*C=Z    ***
14002 REM ******************************************
14040 FOR I=0 TO M : D4=D(I,5) : FOR J=5 TO 10
14060     D(I,J)=D(I,J)/D4 : NEXT J : R(I)=R(I)/D4
14080   J=I+5 : IF J>M  THEN J=M
14100   IF I=M THEN GOTO 14400
14140   FOR K=I+1 TO J
14150     L3=-D(K,5-K+I)
14155       IF L3=0 THEN GOTO 14240
14160       FOR L=0 TO 5
```

```
14180      L1=L+5 : L2=L+5-(K-I)
14200      D(K,L2)=D(K,L2)+L3*D(I,L1)
14220    NEXT L : R(K)=R(K)+L3*R(I)
14240   NEXT K
14400 NEXT I
14480 FOR I=0 TO M : Z(I)= 0 : NEXT I
14500 Z(M)=R(M) : Z(M-1)=R(M-1)-D(M-1,6)*Z(M)
14510 Z(M-2)=R(M-2)-D(M-2,6)*Z(M-1)-D(M-2,7)*Z(M)
14520 Z(M-3)=R(M-3)-D(M-3,6)*Z(M-2)-D(M-3,7)*Z(M-1)
14525 Z(M-3)=Z(M-3)-D(M-3,8)*Z(M)
14530 Z(M-4)=R(M-4)-D(M-4,6)*Z(M-3)-D(M-4,7)*Z(M-2)
14535 Z(M-4)=Z(M-4)-D(M-4,8)*Z(M-1)
14540 Z(M-4)=Z(M-4)-D(M-4,9)*Z(M)
14560 FOR I=M-5 TO 0 STEP -1
14580   Z(I)=R(I)-D(I,6)*Z(I+1)-D(I,7)*Z(I+2)
14590   Z(I)=Z(I)-D(I,8)*Z(I+3)
14600   Z(I)=Z(I)-D(I,9)*Z(I+4)-D(I,10)*Z(I+5)
14620 NEXT I
14900 RETURN
16000 REM *******************************************
16001 REM *** Plotting subroutine              ***
16002 REM *******************************************
16020 KEY OFF : SCREEN 2
16030 VIEW (80,0)-(559,199)
16040 DI=(XM-X0)/20 : X0=X0-DI : XM=XM+DI
16060 DI=(YM-Y0)/20 : Y0=Y0-DI : YM=YM+DI
16080 WINDOW (X0,Y0) - (XM,YM)
16120 FOR I=0 TO N
16140   CIRCLE (X(I),Y(I)),(XM-X0)/100
16180 NEXT I
16200 LINE   (X0,Y0) - (XM,YM),,B
16400 X=X(0) : GOSUB 22000 : LINE (X,Y) - (X,Y)
16420 FOR X=X(0) TO X(N) STEP (X(N)-X(0))/100
16440   GOSUB 22000 : LINE -(X,Y)
16460 NEXT X
16900 RETURN
20000 REM *******************************************
20001 REM *** Calculating the spline function   ***
20002 REM *******************************************
20020 FOR I=0 TO N
20030   X=X(I)
20040   GOSUB 22000 : REM calculating spline value
20100   PRINT I,X(I),Y(I),Y
20200 NEXT I
20990 RETURN
```

```
21000 REM *********************************************
21001 REM *** Determining spline interval j       ***
21002 REM *********************************************
21010 J1=0 : J2=N
21020 IF X<=X(J1) THEN : J=0 : RETURN
21040 IF X>=X(J2) THEN : J=N-1 : RETURN
21060 J=INT((J1+J2)/2)
21080 IF X>X(J) THEN : J1=J : GOTO 21120
21100 J2=J
21120 IF J2-J1>1 THEN GOTO 21060
21140 J=J1
21200 RETURN
22000 REM *********************************************
22001 REM ***  Evaluating the spline function     ***
22002 REM *********************************************
22040 GOSUB 21000 : REM determining spline interval j
22060 T1=X(J+1)-X : T2=X-X(J)
22080 Y=A(J)*T1*T1*T1/6/H(J+1)+A(J+1)*T2*T2*T2/6/H(J+1)
22100 Y=Y+B(J)*(X-X(J))+C(J)
22900 RETURN
60000 DATA 27
60010 DATA    0,    0.5
60020 DATA    1,    1
60030 DATA    2,    3.5
60040 DATA    3,    4.2
60050 DATA    4,    5.5
60060 DATA    5,    8.2
60070 DATA    6,    8.5
60080 DATA    7,   10.2
60085 DATA    8,   10.9
60090 DATA    9,   11.5
60100 DATA   10,   11.6
60110 DATA   11,   11.5
60120 DATA   12,   10.6
60130 DATA   13,    9.5
60140 DATA   14,    7.2
60150 DATA   15,    5.5
60160 DATA   16,    3.2
60170 DATA   17,    2.1
60180 DATA   18,    0.9
60190 DATA   19,    0.0
60200 DATA  0.3 ,   0.6
60210 DATA  4.8 ,   8.6
60220 DATA  8.5 ,  11.6
60230 DATA 12.1 ,  10.5
60240 DATA 12.8 ,  10.0
```

```
60260 DATA 16.7 ,   2.5
60270 DATA 18.2 ,   0.5
63999 END

RUN
sum of squared deviations = 3.337253
Integral on the square of the
2nd derivatives =   1.99107
```

0	0	.5	.2325987
1	.3	.6	.61052
2	1	1	1.532291
3	2	3.5	3.003257
4	3	4.2	4.541642
5	4	5.5	6.229048
6	4.8	8.600001	7.584201
7	5	8.2	7.881948
8	6	8.5	9.113991
9	7	10.2	10.16765
10	8	10.9	11.02115
11	8.5	11.6	11.33793
12	9	11.5	11.55721
13	10	11.6	11.69925
14	11	11.5	11.42135
15	12	10.6	10.64049
16	12.1	10.5	10.5327
17	12.8	10	9.61603
18	13	9.5	9.299307
19	14	7.2	7.447053
20	15	5.5	5.453778
21	16	3.2	3.598735
22	16.7	2.5	2.503192
23	17	2.1	2.084348
24	18	.9	.8564234
25	18.2	.5	.6353283
26	19	0	-.2014265

```
For plotting press any key
Break in 9999
```

The user has to enter data points as DATA-statements beginning in line 60000
and set the parameter PP in line 200 for the weight of the squared deviations.
In the example printed, both parts are weighted by 50%.

The development of the numerical algorithm is outlined here briefly. It is assumed that the *y*-values, through which the spline passes, are already known and the linear equation system for the calculation of the spline parameters can be set up. This equation system is considered as the constraints for the minimizing problem of the above mentioned linear combination of squared deviations and the integral of the squared second derivation function. The search for a minimum with constraints is done using Lagrange's multiplication factors. From this, we again obtain a linear equation system with a band structure. The solution of this equation system supplies the parameters of the spline function, their *y*-values at the grid points, and the Lagrange's multiplicators.

In Fig. 14.4 smoothed splines are shown which are made with the program REGSPL2 and different smoothing factors PP.

If this method is used with weighted data points, a weighting vector W has to be added and the respective weights have to be added in the DATA-statements. The weighting factors are read in the input subroutine 10000 and they must also be used in the sorting algorithm. The changes in the calculation algorithm are limited to the beginning of the subroutine starting with line 13000. The new program part is listed below:

```
13000 REM ******************************************
13001 REM *** Filling band matrix D              ***
13002 REM ******************************************
13020 M=3*N-2
13030 FOR I=0 TO M : FOR J=0 TO 10
13040    D(I,J)=0 : NEXT J : R(I)=0
13050 NEXT I
13080 D(0,5)=PP*W(0) : D(0,8)=-IH(1)
13090 R(0)=PP*Y(0)*W(0)
13100 D(1,5)=PP*W(1) : D(1,7)=IH(1)+IH(2)
13110 D(1,10)=-IH(2) : R(1)=PP*Y(1)*W(1)
13120 D(M-3,4)=-IH(N-1) : D(M-3,5)=PP*W(N-1)
13130 D(M-3,7)=IH(N)+IH(N-1) : R(M-3)=PP*Y(N-1)*W(N-1)
13140 D(M,4)=-IH(N) : D(M,5)=PP*W(N) : R(M)=PP*Y(N)*W(N)
13160 FOR I=2 TO N-2 : J=3*I-2
13180    D(J,5)=PP*W(I) : R(J)=PP*Y(I)*W(I)
13200    D(J,4)=-IH(I) : D(J,7)=IH(I)+IH(I+1)
13210    D(J,10)=-IH(I+1)
13220 NEXT I
```

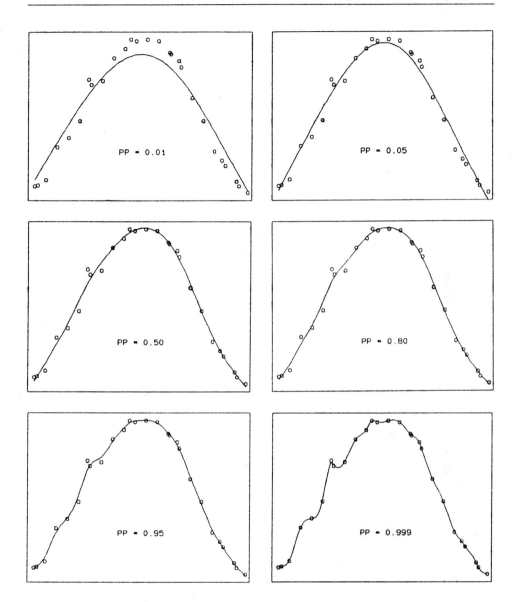

Fig. 14.4. Smoothed splines through experimental points with smoothing factors (PP) 0.999, 0.95, 0.8, 0.5, 0.05 and 0.01

14.4   Stiff Differential Equations

In this excursion into the field of stiff differential equations, a program is shown which is able to solve this type of equation. It can solve problems which are derived from chemical reaction kinetics of complex mechanisms. The system of differential equations for chemical kinetics is usually nonlinear and may have reaction rate constants (parameters for the ODE-system) which differ in many orders of magnitude. The program shown for the numerical solution of stiff ODE-systems contains only a very simple step size control algorithm and it is not optimized. It will help to explain the principle by which stiff ODE's are solved, and the reader is encouraged to study numerical methods to solve such problems by himself.[2,3,4,5]

In trying to solve systems of stiff differential equations with the usual numerical methods, such as the well-known Runge-Kutta methods, completely false results may be obtained, especially for problems of chemical reaction kinetics. Improving this behavior by reducing the step size may result in extremely long computing times, and the chemical conversions can remain very low.

This typical behavior for stiff differential equations shall be demonstrated on a chemical example which also has an analytical solution.[6] How does a numerical algorithm, like that of Euler, behave with a stiff differential equation? This example will show that the problem can be treated with the implicit Euler method.

---

[2] G. Bader, U. Novak, P. Deuflhard, *An advanced Simulation package for Large Chemical Reaction Systems* in: Aiken(ed.), *Stiff Computation*, Oxford University Press (1983)

[3] G. Bader, P. Deuflhard, *A Semi-Implicit Midpoint Rule for Stiff Systems of Ordinary Differential Equations, Numer. Math.*, (1983)

[4] E.M. Chance, A.R. Curtis, I.P. Jones, C.R. Kirby, *FACSIMILE : A Computer Program for Flow and Chemistry Simulations, and General Initial Value Problems*, Harwell, AERE Tech. Rep. R. 8775 (Dec. 1977)

[5] R.J. Kee, J.A. Miller, T.H. Jefferson, *CHEMKIN : A General Purpose, Problem Independent, Transportable, Fortan Chemical Kinetics Code Package*, Sandia National Laboratories, Livermore, Tech. Rep. SAND80-8003 (1980)

[6] To our knowledge the phenomena was first shown in the literature by: C.F. Curtis, J.O. Hirschfelder, *Proc. Nat. Acad. Sci. USA*, **38**, 235 (1952)

Given are the following two differential equations which describe a simple consecutive chemical reaction:

$$A \xrightarrow{\ \ k_1 = 1\ \ } R \xrightarrow{\ \ k_2 = 10^6\ \ } P$$

$$\frac{d[A]}{dt} = -[A] \qquad\qquad [A]_{t=0} = 1$$

$$\frac{d[R]}{dt} = +[A] - 10^6[R] \qquad [R]_{t=0} = 0$$

R is an intermediate compound which reacts much faster in the second reaction than it is formed in the first reaction. The analytical solutions for both substances A and R can be written down using the initial conditions above for determining the integration parameters.

$$[A] = e^{-t} \qquad\qquad [R] = \frac{e^{-t} - e^{-10^6 t}}{10^6 - 1}$$

If [A] is assigned to x and [R] to y, and if the two right hand sides of the ODE's are defined as f_1 and f_2, then the iteration step according to Euler is given by

$$x_{i+1} = x_i + f_1(x_i, y_i) \cdot \Delta t$$

$$y_{i+1} = y_i + f_2(x_i, y_i) \cdot \Delta t$$

If this explicit method is changed into an implicit method by changing the indices on the right hand sides from i to $i+1$, then the following iteration steps are obtained

$$x_{i+1} = x_i + f_1(x_{i+1}, y_{i+1}) \cdot \Delta t$$

$$y_{i+1} = y_i + f_2(x_{i+1}, y_{i+1}) \cdot \Delta t$$

For our chemical reaction this implicit method is written as

$$[A(t + \Delta t)] = [A(t)] - [A(t + \Delta t)] \cdot \Delta t$$

$$[R(t + \Delta t)] = [R(t)] + ([A(t + \Delta t)] - 10^6[R(t + \Delta t)]) \cdot \Delta t$$

If this linear equation system is solved for $[A(t + \Delta t)]$ and $[R(t + \Delta t)]$, then the following iteration steps for the implicit method are obtained.

$$[A(t + \Delta t)] = \frac{[A(t)]}{1 + \Delta t}$$

$$[R(t + \Delta t)] = \frac{[R(t)] + \frac{[A(t)]}{1 + \Delta t} \cdot \Delta t}{1 + 10^6 \cdot \Delta t}$$

The results of this implicit method are shown in Tables 14.2 and 14.3 in the column 'stiff'; they are compared with those of the analytical solution ('exact') and the explicit 'Euler' method.

Table 14.2. Comparison of the explicit Euler method, the implicit ('stiff') Euler method and the exact solution for the calculation of the stiff reaction system: A $\longrightarrow$ R $\longrightarrow$ P. The time step for the different integration methods was $\Delta t = 1 \cdot 10^{-7}$.

Time	A			R		
	Euler	exact	stiff	Euler	exact	stiff
0	1	1	1	0	0	0
$1 \cdot 10^{-7}$	1	1	1	$+1.00 \cdot 10^{-7}$	$9.52 \cdot 10^{-8}$	$9.09 \cdot 10^{-8}$
$2 \cdot 10^{-7}$	1	1	1	$+1.90 \cdot 10^{-5}$	$1.81 \cdot 10^{-5}$	$1.74 \cdot 10^{-5}$
$3 \cdot 10^{-7}$	1	1	1	$+2.71 \cdot 10^{-7}$	$2.59 \cdot 10^{-7}$	$2.49 \cdot 10^{-7}$
$4 \cdot 10^{-7}$	1	1	1	$+3.44 \cdot 10^{-7}$	$3.30 \cdot 10^{-7}$	$3.17 \cdot 10^{-7}$
$5 \cdot 10^{-7}$	1	1	1	$+4.10 \cdot 10^{-7}$	$3.93 \cdot 10^{-7}$	$3.79 \cdot 10^{-7}$
$6 \cdot 10^{-7}$	1	1	1	$+4.69 \cdot 10^{-7}$	$4.51 \cdot 10^{-7}$	$4.36 \cdot 10^{-7}$
$7 \cdot 10^{-7}$	1	1	1	$+5.22 \cdot 10^{-7}$	$5.03 \cdot 10^{-7}$	$4.87 \cdot 10^{-7}$
$8 \cdot 10^{-7}$	1	1	1	$+5.70 \cdot 10^{-7}$	$5.51 \cdot 10^{-7}$	$5.33 \cdot 10^{-7}$
$9 \cdot 10^{-7}$	1	1	1	$+6.13 \cdot 10^{-7}$	$5.93 \cdot 10^{-7}$	$5.76 \cdot 10^{-7}$
$10 \cdot 10^{-7}$	1	1	1	$+6.51 \cdot 10^{-7}$	$6.32 \cdot 10^{-7}$	$6.14 \cdot 10^{-7}$
✱✱✱		✱✱✱		✱✱✱	✱✱✱	✱✱✱
$90 \cdot 10^{-7}$	1	1	1	$+1.00 \cdot 10^{-6}$	$1.00 \cdot 10^{-6}$	$1.00 \cdot 10^{-6}$
$91 \cdot 10^{-7}$	1	1	1	$+1.00 \cdot 10^{-6}$	$1.00 \cdot 10^{-6}$	$1.00 \cdot 10^{-6}$
$92 \cdot 10^{-7}$	1	1	1	$+1.00 \cdot 10^{-6}$	$1.00 \cdot 10^{-6}$	$1.00 \cdot 10^{-6}$
$93 \cdot 10^{-7}$	1	1	1	$+1.00 \cdot 10^{-6}$	$1.00 \cdot 10^{-6}$	$1.00 \cdot 10^{-6}$
$94 \cdot 10^{-7}$	1	1	1	$+1.00 \cdot 10^{-6}$	$1.00 \cdot 10^{-6}$	$1.00 \cdot 10^{-6}$
$95 \cdot 10^{-7}$	1	1	1	$+1.00 \cdot 10^{-6}$	$1.00 \cdot 10^{-6}$	$1.00 \cdot 10^{-6}$
$96 \cdot 10^{-7}$	1	1	1	$+1.00 \cdot 10^{-6}$	$1.00 \cdot 10^{-6}$	$1.00 \cdot 10^{-6}$
$97 \cdot 10^{-7}$	1	1	1	$+1.00 \cdot 10^{-6}$	$1.00 \cdot 10^{-6}$	$1.00 \cdot 10^{-6}$
$98 \cdot 10^{-7}$	1	1	1	$+1.00 \cdot 10^{-6}$	$1.00 \cdot 10^{-6}$	$1.00 \cdot 10^{-6}$
$99 \cdot 10^{-7}$	1	1	1	$+1.00 \cdot 10^{-6}$	$1.00 \cdot 10^{-6}$	$1.00 \cdot 10^{-6}$
$100 \cdot 10^{-7}$	1	1	1	$+1.00 \cdot 10^{-6}$	$1.00 \cdot 10^{-6}$	$1.00 \cdot 10^{-6}$

Table 14.3. Comparison of the explicit Euler method, the implicit ('stiff') Euler method and the exact solution for the calculation of the stiff reaction system: $A \longrightarrow R \longrightarrow P$. A constant time step of $\Delta t = 1 \cdot 10^{-3}$ was used in this table; this is increased considerably compared with Table 14.2.

Time	A			R		
	Euler	exact	stiff	Euler	exact	stiff
0	1	1	1	0	0	0
$1 \cdot 10^{-3}$	0.9990	0.9990	0.9990	$+9.99 \cdot 10^{-2}$	$9.99 \cdot 10^{-5}$	$9.98 \cdot 10^{-5}$
$2 \cdot 10^{-3}$	0.9980	0.9980	0.9980	$-9.97 \cdot 10^{-1}$	$9.98 \cdot 10^{-5}$	$9.98 \cdot 10^{-5}$
$3 \cdot 10^{-3}$	0.9970	0.9970	0.9970	$+9.96 \cdot 10^{2}$	$9.97 \cdot 10^{-5}$	$9.97 \cdot 10^{-5}$
$4 \cdot 10^{-3}$	0.9960	0.9960	0.9960	$-9.95 \cdot 10^{5}$	$9.96 \cdot 10^{-5}$	$9.96 \cdot 10^{-5}$
$5 \cdot 10^{-3}$	0.9950	0.9950	0.9950	$+9.94 \cdot 10^{8}$	$9.95 \cdot 10^{-5}$	$9.95 \cdot 10^{-5}$
$6 \cdot 10^{-3}$	0.9940	0.9940	0.9940	$-9.93 \cdot 10^{11}$	$9.94 \cdot 10^{-5}$	$9.94 \cdot 10^{-5}$
$7 \cdot 10^{-3}$	0.9930	0.9930	0.9930	$+9.92 \cdot 10^{15}$	$9.93 \cdot 10^{-5}$	$9.93 \cdot 10^{-5}$
$8 \cdot 10^{-3}$	0.9920	0.9920	0.9920	$-9.91 \cdot 10^{17}$	$9.92 \cdot 10^{-5}$	$9.92 \cdot 10^{-5}$
$9 \cdot 10^{-3}$	0.9910	0.9910	0.9910	$+9.90 \cdot 10^{20}$	$9.91 \cdot 10^{-5}$	$9.91 \cdot 10^{-5}$
$1 \cdot 10^{-2}$	0.9900	0.9900	0.9901	$-9.89 \cdot 10^{23}$	$9.90 \cdot 10^{-5}$	$9.90 \cdot 10^{-5}$
* * *	* * *	* * *	* * *	* * *	* * *	* * *
$9 \cdot 10^{-2}$	0.9139	0.9139	0.9140	$-1.69 \cdot 10^{35}$	$9.14 \cdot 10^{-5}$	$9.14 \cdot 10^{-5}$
$9.1 \cdot 10^{-2}$	0.9130	0.9130	0.9131	$+9.86 \cdot 10^{32}$	$9.13 \cdot 10^{-5}$	$9.13 \cdot 10^{-5}$
$9.2 \cdot 10^{-2}$	0.9121	0.9121	0.9121	$-1.69 \cdot 10^{35}$	$9.12 \cdot 10^{-5}$	$9.12 \cdot 10^{-5}$
$9.3 \cdot 10^{-2}$	0.9112	0.9112	0.9112	$+9.86 \cdot 10^{32}$	$9.11 \cdot 10^{-6}$	$9.11 \cdot 10^{-5}$
$9.4 \cdot 10^{-2}$	0.9102	0.9103	0.9103	$-1.69 \cdot 10^{35}$	$9.10 \cdot 10^{-5}$	$9.10 \cdot 10^{-5}$
$9.5 \cdot 10^{-2}$	0.9093	0.9094	0.9094	$+9.86 \cdot 10^{32}$	$9.09 \cdot 10^{-5}$	$9.09 \cdot 10^{-5}$
$9.6 \cdot 10^{-2}$	0.9084	0.9085	0.9085	$-1.69 \cdot 10^{35}$	$9.08 \cdot 10^{-5}$	$9.09 \cdot 10^{-5}$
$9.7 \cdot 10^{-2}$	0.9075	0.9076	0.9076	$+9.86 \cdot 10^{32}$	$9.08 \cdot 10^{-5}$	$9.08 \cdot 10^{-5}$
$9.8 \cdot 10^{-2}$	0.9066	0.9066	0.9067	$-1.69 \cdot 10^{35}$	$9.07 \cdot 10^{-5}$	$9.07 \cdot 10^{-5}$
$9.9 \cdot 10^{-2}$	0.9057	0.9057	0.9058	$+9.86 \cdot 10^{32}$	$9.06 \cdot 10^{-5}$	$9.06 \cdot 10^{-5}$
$1 \cdot 10^{-1}$	0.9048	0.9048	0.9049	$-1.69 \cdot 10^{35}$	$9.05 \cdot 10^{-5}$	$9.05 \cdot 10^{-5}$

Beginning with a step size of $\Delta t = 10^{-7}$, [R] is calculated with both methods correctly. After 100 steps the quasi steady state concentration of [R] is nearly obtained. But because [A] is not really changing using such small increments, the step size is increased from 10^{-7} to 10^{-3}. This leads, with the explicit Euler method, to oscillations for the value of [R] and soon to a number overflow. The implicit method, however, yields correct values for [R].

If a general linear system of differential equations is considered it can be written as a matrix A and a vector $\vec{y}$ in a vector notation as

$$\frac{d\vec{y}}{dt} = A \cdot \vec{y}$$

The explicit Euler method for these ODE's reads

$$\vec{y}(t + \Delta t) = \vec{y}(t) + A \cdot \vec{y}(t) \cdot \Delta t$$

The implicit Euler method for these ODE's is expressed by

$$\vec{y}(t + \Delta t) = \vec{y}(t) + A \cdot \vec{y}(t + \Delta t) \cdot \Delta t$$

This equation can be solved for the vector $\vec{y}(t + \Delta t)$

$$\vec{y}(t + \Delta t) = (I - A \cdot \Delta t)^{-1} \cdot \vec{y}(t)$$

$$I = \text{ unity matrix}$$

Using the implicit method it is necessary to invert a matrix for each iteration step to solve a linear system of differential equations. For linear ODE's the matrix A and the Jacobi matrix are identical.

For the general nonlinear case

$$\frac{d\vec{y}}{dt} = \vec{f}(\vec{y})$$

the implicit equation system cannot be solved directly. The first summand of the Taylor series is used as an approximation and the following iteration formula for the semi-implicit Euler method is obtained

$$\vec{y}(t + \Delta t) = (I - A \cdot \Delta t)^{-1} \cdot \vec{y}$$

$$I = \text{ unity matrix}$$

$$A = \frac{\partial \frac{d\vec{y}}{dt}}{\partial \vec{y}} = \frac{\partial \vec{f}(\vec{y})}{\partial \vec{y}}$$

In this case, A is the Jacobian of the right hand side of the differential equation system.

The program ER-KIN can be extended to calculate both the Jacobi matrix and the system of ODE's from the chemical reaction system. As an example the following simple chemical reaction mechanism shall be treated:

$$
\begin{array}{ccccc}
A & & \xrightarrow{k_1} & M + M \\
A & + & M \xrightarrow{k_2} & B + R \\
R & & \xrightarrow{k_3} & M + D \\
M & + & M \xrightarrow{k_4} & E
\end{array}
$$

From this the following differential equations and the elements of the Jacobian, which are not equal to zero, result

$$\frac{d[A]}{dt} = -k_1[A] - k_2[A][M]$$

$$\frac{d[M]}{dt} = 2k_1[A] - k_2[A][M] + k_3[R] - 2k_4[M]^2$$

$$\frac{d[B]}{dt} = k_2[A][M]$$

$$\frac{d[R]}{dt} = k_2[A][M] - k_3[R]$$

$$\frac{d[D]}{dt} = k_3[R]$$

$$\frac{d[E]}{dt} = k_4[M]^2$$

$$\frac{\partial[\dot{A}]}{\partial[A]} = -k_1 - k_2[M] \qquad\qquad \frac{\partial[\dot{B}]}{\partial[M]} = k_2[A]$$

$$\frac{\partial[\dot{A}]}{\partial[M]} = -k_2[A] \qquad\qquad \frac{\partial[\dot{R}]}{\partial[R]} = -k_3$$

$$\frac{\partial[\dot{M}]}{\partial[M]} = -k_2[A] - 4k_4[M] \qquad\qquad \frac{\partial[\dot{R}]}{\partial[A]} = k_2[M]$$

$$\frac{\partial[\dot{M}]}{\partial[A]} = 2k_1 - k_2[M] \qquad \frac{\partial[\dot{R}]}{\partial[M]} = k_2[A]$$

$$\frac{\partial[\dot{M}]}{\partial[R]} = k_3 \qquad \frac{\partial[\dot{D}]}{\partial[R]} = k_3$$

$$\frac{\partial[\dot{B}]}{\partial[A]} = k_2[M] \qquad \frac{\partial[\dot{E}]}{\partial[M]} = 2k_4[M]$$

The program for the numerical simulation of the reaction kinetics of this mechanism is listed below:

```
0 REM "ODESTIFF"          EBERT/EDERER          880506
1 REM ************************************************
2 REM *** Semi-implicit Euler method for solving   ***
3 REM *** systems of stiff differential equations. ***
4 REM *** This is a modification of problem 128.   ***
9 REM ************************************************
20    REM   A         => k1 =>   M  +  M
30    REM   A  +  M   => k2 =>   B  +  R
40    REM   R         => k3 =>   M  +  D
50    REM   M  +  M   => k4 =>   E
100   DIM A(20,40),Y(20),Y1(20),D(20),YA(20)
200   N=6 : REM N is the dimension of the ODE-system
300   FOR I=1 TO N
400      PRINT "Initial value of y(";I;") = ";
500      INPUT Y1(I) : Y(I)=Y1(I) : YA(I)=Y(I) : NEXT I
600   INPUT "x-values start,end  ";XA,XE
620   X=XA : DV=(XE-XA)/50
700   INPUT "Delta-x-value ";DD
710   K1=.01
720   K2=1000
730   K3=100
740   K4=1E+08
750   REM GOTO 999
800   KEY OFF : SCREEN 2
810   WINDOW (XA,0) - (XE,Y(1))
830   LINE (XA,0) - (XE,Y(1)) ,,B
999   GOTO 5000
1000  GOSUB 20000 : REM Output
1100  GOSUB 10000 : REM Right hand sides of ODE's
1120  GOSUB 11000 : REM Jacobimatrix
1200  GOSUB 50000 : REM Inversion of the Jacobi matrix
1300  X=X+DX
```

```
1400    FOR I=1 TO N
1420      DY=0
1440      FOR J=1 TO N : DY=DY+A(I,N+J)*D(J) : NEXT J
1460      Y(I)=Y(I)+DY*DX
1480    NEXT I
1600    RETURN
5000    REM ******************************************
5010    REM ***        Main Program               ***
5020    REM ******************************************
5030    N1=1
5100    GOSUB 20000
5140    DX=DD/N1 : X=XA
5150    FOR ID=1 TO N1 : GOSUB 1100 : NEXT ID
5160    IF N1>50 THEN : DD=DD/1000 : N1=1 : GOTO 5140
5170    FOR I=1 TO N : YA(I)=Y(I) : Y(I)=Y1(I) : NEXT I
5175    N1=N1*2
5180    DX=DD/N1 : X=XA
5190    FOR ID=1 TO N1 : GOSUB 10000 : GOSUB 1300 : NEXT ID
5191    REM FOR ID=1 TO N1 : GOSUB 1100 : NEXT ID
5200    FOR I=1 TO N : YS=ABS(Y(I))+ABS(YA(I))+.000001
5210      REM PRINT X,I,Y(I),YA(I),N1
5240      IF ABS(Y(I)-YA(I))/YS>.05 THEN 5160
5300    NEXT I
5320    GOSUB 15000
5340    XA=XA+DD
5360    FOR I=1 TO N : Y1(I)=Y(I) : NEXT I
5400    IF N1=2 THEN DD=DD*2
5420    N1=INT(N1/5)+1
5430    IF DD>DV THEN DD=DV
5440    IF DD>(XE-XA) THEN DD=(XE-XA)
6200    IF XA<XE THEN 5100
6400    GOSUB 20000
7000    A$="" : A$=INKEY$ : IF A$="" THEN GOTO 7000
7100    SCREEN 0 : KEY ON
9900    STOP
9910    REM ######################################
9920    REM ### Main program ends here.        ###
9930    REM ######################################
10000   REM ******************************************
10002   REM *** Subroutine: RHS                    ***
10004   REM *** The right hand sides of the ODE's   ***
10006   REM *** are calculated.                    ***
10010   REM ******************************************
10100   D(1)=-K1*Y(1)-K2*Y(1)*Y(2)
10200   D(2)=2*K1*Y(1)-K2*Y(1)*Y(2)+K3*Y(4)-2*K4*Y(2)*Y(2)
10300   D(3)=K2*Y(1)*Y(2)
```

```
10400 D(4)=K2*Y(1)*Y(2)-K3*Y(4)
10500 D(5)=K3*Y(4)
10600 D(6)=K4*Y(2)*Y(2)
10999 RETURN
11000 REM *******************************************
11001 REM *** Subroutine: JACOBI                 ***
11002 REM *** Calculation of the Jacobi matrix.  ***
11003 REM *******************************************
11010 FOR I=1 TO 2*N : FOR J=1 TO N
11015    A(J,I)=0 : NEXT J : NEXT I
11020 FOR I=1 TO N : A(I,I)=1 : A(I,N+I)=1 : NEXT I
11100 A(1,1)= A(1,1)-DX*(-K1-K2*Y(2))
11120 A(1,2)= DX*K2*Y(1)
11200 A(2,2)= A(2,2)-DX*(-K2*Y(1)-4*K4*Y(2))
11210 A(2,1)=-DX*(2*K1-K2*Y(2))
11220 A(2,4)=-DX*K3
11300 A(3,1)=-DX*K2*Y(2)
11301 A(3,2)=-DX*K2*Y(1)
11400 A(4,4)= A(4,4)+DX*K3
11410 A(4,1)=-DX*K2*Y(2)
11420 A(4,2)=-DX*K2*Y(1)
11520 A(5,4)=-DX*K3
11620 A(6,2)=-DX*K4*2*Y(2)
14900 RETURN
15000 REM *******************************************
15001 REM *** Subroutine: PLOT                   ***
15002 REM *******************************************
15010 REM GOTO 15990
15020 FOR I=1 TO N
15040    LINE (XA,Y1(I)) - (X,Y(I))
15080 NEXT I
15990 RETURN
20000 REM *******************************************
20001 REM *** Subroutine: OUTPUT
20002 REM *******************************************
20005 GOTO 20200
20010 PRINT
20100 FOR I=1 TO N
20120    PRINT X,I,Y(I)
20140 NEXT I
20200 RETURN
50000 REM *******************************************
50001 REM *** Subroutine: INVERSION              ***
50002 REM *** Inversion of the Jacobi matrix.    ***
50003 REM *******************************************
```

```
50900 FOR S=1 TO N
51000    FOR T=S TO N
51100       IF A(T,S)<>0 THEN 51300
51200    NEXT T : PRINT "No solution" : GOTO 63999
51300    FOR J=1 TO 2*N : B=A(S,J)
51320       A(S,J)=A(T,J) : A(T,J)=B : NEXT J
51400    C=1/A(S,S)
51500    FOR J=1 TO 2*N : A(S,J)=C*A(S,J) : NEXT J
51600    FOR T=1 TO N
51700       IF T=S THEN 52000
51800       C=-A(T,S)
51900       FOR J=1 TO 2*N: A(T,J)=A(T,J)+C*A(S,J): NEXT J
52000    NEXT T
52100 NEXT S
52200 RETURN
63999 END


RUN
Initial value of y( 1 ) = ? 1
Initial value of y( 2 ) = ? 0
Initial value of y( 3 ) = ? 0
Initial value of y( 4 ) = ? 0
Initial value of y( 5 ) = ? 0
Initial value of y( 6 ) = ? 0
x-values start,end ? 0,100
Delta-x-value ? 1e-7
  0              1              1
  0              2              0
  0              3              0
  0              4              0
  0              5              0
  0              6              0

 .0000001        1              1
 .0000001        2              1.99975E-09
 .0000001        3              2.4997E-13
 .0000001        4              2.49967E-13
 .0000001        5              2.999615E-18
 .0000001        6              4.999E-18

 .0000003        1              1
 .0000003        2              5.998349E-09
 .0000003        3              1.64958E-12
 .0000003        4              1.649538E-12
 .0000003        5              4.198878E-17
 .0000003        6              5.247689E-16
```

.0000007	1	1
.0000007	2	1.399194E-08
.0000007	3	8.045795E-12
.0000007	4	8.045375E-12
.0000007	5	4.197719E-16
.0000007	6	7.07846E-15

. . . .

35.35544	1	.530183
35.35544	2	7.282975E-06
35.35544	3	.216709
35.35544	4	3.859662E-05
35.35544	5	.2166704
35.35544	6	.253085

37.35544	1	.5127703
37.35544	2	7.162355E-06
37.35544	3	.2239542
37.35544	4	3.671104E-05
37.35544	5	.2239175
37.35544	6	.2632535

. . . .

97.35544	1	.2049932
97.35544	2	4.528295E-06
97.35544	3	.337368
97.35544	4	9.280215E-06
97.35544	5	.3373587
97.35544	6	.4576316

99.35544	1	.1992773
99.35544	2	4.46471E-06
99.35544	3	.3391272
99.35544	4	8.894802E-06
99.35544	5	.3391183
99.35544	6	.4615885

100	1	.1974441
100	2	4.44377E-06
100	3	.3396907
100	4	8.774442E-06
100	5	.3396819
100	6	.4628583

Break in 9900

In Fig. 14.5 the decay of A and the formation of B, D and E are shown. The x-axis corresponds to 100 seconds. The concentration of the reactive species M and R is very low and coincides with the x-axis in this linear concentration scale.

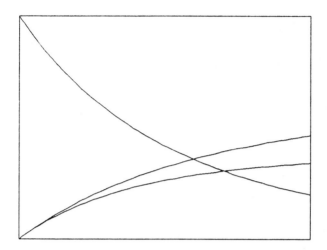

Fig. 14.5. The concentrations of species A, B, D and E (y-axis is scaled from 0–1) as a function of reaction time (0–100) are shown. The concentrations of B and D are nearly identical.

In the subroutine beginning with line 10000 the right hand sides of the ODE-system, which result from the chemical reaction model, are calculated. In subroutine 11000, the matrix $(I - A \cdot \Delta t)$ is evaluated and inverted in subroutine 50000. The graphical output is contained in subroutine 15000, the numerical output in subroutine 20000. The program section through line 999 is used for the input of the initial data.

The semi-implicit Euler steps, according to the formula above, are accomplished in the subroutine beginning with line 1100. In the main program (from line 5000), the total time interval is successively computed by using small time intervals DD which are changed by the step size control algorithm. The Euler iteration runs N1 times for the calculation of the time interval DD using the step size DD/N1. If the accuracy requirement in line 5240 with a doubled N1 is fulfilled, then this integration step is ended. Otherwise N1 is again multiplied by 2. If N1 becomes greater than 50 (line 5160), then the time interval DD is divided by 1000 and the subroutine is restarted. The time interval is doubled if the accuracy is already obtained with N1=2. N1 is reduced after each successful integration step (line 5420). This represents a very simple step size control procedure. Step size control is necessary because the program must start with very small time intervals since there are relatively big changes in the concentrations at the very beginning of the reaction, especially for the reactive species. After the initial phase, however, the relative variation of the concentrations slows down. Lines 5430 and 5440 of the step size control section serve to restrict the maximum step size for the graphical output and to stop the program at a given time.

15 Programs in Pascal

This chapter contains some remarks on the following PASCAL-programs which are filed on the floppy disk attached to this book. The printed programs contain lines with more than one statement, whereas the stored programs use a seperate line for each statement.

During the last several years, interest in programs written in PASCAL, especially in scientific software, has increased more and more. This chapter is not intended to be an introductory guide to the PASCAL-language. Rather, it should interest readers, who have not yet written programs in PASCAL. There are a number of books available containing introductions for working with PASCAL. The programs are written in a PASCAL-dialect known as Turbo-Pascal. Therefore, in studying and analyzing the programs, it is recommended to keep the Turbo-Pascal manual always at hand.

It should be emphasized that all the PASCAL-programs in this chapter were translated from the corresponding BASIC-programs. Therefore, elements of program structures in BASIC are also visible in the PASCAL-version. No further comments are given to explain the PASCAL-programs; however, it should be mentioned, that the logical structures of the PASCAL-programs are similar to the BASIC-programs, which are described in detail.

One of the most significant differences between PASCAL and BASIC is that PASCAL usually is a compiler language, while BASIC is an interpreter language. This means that it is easier to test and debug a BASIC-program, because, after a break, variables can be verified and assigned by an interactive mode. The price which must be paid, is the poor performance of the BASIC-programs. However, the Turbo-Pascal version 5.0, which has recently appeared, includes a built-in debugger, a very powerful program development tool. Programming in PASCAL leads to clearly arranged structures that allowing the user to analyze, check and understand programs more easily and — which is often a problem — to get familiar again with his own programs even after a long time.

A further advantage of PASCAL is the modular structure of the language. A procedure or a function, once programmed, can be used easily in any other program. Furthermore, one has the facility to pool all the procedures and functions, which may also be collected in a personal library. In this way a new vocabulary can be created which is accessible to all programs in the same way as the built-in procedures and functions. Typical examples are: matrix-operations, least squares sums or searches for maxima or minima.

A few, more precise comments concerning the basic structure of the programming language PASCAL should be added. The body of a typical PASCAL-program consists of a set of calls of different procedures and functions. All the built-in features are either procedures or functions. The body of the main program, which contains these calls, must begin with the reserved word BEGIN; after the last call the reserved word END. must be written. Of course these reserved words must not be used as names for user-programmed procedures or functions.

A procedure is a subprogram, which starts with the reserved word PROCEDURE, followed by its name. It can only be called from the main program with its name. Within this subroutine, all the vocabulary of the libraries is also available. The body of the subroutine must start with BEGIN and must end with END;. The semicolon shows the compiler, that this end is the end of a subprogram.

A function is also a subroutine, but, in contrast to a procedure, it returns a value. The function has to be called by the name which follows the reserved word FUNCTION. Its body also starts with BEGIN and ends with END;. The value given back by a function is assigned to a variable of the same name with the possibility to keep the whole vocabulary accessible. The call of a function must always be placed at the right hand side of the assignment symbol.

Another characteristic of PASCAL are the local variables. Variables are defined only within the procedure or function where they are declared. It is possible to use the same name for variables in different independent subroutines and their values are also independent. Only the variables declared before the beginning of the main program are valid in all subroutines and are therefore called global variables. Besides functions and procedures, structure elements can be used to complete programs. There are three different types of loops and one element to control a more complex type of variable, the RECORD. However, no further explanation of RECORD is needed, because RECORD is not used in the programs in this book.

If special variables of a certain subroutine are to be used within other subroutines and they should not be declared global, these variables must be transferred. PASCAL offers two possibilities to do this: the call by value and the call by reference. In case of a call by value, it is possible to calculate with the values of the transferred variables, while by a call by reference they can be assigned anew within the subroutine.

For the programs printed in this book and stored on Disk II, some special details should be discussed. All the programs were developed with the editor of Turbo-Pascal version 4.0 of Borland International, Inc., and tested with XT- and AT-models of IBM as well as with some XT- and AT-clones. This means that the programs should run with the appropriate computers and the corresponding software. Only common types of variables and data structures are used to avoid conflicts with the (eventually installed) math-coprocessors; therefore only two-byte integers and six-byte reals are used for declarations. The example programs on the disk are programmed so that the input of data must be carried out interactively or that the data are defined using CONST. The output of data is directed to the screen either in alphanumeric mode or in the graphics mode.

In case the amount of experimental data is large, it might be too cumbersome to feed into the computer by hand. Likewise one might need to save a large number of results from processed data (i.e. spectra, interpolated data, regression coefficients) on disk. Then the input- and output-routines must be adapted. To be able to read data from a disk, a file must be named using ASSIGN; the following statement RESET puts a pointer to the beginning of the file. Now the data can be read into the corresponding variables via this channel; after reading in the data from this file, the data channel must be closed by the statement CLOSE.

For the output of data to a disk, a similar procedure must be applied. The file has to be named by ASSIGN, followed by the instruction REWRITE. The program creates a new file and puts a pointer to the beginning of that file. Now the computer is ready to transfer the data to the disk. After termination of this process, the file must be closed again by CLOSE.

These instructions to process data on your disk have to be used very cautiously. To avoid unintentional deleting of data or files, the user should consult the manual for details.

To print out data, just add `USES PRINTER;` in the headline of your program and add `LST,` as the first item of the parameter list in all lines starting with `WRITE` or `WRITELN`.

In some of the programs it was necessary to include a switch to hold the results on the screen for a while, to give the user the chance to read them. This switch is programmed using the key `<ENTER>`. Results or graphics remain on screen; the program waits for `<ENTER>` to be pressed before it continues.

Dimensions of the arrays are the same as in the BASIC-programs (see Chapter 7). Closely connected with arrays is the instruction `FILLCHAR`. This shows a further fundamental difference between BASIC and PASCAL: in BASIC, the interpreter zeroes all variables at the start of the program by `RUN`. In PASCAL, the programmer has to assign the initial values of the variables; the declaration of a variable reserves the corresponding bytes in the data segment, but does not zero them. Hence, `FILLCHAR` fills the data segment byte by byte with well-defined values following the order of declaration. For example, `FILLCHAR(x[1], 600, chr($00))` means that in the data segment, exactly 600 bytes, beginning at the address of `x[1]`, will be set to zero.

At this time a very interesting facility of PASCAL shall be mentioned, though it is not used within this book: the type `POINTER`. Sorting, merging and searching routines operate faster and much more effectively using `POINTER`-variables. For further details consult a PASCAL-handbook.

The graphics mode in Turbo-Pascal must be initialized by `INITGRAPH`. Besides specifying the necessary graphics driver and the desired graphics mode, the subdirectory must be identified within which the compiler searches the graphics driver. If Turbo-Pascal software is installed following the instructions of the manual, no special path needs to be specified. In any case, we recommend the user to verify the searching paths before running a graphics program. Turbo-Pascal version 4.0 is able to search the available graphics driver and initialize it. All the graphics programs of this book are designed to make use of this facility. Only black and white colors were used in these programs, because all combinations of graphics cards and screens can handle black and white.

The control sequences discussed in Chapter 14 (data acquisition) are valid only for the analog-digital-converter PCI 6380 of CIL Microsystems Ltd., Worthing, Sussex, United Kingdom, which was used in our laboratory. If other instruments are used, the parameters must be fitted by the user according to the individual analog-digital-converter. Consult the appropriate manual for more detailed information.

In contrast to BASIC-programs, in PASCAL-programs there is a possibility to configurate the serial port before starting data acquisition. Two alternatives are shown on how to program the serial port COM1 directly. Within the code LABDATA1, the unit AUXINOUT is used, which is available in the Turbo-Pascal version 4.0; the program GPCDATA uses the direct statement PORT. The latter program needs a number of parameters to program the I/O-port correctly, but offers a greater variety of configurations. Details should be taken from the Reference Handbook of Turbo-Pascal 4.0 and advanced literature of MS-DOS.[1]

The special details mentioned above may produce the impression that programming in PASCAL seems to be rather difficult. It should be emphasized, however, that these apparent difficulties are more than compensated for by the advantages of PASCAL which are summarized as follows:

- structured, modular programming
- local variables in the subroutines
- function definition keeping the whole vocabulary accessible
- complex data structures
- recursive progamming
- high performance during execution
- private libraries
- crossover from compiled subroutines to and from Assembler, C and FORTRAN
- program code is relatively easy to transfer between computers, e.g. from PC to main frames

By programming and gathering experience, the user will recognize that there are additional advantages to getting more familiar with a programming language like PASCAL; this language may be one of the most suitable for scientific computing.

---

[1] R. Duncan, *Advanced MS-DOS*, Microsoft Press (1986)

```
                              TRIANGLE
```

```
program triangle;
    { J. Kinkel, January 1989, based on Ebert/Ederer }

uses   crt;

var  a,b,c,area : real;

 procedure Input;
{----------------}
 begin
    clrscr;   writeln;
    write (' Side a in cm ? ');   readln (a);
    write (' Side b in cm ? ');   readln (b);
    write (' Side c in cm ? ');   readln (c)
 end;   { Input }

 procedure Calc;
{----------------}
 var   s : real;

 begin
    s := 0.5*(a+b+c);   area := sqrt(s*(s-a)*(s-b)*(s-c))
 end;   { Calc }

 procedure Output;
{----------------}
 begin
    writeln;
    writeln (' Area of the triangle = ',area:12:7,' cm^2')
 end;   { Output }

begin
   Input;
   Calc;
   Output
end.
```

```
                              FREEPATH
```

```
program freepath;
    { J. Kinkel, January 1989, based on Ebert/Ederer }

uses   crt;

const   avogadro = 6.022e23;   { particles/m^3 }
        gas_constant = 8.2056e-5;   { m^3*atm/(mol*K) }

var   press,temp,sigma,fpl : real;
```

```
 procedure Input;
{----------------}
 begin
    clrscr;   writeln;
    write (' Pressure in atm ? ');                readln (press);
    write (' Temperature in K ? ');               readln (temp);
    write (' Diameter of the molecule in m ? ');  readln (sigma)
 end;   { Input }

 procedure Calc;
{---------------}
 var   n,pi : real;

 begin
    pi := 4*arctan(1);   n := avogadro*press/(temp*gas_constant);
    fpl := 1/(pi*sqrt(2)*sqr(sigma)*n)
 end;   { Calc }

 procedure Output;
{-----------------}
 begin
    writeln;   writeln (' Mean free path length = ',fpl,' m')
 end;   { Output }

begin
   Input;
   Calc;
   Output
end.
```

```
                              MAXWELL
```

```
program maxwell;
   { J. Kinkel, January 1989, based on Ebert/Ederer }

uses   crt;

const   r = 8.31441;   { J/(K*mol) }

var   temp,mw,u,pd : real;

 procedure Input;
{----------------}
 begin
    clrscr;   writeln;
    write (' Temperature in K ? ');               readln (temp);
    write (' Molecular weight ? ');               readln (mw);
    write (' Molecular velocity in m/sec ? ');    readln (u)
 end;   { Input }

 procedure Calc;
{---------------}
 var   pi,uh,rh,fac1,fac2,fac3 : real;
```

```
begin
   pi := 4*arctan(1);   rh := r*1e7;   uh := u*100;
   fac1 := 4*pi*sqr(uh);
   fac2 := sqrt(mw/(2*pi*rh*temp)*sqr(mw/(2*pi*rh*temp)));
   fac3 := exp(-0.5*mw*sqr(uh)/(rh*temp));
   pd := 100*fac1*fac2*fac3
end;   { Calc }

procedure Output;
{------------------}
begin
   writeln (' Temperature        = ',temp:4:0,' K');
   writeln (' Molecular weight   = ',mw:4:0,' amu');
   writeln (' Molecular velocity = ',u:4:0,' m/sec');
   writeln;   writeln (' Particle density   = ',pd,' mol/(m/sec)')
end;   { Output }

begin
   Input;
   Calc;
   Output
end.
```

PLANCK

```
program planck;
   { J. Kinkel, January 1989, based on Ebert/Ederer }

uses   crt;

const   h = 6.6262824e-34;   { J*sec }
        k = 1.38054e-23;     { J/K }
        c = 2.9979245812e10;   { cm/sec }

var   freq,temp,u : real;

 procedure Input;
{----------------}
 begin
    clrscr;   writeln;
    write (' Frequency in 1/sec ? ');   readln (freq);
    write (' Temperature in K ? ');     readln (temp)
 end;   { Input }

 procedure Calc;
{---------------}
 var   pi,num,den : real;

 begin
    pi := 4*arctan(1);   num := 8*pi*h*sqr(freq/c)*(freq/c);
    den := exp(h*freq/(k*temp))-1;   u := num/den
 end;   { Calc }
```

```
 procedure Output;
{-----------------}
 begin
    writeln;
    writeln (' Volume density of the energy of radiation');
    writeln (' at frequency ',freq,' 1/sec and');
    writeln (' at temperature ',temp,' K');
    writeln (' = ',u,'J*sec/cm^3')
 end;    { Output }

begin
   Input;
   Calc;
   Output
end.
```

```
                            C_F
```

```
program c_f;
   { J. Kinkel, January 1989, based on Ebert/Ederer }

uses   crt;

var   celsius,fahrenheit : real;

 procedure Input;
{----------------}
 begin
    clrscr;   writeln;
    write (' Temperature in Celsius ? ');   readln (celsius)
 end;   { Input }

 procedure Calc;
{--------------}
 begin
    fahrenheit := celsius*9/5+32
 end;   { Calc }

 procedure Output;
{-----------------}
 begin
    writeln;
    writeln (celsius:7:2,' C corresponds to ',fahrenheit:7:2,' F')
 end;   { Output }

begin
   Input;
   Calc;
   Output
end.
```

```
┌──────────────────────────────────────────────────────────────┐
│                           DIFF_POT                             │
└──────────────────────────────────────────────────────────────┘
```

```pascal
program diff_pot;
   { J. Kinkel, January 1989, based on Ebert/Ederer }

uses   crt;

var    hitt,n,conc1,conc2,diffpot : real;

 procedure Input;
{----------------}
 begin
    clrscr;   writeln;
    write (' Hittorf number ? ');              readln (hitt);
    write (' Electrochemical valence ? ');     readln (n);
    write (' Concentration 1 ? ');             readln (conc1);
    write (' Concentration 2 ? ');             readln (conc2)
 end;   { Input }

 procedure Calc;
{---------------}
 begin
    diffpot := (2*hitt-1)*0.058/n*ln(conc1/conc2)
 end;   { Calc }

 procedure Output;
{-----------------}
 begin
    writeln;
    writeln (' Diff.pot. = ',diffpot:10:8,' Volt')
 end;   { Output }

begin
   Input;
   Calc;
   Output
end.
```

```
┌──────────────────────────────────────────────────────────────┐
│                           EFFUSION                             │
└──────────────────────────────────────────────────────────────┘
```

```pascal
program effusion;
   { J. Kinkel, January 1989, based on Ebert/Ederer }

uses   crt;

const   r = 8.31441e7;

var    temp,mw,pex,pin,kappa,u : real;
```

```
procedure Input;
{----------------}
 begin
    clrscr;   writeln;
    write (' Temperature in K ? ');          readln (temp);
    write (' Molecular weight ? ');          readln (mw);
    write (' Cp/Cv ? ');                      readln (kappa);
    write (' External pressure ad. lib. ? ');  readln (pex);
    write (' Internal pressure ad. lib. ? ');  readln (pin)
 end;   { Input }

 procedure Calc;
{---------------}
 var   quot,pr,fac : real;

 begin
    quot := (kappa-1)/kappa;   pr := pex/pin;   fac := r*temp/mw;
    u := sqrt((2*fac/quot)*(1-exp(quot*ln(pr))))/100
 end;   { Calc }

 procedure Output;
{-----------------}
 begin
    writeln;   writeln (' Effusion velocity = ',u:10:5,' m/sec')
 end;   { Output }

begin
    Input;
    Calc;
    Output
end.
```

```
                            GEO_SER
```

```
program geo_ser;
    { J. Kinkel, January 1989, based on Ebert/Ederer }

uses   crt;

var   a,q,sum,h : real;
      count : integer;
 procedure Input;
{----------------}
 begin
    clrscr;   writeln;
    write (' Initial term of the series ? ');   readln (a);
    write (' Multiplication factor ? ');         readln (q)
 end;   { Input }
 procedure Output;
{-----------------}
 begin
    writeln (count:5,h:35,sum:20:8)
 end;   { Output }
```

```
procedure Calc;
{---------------}
begin
   count := 0;   sum := 0;   h := a;
   repeat
    sum := sum + h;   h := h*q;
    Output;
    count := count + 1
   until (count = 20)
end;   { Calc }

begin
   Input;
   Calc
end.
```

```
                              FOU_SER
```

```
program fou_ser;
{ J. Kinkel, January 1989, based on Ebert/Ederer }
uses   crt;

var   x,f,h : real;
      count : integer;

procedure Input;
{----------------}
begin
   clrscr;   writeln;
   write (' Which x shall be tried ? ');   readln (x)
end;   { Input }

procedure Output;
{-----------------}
begin
   writeln (count:5,x:8:4,h:35,f:20:8)
end;   { Output }

procedure Calc;
{---------------}
var   pi : real;

begin
   pi := 4*arctan(1);   f := 1;   count := 1;
   repeat
    h := 4/sqr(count)/sqr(pi)*(cos(count*pi)-1);
    h := h*cos(count*pi*x/2);   f := f+h;
    Output;
    count := count + 1
   until (count > 20)
end;   { Calc }

begin
   Input;
   Calc
end.
```

```
┌─────────────────────────────────────────────────────────────────┐
│                          HEATCOND                                 │
└─────────────────────────────────────────────────────────────────┘
```

```pascal
program heatcond;
   { J. Kinkel, January 1989, based on Ebert/Ederer }

uses   crt;

var    x,z,h,temp : real;
       count : integer;

 procedure Input;
{----------------}
 begin
    clrscr;   writeln;   writeln (' Calculation of temperature ');
    write (' at which position x ? ');   readln (x);
    write (' at which time t ? ');       readln (z)
 end;   { Input }

 procedure Output;
{-----------------}
 begin
    writeln (count:5,h:35,temp:20:8)
 end;   { Output }

 procedure Calc;
{---------------}
 var   pi : real;

 begin
    pi := 4*arctan(1);   temp := 10*x + 10;   count := 0;
    repeat
     count := count + 1;
     h := 30/count/pi*(cos(count*pi)-1);
     h := h*exp(-2*sqr(count)*sqr(pi)*z/9/100);
     h := h*sin(count*pi*x/3);
     temp := temp + h;
     Output
    until (count > 25)
 end;   { Calc }

begin
   Input;
   Calc
end.
```

```
┌─────────────────────────────────────────────────────────────────┐
│                          ROTATION                                 │
└─────────────────────────────────────────────────────────────────┘
```

```pascal
program rotation;
   { J. Kinkel, January 1989, based on Ebert/Ederer }

uses   crt;

const  h = 6.626176e-34;   { J/sec }
       lo = 6.022e23;   { 1/mol }
       k = 1.38066e-23;   { J/K }
```

```
var    temp,aw1,aw2,dist,en,q,zsum : real;
       j : integer;

procedure Input;
{----------------}
 begin
    clrscr;   writeln;
    write (' Temperature in K ? ');          readln (temp);
    write (' Atomic weight 1 in amu ? ');    readln (aw1);
    write (' Atomic weight 2 in amu ? ');    readln (aw2);
    write (' Distance in Angstroem ? ');     readln (dist)
 end;    { Input }

procedure Output;
{-----------------}
 begin
    writeln;
    writeln (' Number of the level      : ',j:5);
    writeln (' Energy level             : ',en,' J');
    writeln (' Frequency of occupation : ',q);
    writeln (' Partition function       : ',zsum)
 end;    { Output }

procedure Calc;
{---------------}
 var    pi,mue,tm : real;

 begin
    pi := 4*arctan(1);    mue := aw1*aw2/(aw1+aw2);
    tm := mue/1000*sqr(dist*1e-10);    zsum := 0;    j := 0;
    repeat
     en := j*(j+1)*h*lo/(8*sqr(pi)*tm)*h;
     q := (2*j+1)*exp(-en/k/temp);    zsum := zsum + q;
     Output;
     j := j + 1
    until (j > 100)
 end;   { Calc }

begin
   Input;
   Calc
end.
```

```
                                    PI
```

```
program pi;
   { J. Kinkel, January 1989, based on Ebert/Ederer }

uses   crt;

var    count,sum : integer;
```

```
 procedure Input;
{----------------}
 begin
    clrscr;   writeln;   randomize
 end;   { Input }

 procedure Output;
{------------------}
 begin
    writeln (count:5,sum:8,sum*4/count:20:10)
 end;   { Output }

 procedure Calc;
{---------------}
 var   x,y : real;

 begin
    sum := 0;   count := 0;
    repeat
     x := random;   y := random;   count := count+1;
     if (sqr(x)+sqr(y) <= 1) then sum := sum + 1;
     Output
    until (count >= maxint)
 end;   { Calc }

begin
   Input;
   Calc
end.
```

```
                              SUM
```

```
program sum;
   { J. Kinkel, January 1989, based on Ebert/Ederer }

uses   crt;

var   sum,count,max : real;

 procedure Input;
{----------------}
 begin
    clrscr;   writeln;
    write (' Calculation of the sum of the first n squared');
    writeln (' integer numbers');
    write (' n = ? ');   readln (max)
 end;   { Input }

 procedure Calc;
{---------------}
 begin
    sum := 0;   count := 0;
    while (count < max) do
    begin
       count := count + 1;   sum := sum + sqr(count)
    end
 end;   { Calc }
```

```
 procedure Output;
{------------------}
 begin
    writeln;   writeln (' Sum = ',sum:20:0)
 end;   { Output }

begin
   Input;
   Calc;
   Output
end.
```

```
┌─────────────────────────────────────────────────────────────────┐
│                           MOL_VEL                                 │
└─────────────────────────────────────────────────────────────────┘
```

```
program mol_vel;
   { J. Kinkel, January 1989, based on Ebert/Ederer }

uses   crt;

const   h2 = 'Hydrogen';   c6 = 'Benzene';
        mwh2 = 2;          mwc6 = 78;         r = 8314.34;

 procedure Calc;
{---------------}
 var   vh2,vc6 : real;
       temp : integer;

 begin
    temp := 150;
    repeat
     vh2 := sqrt(3*r*temp/mwh2);   vc6 := sqrt(3*r*temp/mwc6);
     writeln (temp:6,vh2:25:8,vc6:18:8);
     temp := temp + 50
    until (temp > 600)
 end;   { Calc }

 procedure Output;
{------------------}
 begin
    clrscr;   writeln;
    writeln (' Mean molecular velocity in m/sec');   writeln;
    writeln ('Temperature':10,h2:17,c6:18)
 end;   { Output }

begin
   Output;
   Calc
end.
```

```
                              MEAN

program mean;
    { J. Kinkel, January 1989, based on Ebert/Ederer }

uses  crt;

const   ndp = 9;
        x : array[1..9] of real = (5,5.5,3,7.2,8.1,5.95,4.7,4,6.05);

var   sum,sdev : real;

 procedure Calc;
{---------------}
 var   cnt : integer;

 begin
    sum := 0;    for cnt := 1 to ndp do sum := sum + x[cnt];
    sum := sum/ndp;    sdev := 0;
    for cnt := 1 to ndp do sdev := sdev + sqr(sum - x[cnt]);
    sdev := sqrt(sdev/(ndp-1))
 end;    { Calc }

 procedure Output;
{-----------------}
 begin
    clrscr;    writeln;
    writeln (' Mean value           =',sum:10:6);
    writeln (' Standard deviation =',sdev:10:6)
 end;    { Output }

begin
   Calc;
   Output
end.
```

```
                             MC_INT

program mc_int;
    { J. Kinkel, January 1989, based on Ebert/Ederer }

uses   crt;

var   xa,xe,ya,ye,za,ze,v,sum : real;
      count : integer;

 procedure Input;
{------------------}
 begin
    clrscr;    writeln;
    write (' x - interval from ');    readln (xa);
```

```
      write ('                  to ');    readln (xe);
      write (' y - interval from ');      readln (ya);
      write ('                  to ');    readln (ye);
      write (' z - interval from ');      readln (za);
      write ('                  to ');    readln (ze)
   end;   { Input }

 procedure Output;
{-----------------}
 begin
      writeln (' Estimated integral = ',(sum/count*v):20:8,count:10)
   end;   { Output }

 procedure Calc;
{---------------}
 var   x,y,z,xm,ym,zm,f : real;

 begin
      xm := xe-xa;    ym := ye-ya;    zm := ze-za;
      v := xm*ym*zm;    count := 0;    sum := 0;
      randomize;
      repeat
         count := count + 1;
         x := random*xm+xa;    y := random*ym+ya;    z := random*zm+za;
         f := sqr(x)+sqr(y)+sqr(z);    sum := sum + f;
         Output
      until (count >= maxint)
   end;   { Calc }

begin
   Input;
   Calc
end.
```

```
+----------------------------------------------------------------+
|                          C_H_N_O                               |
+----------------------------------------------------------------+
```

```
program c_h_n_o;
   { J. Kinkel, January 1989, based on Ebert/Ederer }

uses   crt;

var   c,h,n,o,ccalc,hcalc,ncalc,ocalc : real;

 procedure Input;
{-----------------}
 begin
      clrscr;   writeln;
      write (' Weight% C ? ');    readln (c);
      write (' Weight% H ? ');    readln (h);
      write (' Weight% N ? ');    readln (n);
      o := 100 - h - n - c;       writeln (' Weight% O   ',o:5:2)
   end;   { Input }
```

```
procedure Output;
{-----------------}
 begin
    writeln;   writeln ('C':8,'H':10,'N':10,'O':10)
 end;   { Output }

procedure Calc;
{---------------}
 var   cz,hz,nz,oz,x : real;
       cnt : integer;

 begin
    cz := c/12.011;   hz := h/1.008;   nz := n/14.007;   oz := o/15.999;
    Output;
    for cnt := 1 to 15 do
    begin
       x := cnt/cz;   ccalc := x*cz;   hcalc := int(hz*x*100)/100;
       ncalc := int(nz*x*100)/100;   ocalc := int(oz*x*100)/100;
       writeln (ccalc:9:2,hcalc:10:2,ncalc:10:2,ocalc:10:2)
    end
 end;   { Calc }


begin
   Input;
   Calc
end.
```

```
                              MS
```

```
program ms;
   { J. Kinkel, January 1989, based on Ebert/Ederer }

uses   crt;

const   m1 = 1.0078;   m2 = 31.9721;   m3 = 15.9949;
        m4 = 14.0031;   m5 = 12.0000;

var    mba,mbe,sum : real;
       zs,zo,zn,zc,zh : integer;

 procedure Input;
{----------------}
 begin
    clrscr;   writeln;
    write (' Mass interval from ');   readln (mba);
    write ('                 to ');   readln (mbe)
 end;   { Input }

 procedure Output;
{-----------------}
 begin
    writeln (' S ',zs,' O ',zo, ' N ',zn,' C ',zc,' H ',zh,' = ',sum:12:8)
 end;   { Output }
```

```pascal
procedure Calc;
{---------------}
var   cnt1,cnt2,cnt3,cnt4,cnt5,max : real;
      zw1,zw2 : integer;

begin
   cnt1 := 0;   cnt2 := 0;   cnt3 := 0;   cnt4 := 0;   cnt5 := 0;
   max := mbe;
   while (cnt2 <= max) do
   begin
      cnt3 := cnt2;
      while (cnt3 <= max) do
      begin
         cnt4 := cnt3;
         while (cnt4 <= max) do
         begin
            cnt5 := cnt4;
            while (cnt5 <= max) do
            begin
               zw1 := trunc((mba-cnt5)/m1+1);
               zw2 := trunc((mbe-cnt5)/m1);
               if (zw1 <= zw2) then
               begin
                  cnt1 := zw1;
                  while (cnt1 <= zw2) do
                  begin
                     if (cnt1 >= 0) then
                     begin
                        if (cnt1 <= max/4+2) then
                        begin
                           sum := cnt5 + cnt1*m1;
                           zs := trunc(cnt2/m2 + 0.1);
                           zo := trunc((cnt3-cnt2)/m3 + 0.1);
                           zn := trunc((cnt4-cnt3)/m4 + 0.1);
                           zc := trunc((cnt5-cnt4)/m5 + 0.1);
                           zh := trunc(cnt1/m1 + 0.1);
                           Output
                        end
                        else cnt1 := zw2 + 1
                     end;
                     cnt1 := cnt1 + 1
                  end
               end;
               cnt5 := cnt5 + m5
            end;
            cnt4 := cnt4 + m4
         end;
         cnt3 := cnt3 + m3
      end;
      cnt2 := cnt2 + m2
   end
end;   { Calc }

begin
   Input;
   Calc
end.
```

```
┌────────────────────────────────────────────────────────────┐
│                            EULER                             │
└────────────────────────────────────────────────────────────┘
```

```pascal
program euler;
   { J. Kinkel, January 1989, based on Ebert/Ederer }

uses   crt;

var    ll,ul,sum : real;
       subi : integer;

 procedure Input;
{----------------}
 begin
    clrscr;   writeln;
    write (' Lower integration limit ? ');   readln (ll);
    write (' Upper integration limit ? ');   readln (ul)
 end;   { Input }

 procedure Calc;
{---------------}
 var   help,x,y : real;
       cnt : integer;

 begin
    sum := 0;   help := (ul-ll)/subi;
    x := ll;   y := exp(-sqr(x));   sum := sum + y/2;
    for cnt := 1 to subi-1 do
    begin
       x := ll + cnt*help;   y := exp(-sqr(x));   sum := sum + y
    end;
    x := ul;   y := exp(-sqr(x));   sum := sum + y/2;
    sum := sum*help
 end;   { Calc }

 procedure Output;
{-----------------}
 begin
    writeln;
    writeln (' Integral from ',ll:5:2,' to ',ul:5:2);
    writeln (' with ',subi,' subintervals = ',sum:12:8)
 end;   { Output }

begin
   Input;
   subi := 20;
   repeat
   Calc;
   Output;
   subi := subi*2
   until (subi > 21000)
end.
```

EULER2

```pascal
program euler2;
   { J. Kinkel, January 1989, based on Ebert/Ederer }

uses   crt;

var   ll,ul,sum : real;
      subi : integer;

 procedure Input;
{----------------}
 begin
    clrscr;   writeln;
    write (' Lower integration limit ? ');   readln (ll);
    write (' Upper integration limit ? ');   readln (ul)
 end;   { Input }

 procedure Calc;
{---------------}
 var   help,x,y : real;
       cnt : integer;

    function fnf(xh : real) : real;
   {----------------------------}
    begin
       fnf := exp(-sqr(xh))
    end;   { fnf }

 begin
    sum := 0;   help := (ul-ll)/subi;   sum := sum + fnf(ll)/2;
    x := ll;
    for cnt := 1 to subi-1 do
    begin
       x := ll + cnt*help;   sum := sum + fnf(x)
    end;
    sum := sum + fnf(ul)/2;   sum := sum*help
 end;   { Calc }

 procedure Output;
{-----------------}
 begin
    writeln;
    writeln (' Integral from ',ll:5:2,' to ',ul:5:2);
    writeln (' with ',subi,' subintervals = ',sum:12:8)
 end;   { Output }

begin
   Input;
   subi := 20;
   repeat
    Calc;
    Output;
    subi := subi*2
   until (subi > 21000)
end.
```

```
┌─────────────────────────────────────────────────────────────────┐
│                            ELLIPSE                                │
└─────────────────────────────────────────────────────────────────┘
```

```pascal
program ellipse;
   { J. Kinkel, January 1989, based on Ebert/Ederer }

uses   crt;

var   ha1,ha2,sum : real;
      max : integer;

 procedure Input;
{-----------------}
 begin
    clrscr;   writeln;
    write (' Length of the first half axis ? ');    readln (ha1);
    write (' Length of the second half axis ? ');   readln (ha2)
 end;   { Input }

 procedure Calc;
{---------------}
 var   pi,ll,ul,help : real;
       cnt : integer;

    function fnf(x : real) : real;
    {-----------------------------}
    begin
       fnf := sqrt(sqr(ha1*sin(x))+sqr(ha2*cos(x)))
    end;   { fnf }

 begin
    pi := 4*arctan(1);
    ll := 0;   ul := 2*pi;   help := (ul-ll)/max;
    sum := 0;   sum := sum + fnf(ll)/2;
    for cnt := 1 to max-1 do sum := sum + fnf(cnt*help);
    sum := sum + fnf(ul)/2;   sum := sum*help
 end;   { Calc }

 procedure Output;
{------------------}
 begin
    writeln;   writeln (' Circumference of the ellipse');
    writeln (' with ',max,' subintervals = ',sum:12:8)
 end;   { Output }

begin
   Input;
   max := 10;
   repeat
    Calc;
    Output;
    max := max*2
   until (max > 21000)
end.
```

```
                              CO_POL
```

```
program co_pol;
   { J. Kinkel, January 1989, based on Ebert/Ederer }

uses   crt;

var    paa,pbb,x : real;
       mon : char;
       cnt : integer;

 procedure Input;
{----------------}
 begin
    clrscr;   writeln;
    writeln (' What is the probability for a polymer with');
    write (' an end group A to react with the monomer molecule A ? ');
    readln (paa);   writeln;
    writeln (' What is the probability for a polymer with');
    write (' an end group B to react with the monomer molecule B ? ');
    readln (pbb);   writeln;
    write (' Do you want to start with A or B ? ');   readln (mon)
 end;   { Input }

 procedure Calc;
{---------------}
 begin
    randomize;   cnt := 0;
    repeat
     x := random;   write (mon);
     if (mon = 'b') then
     begin
        if (x >= pbb) then mon := 'a'
     end
     else if (x >= paa) then mon := 'b';
     cnt := cnt + 1
    until (cnt > 1000)
 end;   { Calc }

begin
   Input;
   Calc
end.
```

```
                             SIMPSON1
```

```
program simpson1;
   { J. Kinkel, January 1989, based on Ebert/Ederer }

uses   crt;

var    ll,ul,sum : real;
       subi : integer;
```

```
procedure Input;
{----------------}
begin
   clrscr;   writeln;
   write (' Lower integration limit ? ');   readln (ll);
   write (' Upper integration limit ? ');   readln (ul)
end;   { Input }

procedure Calc;
{---------------}
var   help: real;
      cnt : integer;

   function fnf(xh : real) : real;
   {----------------------------}
   begin
      fnf := sqr(xh)*exp(-sqr(xh))
   end;   { fnf }

begin
   sum := fnf(ll);   help := (ul-ll)/subi;   cnt := 1;
   repeat
    sum := sum + 4*fnf(ll+cnt*help);   cnt := cnt + 2
   until (cnt > subi-1);
   cnt := 2;
   repeat
    sum := sum + 2*fnf(ll+cnt*help);   cnt := cnt + 2
   until (cnt > subi-2);
   sum := sum + fnf(ul);   sum := sum*help/3
end;   { Calc }

procedure Output;
{-----------------}
begin
   writeln;
   writeln (' Integral from ',ll:6:3,' to ',ul:6:3);
   writeln (' with ',subi,' subintervals = ',sum:12:8)
end;   { Output }

begin
   Input;
   subi := 4;
   repeat
    Calc;
    Output;
    subi := subi*2
   until (subi > 1000)
end.
```

```
                              SIMPSON2

program simpson2;
    { J. Kinkel, January 1989, based on Ebert/Ederer }

uses   crt;

var    ll,ul,sum,areq,acal : real;
       subi : integer;
       flag : boolean;

 procedure Input;
{----------------}
 begin
    flag := true;   clrscr;   writeln;
    write (' Lower integration limit ? ');   readln (ll);
    write (' Upper integration limit ? ');   readln (ul);
    write (' Accuracy in % ? ');             readln (areq)
 end;   { Input }

 procedure Calc;
{---------------}
 var    isum,help : real;
        cnt : integer;

    function fnf(xh : real) : real;
    {---------------------------}
    begin
       fnf := sqr(xh)*exp(-sqr(xh))
    end;   { fnf }

 begin
    isum := 0;   subi := 4;
    repeat
     sum := fnf(ll);   help := (ul-ll)/subi;   cnt := 1;
     repeat
      sum := sum + 4*fnf(ll+cnt*help);   cnt := cnt + 2
     until (cnt > subi-1);
     cnt := 2;
     repeat
      sum := sum + 2*fnf(ll+cnt*help);   cnt := cnt + 2
     until (cnt > subi-2);
     sum := sum + fnf(ul);    sum := sum*help/3;
     if (subi > 2500) then flag := false;
     subi := subi*2;   acal := abs(sum-isum);   isum := sum;
    until (acal <= abs(sum)*areq/100) or not(flag)
 end;   { Calc }

 procedure Output;
{-----------------}
 begin
    writeln;
    if flag then
```

```
      begin
         writeln (' Integral from ',ll:6:3,' to ',ul:6:3);
         writeln (' with ',round(subi/2),' subintervals = ',sum:12:8);
         writeln;
         writeln (' Accuracy required  : ',abs(areq)*sum/100);
         writeln (' Accuracy calculated: ',acal)
      end
      else
      begin
         writeln (' The required accuracy is not possible');
         writeln (' Last estimation of the integral = ',sum:12:8)
      end
  end;   { Output }

begin
   Input;
   Calc;
   Output
end.
```

```
                              DEBYE
```

```
program debye;
   { J. Kinkel, January 1989, based on Ebert/Ederer }

uses   crt;

var    ul,sum : real;
       flag : boolean;

 procedure Input;
{----------------}
 begin
    flag := true;  clrscr;  writeln;
    write (' 0/T ? ');  readln (ul)
 end;   { Input }

 procedure Calc;
{----------------}
 var    ll,isum,help,areq,acal : real;
        cnt,subi : integer;

    function fnf(xh : real) : real;
   {-----------------------------}
    begin
       fnf := sqr(xh)*sqr(xh)*exp(xh)/sqr(exp(xh)-1)
    end;   { fnf }

 begin
    ll := 1e-5;  areq := 1e-2;  isum := 0;  subi := 4;
    repeat
     sum := fnf(ll);  help := (ul-ll)/subi;  cnt := 1;
     repeat
      sum := sum + 4*fnf(ll+cnt*help);  cnt := cnt + 2
     until (cnt > subi-1);
```

```
      cnt := 2;
      repeat
       sum := sum + 2*fnf(ll+cnt*help);   cnt := cnt + 2
      until (cnt > subi-2);
      sum := sum + fnf(ul);   sum := sum*help/3;
      sum := sum*9/ul/ul/ul;
      if (subi > 2500) then flag := false;
      subi := subi*2;   acal := abs(sum-isum);   isum := sum;
    until (acal <= abs(sum)*areq/100) or not(flag)
  end;   { Calc }

procedure Output;
{----------------}
 begin
    writeln;
    if flag then
    begin
       writeln (' Spec. molar heat Cv of a metal');
       writeln (' with Q/T of ',ul:8:3,' = ',sum:12:8,'*R')
    end
    else
    begin
       writeln (' The required accuracy is not possible');
       writeln (' Last estimation for the integral ',sum:12:8)
    end
  end;   { Output }

begin
   Input;
   Calc;
   Output
end.
```

```
┌─────────────────────────────────────────────────────────┐
│                        VIRIAL                           │
└─────────────────────────────────────────────────────────┘
```

```
program virial;
   { J. Kinkel, January 1989, based on Ebert/Ederer }

uses   crt;

var    ll,ul,sigma,eps,temp,sum,areq,acal : real;
       subi : integer;
       flag : boolean;

procedure Input;
{----------------}
 begin
    flag := true;   clrscr;   writeln;
    writeln (' First parameter of the Lennard-Jones pot.');
    write (' Sigma ? ');                readln (sigma);
    writeln (' Second parameter of the Lennard-Jones pot.');
    write (' Epsilon/k(Boltzmann) ? ');  readln (eps);
    write (' Temperature in K ? ');      readln (temp);
    write (' Accuracy in % ? ');         readln (areq)
  end;   { Input }
```

```
procedure Calc;
{--------------}
var   isum,help : real;
      cnt : integer;

   function fnf(x : real) : real;
   {---------------------------}
   var    diff,arg : real;

   begin
      diff := exp(12*ln(sigma/x))-exp(6*ln(sigma/x));
      arg := 4*eps/temp*diff;
      if (arg > 1419) then arg := 1419;
      fnf := (1-exp(-arg))*sqr(x)
   end;   { fnf }

begin
   ll := sigma*1e-3;   ul := sigma*10;   isum := 0;   subi := 4;
   repeat
    sum := fnf(ll);   help := (ul-ll)/subi;   cnt := 1;
    repeat
     sum := sum + 4*fnf(ll+cnt*help);   cnt := cnt + 2
    until (cnt > subi-1);
    cnt := 2;
    repeat
     sum := sum + 2*fnf(ll+cnt*help);   cnt := cnt + 2
    until (cnt > subi-2);
    sum := sum + fnf(ul);   sum := sum*help/3;
    sum := sum*8*arctan(1)*6.022e23*1e-24;
    if (subi > 2500) then flag := false;
    subi := subi*2;   acal := abs(sum-isum);   isum := sum;
   until (acal <= abs(sum)*areq/100) or not(flag)
end;   { Calc }

procedure Output;
{---------------}
begin
   clrscr;   writeln;
   if flag then
   begin
      writeln (' Integral from ',ll:6:3,' to ',ul:6:3,' Angstroem');
      writeln (' with ',round(subi/2),' subintervals = ',sum:12:8);
      writeln;
      writeln (' Accuracy required   : ',abs(sum)*areq/100);
      writeln (' Accuracy calculated : ',acal);   writeln;
      writeln (' Sigma               = ',sigma:8:4,' Angstroem');
      writeln (' Epsilon/k(Boltzmann) = ',eps:8:4,' K');
      writeln (' Temperature         = ',temp:8:4,' K');
      writeln (' 2. Virialcoeff.     = ',sum:8:4,' cm^3/mol')
   end
   else
   begin
      writeln (' The required accuracy is not possible');
      writeln (' Last estimation of the integral : ',sum:12:8)
   end
end;   { Output }
```

```
begin
   Input;
   Calc;
```

```
                              BISECT
```

```
program bisect;
   { J. Kinkel, January 1989, based on Ebert/Ederer }

uses   crt;

var    ub,lb,eps : real;

 procedure Input;
{------------------}
 begin
    clrscr;    writeln;
    write (' Boundaries of the searching interval (beg end) ? ');
    readln (lb,ub);
    write (' Absolute accuracy of the root ? ');    readln (eps)
 end;    { Input }

 procedure Calc;
{--------------------}
 var    sgn1,sgn2,sgn3 : char;
        max : integer;
        y1,y2,middle,ym : real;

    function fnf(x:real) : real;
   {--------------------------}
    begin
       fnf := 23+x*(-17+x*(2+x*(-3+x*(1+0.1*x))))
    end;    { fnf }

    procedure Search;
   {----------------}
    var   cnt: integer;
          help : real;

    begin
       while (max <= 1000) and (sgn1 = sgn3) do
       begin
          help := (ub-lb)/max;
          for cnt := 1 to max-1 do
          begin
             ym := fnf(lb+cnt*help);
             if (ym < 0) then sgn3 := '-' else sgn3 := '+';
             if not(sgn1 = sgn3) then ub := lb+cnt*help
          end;
          max := max*10
       end
    end;    { Search }
```

```
      procedure Output;
    {------------------}
    begin
       writeln;
       writeln (' Root = ',(ub+lb)/2:12:8,' +/- ',abs(ub-lb)/2:12:8)
    end;   { Output }

  begin
     sgn1 := '+';   sgn2 := sgn1;    sgn3 := sgn1;
     max := 10;   y1 := fnf(lb);
     if (y1 < 0) then sgn1 := '-';
     y2 := fnf(ub);
     if (y2 < 0) then sgn2 := '-';
     sgn3 := sgn2;
     if (sgn1 = sgn2) then Search;
     y2 := fnf(ub);
     if (y2 < 0) then sgn2 := '-' else sgn2 := '+';
     if (max > 1000) then
      writeln (' No root of the function has been found !')
     else
     begin
        while (abs(ub-lb) >= eps) do
        begin
           if not(sgn1 = sgn2) then
           begin
              writeln (' y (',lb:6:6,') = ',y1:12:8);
              writeln (' y (',ub:6:6,') = ',y2:12:8);
              middle := (lb+ub)/2;   ym := fnf(middle);
              if (ym < 0) then sgn3 := '-' else sgn3 :='+';
              if (sgn3 = sgn1) then lb := middle else ub := middle;
              y1 := fnf(lb);
              if (y1 < 0) then sgn1 := '-' else sgn1 := '+';
              y2 := fnf(ub);
              if (y2 < 0) then sgn2 := '-' else sgn2 := '+'
           end
        end;
        Output
     end
  end;   { Calc }

begin
   Input;
   Calc
end.
```

```
+------------------------------------------------------------+
|                         NEWTON                             |
+------------------------------------------------------------+
```

```
program newton;
   { J. Kinkel, January 1989, based on Ebert/Ederer }

uses   crt;

var   beg,tmp,y : real;
```

```
   procedure Input;
{----------------}
 begin
    clrscr;   writeln;
    write (' Estimation of the root ? ');   readln (beg)
 end;   { Input }

 procedure Calc;
{---------------}
 var   yf,yd : real;
       flag  : boolean;

    function fnf(x : real) : real;
   {----------------------------}
    begin
       fnf := 10*x*sqr(sqr(x))-exp(sqr(x))
    end;   { fnf }

    function fnd(x : real) : real;
   {----------------------------}
    begin
       fnd := 50*sqr(sqr(x))-2*x*exp(sqr(x))
    end;   { fnd }

 begin
    tmp := 0;
    flag := false;
    repeat
       yf := fnf(beg);   yd := fnd(beg);
       tmp := beg - yf/yd;
       if (abs(beg-tmp) < abs(beg*1e-7)) then flag := true;
       beg := tmp
    until flag;
    y := yf
 end;   { Calc }

 procedure Output;
{-----------------}
 begin
    writeln;
    writeln (' Root          = ',(beg+tmp)/2:12:8);
    writeln (' Function value = ',y)
 end;   { Output }

begin
   Input;
   Calc;
   Output
end.
```

```
                              REG_FAL


program reg_fal;
   { J. Kinkel, January 1989, based on Ebert/Ederer }

uses   crt;

var    lb,ub,fv : real;
       answer : char;
       flag : boolean;

 procedure Input;
{----------------}
 begin
    clrscr;   writeln;
    write (' Starting interval (beg end) ? ');   readln (lb,ub);
    writeln;
    write (' Intermediate results on the screen (y/n) ? ');
    readln (answer);
    if (answer in ['j','J','y','Y']) then flag := true else flag := false
 end;   { Input }

 procedure Calc;
{---------------}
 var    y0,y1,xz : real;

    function fnf(x:real) : real;
   {--------------------------}
    begin
       fnf := (x-4)*(x-2.3)*(x+2.1)*(x+3.34)
    end;   { fnf }

 begin
    repeat
       y0 := fnf(lb);   y1 := fnf(ub);   xz := ub-(ub-lb)/(y1-y0)*y1;
       if flag then writeln (lb:10:8,ub:15:8,y1:15:8,y0:15:8);
       lb := ub;   ub := xz
    until (abs(lb-ub) <= abs(lb*1e-7));
    fv := (y1+y0)/2
 end;   { Calc }

 procedure Output;
{-----------------}
 begin
    writeln;
    writeln (' Root          = ',(ub+lb)/2:12:8);
    writeln (' Function value = ',fv:12:8)
 end;   { Output }

begin
   Input;
   Calc;
   Output
end.
```

```
┌──────────────────────────────────────────────────────────────────┐
│                              NI_CRNI                               │
└──────────────────────────────────────────────────────────────────┘
```

```pascal
program ni_crni;
    { J. Kinkel, January 1989, based on Ebert/Ederer }

uses    crt;

const   k1 = 25.4497;      k2 = -0.559195;    k3 = 0.10452439;
        k4 = -8.776153e-3;  k5 = 3.76041e-4;  k6 = -8.64943e-6;
        k7 = 1.021005e-7;   k8 = -4.891009e-10;

var    temp,ub,lb : real;

 procedure Input;
{----------------}
 begin
    clrscr;    writeln;
    write (' Temperature in Celsius ? ');    readln (temp)
 end;    { Input }

 procedure Calc;
{---------------}
 var    sgn1,sgn2,sgn3 : char;
        max : integer;
        y1,y2,middle,ym,eps : real;

    function fnf(x,y : real) : real;
    {------------------------------}
    var    h : real;

    begin
        h := k1*x+k2*sqr(x)+k3*x*sqr(x)+k4*sqr(sqr(x))+k5*x*sqr(sqr(x));
        h := h+k6*sqr(x)*sqr(sqr(x))+k7*x*sqr(x)*sqr(sqr(x))
            +k8*sqr(sqr(sqr(x)));
        fnf := h-y
    end;    { fnf }

    procedure Search;
    {-----------------}
    var    cnt: integer;
           help : real;

    begin
        while (max <= 1000) and (sgn1 = sgn3) do
        begin
            help := (ub-lb)/max;
            for cnt := 1 to max-1 do
            begin
                ym := fnf(lb+cnt*help,temp);
                if (ym < 0) then sgn3 := '-' else sgn3 := '+';
                if not(sgn1 = sgn3) then ub := lb+cnt*help
            end;
            max := max*10
        end
    end;    { Search }
```

```
     procedure Output;
    {------------------}
     begin
        writeln;
        writeln (' The voltage of the Ni-CrNi TC');
        writeln (' at ',temp:6:2,' Celsius');
        writeln (' is ',int(50*(ub+lb)+0.5)/100:6:2,' mV')
     end;   { Output }

  begin
     sgn1 := '+';   sgn2 := sgn1;   sgn3 := sgn1;
     lb := -6;   ub := 55;   eps := 1e-3;   max := 10;
     y1 := fnf(lb,temp);
     if (y1 < 0) then sgn1 := '-';
     y2 := fnf(ub,temp);
     if (y2 < 0) then sgn2 := '-';
     if (sgn1 = sgn2) then Search;
     y2 := fnf(ub,temp);
     if (y2 < 0) then sgn2 := '-' else sgn2 := '+';
     if (max > 1000) then
      writeln (' No root of the function has been found !')
     else
     begin
        while (abs(ub-lb) >= eps) do
        begin
           if not(sgn1 = sgn2) then
           begin
              middle := (lb+ub)/2;   ym := fnf(middle,temp);
              if (ym < 0) then sgn3 := '-' else sgn3 :='+';
              if (sgn3 = sgn1) then lb := middle else ub := middle;
              y1 := fnf(lb,temp);
              if (y1 < 0) then sgn1 := '-' else sgn1 := '+';
              y2 := fnf(ub,temp);
              if (y2 < 0) then sgn2 := '-' else sgn2 := '+'
           end
        end;
        Output
     end
  end;   { Calc }

begin
   Input;
   Calc
end.
```

PERSIST

```
program persist;
   { J. Kinkel, January 1989, based on Ebert/Ederer }

uses   crt;

var   rg,cl : real;
```

```pascal
procedure Input;
{----------------}
begin
   clrscr;   writeln;
   write (' Radius of gyration in Angstroem ? ');     readln (rg);
   write (' Contour length in Angstroem ? ');     readln (cl)
end;    { Input }

procedure Calc;
{--------------}
var   sgn1,sgn2,sgn3 : char;
      max : integer;
      ub,lb,eps,y1,y2,middle,ym : real;

   function fnf(x : real) : real;
   {----------------------------}
   var   arg : real;

   begin
      arg := cl/x;
      if (arg > 1419) then arg := 1419;
      fnf := sqr(x)*(cl/3/x-1+2*x/cl-2*(1-exp(-arg))*sqr(x/cl))-sqr(rg)
   end;    { fnf }

   procedure Search;
   {----------------}
   var   cnt: integer;
         help : real;

   begin
      while (max <= 1000) and (sgn1 = sgn3) do
      begin
         help := (ub-lb)/max;
         for cnt := 1 to max-1 do
         begin
            ym := fnf(lb+cnt*help);
            if (ym < 0) then sgn3 := '-' else sgn3 := '+';
            if not(sgn1 = sgn3) then ub := lb+cnt*help
         end;
         max := max*10
      end
   end;    { Search }

   procedure Output;
   {----------------}
   begin
      writeln;
      writeln (' Persistence length in Angstroem :');
      writeln ((ub+lb)/2:12:8,' +/- ',abs(ub-lb)/2:12:8)
   end;    { Output }

begin
   sgn1 := '+';   sgn2 := sgn1;   sgn3 := sgn1;
   lb := 1e-4;   ub := 1e5;   eps := 1e-4;   max := 10;
   y1 := fnf(lb);
```

```
      if (y1 < 0) then sgn1 := '-';
      y2 := fnf(ub);
      if (y2 < 0) then sgn2 := '-';
      sgn3 := sgn2;
      if (sgn1 = sgn2) then Search;
      y2 := fnf(ub);
      if (y2 < 0) then sgn2 := '-' else sgn2 := '+';
      if (max > 1000) then
       writeln (' No root of the function has been found !')
      else
      begin
         while (abs(ub-lb) >= eps) do
         begin
            if not(sgn1 = sgn2) then
            begin
               middle := (lb+ub)/2;   ym := fnf(middle);
               if (ym < 0) then sgn3 := '-' else sgn3 :='+';
               if (sgn3 = sgn1) then lb := middle else ub := middle;
               y1 := fnf(lb);
               if (y1 < 0) then sgn1 := '-' else sgn1 := '+';
               y2 := fnf(ub);
               if (y2 < 0) then sgn2 := '-' else sgn2 := '+'
            end
         end;
         Output
      end
  end;   { Calc }

begin
   Input;
   Calc
end.
```

```
┌─────────────────────────────────────────────────────────────────┐
│                            DISTILL                                │
└─────────────────────────────────────────────────────────────────┘
```

```
program distill;
   { J. Kinkel, January 1989, based on Ebert/Ederer }

uses   crt;

var   a1,a2,b1,b2,x1 : real;
      tp : integer;

 procedure Input;
{----------------}
 begin
    clrscr;   writeln;
    write (' a1 b1 ? ');   readln (a1,b1);
    write (' a2 b2 ? ');   readln (a2,b2);
    write (' Mole fraction of the lower boiling substance ? ');
    readln (x1);
    write (' How many theoretical plates ? ');   readln (tp)
  end;   { Input }
```

```
procedure Calc;
{---------------}
var    sgn1,sgn2,sgn3 : char;
       st,max : integer;
       y1,y2,middle,ym,lb,ub,xm : real;

    procedure Exchange;
    {------------------}
    var   h : real;

    begin
       h := ub;   ub := lb+1;   lb := h-1;
       h := a1;   a1 := a2;     a2 := h;
       h := b1;   b1 := b2;     b2 := h
    end;    { Exchange }

    function fnf(x : real) : real;
    {----------------------------}
    begin
       fnf := x1*exp(a1/x+b1)+(1-x1)*exp(a2/x+b2)-760
    end;    { fnf }

    procedure Search;
    {----------------}
    var   cnt: integer;
          help : real;

    begin
       while (max <= 1000) and (sgn1 = sgn3) do
       begin
          help := (ub-lb)/max;
          for cnt := 1 to max-1 do
          begin
             ym := fnf(lb+cnt*help);
             if (ym < 0) then sgn3 := '-' else sgn3 := '+';
             if not(sgn1 = sgn3) then ub := lb+cnt*help
          end;
          max := max*10
       end
    end;    { Search }

    procedure Output;
    {----------------}
    begin
       writeln (st:9,int(x1*1e4)*1e-4:12:5,int(xm*1e4)*1e-4:10:5,
                int((lb-273.15)*1e4)/1e4:10:2)
    end;    { Output }

begin
   writeln;
   writeln ('Plate':10,'x-liq':10,'x-vap':10,'Bp in C':12);
   st := 1;
   repeat
     sgn1 := '+';   sgn2 := sgn1;   sgn3 := sgn1;
     lb := a1/(ln(760)-b1);    ub := a2/(ln(760)-b2);
```

```
        if (ub <= lb) then Exchange;
        max := 10;
        y1 := fnf(lb);
        if (y1 < 0) then sgn1 := '-';
        y2 := fnf(ub);
        if (y2 < 0) then sgn2 := '-';
        sgn3 := sgn2;
        if (sgn1 = sgn2) then Search;
        y2 := fnf(ub);
        if (y2 < 0) then sgn2 := '-' else sgn2 := '+';
        while (abs(ub-lb) >= abs(lb*1e-7)) do
        begin
           if not(sgn1 = sgn2) then
           begin
              middle := (lb+ub)/2;   ym := fnf(middle);
              if (ym < 0) then sgn3 := '-' else sgn3 :='+';
              if (sgn3 = sgn1) then lb := middle else ub := middle;
              y1 := fnf(lb);
              if (y1 < 0) then sgn1 := '-' else sgn1 := '+';
              y2 := fnf(ub);
              if (y2 < 0) then sgn2 := '-' else sgn2 := '+'
           end
        end;
        xm := x1*exp(a1/lb+b1)/760;
        Output;
        x1 := xm;   st := st+1
     until (st > tp)
  end;   { Calc }

begin
   Input;
   Calc
end.
```

```
                              ACID
```

```
program acid;
   { J. Kinkel, January 1989, based on Ebert/Ederer }

uses   crt;

var   conc,ka,kw : real;

 procedure Input;
{----------------}
 begin
    clrscr;   writeln;
    write (' Concentration of the acid in mol/l ? ');    readln (conc)
    write (' Dissociation constant Ka ? ');              readln (ka);
    write (' Dissociation constant Kw of the water ? '); readln (kw)
  end;   { Input }
```

```pascal
procedure Calc;
{---------------}
var    sgn1,sgn2,sgn3 : char;
       max : integer;
       y1,y2,middle,ym,ub,lb,eps : real;

   function fnf(x : real) : real;
   {-----------------------------}
   begin
       fnf := x*(x-kw/x)/ka+x-kw/x-conc
   end;   { fnf }

   procedure Search;
   {-----------------}
   var   cnt: integer;
         help : real;

   begin
       while (max <= 1000) and (sgn1 = sgn3) do
       begin
          help := (ub-lb)/max;
          for cnt := 1 to max-1 do
          begin
             ym := fnf(lb+cnt*help);
             if (ym < 0) then sgn3 := '-' else sgn3 := '+';
             if not(sgn1 = sgn3) then ub := lb+cnt*help
          end;
          max := max*10
       end
   end;   { Search }

   procedure Output;
   {-----------------}
   begin
       writeln;   writeln (' pH = ',-ln((ub+lb)/2)/ln(10):12:8)
   end;   { Output }

begin
    sgn1 := '+';    sgn2 := sgn1;    sgn3 := sgn1;
    lb := 1e-10;   ub := 1;
    eps := (ub+lb)*1e-4;   max := 10;
    y1 := fnf(lb);
    if (y1 < 0) then sgn1 := '-';
    y2 := fnf(ub);
    if (y2 < 0) then sgn2 := '-';
    sgn3 := sgn2;
    if (sgn1 = sgn2) then Search;
    y2 := fnf(ub);
    if (y2 < 0) then sgn2 := '-' else sgn2 := '+';
    if (max >= 1000) then
      writeln (' No root of the function has been found')
    else
    begin
       while (abs(ub-lb) >= eps) do
```

```
      begin
        if not(sgn1 = sgn2) then
        begin
           middle := (lb+ub)/2;    ym := fnf(middle);
           if (ym < 0) then sgn3 := '-' else sgn3 :='+';
           if (sgn3 = sgn1) then lb := middle else ub := middle;
           y1 := fnf(lb);
           if (y1 < 0) then sgn1 := '-' else sgn1 := '+';
           y2 := fnf(ub);
           if (y2 < 0) then sgn2 := '-' else sgn2 := '+'
        end
      end;
      Output
    end
 end;   { Calc }

begin
   Input;
   Calc
end.
```

```
                              SORTNUM
```

```
program sortnum;
   { J. Kinkel, January 1989, based on Ebert/Ederer }

uses   crt;

var   x,y : array[1..100] of real;
      cnt,max : integer;

 procedure Input;
{----------------}
 begin
    clrscr;   writeln;
    write (' Number of values ? ');    readln (max);
    for cnt := 1 to max do
    begin
        write (cnt:6,'. ? ');    readln (x[cnt]);    y[cnt] := x[cnt]
    end
 end;   { Input }

 procedure Calc;
{----------------}
 var   help1 : real;
       help2,cntb : integer;

 begin
    for cnt := 1 to max-1 do
    begin
        help1 := y[cnt];   help2 := cnt;
        for cntb := cnt+1 to max do
```

```
          begin
             if (help1 <= y[cntb]) then
             begin
                help1 := y[cntb];   help2 := cntb
             end
          end;
          y[help2] := y[cnt];   y[cnt] := help1
       end
 end;   { Calc }

 procedure Output;
{-----------------}
 begin
    writeln;   writeln (' Original values');
    for cnt := 1 to max do write (x[cnt]:10:1);
    writeln;   writeln (' Sorted values');
    for cnt := 1 to max do write (y[cnt]:10:1)
 end;   { Output }

begin
   Input;
   Calc;
   Output
end.
```

                              SORTTEXT

```
program sorttext;
   { J. Kinkel, January 1989, based on Ebert/Ederer }

uses   crt;

var    x,y : array[1..100] of string[20];
       cnt,max : integer;

 procedure Input;
{----------------}
 begin
    clrscr;   writeln;
    write (' Number of strings ? ');   readln (max);
    for cnt := 1 to max do
    begin
       write (cnt:6,'. ? ');   readln (x[cnt]);   y[cnt] := x[cnt]
    end
 end;   { Input }

 procedure Calc;
{---------------}
 var   help1 : string[20];
       help2,cntb : integer;

 begin
    for cnt := 1 to max-1 do
```

```
     begin
        help1 := y[cnt];    help2 := cnt;
        for cntb := cnt+1 to max do
        begin
           if (help1 <= y[cntb]) then
           begin
              help1 := y[cntb];    help2 := cntb
           end
        end;
        y[help2] := y[cnt];    y[cnt] := help1
     end
 end;    { Calc }

procedure Output;
{-----------------}
begin
   writeln;   writeln (' Original strings');
   for cnt := 1 to max do write (x[cnt]:20);
   writeln;   writeln (' Sorted strings');
   for cnt := 1 to max do write (y[cnt]:20)
end;    { Output }

begin
   Input;
   Calc;
   Output
end.
```

```
                              KIN_1ORD
```

```
program kin_1ord;
   { J. Kinkel, January 1989, based on Ebert/Ederer }

uses   crt;

var   x : array[1..100] of real;
      cnt,t,n : integer;

procedure Output;
{-----------------}
begin
   write (t:6,n:6,exp(-0.01*t):12:8);   writeln (' Stop : <ENTER>':55)
end;    { Output }

procedure Calc;
{---------------}
var   h : integer;
      abb : char;

begin
   clrscr;   randomize;
   for cnt := 1 to 100 do x[cnt] := 1;
   t := 0;    n := 100;
```

```
      repeat
      for cnt := 1 to 10 do
      begin
          t := t+1;   h := random(100);
          if (x[h] <> 0) then
          begin
             x[h] := 0;    n := n-1
          end
      end;
      Output;
      abb := readkey
      until (abb = chr($0D))
   end;   { Calc }

begin
   Calc
end.
```

MATMULT

```
program matmult;
   { J. Kinkel, January 1989, based on Ebert/Ederer }

uses   crt;

var   a,b,c : array[1..15,1..15] of real;
      dim : integer;

 procedure Input;
{----------------}
 var   i,j : integer;

 begin
    fillchar (a[1,1],1350,chr($00));   fillchar (b[1,1],1350,chr($00));
    clrscr;    writeln;
    write (' Dimension of the matrices ? ');   readln (dim);
    writeln;
    for i := 1 to dim do
    begin
       for j := 1 to dim do
       begin
          write (i:6,'. row, ',j,'. column ? ');   readln (a[i,j],b[i,j])
       end
    end
 end;   { Input }

 procedure Calc;
{----------------}
 var   i,j,k : integer;

 begin
    fillchar(c[1,1],1350,chr($00));
    for i := 1 to dim do
```

```pascal
   begin
     for j := 1 to dim do
     begin
        for k := 1 to dim do c[i,j] := c[i,j]+a[i,k]*b[k,j]
     end
   end
end;   { Calc }

procedure Output;
{-----------------}
 var   i,j : integer;

 begin
    writeln;
    for i := 1 to dim do
    begin
      for j := 1 to dim do write (a[i,j]:20);   writeln
    end;   writeln;
    for i := 1 to dim do
    begin
      for j := 1 to dim do write (b[i,j]:20);   writeln
    end;   writeln;
    for i := 1 to dim do
    begin
      for j := 1 to dim do write (c[i,j]:20);   writeln
    end
 end;   { Output }

begin
   Input;
   Calc;
   Output
end.
```

```
┌─────────────────────────────────────────────────────────────┐
│                          COMPLEX                            │
└─────────────────────────────────────────────────────────────┘
```

```pascal
program complex;
   { J. Kinkel, January 1989, based on Ebert/Ederer }

uses   crt;

var   a,b,c : array[1..2] of real;
      op : char;

 procedure Input;
 {----------------}
 begin
    clrscr;   writeln;
    writeln (' Input of the complex number (a + b*i) as a b');   writeln;
    write (' 1st complex number ? ');          readln (a[1],a[2]);
    write (' 2nd complex number ? ');          readln (b[1],b[2]);
    write (' Which operation (+ - * /) ? ');   readln (op)
 end;   { Input }
```

```
   procedure Calc;
{---------------}

   procedure Addition;
   {-----------------}
   begin
      c[1] := a[1] + b[1];
      c[2] := a[2] + b[2]
   end;   { Addition }

   procedure Subtraction;
   {--------------------}
   begin
      c[1] := a[1] - b[1];
      c[2] := a[2] - b[2]
   end;   { Subtraction }

   procedure Multiplication;
   {-----------------------}
   begin
      c[1] := a[1]*b[1] - a[2]*b[2];
      c[2] := a[1]*b[2] + a[2]*b[1]
   end;   { Multiplication }

   procedure Division;
   {-----------------}
   var   h : real;

   begin
      h := sqr(b[1])+sqr(b[2]);
      c[1] := (a[1]*b[1] + a[2]*b[2])/h;
      c[2] := (a[2]*b[1] - a[1]*b[2])/h
   end;   { Division }

begin
   case op of
   '+': Addition;
   '-': Subtraction;
   '*': Multiplication;
   '/': Division
   end
end;   { Calc }

procedure Output;
{---------------}
begin
   writeln;
   writeln ('(',a[1]:10:6,'+',a[2]:10:6,'*i) ',op,' (',b[1]:10:6,'+',
            b[2]:10:6,'*i) =');
   writeln ('(',c[1]:10:6,'+',c[2]:10:6,'*i)')
end;   { Output }

begin
   Input;
   Calc;
   Output
end.
```

```
                              BRAGG

program bragg;
   { J. Kinkel, January 1989, based on Ebert/Ederer }

uses   crt;

type   dim = array[1..3] of real;

var    axis,angle : dim;
       lambda : real;

 procedure Input;
{----------------}
 begin
    clrscr;   writeln;
    writeln (' Each length in Angstroem !!!');   writeln;
    write (' Wavelength of the X-rays ? ');     readln (lambda);
    write (' Length of the 3 axes of the unit cell ? ');
    readln (axis[1],axis[2],axis[3]);
    write (' The three angles of the unit cell ? ');
    readln (angle[1],angle[2],angle[3])
 end;   { Input }

 procedure Calc;
{---------------}
 var    h,k,l,cnt1,cnt2 : integer;
        maxis,mc : dim;
        pi,dist : real;
        wait : char;

    procedure Parameter;
    {-------------------}
    var   s,c,ms : dim;
          cnt : byte;

    begin
       fillchar (s[1],18,chr($00));   fillchar (c[1],18,chr($00));
       fillchar (ms[1],18,chr($00));
       pi := 4*arctan(1);
       for cnt := 1 to 3 do
       begin
          s[cnt] := sin(angle[cnt]/180*pi);
          c[cnt] := cos(angle[cnt]/180*pi)
       end;
       mc[1] := (c[2]*c[3]-c[1])/(s[2]*s[3]);
       mc[2] := (c[1]*c[3]-c[2])/(s[1]*s[3]);
       mc[3] := (c[1]*c[2]-c[3])/(s[1]*s[2]);
       for cnt := 1 to 3 do ms[cnt] := sqrt(1-sqr(mc[cnt]));
       maxis[1] := 1/(axis[1]*s[2]*ms[3]);
       maxis[2] := 1/(axis[2]*s[3]*ms[1]);
       maxis[3] := 1/(axis[3]*s[1]*ms[2])
    end;   { Parameter }
```

```pascal
     function zw : real;
   {-------------------}
     var   hv : real;

     begin
        hv := sqr(h*maxis[1]) + sqr(k*maxis[2]) + sqr(l*maxis[3]);
        hv := hv + 2*k*l*maxis[2]*maxis[3]*mc[1]
                 + 2*l*h*maxis[3]*maxis[1]*mc[2];
        zw := hv + 2*h*k*maxis[1]*maxis[2]*mc[3]
     end;   { zw }

     function bw : real;
   {-------------------}
     var   hv : real;

     begin
        hv := lambda/2/dist;   hv := hv/sqrt(1-sqr(hv));
        bw := arctan(hv)*180/pi
     end;   { bw }

   begin
       fillchar (maxis[1],18,chr($00));   fillchar (mc[1],18,chr($00));
       Parameter;
       cnt1 := 1;   cnt2 := 0;
       repeat
        for l := 0 to cnt1 do
        begin
           for k := 0 to (cnt1-1) do
           begin
              h := cnt1 - l - k;   dist := sqrt(1/zw);
              if (cnt2/12 = round(cnt2/12)) then
              begin
                 writeln ('***');   wait := readkey;
                 writeln ('H':4,'K':4,'L':4,'incidence angle':25);
                 writeln
              end;
              if (lambda/2/dist <= 1) then writeln (h:4,k:4,l:4,bw:25:8)
              else writeln (h:4,k:4,l:4,'no incidence angle':25);
              cnt2 := cnt2 + 1
           end
        end;
        cnt1 := cnt1 + 1
       until ( cnt1 > 9)
   end;   { Calc }

begin
   Input;
   Calc
end.
```

```
┌─────────────────────────────────────────────────────────────┐
│                              GC                               │
└─────────────────────────────────────────────────────────────┘
```

```pascal
program gc;
    { J. Kinkel, January 1989, based on Ebert/Ederer }

uses    crt;

var   ntp,rt,hw : integer;
      pc : real;

 procedure Input;
{----------------}
 begin
    clrscr;   writeln;
    write (' Number of theoretical plates ? ');    readln (ntp);
    write (' Partition coefficient ? ');           readln (pc)
 end;    { Input }

 procedure Calc;
{---------------}
 var   gas,liq : array[1..250] of real;
       p1,p2,max : real;
       th : integer;

    procedure Equilibrium;
    {---------------------}
    var   sum : real;
          cnt : integer;

    begin
       for cnt := 1 to ntp do
       begin
          sum := gas[cnt] + liq[cnt];
          gas[cnt] := sum*p1;   liq[cnt] := sum*p2
       end
    end;    { Equilibrium }

    procedure Gasflow;
    {-----------------}
    var   cnt : integer;

    begin
       cnt := ntp;
       repeat
        gas[cnt] := gas[cnt-1];   cnt := cnt - 1
       until (cnt < 2);
       gas[1] := 0
    end;    { Gasflow }

begin
    fillchar (gas[1],1500,chr($00));   fillchar (liq[1],1500,chr($00));
    p1 := pc/(1+pc);   p2 := 1/(1+pc);   rt := 0;
    gas[1] := 1;   max := 0;   hw := 0;   th := 1;
    repeat
```

```
         if (rt <= 0) then
         begin
            if (gas[ntp] > gas[ntp-1]) then
            begin
               rt := th+1;   max := gas[ntp]
            end
         end
         else
          if (gas[ntp] <= max/2) then hw := th - rt;
         Equilibrium;
         Gasflow;
         th := th + 1
      until (hw > 0);
      rt := rt - ntp
   end;   { Calc }

 procedure Output;
{-----------------}
 begin
      writeln;
      writeln ('              Retention time = ',rt);
      writeln (' Width at half peak height = ',hw)
   end;   { Output }

begin
   Input;
   Calc;
   Output
end.
```

REGLIN

```
program reglin;
   { J. Kinkel, January 1989, based on Ebert/Ederer }

uses    crt;

var    x,y : array[1..100] of real;
       sl,ic,ss : real;
       ndp : byte;

 procedure Input;
{----------------}
 var   cnt : byte;

 begin
      clrscr;   writeln;
      write (' How many data points ? ');   readln (ndp);
      for cnt := 1 to ndp do
      begin
         write ('x(',cnt,')  y(',cnt,') ? ');   readln (x[cnt],y[cnt])
      end
   end;   { Input }
```

```
procedure Calc;
{---------------}
var   sum1,sum2,sum3,sum4,sum5,dif1,dif2,den : real;
      cnt : byte;

   function ssd : real;
   {--------------------}
   var  zws : real;
        cnt : byte;

   begin
     zws := 0;
     for cnt := 1 to ndp do zws := zws + sqr(y[cnt] - ic - sl*x[cnt]);
     ssd := zws
   end;   { ssd }

begin
   sum1 := ndp;   sum2 := 0;   sum3 := 0;   sum4 := 0;   sum5 := 0;
   for cnt := 1 to ndp do
   begin
     sum2 := sum2 + x[cnt];   sum3 := sum3 + y[cnt];
     sum4 := sum4 + sqr(x[cnt]);   sum5 := sum5 + x[cnt]*y[cnt]
   end;
   den := sum1*sum4 - sqr(sum2);
   dif1 := sum3*sum4 - sum5*sum2;   dif2 := sum1*sum5 - sum2*sum3;
   ic := dif1/den;   sl := dif2/den;   ss := ssd
end;   { Calc }

procedure Output;
{----------------}
begin
   writeln;   writeln (' The regression line for the data is');
   writeln ('      y = ',ic:12:8,' + ',sl:12:8,' * x');   writeln;
   writeln (' Sum of squared deviations = ',ss:12:8)
end;   { Output }

begin
   Input;
   Calc;
   Output
end.
```

REGLINW

```
program reglinw;
   { J. Kinkel, January 1989, based on Ebert/Ederer }

uses   crt;

var   x,y,w : array[1..100] of real;
      sl,ic,ss,dsl,dic,dev : real;
      ndp : byte;
```

```
procedure Input;
{----------------}
var    cnt : byte;
       answer : char;

begin
   clrscr;   writeln;
   write (' How many data points ? ');   readln (ndp);
   for cnt := 1 to ndp do w[cnt] := 1;
   write (' Do you want to weight your data points (y/n) ? ');
   readln (answer);
   for cnt := 1 to ndp do
   begin
      write (' x(',cnt,') = ');   readln (x[cnt]);
      write (' y(',cnt,') = ');   readln (y[cnt]);
      if (answer in ['j','J','y','Y']) then
      begin
         write (' w(',cnt,') = ');   readln (w[cnt])
      end
      else w[cnt] := 1
   end
end;   { Input }

procedure Calc;
{---------------}
var    sum1,sum2,sum3,sum4,sum5,dif1,dif2,den : real;
       cnt : byte;

   function ssd : real;
   {--------------------}
   var   zws : real;
         cnt : byte;

   begin
      zws := 0;
      for cnt := 1 to ndp do
       zws := zws + w[cnt]*sqr(y[cnt] - ic - sl*x[cnt]);
      ssd := zws
   end;   { ssd }

begin
   sum1 := 0;   sum2 := 0;   sum3 := 0;   sum4 := 0;   sum5 := 0;
   for cnt := 1 to ndp do
   begin
      sum1 := sum1 + w[cnt];   sum2 := sum2 + x[cnt]*w[cnt];
      sum3 := sum3 + y[cnt]*w[cnt];   sum4 := sum4 + sqr(x[cnt])*w[cnt];
      sum5 := sum5 + x[cnt]*y[cnt]*w[cnt]
   end;
   den := sum1*sum4 - sqr(sum2);
   dif1 := sum3*sum4 - sum5*sum2;   dif2 := sum1*sum5 - sum2*sum3;
   ic := dif1/den;   sl := dif2/den;   ss := ssd;
   dev := sqrt(ss/(ndp-2));   dic := dev*sqrt(sum4/den);
   dsl := dev*sqrt(sum1/den)
end;   { Calc }
```

```
 procedure Output;
{-----------------}
 var    answer : char;

    procedure Compare;
    {-----------------}
    var    cnt : byte;

    begin
       writeln;
       writeln('x':7,'y(input)':21,'y(regression)':20);
       for cnt := 1 to ndp do
          writeln (x[cnt]:12:5,y[cnt]:14:5,ic+sl*x[cnt]:20:5)
    end;   { Compare }

    procedure Search;
    {----------------}
    var    xvalue : real;

    begin
       repeat
       write (' x-value (stop if x > 99) ? ');   readln (xvalue);
       writeln (' y(',xvalue:5:3,') = ',ic+sl*xvalue:12:8)
       until (xvalue > 99)
    end;   { Search }

begin
    writeln;
    writeln (' The regression line for the data is');
    writeln ('      y = ',ic:12:8,' + ',sl:12:8,' * x');   writeln;
    writeln (' Sum of squared deviations = ',ss:12:8);   writeln;
    writeln (' Standard deviation = ',dev:12:8);   writeln;
    writeln (' Number of data points = ',ndp);   writeln;
    writeln ('      slope = ',sl:12:8,' +/- ',dsl:12:8);
    writeln (' intercept = ',ic:12:8,' +/- ',dic:12:8);   writeln;
    writeln (' Do you want to compare the input values');
    write (' with the regression values (y/n) ? ');   readln (answer);
    if (answer in ['j','J','y','Y']) then Compare;   writeln;
    writeln (' Do you want to interpolate data');
    write (' using the regression line (y/n) ? ');   readln (answer);
    if (answer in ['j','J','y','Y']) then Search
end;   { Output }

begin
    Input;
    Calc;
    Output
end.
```

```
                              ISOMASS
```

```
program isomass;
   { J. Kinkel, January 1989, based on Ebert/Ederer }

uses   crt;

type   iso = record
                 sym : string[2];
                 lm,noi : integer;
                 freq : array[1..3] of real
             end;

const  numel = 6;
       ele : array[1..6] of iso =
                 ((sym:'C';lm:12;noi:2;freq:(0.989,0.011,0)),
                  (sym:'H';lm:1;noi:2;freq:(0.99985,0.00015,0)),
                  (sym:'N';lm:14;noi:2;freq:(0.9963,0.0037,0)),
                  (sym:'O';lm:16;noi:3;freq:(0.99762,0.00038,0.002)),
                  (sym:'Cl';lm:35;noi:3;freq:(0.7577,0,0.2423)),
                  (sym:'Br';lm:79;noi:3;freq:(0.5069,0,0.4931)));

var    fr : array[1..250] of real;
       cnt,numat,nmax,mmin,mmax : integer;

procedure Input;
{----------------}
begin
   write (' How many * ',ele[cnt].sym,' * atoms ? ');   readln (numat)
end;   { Input }

procedure Calcfm;
{-----------------}
var    fraux : array[1..250] of real;
       c1,c2,c3,hmax : integer;

begin
   with ele[cnt] do
   begin
      for c1 := 1 to numat do
      begin
         for c2 := 1 to nmax do fraux[c2] := fr[c2];
         hmax := nmax;
         mmin := mmin+lm;    mmax := mmax+lm+noi-1;    nmax := nmax+noi-1;
         for c2 := 1 to nmax do fr[c2] := 0;
         for c2 := 1 to hmax do
         begin
            for c3 := 1 to noi do
              fr[c2+c3-1] := fr[c2+c3-1]+fraux[c2]*freq[c3]
         end
      end
   end
end;   { Calcfm }
```

```
procedure Output;
{-----------------}
var    avm : real;
       cout : integer;

begin
   avm := 0;   writeln;
   for cout := 1 to nmax do
   begin
      avm := avm+fr[cout]*(mmin+cout-1);
      if (fr[cout] >= 1e-10) then
        writeln (' Frequency of mass ',mmin+cout-1,' = ',fr[cout]:15:14)
   end;   writeln;
   writeln (' Average molecular mass = ',avm:12:5)
end;    { Output }

begin
   clrscr;   writeln;
   nmax := 1;   mmin := 0;   mmax := 0;   fr[1] := 1;
   for cnt := 1 to numel do
   begin
      Input;
      if (numat > 0) then Calcfm
   end;
   Output
end.
```

```
                              GJ
```

```
program gj;
   { J. Kinkel, January 1989, based on Ebert/Ederer }

uses   crt;

var    mat : array[1..20,1..21] of real;
       dim : integer;
       flag : boolean;

procedure Input;
{----------------}
var    row,col : integer;

begin
   clrscr;    writeln;
   write (' Number of equations ? ');   readln (dim);
   writeln (' Input of the extended coefficient matrix :');
   for row := 1 to dim do
   begin
      for col := 1 to dim do
      begin
         write (' a(',row,',',col,') = ');   readln (mat[row,col])
      end;
      write (' b(',row,') = ');   readln (mat[row,dim+1])
   end
end;    { Input }
```

```
procedure Calc;
{---------------}
var   row,col,i,hlp : integer;

   procedure Exchange;
   {------------------}
   var   cnt : integer;
         zw : real;

   begin
      for cnt := 1 to dim+1 do
      begin
         zw := mat[row,cnt];
         mat[row,cnt] := mat[hlp,cnt];
         mat[hlp,cnt] := zw
      end
   end;   { Exchange }

   procedure Normalize;
   {--------------------}
   var   cnt : integer;
         fac : real;

   begin
      fac := 1/mat[row,row];
      for cnt := 1 to dim+1 do mat[row,cnt] := fac*mat[row,cnt]
   end;   { Normalize }

   procedure Subtract;
   {-------------------}
   var   cnt : integer;
         fac : real;

   begin
      fac := -mat[col,row];
      for cnt := 1 to dim+1 do
       mat[col,cnt] := mat[col,cnt] + fac*mat[row,cnt]
   end;   { Subtract }

begin
   flag := true;   i := 1;
   for row := 1 to dim do
   begin
      hlp := row;
      repeat
         if not(mat[hlp,row] = 0) then
         begin
            Exchange;
            Normalize;
            for col := 1 to dim do
             if not(col = row) then Subtract;
             hlp := dim
         end
         else i := hlp;
         hlp := hlp+1
      until (hlp >= dim);
      if (i = dim) then
```

```
        begin
           flag := false;
           exit
         end
      end
 end;   { Calc }

 procedure Output;
 {------------------}
 var   row : integer;

 begin
    writeln;
    if flag then
      for row := 1 to dim do writeln (' x(',row,') = ',mat[row,dim+1]:12:8)
    else writeln (' No solution is possible !')
 end;   { Output }

begin
   Input;
   Calc;
   Output
end.
```

```
┌─────────────────────────────────────────────────────────────┐
│                          REGLING                            │
└─────────────────────────────────────────────────────────────┘
```

```
program regling;
   { J. Kinkel, January 1989, based on Ebert/Ederer }

uses   crt;

var    x,y : array[1..100] of real;
       coeff : array[1..20] of real;
       ssd : real;
       flag : boolean;
       fct,ndp: integer;

 procedure Input;
 {----------------}
 var   cnt : integer;

 begin
    fillchar (x[1],600,chr($00));   fillchar (y[1],600,chr($00));
    clrscr;   writeln;
    write (' How many functions ? ');     readln (fct);
    write (' How many data points ? ');   readln (ndp);
    for cnt := 1 to ndp do
    begin
       write (' x(',cnt,') ');   readln (x[cnt]);
       write (' y(',cnt,') ');   readln (y[cnt])
    end
 end;   { Input }
```

```pascal
procedure Calc;
{----------------}
var   mat : array[1..20,1..21] of real;
      fnf : array[1..20] of real;
      cnt1,cnt2,cnt3 : integer;

   procedure allfcts(x1 : real);
   {--------------------------}
   begin
      fnf[1] := sqr(x1);   fnf[2] := exp(x1);   fnf[3] := 1/x1
   end;   { allfcts}

   procedure Gauss_Jordan;
   {---------------------}
   var   dim,row,col,i,hlp : integer;

      procedure Exchange;
      {-----------------}
      var   cnt : integer;
            zw : real;

      begin
         for cnt := 1 to dim+1 do
         begin
            zw := mat[row,cnt];
            mat[row,cnt] := mat[hlp,cnt];
            mat[hlp,cnt] := zw
         end
      end;   { Exchange }

      procedure Normalize;
      {------------------}
      var   cnt : integer;
            fac : real;

      begin
         fac := 1/mat[row,row];
         for cnt := 1 to dim+1 do mat[row,cnt] := fac*mat[row,cnt]
      end;   { Normalize }

      procedure Subtract;
      {-----------------}
      var   cnt : integer;
            fac : real;

      begin
         fac := -mat[col,row];
         for cnt := 1 to dim+1 do
           mat[col,cnt] := mat[col,cnt] + fac*mat[row,cnt]
      end;   { Subtract }

   begin
      flag := true;   dim := fct;   i := 1;
      for row := 1 to dim do
      begin
         hlp := row;
         repeat
```

```
                   if not(mat[hlp,row] = 0) then
                   begin
                       Exchange;
                       Normalize;
                       for col := 1 to dim do
                        if not(col = row) then Subtract;
                        hlp := dim
                   end
                   else i := hlp;
                   hlp := hlp+1
               until (hlp >= dim);
               if (i = dim) then
               begin
                   flag := false;
                   exit
               end
           end
       end
   end;   { Gauss_Jordan }

   function SquareSum : real;
   {--------------------------}
   var   cnt1,cntf : integer;
         zws1,zws2 : real;

   begin
       SquareSum := 0;   zws2 := 0;
       for cnt1 := 1 to ndp do
       begin
           zws1 := 0;
           allfcts(x[cnt1]);
           for cntf := 1 to fct do zws1 := zws1 + coeff[cntf]*fnf[cntf];
           zws2 := zws2 + sqr(y[cnt1] - zws1)
       end;
       SquareSum := zws2
   end;   { SquareSum }

begin
   fillchar (mat[1,1],2520,chr($00));
   for cnt1 := 1 to ndp do
   begin
       allfcts(x[cnt1]);
       for cnt2 := 1 to fct do
       begin
           for cnt3 := 1 to cnt2 do
           begin
               mat[cnt2,cnt3] := mat[cnt2,cnt3] + fnf[cnt2]*fnf[cnt3];
               mat[cnt3,cnt2] := mat[cnt2,cnt3]
           end;
           mat[cnt2,fct+1] := mat[cnt2,fct+1] + y[cnt1]*fnf[cnt2]
       end
   end;
   Gauss_Jordan;
   for cnt2 := 1 to fct do coeff[cnt2] := mat[cnt2,fct+1];
   ssd := SquareSum
end;   { Calc }
```

```pascal
 procedure Output;
{------------------}
 var   row : integer;

 begin
    writeln;
    if flag then
    begin
      writeln (' Coefficients of regression :');
      for row := 1 to fct do writeln (' k(',row,') = ',coeff[row]:12:8);
      writeln;   writeln (' Sum of squared deviations = ',ssd)
    end
    else
     writeln (' No solution is possible !')
 end;   { Output }

begin
   Input;
   Calc;
   Output
end.
```

```
+-------------------------------------------------------------+
|                          INV_GJ                             |
+-------------------------------------------------------------+
```

```pascal
program inv_gj;
   { J. Kinkel, January 1989, based on Ebert/Ederer }

uses   crt;

const   dim = 4;
        mat : array[1..4,1..4] of real = ((7,5,3,1),
                                          (0,1,1,1),
                                          (4,2,1,3),
                                          (8,1,9,5));

var   omat,imat,cmat : array[1..4,1..4] of real;
      flag : boolean;

 procedure Calc;
{---------------}
 var   row,col,i,hlp : integer;

    procedure Exchange;
    {------------------}
    var   cnt : integer;
          zw : real;

    begin
       for cnt := 1 to dim do
       begin
         zw := omat[row,cnt];
         omat[row,cnt] := omat[hlp,cnt];   omat[hlp,cnt] := zw;
         zw := imat[row,cnt];
         imat[row,cnt] := imat[hlp,cnt];   imat[hlp,cnt] := zw
       end
    end;   { Exchange }
```

```
 procedure Normalize;
 {--------------------}
 var   cnt : integer;
       fac : real;

 begin
    fac := 1/omat[row,row];
    for cnt := 1 to dim do
    begin
       omat[row,cnt] := fac*omat[row,cnt];
       imat[row,cnt] := fac*imat[row,cnt]
    end
 end;   { Normalize }

 procedure Subtract;
 {------------------}
 var   cnt : integer;
       fac : real;

 begin
    fac := -omat[col,row];
    for cnt := 1 to dim do
    begin
       omat[col,cnt] := omat[col,cnt] + fac*omat[row,cnt];
       imat[col,cnt] := imat[col,cnt] + fac*imat[row,cnt]
    end
 end;    { Subtract }

 procedure Control;
 {-----------------}
 var   rcnt,ccnt,hcnt : integer;

 begin
    for rcnt := 1 to dim do
    begin
       for ccnt := 1 to dim do
       begin
          cmat[rcnt,ccnt] := 0;
          for hcnt := 1 to dim do
           cmat[rcnt,ccnt] := cmat[rcnt,ccnt]+mat[rcnt,hcnt]
                               *imat[hcnt,ccnt]
       end
    end
 end;    { Control }

begin
   for row := 1 to dim do
   begin
      for col := 1 to dim do
      begin
         omat[row,col] := mat[row,col];
         if (col = row) then imat[row,col] := 1 else imat[row,col] := 0
      end
   end;
```

```
      flag := true;    i := 1;
      for row := 1 to dim do
      begin
         hlp := row;
         repeat
            if not(omat[hlp,row] = 0) then
            begin
               Exchange;
               Normalize;
               for col := 1 to dim do
                if not(col = row) then Subtract;
                hlp := dim
            end
            else i := hlp;
            hlp := hlp+1
         until (hlp >= dim);
         if (i = dim) then
         begin
            flag := false;
            exit
         end
      end;
      Control
   end;    { Calc }

 procedure Output;
 {-----------------}
 var   row,col : integer;

 begin
    clrscr;   writeln;   writeln (' Matrix :');
    for row := 1 to dim do
    begin
       for col := 1 to dim do write (mat[row,col]:15:10);    writeln
    end;   writeln;        writeln (' Inverted matrix :');
    for row := 1 to dim do
    begin
       for col := 1 to dim do
       write (imat[row,col]:15:10);    writeln
    end;   writeln;        writeln (' Control :');
    for row := 1 to dim do
    begin
       for col := 1 to dim do
       write (cmat[row,col]:15:10);    writeln
    end
 end;    { Output }

 begin
    Calc;
    Output
 end.
```

```
                                    INVERS
```

```
program invers;
    { J. Kinkel, January 1989, based on Ebert/Ederer }

uses    crt;

const    dim = 4;
         omat : array[1..4,1..4] of real = ((7,5,3,1),
                                            (0,1,1,1),
                                            (4,2,1,3),
                                            (8,1,9,5));

var     mata,matb,matu,matv,matw : array[1..4,1..4] of real;

 procedure Fill;
{---------------}
 var    row,col : integer;

 begin
    for row := 1 to dim do
    begin
        for col := 1 to dim do
        begin
            mata[row,col] := omat[row,col];
            if (col = row) then
            begin
               matb[row,col] := 1;    matv[row,col] := 1
            end
            else
            begin
               matb[row,col] := 0;    matv[row,col] := 0
            end
        end
    end
end;    { Fill }

 procedure Pivot;
{----------------}
 var   pcnt,cnt,t : integer;
         s : real;

    procedure Exchange;
   {------------------}
    var    tcnt : integer;

    begin
        for tcnt := 1 to dim do
        begin
            s:= mata[pcnt,tcnt];
            mata[pcnt,tcnt] := mata[t,tcnt];
            mata[t,tcnt] := s
        end;
        if (abs(mata[pcnt,pcnt]) <= 1e-30) then
        begin
            writeln (' No inversion is possible !');    Halt
        end
```

```
        else
        begin
           matv[pcnt,pcnt] := 0;    matv[t,t] := 0;
           matv[pcnt,t] := 1;    matv[t,pcnt] := 1
        end
   end;    { Exchange }

   procedure GJ_Elimination;
   {-------------------------}
   var    cnt1,cnt2 : integer;

   begin
      for cnt1 := 1 to dim do
      begin
         for cnt2 := 1 to dim do
         begin
            if (cnt1 = pcnt) then
             if (cnt2 = cnt1) then matu[pcnt,pcnt] := 1/mata[pcnt,pcnt]
             else matu[cnt1,cnt2] := -mata[cnt1,cnt2]/mata[pcnt,pcnt]
            else
            begin
               if (cnt2 = pcnt) then
                matu[cnt1,pcnt] := mata[cnt1,pcnt]/mata[pcnt,pcnt]
               else matu[cnt1,cnt2] := mata[cnt1,cnt2]-mata[pcnt,cnt2]
                                  *mata[cnt1,pcnt]/mata[pcnt,pcnt]
            end
         end
      end
   end;    { GJ_Elimination }

   procedure Multiplication;
   {-------------------------}
   var    cnt1,cnt2,cnt3 : integer;

   begin
      for cnt1 := 1 to dim do
      begin
         for cnt2 := 1 to dim do
         begin
            matw[cnt1,cnt2] := 0;
            for cnt3 := 1 to dim do
             matw[cnt1,cnt2] := matw[cnt1,cnt2]+matw[cnt1,cnt3]
                                  *matb[cnt3,cnt2]
         end
      end;
      matb := matw;
      for cnt1 := 1 to dim do
      begin
         for cnt2 := 1 to dim do
         begin
            mata[cnt1,cnt2] := matu[cnt1,cnt2];
            if (cnt1 = cnt2) then matv[cnt1,cnt2] := 1
            else matv[cnt1,cnt2] := 0
         end
      end
   end; { Multiplication }
```

```
begin
   for pcnt := 1 to dim do
   begin
      s := 0;
      for cnt := pcnt to dim do
       if (s <= abs(mata[cnt,pcnt])) then
       begin
          s := abs(mata[cnt,pcnt]);   t := cnt
       end;
      Exchange;
      GJ_Elimination;
      Multiplication
   end
end;   { Pivot }

procedure Result;
{-----------------}
 var    cnt1,cnt2,cnt3 : integer;

begin
   for cnt1 := 1 to dim do
   begin
      for cnt2 := 1 to dim do
      begin
         matw[cnt1,cnt2] := 0;
         for cnt3 := 1 to dim do
         matw[cnt1,cnt2] := matw[cnt1,cnt2]+mata[cnt1,cnt3]
                             *matb[cnt3,cnt2]
      end
   end
end;   { Result }

procedure Control;
{------------------}
 var    cnt1,cnt2,cnt3 : integer;

begin
   for cnt1 := 1 to dim do
   begin
      for cnt2 := 1 to dim do
      begin
         matb[cnt1,cnt2] := 0;
         for cnt3 := 1 to dim do
         matb[cnt1,cnt2] := matb[cnt1,cnt2]+omat[cnt1,cnt3]
                             *matw[cnt3,cnt2]
      end
   end
end;   { Control }

procedure Output;
{-----------------}
 var    row,col : integer;

begin
   clrscr;   writeln;   writeln (' Matrix :');
```

```
      for row := 1 to dim do
      begin
         for col := 1 to dim do write (omat[row,col]:15:10);    writeln
      end;  writeln;       writeln (' Inverted matrix :');
      for row := 1 to dim do
      begin
         for col := 1 to dim do write (matw[row,col]:15:10);    writeln
      end;  writeln;       writeln (' Product:');
      for row := 1 to dim do
      begin
         for col := 1 to dim do write (matb[row,col]:15:10);    writeln
      end
end;    { Output }

begin
   Fill;
   Pivot;
   Result;
   Control;
   Output
end.
```

```
                                EIGEN
```

```
program eigen;
   { J. Kinkel, January 1989, based on Ebert/Ederer }

uses   crt;

const   dim = 6;
        omat : array[1..dim,1..dim] of real = ((0,1,0,0,1,0),
                                                (1,0,1,0,0,0),
                                                (0,1,0,1,0,0),
                                                (0,0,1,0,1,0),
                                                (1,0,0,1,0,1),
                                                (0,0,0,0,1,0));

        eps = 1e-8;

var    mat : array[1..dim,1..dim] of real;
       s : real;
       z,t1,t2 : integer;

 procedure Fill;
 {---------------}
 var   row,col : integer;

 begin
    for row := 1 to dim do
    begin
       for col := 1 to dim do mat[row,col] := omat[row,col];
    end
 end;    { Fill }
```

```
procedure Pivot;
{----------------}
var    cnt1,cnt2 : integer;

begin
   s := 0;
   for cnt1 := 1 to dim-1 do
   begin
      for cnt2 := cnt1+1 to dim do
      if (s <= abs(mat[cnt1,cnt2])) then
      begin
         s := abs(mat[cnt1,cnt2]);    t1 := cnt1;    t2 := cnt2
      end
   end
end;    { Pivot }

procedure Matrix;
{----------------}
var    ang,h1,h2,h3,h4 : real;
       cnt : integer;

   function Angle : real;
   {---------------------}
   var    sgn : integer;
          p : real;

   begin
      if not(mat[t1,t1] = mat[t2,t2]) then
      begin
         p := 2*mat[t1,t2]/(mat[t2,t2]-mat[t1,t1]);
         Angle := 0.5*arctan(p)
      end
      else
      begin
         if (mat[t1,t2] >= 0) then sgn := 1 else sgn := -1;
         Angle := sgn*arctan(1)
      end
   end;    { Angle }

begin
   ang := Angle;
   for cnt := 1 to dim do
   begin
      if not((cnt = t1) or (cnt = t2)) then
      begin
         h1 := mat[cnt,t1];
         mat[cnt,t1] := mat[cnt,t1]*cos(ang)-mat[cnt,t2]*sin(ang);
         mat[t1,cnt] := mat[cnt,t1];
         mat[cnt,t2] := h1*sin(ang)+mat[cnt,t2]*cos(ang);
         mat[t2,cnt] := mat[cnt,t2]
      end
   end;
   h2 := mat[t1,t1];    h3 := mat[t2,t2];    h4 := mat[t1,t2];
   mat[t1,t2] := 0;    mat[t2,t1] := 0;
   mat[t1,t1] := h2*sqr(cos(ang))-h4*sin(2*ang)+h3*sqr(sin(ang));
   mat[t2,t2] := h2*sqr(sin(ang))+h4*sin(2*ang)+h3*sqr(cos(ang))
end;    { Matrix }
```

```
   procedure Output;
{-----------------}
 var   cnt : integer;

 begin
    clrscr;   writeln;
    for cnt := 1 to dim do writeln (' Eigenvalue = ',mat[cnt,cnt]:12:8)
 end;   { Output }

begin
   Fill;
   for z := 1 to 400 do
   begin
      Pivot;
      if (s >= eps) then Matrix
   end;
   Output
end.
```

┌───┐
│   ISOMERS   │
└───┘

```
program isomers;
   { J. Kinkel, January 1989, based on Ebert/Ederer }

   { if you want to calculate the number of isomers of molecules
     containing more than 11 C-atoms, you have to change the data
     structure and the OutPut-routine, because we used only integers
     for the number of isomers }

uses   crt;

var    alkan,alkyl,alkenon,aldehyd,
       keton,alkct,aldct,ketct,
       x,y,z,fnf,p,q,e,a2,a3,a4,ak,
       b2,b3,b4,c2 : array[1..50] of integer;
       nc,max : integer;

 procedure Input;
{----------------}
 begin
    clrscr;   writeln;
    write (' Maximum number of C-atoms (max. 11) ? ');   readln (nc)
 end;   { Input }

 procedure Calc;
{---------------}

    procedure allfcts;
   {------------------}
    var   cnt1,cnt2 :integer;

    begin
       for cnt1 := 1 to max do
```

```
       begin
          fnf[cnt1] := 0;
          for cnt2 := 1 to cnt1 do
          begin
             fnf[cnt1] := fnf[cnt1]+x[cnt2]*y[cnt1-cnt2+1]
          end
       end
  end;   { allfcts }

  procedure Alkyles;
  {------------------}
  var   i,j,k,l,sum : integer;

  begin
     for i := 1 to 3 do alkyl[i] := 1;
     alkyl[4] := 2;
     for i := 4 to nc do
     begin
        sum := 0;    j := i;
        repeat
         for k := 1 to i-j+1 do
         begin
            l := i-j-k+1;    sum := sum+alkyl[j]*alkyl[k]*alkyl[l+1]
         end;
         j := j-1
        until (j < 1);
        j := 0;
        repeat
         k := i-j;    sum := sum+3*alkyl[round(j/2)+1]*alkyl[k];
         j := j+2
        until (j > i-1);
        if (round((i-1)/3) = (i-1)/3) then
         sum := sum+2*alkyl[round((i-1)/3)];
        sum := round(sum/6);   alkyl[i+1] := sum
     end
  end;   { Alkyles }

  procedure Atwo;
  {---------------}
  begin
     max := nc+1;   x := alkyl;   y := alkyl;
     allfcts;   a2 := fnf
  end;    { Atwo }

  procedure Athree;
  {-----------------}
  begin
     max := nc+1;   x := a2;   y := alkyl;
     allfcts;   a3 := fnf
  end;    { Athree }

  procedure Afour;
  {----------------}
  begin
     max := nc+1;   x := a2;   y := a2;
     allfcts;   a4 := fnf
  end;    { Afour }
```

```
procedure Polynom;
{------------------}
 var   cnt : integer;

 begin
    fillchar(ak[1],100,chr($00));
    for cnt := 0 to round((nc)/2) do ak[2*cnt+1] := alkyl[cnt+1];
    for cnt := 1 to nc+1 do ak[cnt] := ak[cnt]+a2[cnt];
    cnt := nc+2;
    repeat
     ak[cnt] := round(ak[cnt-1]/2);
     cnt := cnt-1
    until (cnt < 2);
    ak[1] := 0
 end;   { Polynom }

procedure Sum;
{--------------}
 var   i,j,k,l,m : integer;

 begin
    fillchar(z[1],100,chr($00));
    z[1] := 1;   p := ak;
    for i:= 1 to nc+1 do z[i] := z[i]+ak[i];
    for m := 3 to nc+1 do
    begin
       fillchar(e[1],100,chr($00));
       for j := 1 to nc+1 do
       begin
          for k := 1 to j do
          begin
             l := j-k+1;   e[j] := e[j]+ak[k]*p[l]
          end
       end;
       p := e;
       for i := 1 to nc+1 do z[i] := z[i]+p[i]
    end
 end;    { Sum }

procedure Alkenonesct;
{----------------------}
 var   i,j,k,l : integer;

 begin
    fillchar(alkct[1],100,chr($00));
    for j := 1 to nc+1 do
    begin
       for k := 1 to j do
       begin
          l := j-k+1;   alkct[j+3] := alkct[j+3]+z[k]*a4[l]
       end
    end
 end;   { Alkenonesct }
```

```
 procedure Aldehydesct;
{---------------------}
 var    cnt : integer;

 begin
    max := nc+4;    x := a3;    y := z;    allfcts;
    for cnt := 1 to nc+1 do aldct[cnt+3] := fnf[cnt]
 end;    { Aldehydesct }

 procedure Ketonesct;
{-------------------}
 begin
    max := nc+4;    x := aldcz;    y := alkyl;
    y[1] := 0;    allfcts;    ketct := fnf
 end;    { Ketonesct }

 procedure Alkenones;
{-------------------}
 var    cnt : integer;

 begin
    for cnt := 1 to nc+1 do x[cnt] := ak[cnt+1];
    y := a2;    allfcts;
    x := fnf;    y := z;    allfcts;
    for cnt := 1 to nc+1 do alkenon[cnt+3] := fnf[cnt]
 end;    { Alkenones }

 procedure Aldehydes;
{-------------------}
 var    cnt : integer;

 begin
    max := nc+4;
    for cnt := 1 to max do x[cnt] := ak[cnt+1];
    y := alkyl;    allfcts;
    x := fnf;    y := z;    allfcts;
    for cnt := 1 to max do aldehyd[cnt+3] := fnf[cnt]
 end;    { Aldehydes }

 procedure Ketones;
{-----------------}
 begin
    max := nc+4;    x := aldehyd;    y := alkyl;
    y[1] := 0;    allfcts;    keton := fnf
 end;    { Ketones }

 procedure Btwo;
{--------------}
 var    cnt : integer;

 begin
    fillchar(b2[1],100,chr($00));
    for cnt := 0 to round(nc/2) do b2[2*cnt+1] := alkyl[cnt+1]
 end;    { Btwo }
```

```
procedure Bthree;
{-----------------}
 var    cnt : integer;

 begin
    fillchar(b3[1],100,chr($00));
    for cnt := 0 to round(nc/3) do b3[3*cnt+1] := alkyl[cnt+1]
 end;   { Bthree }

procedure Bfour;
{---------------}
 var    cnt : integer;

 begin
    fillchar(b4[1],100,chr($00));
    for cnt := 0 to round(nc/4) do b4[4*cnt+1] := alkyl[cnt+1]
 end;   { Btwo }

procedure Ctwo;
{--------------}
 begin
    max := nc+1;   x := b2;   y := b2;
    allfcts;   c2 := fnf
 end;   { Ctwo }

procedure Calcp;
{---------------}
 var    cnt : integer;

 begin
    max := nc+1;   x := a2;   y := b2;   allfcts;
    for cnt := 1 to max do p[cnt] := a4[cnt]+6*fnf[cnt]+3*c2[cnt];
    x := alkyl;   y := b3;   allfcts;
    for cnt := 1 to max do
     p[cnt] := round((p[cnt]+8*fnf[cnt]+6*b4[cnt])/24);
    cnt := max;
    repeat
       p[cnt+1] := p[cnt];
       cnt := cnt -1
    until (cnt < 2);
    p[1] := 0
 end;   { Calcp }

procedure Calcq;
{---------------}
 var    cnt : integer;

 begin
    max := nc+1;   x := alkyl;   y := alkyl;
    x[1] := 0;   y[1] := 0;   allfcts;
    for cnt := 1 to max do q[cnt] := round((fnf[cnt]+b2[cnt])/2);
    q[1] := round(q[1]-0.5)
 end;   { Calcq }
```

```
   procedure Alkanes;
   {------------------}
   var   cnt : integer;

   begin
      for cnt := 1 to nc+1 do alkan[cnt] := p[cnt]-q[cnt]+b2[cnt]
   end;   { Alkanes }

begin
   fillchar(alkan[1],100,chr($00));    fillchar(alkyl[1],100,chr($00));
   fillchar(alkenon[1],100,chr($00)); fillchar(aldehyd[1],100,chr($00));
   fillchar(keton[1],100,chr($00));    fillchar(alkct[1],100,chr($00));
   fillchar(aldct[1],100,chr($00));    fillchar(ketct[1],100,chr($00));
   Alkyles;
   Atwo;
   Athree;
   Afour;
   Polynom;
   Sum;
   Alkenonesct;
   Aldehydesct;
   Ketonesct;
   Alkenones;
   Aldehydes;
   Ketones;
   Btwo;
   Bthree;
   Bfour;
   Ctwo;
   Calcp;
   Calcq;
   Alkanes
 end;   { Calc }

procedure Output;
{------------------}
 var   cnt : integer;

 begin
   writeln;
   writeln ('C-atoms':10,'alkanes':9,'alkyls':9,'alkenones':11,
           'aldehydes':15,'ketones':13);   writeln;
   for cnt := 1 to nc+1 do
   begin
      write(cnt-1:7,alkan[cnt]:9,alkyl[cnt]:9,alkct[cnt]:9,
           aldct[cnt]:15,ketct[cnt]:13);   writeln;
      write(alkenon[cnt]:36,aldehyd[cnt]:15,keton[cnt]:13);   writeln
   end
end;   { Output }

begin
   Input;
   Calc;
   Output
end.
```

```
                              EUL_ODE
```

```pascal
program eul_ode;
    { J. Kinkel, January 1989, based on Ebert/Ederer }

uses   crt;

var    ya,xb,xe : real;

 procedure Input;
{----------------}
 begin
    clrscr;   writeln;
    write (' Initial value of y ? ');    readln (ya);
    write (' x-values  (beg end) ? ');   readln (xb,xe)
 end;   { Input }

 procedure Calc;
{---------------}
 var   x,y,h : real;
       cnt,subi : integer;

    procedure Output;
   {-----------------}
    begin
       writeln (subi:5,' sub-intervals :     y = ',y:12:8)
    end;   { Output }

    function ode : real;
   {--------------------}
    begin
       ode := x*y
    end;  { ode }

 begin
    subi := 2;
    repeat
    h := (xe-xb)/subi;   x := xb;   y := ya;
    for cnt :=1 to subi-1 do
    begin
       x := xb+cnt*h;   y := y+ode*h
    end;
    Output;
    subi := subi*2
    until (subi >= 8192)
 end;   { Calc }

begin
   Input;
   Calc
end.
```

```
┌──────────────────────────────────────────────────────────────┐
│                           EUL_IMPR                             │
└──────────────────────────────────────────────────────────────┘
```

```pascal
program eul_impr;
   { J.Kinkel, January 1989, based on Ebert/Ederer }

uses   crt;

var    ya,xb,xe : real;

 procedure Input;
{----------------}
 begin
    clrscr;   writeln;
    write (' Initial value of y ? ');   readln (ya);
    write (' x-values (beg end) ? ');   readln (xb,xe)
 end;   { Input }

 procedure Calc;
{---------------}
 var   x,y,xh,yh,yh1,yh2,h : real;
       cnt,subi : integer;

    procedure Output;
   {-----------------}
    begin
       writeln (subi:5,' sub-intervals :     y = ',y:12:8)
    end;   { Output }

    function ode : real;
   {--------------------}
    begin
       ode := xh*yh
    end;  { ode }

 begin
    subi := 2;
    repeat
    h := (xe-xb)/subi;   y := ya;
    for cnt := 0 to subi-1 do
    begin
       x := xb+cnt*h;   xh := x;   yh := y;
       yh1 := y+ode*h;   xh := x+h;   yh := yh1;
       yh2 := y+ode*h;   y := (yh1+yh2)/2
    end;
    Output;
    subi := subi*2
    until (subi >= 8192)
 end;   { Calc }

begin
   Input;
   Calc
end.
```

```
┌─────────────────────────────────────────────────────────────────┐
│                             R_K_N                                 │
└─────────────────────────────────────────────────────────────────┘
```

```pascal
program r_k_n;
   { J. Kinkel, January 1989, based on Ebert/Ederer }

uses   crt;

var   ya,xb,xe : real;

 procedure Input;
{----------------}
 begin
    clrscr;   writeln;
    write (' Initial value of y ? ');   readln (ya);
    write (' x-value (beg end) ? ');   readln (xb,xe)
 end;   { Input }

 procedure Calc;
{---------------}
 var   x,y,xh,yh,h,k1,k2,k3,k4 : real;
       cnt,subi : integer;

    procedure Output;
   {------------------}
    begin
       writeln (subi:5,' sub-intervals :     y = ',y:12:8)
    end;   { Output }

    function ode : real;
   {--------------------}
    begin
       ode := xh*yh
    end;  { ode }

 begin
    subi := 2;
    repeat
    h := (xe-xb)/subi;   y := ya;
    for cnt := 0 to subi-1 do
    begin
       x := xb+cnt*h;   xh := x;   yh := y;
       k1 := ode*h;   xh := x+h/2;   yh := y+k1/2;
       k2 := ode*h;   yh := y+k2/2;
       k3 := ode*h;   xh := x+h;   yh := y+k3;
       k4 := ode*h;   y := y+(k1+2*k2+2*k3+k4)/6
    end;
    Output;
    subi := subi*2
    until (subi >= 8192)
 end;   { Calc }

begin
   Input;
   Calc
end.
```

```
                              SYS_EUL
```

```
program sys_eul;
   { J. Kinkel, January 1989, based on Ebert/Ederer }

uses   crt;

const   dim = 2;

var   ya : array[1..20] of real;
      xa,xe : real;

 procedure Input;
{----------------}
 var   cnt : integer;

 begin
    clrscr;   writeln;
    for cnt := 1 to dim do
    begin
       write (' Initial value of y(',cnt,') ? ');   readln (ya[cnt])
    end;
    write (' x-values (beg end) ? ');   readln (xa,xe)
 end;   { Input }

 procedure Calc;
{----------------}
 var   y,ode : array[1..20] of real;
       x,h : real;
       ccnt,subi,z : integer;

    procedure Output;
   {------------------}
    var   cnt : integer;

    begin
       writeln;   writeln (subi:5,' sub-intervals :');
       for cnt := 1 to dim do writeln ('   y(',cnt,') = ',y[cnt]:12:8)
    end;   { Output }

    procedure diff;
   {---------------}
    begin
       ode[1] := -3*sqr(y[1]) + 2*y[2] + exp(-x);
       ode[2] := 3*sqr(y[1]) - 12*y[2] + exp(-x)
    end;   { diff }

 begin
    subi := 4;
    repeat
     h := (xe-xa)/subi;
     for z := 1 to dim do y[z] := ya[z];
     for ccnt := 0 to subi-1 do
```

```
      begin
         x := xa+ccnt*h;
         diff;
         for z := 1 to dim do y[z] := y[z]+ode[z]*h
      end;
      Output;
      subi := subi*2
    until (subi >= 8192)
  end;    { Calc }

begin
   Input;
   Calc
end.
```

SYS_RKN

```
program sys_rkn;
    { J. Kinkel, January 1989, based on Ebert/Ederer }

uses    crt;

const    dim = 2;

var    ya : array[1..20] of real;
       xb,xe : real;

 procedure Input;
{----------------}
 var    cnt : integer;

 begin
    clrscr;    writeln;
    for cnt := 1 to dim do
    begin
       write (' Initial value of y (',cnt,') ? ');    readln (ya[cnt])
    end;
    write (' x-values (beg end) ? ');    readln (xb,xe)
 end;    { Input }

 procedure Calc;
{----------------}
 var    y,yh,ode,k1,k2,k3,k4 : array[1..20] of real;
        x,xh,h : real;
        ccnt,subi,z : integer;

    procedure Output;
   {------------------}
    var    cnt : integer;

    begin
       writeln;    writeln (subi:5,' sub-intervals :');
       for cnt := 1 to dim do writeln ('    y(',cnt,') = ',y[cnt]:12:8)
    end;    { Output }
```

```
      procedure diff;
   {---------------}
   begin
      ode[1] := -3*sqr(yh[1]) + 2*yh[2] + exp(-xh);
      ode[2] := 3*sqr(yh[1]) - 12*yh[2] + exp(-xh)
   end;  { diff }

begin
   subi := 4;
   repeat
   h := (xe-xb)/subi;
   for z := 1 to dim do y[z] := ya[z];
   for ccnt := 0 to subi-1 do
   begin
      x := xb+ccnt*h;    xh := x;
      for z := 1 to dim do yh[z] := y[z];
      diff;
      for z := 1 to dim do k1[z] := ode[z]*h;
      xh := x+h/2;
      for z := 1 to dim do yh[z] := y[z]+k1[z]/2;
      diff;
      for z := 1 to dim do k2[z] := ode[z]*h;
      for z := 1 to dim do yh[z] := y[z]+k2[z]/2;
      diff;
      for z := 1 to dim do k3[z] := ode[z]*h;
      xh := x+h;
      for z := 1 to dim do yh[z] := y[z]+k3[z];
      diff;
      for z := 1 to dim do k4[z] := ode[z]*h;
      for z := 1 to dim do
       y[z] := y[z]+(k1[z]+2*k2[z]+2*k3[z]+k4[z])/6
   end;
   Output;
   subi := subi*2
   until (subi >= 8192)
 end;   { Calc }

begin
   Input;
   Calc
end.
```

BOUNDARY

```
program boundary;
   { J. Kinkel, January 1989, based on Ebert/Ederer }

uses   crt;

const   dim = 2;

var    ya,y : array[1..20] of real;
       y1,y2,yv,xb,xe : real;
```

```
procedure SetPar;
{------------------}
begin
   clrscr;
   ya[1] := 1;   xb := 0;   xe := 1;   y1 := 0;   y2 := 5;   yv := 0.75
end;   { SetPar }

procedure Calc;
{--------------------}
var   yh,ode,k1,k2,k3,k4 : array[1..20] of real;
      x,xh,h : real;
      ccnt,subi,z : integer;

   procedure diff;
   {---------------}
   begin
      ode[1] := -0.1*yh[1] - 0.1*yh[1]*yh[2];
      ode[2] := 0.1*yh[1] - 0.1*yh[1]*yh[2]
   end;   { diff }

begin
   subi := 16;   h := (xe-xb)/subi;
   for z := 1 to dim do y[z] := ya[z];
   for ccnt := 0 to subi-1 do
   begin
      x := xb+ccnt*h;   xh := x;
      for z := 1 to dim do yh[z] := y[z];
      diff;
      for z := 1 to dim do k1[z] := ode[z]*h;
      xh := x+h/2;
      for z := 1 to dim do yh[z] := y[z]+k1[z]/2;
      diff;
      for z := 1 to dim do k2[z] := ode[z]*h;
      for z := 1 to dim do yh[z] := y[z]+k2[z]/2;
      diff;
      for z := 1 to dim do k3[z] := ode[z]*h;
      xh := x+h;
      for z := 1 to dim do yh[z] := y[z]+k3[z];
      diff;
      for z := 1 to dim do k4[z] := ode[z]*h;
      for z := 1 to dim do
      y[z] := y[z]+(k1[z]+2*k2[z]+2*k3[z]+k4[z])/6
   end
end;   { Calc }

procedure Search;
{------------------}
var   ym,fcta,fcte,fctm : real;
      cnt : integer;

   function sgn(f : real) : integer;
   {---------------------------------}
   begin
      if (f >= 0) then sgn := 1 else sgn := -1
   end;   { sgn }
```

```
   procedure Output;
   {-----------------}
   var   cnt : integer;

   begin
      writeln;
      writeln (' T = ',xb:1:0,' :        y(1) = ',ya[1]:12:8,
               '   y(2) = ',ya[2]:12:8);
      writeln (' T = ',xe:1:0,' :        y(1) = ',y[1]:12:8,
               '   y(2) = ',y[2]:12:8)
   end;   { Output }

begin
   ya[2] := y1;
   Calc;
   fcta := y[1];
   Output;
   ya[2] := y2;
   Calc;
   fcte := y[1];
   Output;
   cnt := 0;
   repeat
    if (sgn(fcta-yv) = sgn(fcte-yv)) then
    begin
      writeln (' Make a better choice of the interval !');
      Halt
    end;
    ym := (y1+y2)/2;   ya[2] := ym;
    Calc;
    fctm := y[1];
    Output;
    if (sgn(fcta-yv) = sgn(fctm-yv)) then
    begin
       y1 := ym;   fcta := fctm
    end
    else
    begin
       y2 := ym;   fcte := fctm
    end;
    cnt := cnt + 1
   until (cnt > 10)
end;   { Search }

begin
   SetPar;
   Search
end.
```

```
┌─────────────────────────────────────────────────────────────────────┐
│                              HARM                                     │
└─────────────────────────────────────────────────────────────────────┘
```

```pascal
program harm;
   { J. Kinkel, January 1989, based on Ebert/Ederer }

uses   crt;

const   dim = 2;    nl = 6.022e23;   rg = 0.7412e-8;
        dg = 7.6064e-12;    hh = 6.626176e-27;

var   ya : array[1..20] of real;
      eg : real;
      sgn : boolean;

 procedure Input;
{----------------}
 begin
    clrscr;    writeln;
    write (' Wave function at equilibrium distance ? ');    readln (ya[1]);
    write (' Derivation of the wave function at equilibrium distance ? ');
    readln (ya[2]);
    write (' Total energy in erg ? ');    readln (eg)
 end;   { Input }

 procedure Calc;
{---------------}
 var   y,yh,ode,k1,k2,k3,k4 : array[1..20] of real;
       x,xb,xe,xh,h,my,aa : real;
       ccnt,z : integer;

    procedure Output;
   {-----------------}
    begin
       writeln;
       writeln (' Psi(',x:12:8,') = ',y[1]:12,eg:15,y[2]:12)
    end;   { Output }

    function fnv (r : real) : real;
   {------------------------------}
    begin
       fnv := aa*sqr((r-rg))
    end;   { fnv }

    procedure diff;
   {---------------}
    begin
       ode[1] := yh[2];
       ode[2] := -(eg-fnv(xh))*78.952*my/hh*yh[1]/hh
    end;  { diff }

 begin
    xb := rg;    xe := 5*rg;    my := 0.504/nl;
    aa := sqr(8*arctan(1))*1.3192e14*my/2*1.3192e14;
    h := (xe-xb)/100;
```

```
      for z := 1 to dim do y[z] := ya[z];
      for ccnt := 0 to 99 do
      begin
         x := xb+ccnt*h;    xh := x;
         for z := 1 to dim do yh[z] := y[z];
         diff;
         for z := 1 to dim do k1[z] := ode[z]*h;
         xh := x+h/2;
         for z := 1 to dim do yh[z] := y[z]+k1[z]/2;
         diff;
         for z := 1 to dim do k2[z] := ode[z]*h;
         for z := 1 to dim do yh[z] := y[z]+k2[z]/2;
         diff;
         for z := 1 to dim do k3[z] := ode[z]*h;
         xh := x+h;
         for z := 1 to dim do yh[z] := y[z]+k3[z];
         diff;
         for z := 1 to dim do k4[z] := ode[z]*h;
         for z := 1 to dim do
         y[z] := y[z]+(k1[z]+2*k2[z]+2*k3[z]+k4[z])/6;
         if (abs(y[dim]) > 1e27) then
         begin
            ccnt := 99;
            if (y[dim] > 0) then sgn := true
            else sgn := false
         end;
         Output
      end
   end;   { Calc }

begin
   Input;
   Calc;
   writeln;   write (' Wave function diverges to');
   if (sgn) then write (' positive ') else write (' negative ');
   writeln ('infinity !');   writeln
end.
```

```
+--------------------------------------------------+
|                      PDE                         |
+--------------------------------------------------+
```

```
program pde;
   { J. Kinkel, January 1989, based on Ebert/Ederer }

uses   crt;

const   dim = 11;

var   ya : array[1..20] of real;
      xb,xe : real;

 procedure Input;
{----------------}
 var   cnt : integer;
```

```
begin
   clrscr;   writeln;
   ya[1] := 0;   ya[dim] := 0;
   for cnt := 2 to dim-1 do ya[cnt] := (cnt-1)*10;
   xb := 0;
   writeln (' At which time do you want to calculate');
   write (' the temperature distribution ? ');   readln (xe)
end;   { Input }

procedure Calc;
{---------------}
var   y,yh,ode,k1,k2,k3,k4 : array[1..20] of real;
      x,xh,h : real;
      ccnt,subi,z : integer;

   procedure Output;
   {------------------}
   var   cnt : integer;

   begin
      writeln;
      writeln (' Temperature distribution at time = ',xe:6:2);
      writeln (' in degree C with ',subi:5,' time intervals :');
      for cnt := 1 to dim do write ((cnt-1)*10:4,' cm');   writeln;
      for cnt := 1 to dim do write (int(y[cnt]*10+0.5)/10:7:1);
      writeln
   end;    { Output }

   procedure diff;
   {---------------}
   var   cnt : integer;

   begin
      ode[1] := 0;
      for cnt := 2 to 10 do
      ode[cnt] := (yh[cnt+1]-2*yh[cnt]+yh[cnt-1])*0.0023;
      ode[11] := 0
   end;  { diff }

begin
   subi := 4;
   repeat
   h := (xe-xb)/subi;
   for z := 1 to dim do y[z] := ya[z];
   for ccnt := 0 to subi-1 do
   begin
      x := xb+ccnt*h;   xh := x;
      for z := 1 to dim do yh[z] := y[z];
      diff;
      for z := 1 to dim do k1[z] := ode[z]*h;
      xh := x+h/2;
      for z := 1 to dim do yh[z] := y[z]+k1[z]/2;
      diff;
      for z := 1 to dim do k2[z] := ode[z]*h;
      for z := 1 to dim do yh[z] := y[z]+k2[z]/2;
```

```
         diff;
         for z := 1 to dim do k3[z] := ode[z]*h;
         xh := x+h;
         for z := 1 to dim do yh[z] := y[z]+k3[z];
         diff;
         for z := 1 to dim do k4[z] := ode[z]*h;
         for z := 1 to dim do
           y[z] := y[z]+(k1[z]+2*k2[z]+2*k3[z]+k4[z])/6
       end;
       Output;
       subi := subi*2
     until (subi >= 8192)
  end;   { Calc }

begin
   Input;
   Calc
end.
```

STAT

```
program stat;
   { J. Kinkel, January 1989, based on Ebert/Ederer }

uses   crt;

var    side : array[1..4] of real;
       t0,tn : array[1..30,1..30] of real;
       iv : integer;

 procedure Input;
{----------------}
 var   cnt : integer;

 begin
    clrscr;   writeln;
    write (' Into how many sub-intervals shall a side be subdivided ? ');
    readln (iv);
    for cnt := 1 to 4 do
    begin
       write (' Temperature of the ',cnt,'. side ? ');
       readln(side[cnt])
    end
 end;   { Input }

 procedure Calc;
{---------------}
 var   eps : real;
       cnt1,cnt2 : integer;
       abb : boolean;

    procedure SetMat0;
   {------------------}
    var   tm : real;
          cnt1,cnt2 : integer;
```

```pascal
   begin
      iv := iv + 1;    tm := (side[1]+side[2]+side[3]+side[4])/4;
      for cnt1 := 2 to iv-1 do
      begin
         t0[cnt1,1] := side[1];    t0[iv,cnt1] := side[2];
         t0[cnt1,iv] := side[3];    t0[1,cnt1] := side[4]
      end;
      t0[1,1] := (side[1]+side[4])/2;
      t0[iv,1] := (side[1]+side[2])/2;
      t0[iv,iv] := (side[2]+side[3])/2;
      t0[1,iv] := (side[3]+side[4])/2;
      for cnt1 := 2 to iv-1 do
       for cnt2 := 2 to iv-1 do t0[cnt1,cnt2] := tm
   end;    { SetMat0 }

   procedure SetMatn;
   {------------------}
   var   cnt1,cnt2 : integer;

   begin
      tn := t0
   end;    { SetMatn }

begin
   fillchar(t0[1,1],5400,chr($00));
   SetMat0;
   SetMatn;
   repeat
    eps := 0.1;    abb := true;
    for cnt1 := 2 to iv-1 do
     for cnt2 := 2 to iv-1 do
     begin
        t0[cnt1,cnt2] := (tn[cnt1+1,cnt2]+tn[cnt1-1,cnt2]
                          +tn[cnt1,cnt2+1]+tn[cnt1,cnt2-1])/4;
        if abs(t0[cnt1,cnt2]-tn[cnt1,cnt2]) > eps then abb := false
     end;
    SetMatn
   until abb
end;    { Calc }

procedure Output;
{-----------------}
var   cnt1,cnt2 : integer;

begin
   writeln;    cnt2 := iv;
   repeat
    for cnt1 := 1 to iv do write (int(t0[cnt1,cnt2]+0.5):6:0);
    cnt2 := cnt2-1;    writeln;    writeln;
   until (cnt2 < 1)
end;    { Output }

begin
   Input;
   Calc;
   Output
end.
```

```
                          LAGRANGE
```

```
program lagrange;
   { J. Kinkel, January 1989, based on Ebert/Ederer }

uses   crt;

var   x,y : array[1..70] of real;
      ndp : integer;

 procedure Input;
{----------------}
 var   cnt : integer;

 begin
    clrscr;   writeln;
    write (' Number of data points ? ');   readln (ndp);
    for cnt := 1 to ndp do
    begin
       write (' x(',cnt,')  y(',cnt,') ? ');   readln (x[cnt],y[cnt])
    end
 end;   { Input }

 procedure Calc;
{---------------}
 var   cnt,ocnt,icnt : integer;
       pol,xi,yi : real;

    procedure Output;
    {-----------------}
    begin
       writeln (' x = ',xi:12:8);   writeln (' y = ',yi:12:8)
    end;   { Output }

 begin
    cnt := 1;
    repeat
     writeln;
     write (' At which x-value shall an interpolation value be
              calculated?');
     readln (xi);   yi := 0;
     for ocnt := 1 to ndp do
     begin
        pol := 1;
        for icnt := 1 to ndp do
         if not(icnt = ocnt) then
          pol := pol*(xi-x[icnt])/(x[ocnt]-x[icnt]);
        yi := yi + pol*y[ocnt]
     end;
     Output;
     cnt := cnt + 1
    until (cnt > 100)
 end;   { Calc }

begin
   Input;
   Calc
end.
```

```
┌─────────────────────────────────────────────────────────────┐
│                          SPLINE                             │
└─────────────────────────────────────────────────────────────┘
```

```pascal
program spline;
   { J. Kinkel, January 1989, based on Ebert/Ederer }

uses   crt;

type   region = array[1..25] of real;

var    mat : array[1..25,1..26] of real;
       h,d,k,x,y : region;
       ndp : byte;
       goon : char;

 procedure Input;
{------------------}
 var   cnt : byte;

 begin
    clrscr;   writeln;
    write (' Number of data points (max. 25) ? ');   readln (ndp);
    for cnt := 1 to ndp do
    begin
       write (cnt:3,' : ');   readln (x[cnt],y[cnt])
    end
 end;   { Input }

 procedure Sort;
{---------------}
 var   hx,tmp : real;
       hcnt,cnt1,cnt2 : integer;

 begin
    for cnt1 := 1 to ndp-1 do
    begin
       hx := x[cnt1];   hcnt := cnt1;
       for cnt2 := cnt1+1 to ndp do
       begin
          if (x[cnt2] <= hx) then
          begin
             hx := x[cnt2];   hcnt := cnt2
          end
       end;
       x[hcnt] := x[cnt1];   x[cnt1] := hx;
       tmp := y[hcnt];   y[hcnt] := y[cnt1];   y[cnt1] := tmp
    end
 end;   { Sort }

 procedure FillMatrix;
{--------------------}
 var   cnt,i,j : byte;

 begin
    for cnt := 1 to ndp-1 do
```

```
  begin
    h[cnt] := x[cnt+1] - x[cnt];
    d[cnt] := (y[cnt+1] - y[cnt])/h[cnt]
  end;
  for i := 1 to ndp do
   for j := 1 to ndp do mat[i,j] := 0;
  mat[1,1] := 2;   mat[ndp,ndp] := 2;    mat[1,2] := 1;
  mat[ndp,ndp-1] := 1;    mat[1,ndp+1] := 3*d[1];
  mat[ndp,ndp+1] := 3*d[ndp-1];
  for cnt := 2 to ndp-1 do
  begin
    mat[cnt,ndp+1] := 3*(d[cnt]*h[cnt-1]+d[cnt-1]*h[cnt]);
    mat[cnt,cnt-1] := h[cnt];    mat[cnt,cnt+1] := h[cnt-1];
    mat[cnt,cnt] := 2*(h[cnt]+h[cnt-1])
  end
end;   { FillMatrix }

procedure Gauss_Jordan;
{---------------------}
var   cnt1,cnt2,hlp : byte;
      faktor : real;

  procedure Exchange;
  {------------------}
  var   i : byte;
        tmp : real;

  begin
    for i := 1 to ndp+1 do
    begin
      tmp := mat[cnt1,i];
      mat[cnt1,i] := mat[hlp,i];
      mat[hlp,i] := tmp
    end
  end;   { Exchange }

  procedure Normalize;
  {-------------------}
  var   i : byte;

  begin
    faktor := 1/mat[cnt1,cnt1];
    for i := 1 to ndp+1 do mat[cnt1,i] := faktor*mat[cnt1,i]
  end;   { Normalize }

  procedure Subtract;
  {------------------}
  var   i : byte;

  begin
    faktor := -mat[cnt2,cnt1];
    for i := 1 to ndp+1 do
     mat[cnt2,i] := mat[cnt2,i] + faktor*mat[cnt1,i]
  end;   { Subtract }
```

```pascal
   procedure Solutions;
  {--------------------}
   var   cnt : byte;

   begin
      for cnt := 1 to ndp do k[cnt] := mat[cnt,ndp+1]
   end;   { Solutions }

begin
   for cnt1 := 1 to ndp do
   begin
      hlp := cnt1;
      repeat
         if not(mat[hlp,cnt1] = 0) then
         begin
            Exchange;
            Normalize;
            for cnt2 := 1 to ndp do
             if not(cnt2 = cnt1) then Subtract;
             hlp := ndp
         end;
         hlp := hlp+1
      until (hlp >= ndp)
   end;
   Solutions
end;   { Gauss_Jordan }

procedure Output;
{-----------------}
var   xgiv,yser : real;
      found  : byte;

   procedure Search;
  {-----------------}
   begin
      found := 0;
      if (xgiv <= x[1]) then found := 1
      else
      if (xgiv > x[ndp]) then found := ndp-1
      else
      begin
         repeat  found := found + 1 until (xgiv <= x[found]);
         found := found - 1
      end
   end;   { Search }

   procedure Interpolation;
  {-----------------------}
   var   tmp : real;

   begin
      tmp := (xgiv - x[found])/h[found];
      yser := tmp*y[found+1]+(1-tmp)*y[found];
      yser := yser+h[found]*(1-tmp)*tmp*((k[found]-d[found])*(1-tmp)
             -(k[found+1]-d[found])*tmp)
   end;   { Interpolation }
```

```
begin
  clrscr;   writeln;
  write (' Which x-value ? ');   readln (xgiv);
  Search;
  Interpolation;   writeln;
  writeln (' Spline interpolation at position');
  writeln (' x = ',xgiv:12:8,' : ',yser:10:6)
end;   { Output }

begin
  fillchar (h[1],150,chr($00));   fillchar (d[1],150,chr($00));
  fillchar (k[1],150,chr($00));   fillchar (x[1],150,chr($00));
  fillchar (y[1],150,chr($00));
  Input;
  Sort;
  FillMatrix;
  Gauss_Jordan;
  repeat
  Output;
  goon := readkey until (goon = chr($0D))
end.
```

```
                         MNEWTON
```

```
program mnewton;
  { J. Kinkel, January 1989, based on Ebert/Ederer }

uses    crt;

const   dim = 3;

var     mata,matb,matu,matv,matw : array[1..dim,1..dim] of real;
        x,y,fnf,xh,fnfh : array[1..dim] of real;
        z : integer;

 procedure Input;
{----------------}
 var   cnt : integer;

 begin
   clrscr;   writeln;
   writeln (' Input of the estimated x-values of the roots :');
   writeln;
   for cnt := 1 to dim do
   begin
     write (' x(',cnt,') = ? ');   readln (x[cnt])
   end;
   xh := x
 end;   { Input }

 procedure InvertMatrix;
{-----------------------}

   procedure Fill;
  {---------------}
   var   row,col : integer;
```

```
begin
   for row := 1 to dim do
   begin
      for col := 1 to dim do
      begin
         if (col = row) then
         begin
            matb[row,col] := 1;   matv[row,col] := 1
         end
         else
         begin
            matb[row,col] := 0;   matv[row,col] := 0
         end
      end
   end
end;   { Fill }

procedure Pivot;
{----------------}
var   pcnt,cnt,t : integer;
      s : real;

   procedure Exchange;
   {------------------}
   var   tcnt : integer;

   begin
      for tcnt := 1 to dim do
      begin
         s:= mata[pcnt,tcnt];
         mata[pcnt,tcnt] := mata[t,tcnt];
         mata[t,tcnt] := s
      end;
      if (abs(mata[pcnt,pcnt]) <= 1e-30) then
      begin
         writeln (' Inversion is not possible !');   Halt
      end
      else
      begin
         matv[pcnt,pcnt] := 0;   matv[t,t] := 0;
         matv[pcnt,t] := 1;   matv[t,pcnt] := 1
      end
   end;   { Exchange }

   procedure GJ_Elimination;
   {------------------------}
   var   cnt1,cnt2 : integer;

   begin
      for cnt1 := 1 to dim do
      begin
         for cnt2 := 1 to dim do
         begin
            if (cnt1 = pcnt) then
             if (cnt2 = cnt1) then
              matu[pcnt,pcnt] := 1/mata[pcnt,pcnt]
             else
```

```
                       matu[cnt1,cnt2] := -mata[cnt1,cnt2]/mata[pcnt,pcnt]
                   else
                   begin
                       if (cnt2 = pcnt) then
                         matu[cnt1,pcnt] := mata[cnt1,pcnt]/mata[pcnt,pcnt]
                       else
                         matu[cnt1,cnt2] := mata[cnt1,cnt2]-mata[pcnt,cnt2]
                                         *mata[cnt1,pcnt]/mata[pcnt,pcnt]
                   end
               end
       end
   end;   { GJ_Elimination }

 procedure Multiplication;
 {-----------------------}
 var    cnt1,cnt2,cnt3 : integer;

 begin
     for cnt1 := 1 to dim do
     begin
         for cnt2 := 1 to dim do
         begin
             matw[cnt1,cnt2] := 0;
             for cnt3 := 1 to dim do
               matw[cnt1,cnt2] := matw[cnt1,cnt2]+matv[cnt1,cnt3]
                                 *matb[cnt3,cnt2]
         end
     end;
     matb := matw;
     for cnt1 := 1 to dim do
     begin
         for cnt2 :=  1 to dim do
         begin
             mata[cnt1,cnt2] := matu[cnt1,cnt2];
             if (cnt1 = cnt2) then matv[cnt1,cnt2] := 1
             else matv[cnt1,cnt2] := 0
         end
     end
   end; { Multiplication }

begin
   for pcnt := 1 to dim do
   begin
       s := 0;
       for cnt := pcnt to dim do
        if (s <= abs(mata[cnt,pcnt])) then
        begin
           s := abs(mata[cnt,pcnt]);   t := cnt
        end;
       Exchange;
       GJ_Elimination;
       Multiplication
   end
end;   { Pivot }
```

```pascal
   procedure Result;
   {-----------------}
   var   cnt1,cnt2,cnt3 : integer;

   begin
      for cnt1 := 1 to dim do
      begin
         for cnt2 := 1 to dim do
         begin
            matw[cnt1,cnt2] := 0;
            for cnt3 := 1 to dim do
            matw[cnt1,cnt2] := matw[cnt1,cnt2]+mata[cnt1,cnt3]
                                    *matb[cnt3,cnt2]
         end
      end
   end;   { Result }

begin
   Fill;
   Pivot;
   Result
end;   { InvertMatrix }

procedure Allfcts;
{------------------}
begin
   fnf[1] := sqr(x[1])-x[1]*x[2]-3;
   fnf[2] := x[3]*x[2]*x[1]-sqr(x[2])*1.5;
   fnf[3] := x[1]*sqr(x[1])+x[2]*sqr(x[2])+x[3]*sqr(x[3])-36
end;   { Allfcts }

procedure PartDeriv;
{--------------------}
var   cnt1,cnt2 : integer;
      diff : real;

begin
   x := xh;
   for cnt1 := 1 to dim do
   begin
      diff := abs(x[cnt1]/1e5)+1e-9;   x[cnt1] := x[cnt1]+diff;
      Allfcts;
      for cnt2 := 1 to dim do mata[cnt2,cnt1] := fnf[cnt2];
      x[cnt1] := xh[cnt1]
   end;
   for cnt1 := 1 to dim do
   begin
      diff := abs(x[cnt1]/1e5)+1e-9;   x[cnt1] := x[cnt1]-diff;
      Allfcts;
      for cnt2 := 1 to dim do
       mata[cnt2,cnt1] := (mata[cnt2,cnt1]-fnf[cnt2])/2/diff;
      x[cnt1] := xh[cnt1]
   end;
   Allfcts;
   fnfh := fnf
end;   { PartDeriv }
```

```
   procedure Optfac;
   {----------------}
   var    help : array[1..dim] of real;
          df,opf,ff,sum1,sum2,sum3 : real;
          cnt1,cnt2 : integer;

   begin
      for cnt1 := 1 to dim do
      begin
         help[cnt1] := 0;
         for cnt2 := 1 to dim do
           help[cnt1] := help[cnt1]+matw[cnt1,cnt2]*fnfh[cnt2]
      end;
      df := 0.6;   opf := 1;   ff := opf;   sum1 := 0;
      for cnt1 := 1 to dim do x[cnt1] := xh[cnt1]-ff*help[cnt1];
      Allfcts;
      for cnt1 := 1 to dim do sum1 := sum1+sqr(fnf[cnt1]);
      for cnt1 := 0 to dim  do
      begin
         df := df*exp(cnt1*ln(0.5));
         repeat
          if not(cnt1/2 = int(cnt1/2)) then df := -df;
          ff := opf+df;   sum3 := sum1;   sum2 := 0;
          for cnt2 := 1 to dim do x[cnt2] := xh[cnt2]-ff*help[cnt2];
          Allfcts;
          for cnt2 := 1 to dim do sum2 := sum2+sqr(fnf[cnt2]);
          if (sum2 < sum1) then
          begin
             sum1 := sum2;   opf := ff
          end
         until (sum2 >= 0.9*sum3)
      end;
      for cnt1 := 1 to dim do xh[cnt1] := xh[cnt1]-opf*help[cnt1]
   end;   { Optfac }

   procedure Output;
   {----------------}
   var    cnt : integer;

   begin
      writeln;
      for cnt := 1 to dim do
       writeln (' x(',cnt,') = ',xh[cnt]:12:8,
                            '   f(',cnt,') = ',fnfh[cnt]:12:8)
   end;   { Output }

begin
   Input;
   z := 1;
   repeat
    PartDeriv;
    Output;
    InvertMatrix;
    Optfac;
    z := z+1
   until (z > maxint)
end.
```

```
                              NL_REGR
```

```
program nl_regr;
   { J. Kinkel, January 1989, based on Ebert/Ederer }

uses    crt;

const   dim = 2;    mp = 6;
        xx : array[1..6] of real = (0,1,2,3,0.5,1.5);
        yy : array[1..6] of real = (2,1.368,1.1353,1.04979,1.6065,1.2231);

var     mata,matb,matu,matv,
                   matw : array[1..dim,1..dim] of real;
        x,y,k,fnf,xh,fnfh : array[1..dim] of real;
        sm : real;
        z : integer;

 procedure Input;
{----------------}
 var   cnt : integer;

 begin
    clrscr;    writeln;
    writeln (' Input of the estimated parameters :');    writeln;
    for cnt := 1 to dim do
    begin
       write (' k(',cnt,') = ? ');    readln (x[cnt])
    end;
    xh := x;    sm := 1e30
 end;    { Input }

 procedure InvertMatrix;
{-----------------------}

    procedure Fill;
    {----------------}
    var    row,col : integer;

    begin
       for row := 1 to dim do
       begin
          for col := 1 to dim do
          begin
             if (col = row) then
             begin
                matb[row,col] := 1;    matv[row,col] := 1
             end
             else
             begin
                matb[row,col] := 0;    matv[row,col] := 0
             end
          end
       end
    end;    { Fill }
```

```
procedure Pivot;
{----------------}
var   pcnt,cnt,t : integer;
      s : real;

   procedure Exchange;
   {------------------}
   var   tcnt : integer;

   begin
      for tcnt := 1 to dim do
      begin
         s:= mata[pcnt,tcnt];
         mata[pcnt,tcnt] := mata[t,tcnt];
         mata[t,tcnt] := s
      end;
      if (abs(mata[pcnt,pcnt]) <= 1e-30) then
      begin
         writeln (' No inversion is possible !');    Halt
      end
      else
      begin
         matv[pcnt,pcnt] := 0;    matv[t,t] := 0;
         matv[pcnt,t] := 1;    matv[t,pcnt] := 1
      end
   end;    { Exchange }

   procedure GJ_Elimination;
   {------------------------}
   var   cnt1,cnt2 : integer;

   begin
      for cnt1 := 1 to dim do
      begin
         for cnt2 := 1 to dim do
         begin
            if (cnt1 = pcnt) then
             if (cnt2 = cnt1) then
              matu[pcnt,pcnt] := 1/mata[pcnt,pcnt]
             else
              matu[cnt1,cnt2] := -mata[cnt1,cnt2]/mata[pcnt,pcnt]
            else
            begin
               if (cnt2 = pcnt) then
                matu[cnt1,pcnt] := mata[cnt1,pcnt]/mata[pcnt,pcnt]
               else
                matu[cnt1,cnt2] := mata[cnt1,cnt2]-mata[pcnt,cnt2]
                                   *mata[cnt1,pcnt]/mata[pcnt,pcnt]
            end
         end
      end
   end;    { GJ_Elimination }
```

```
procedure Multiplication;
{------------------------}
var    cnt1,cnt2,cnt3 : integer;
begin
   for cnt1 := 1 to dim do
   begin
      for cnt2 := 1 to dim do
      begin
         matw[cnt1,cnt2] := 0;
         for cnt3 := 1 to dim do
         matw[cnt1,cnt2] := matw[cnt1,cnt2]+matv[cnt1,cnt3]
                           *matb[cnt3,cnt2]
      end
   end;
   matb := matw;
   for cnt1 := 1 to dim do
   begin
      for cnt2 :=  1 to dim do
      begin
         mata[cnt1,cnt2] := matu[cnt1,cnt2];
         if (cnt1 = cnt2) then matv[cnt1,cnt2] := 1
         else matv[cnt1,cnt2] := 0
      end
   end
end;  { Multiplication }
begin
   for pcnt := 1 to dim do
   begin
      s := 0;
      for cnt := pcnt to dim do
       if (s <= abs(mata[cnt,pcnt])) then
       begin
          s := abs(mata[cnt,pcnt]);    t := cnt
       end;
      Exchange;
      GJ_Elimination;
      Multiplication
   end
end;   { Pivot }

procedure Result;
{-----------------}
var    cnt1,cnt2,cnt3 : integer;
begin
   for cnt1 := 1 to dim do
   begin
      for cnt2 := 1 to dim do
      begin
         matw[cnt1,cnt2] := 0;
         for cnt3 := 1 to dim do
         matw[cnt1,cnt2] := matw[cnt1,cnt2]+mata[cnt1,cnt3]
                           *matb[cnt3,cnt2]
      end
   end
end;   { Result }
```

```
begin
   Fill;
   Pivot;
   Result
end;   { InvertMatrix }

function sqsum (ko1,ko2 : real) : real;
{---------------------------------------}
var   hy1,hy2,zs : real;
      i : integer;

begin
   zs := 0;
   for i := 1 to mp do
   begin
      hy1 := ko1*exp(-ko2*xx[i])+ko1*ko2;   hy2 := hy1-yy[i];
      zs := zs+sqr(hy2)
   end;
   sqsum := zs
end;   { sqsum }

procedure PartDeriv;
{-------------------}
var   cnt1,cnt2 : integer;
      diff : real;

   procedure Allfcts;
   {-----------------}
   var   dj : real;
         cnt : integer;

   begin
      k := x;
      for cnt := 1 to dim do
      begin
         dj := abs(x[cnt]/1e4)+1e-8;   k[cnt] := k[cnt]+dj;
         fnf[cnt] := sqsum(k[1],k[2]);   k[cnt] := k[cnt]-2*dj;
         fnf[cnt] := (fnf[cnt]-sqsum(k[1],k[2]))/2/dj;
         k[cnt] := x[cnt]
      end
   end;   { Allfcts }

begin
   x := xh;
   Allfcts;
   fnfh := fnf;
   for cnt1 := 1 to dim do
   begin
      diff := abs(x[cnt1]/1e4)+1e-8;   x[cnt1] := x[cnt1]+diff;
      Allfcts;
      for cnt2 := 1 to dim do mata[cnt2,cnt1] := fnf[cnt2];
      x[cnt1] := xh[cnt1]
   end;
   for cnt1 := 1 to dim do
   begin
      diff := abs(x[cnt1]/1e4)+1e-8;   x[cnt1] := x[cnt1]-diff;
```

```
          Allfcts;
          for cnt2 := 1 to dim do
           mata[cnt2,cnt1] := (mata[cnt2,cnt1]-fnf[cnt2])/2/diff;
          x[cnt1] := xh[cnt1]
       end
  end;    { PartDeriv }

  procedure Optfac;
  {-----------------}
  var    help : array[1..dim] of real;
         ff,df,fm : real;
         cnt1,cnt2 : integer;

  begin
     for cnt1 := 1 to dim do
     begin
        help[cnt1] := 0;
        for cnt2 := 1 to dim do
         help[cnt1] := help[cnt1]+matw[cnt1,cnt2]*fnfh[cnt2];
        k[cnt1] := xh[cnt1]-ff*help[cnt1]
     end;
     ff := 0.6;    fm := 0;    df := 0.5;
     while (sqsum(k[1],k[2]) < sm) do
     begin
      sm := sqsum(k[1],k[2]);    fm := ff;    ff := ff+df;
      for cnt1 := 1 to dim do k[cnt1] := xh[cnt1]-ff*help[cnt1]
     end;
     ff := fm-df;
     for cnt1 := 1 to dim do k[cnt1] := xh[cnt1]-ff*help[cnt1];
     while (sqsum(k[1],k[2]) < sm) do
     begin
      sm := sqsum(k[1],k[2]);    fm := ff;    ff := ff-df;
      for cnt1 := 1 to dim do k[cnt1] := xh[cnt1]-ff*help[cnt1]
     end;
     for cnt1 := 1 to mp+1 do
     begin
        df := df/2;    ff := fm+df;
        for cnt2 := 1 to dim do k[cnt2] := xh[cnt2]-ff*help[cnt2];
        if (sqsum(k[1],k[2]) < sm) then
        begin
           sm := sqsum(k[1],k[2]);    fm := ff
        end
        else
        begin
           ff := fm-df;
           for cnt2 := 1 to dim do k[cnt2] := xh[cnt2]-ff*help[cnt2];
           if (sqsum(k[1],k[2]) < sm) then
           begin
              sm := sqsum(k[1],k[2]);    fm := ff
           end
        end
     end;
     ff := fm;
     for cnt1 := 1 to dim do xh[cnt1] := xh[cnt1]-ff*help[cnt1]
  end;    { Optfac }
```

```
procedure Output;
{-----------------}
var   ss : real;
      cnt : integer;

begin
   ss := sqsum(k[1],k[2]);   writeln;
   for cnt := 1 to dim do
     writeln (' k(',cnt,') = ',xh[cnt]:12:8,' +/- '
                ,sqr(abs(matw[cnt,cnt]*ss/(mp-dim))));
   writeln (' Sum of squared errors = ',ss)
end;   { Output }

begin
   Input;
   z := 1;
   repeat
    PartDeriv;
    InvertMatrix;
    Output;
    Optfac;
    z := z+1
   until (z > 15)
end.
```

```
                         NL_REGRA
```

```
program nl_regra;
   { J. Kinkel, January 1989, based on Ebert/Ederer }

uses   crt;

const   dim = 3;   mp = 64;
        xx : array[1..64] of real
               = (0.000,0.005,0.010,0.015,0.020,0.025,0.030,0.035,
                  0.040,0.045,0.050,0.055,0.060,0.065,0.070,0.075,
                  0.080,0.085,0.090,0.095,0.100,0.105,0.110,0.115,
                  0.120,0.125,0.130,0.135,0.140,0.145,0.150,0.155,
                  0.160,0.165,0.170,0.175,0.180,0.185,0.190,0.195,
                  0.200,0.205,0.210,0.215,0.220,0.225,0.230,0.235,
                  0.240,0.245,0.250,0.255,0.260,0.265,0.270,0.275,
                  0.600,0.605,0.610,0.615,0.620,0.625,0.630,0.635);
        yy : array[1..64] of real
               = (1.000000,0.854415,0.734266,0.627988,0.535742,0.459174,
                  0.395416,0.337545,0.288678,0.249475,0.213886,0.185657,
                  0.157447,0.135157,0.118451,0.099275,0.085432,0.074251,
                  0.063035,0.054768,0.047787,0.039426,0.034518,0.030223,
                  0.026509,0.023732,0.019806,0.014693,0.011894,0.009542,
                  0.006593,0.005535,0.005540,0.007713,0.003393,0.004965,
                  0.001981,0.003614,0.000934,0.0000215,0.0002637,0.0013670,
                  0.0011571,-0.0006458,0.0019994,-0.0035763,-0.0026426,
                  -0.0023520,-0.0026480,-0.0054493,-0.0026830,-0.0028068,
                  -0.0018326,-0.0008719,-0.0006889,-0.0008531,-0.0052879,
                  -0.0058611,-0.0047900,-0.0052341,-0.0022847,-0.0024500,
                  -0.0030974,-0.0);
```

```
var    mata,matb,matu,matv,matw,fnf2 : array[1..dim,1..dim] of real;
       x,y,k,fnf1,fnfh,xh : array[1..dim] of real;
       sm : real;
       z : integer;

procedure Input;
{----------------}
var    cnt : integer;

begin
   clrscr;   writeln;
   writeln (' Input of the estimated parameters :');   writeln;
   for cnt := 1 to dim do
   begin
       write (' k(',cnt,') = ? ');   readln (x[cnt])
   end;
   xh := x;   sm := 1e30
end;    { Input }

procedure InvertMatrix;
{-----------------------}

   procedure Fill;
   {---------------}
   var    row,col : integer;

   begin
      for row := 1 to dim do
      begin
         for col := 1 to dim do
         begin
            if (col = row) then
            begin
               matb[row,col] := 1;   matv[row,col] := 1
            end
            else
            begin
               matb[row,col] := 0;   matv[row,col] := 0
            end
         end
      end
   end;    { Fill }

   procedure Pivot;
   {----------------}
   var    pcnt,cnt,t : integer;
          s : real;

      procedure Exchange;
      {-------------------}
      var    tcnt : integer;

      begin
         for tcnt := 1 to dim do
         begin
            s:= mata[pcnt,tcnt];
```

```
            mata[pcnt,tcnt] := mata[t,tcnt];
            mata[t,tcnt] := s
         end;
         if (abs(mata[pcnt,pcnt]) <= 1e-30) then
         begin
            writeln (' No inversion is possible !');    Halt
         end
         else
         begin
            matv[pcnt,pcnt] := 0;    matv[t,t] := 0;
            matv[pcnt,t] := 1;    matv[t,pcnt] := 1
         end
   end;    { Exchange }

   procedure GJ_Elimination;
   {-----------------------}
   var   cnt1,cnt2 : integer;

   begin
      for cnt1 := 1 to dim do
      begin
         for cnt2 := 1 to dim do
         begin
            if (cnt1 = pcnt) then
             if (cnt2 = cnt1) then
              matu[pcnt,pcnt] := 1/mata[pcnt,pcnt]
             else
              matu[cnt1,cnt2] := -mata[cnt1,cnt2]/mata[pcnt,pcnt]
            else
            begin
               if (cnt2 = pcnt) then
                matu[cnt1,pcnt] := mata[cnt1,pcnt]/mata[pcnt,pcnt]
               else
                matu[cnt1,cnt2] := mata[cnt1,cnt2]-mata[pcnt,cnt2]
                                 *mata[cnt1,pcnt]/mata[pcnt,pcnt]
            end
         end
      end
   end;    { GJ_Elimination }

   procedure Multiplication;
   {------------------------}
   var   cnt1,cnt2,cnt3 : integer;

   begin
      for cnt1 := 1 to dim do
      begin
         for cnt2 := 1 to dim do
         begin
            matw[cnt1,cnt2] := 0;
            for cnt3 := 1 to dim do
             matw[cnt1,cnt2] := matw[cnt1,cnt2]+matv[cnt1,cnt3]
                              *matb[cnt3,cnt2]
         end
      end;
```

```
         matb := matw;
         for cnt1 := 1 to dim do
         begin
            for cnt2 :=  1 to dim do
            begin
               mata[cnt1,cnt2] := matu[cnt1,cnt2];
               if (cnt1 = cnt2) then matv[cnt1,cnt2] := 1
               else matv[cnt1,cnt2] := 0
            end
         end
      end;  { Multiplication }

   begin
      for pcnt := 1 to dim do
      begin
         s := 0;
         for cnt := pcnt to dim do
          if (s <= abs(mata[cnt,pcnt])) then
          begin
             s := abs(mata[cnt,pcnt]);   t := cnt
          end;
         Exchange;
         GJ_Elimination;
         Multiplication
      end
   end;   { Pivot }

   procedure Result;
   {-----------------}
   var   cnt1,cnt2,cnt3 : integer;

   begin
      for cnt1 := 1 to dim do
      begin
         for cnt2 := 1 to dim do
         begin
            matw[cnt1,cnt2] := 0;
            for cnt3 := 1 to dim do
            matw[cnt1,cnt2] := matw[cnt1,cnt2]+mata[cnt1,cnt3]
                                  *matb[cnt3,cnt2]
         end
      end
   end;   { Result }

begin
   Fill;
   Pivot;
   Result
end;   { InvertMatrix }

function sqsum(x1,x2,x3 : real) : real;
{---------------------------------------}
var   hy1,hy2,zs : real;
      i : integer;
```

```
begin
   zs := 0;
   for i := 1 to mp do
   begin
      hy1 := x1+x2*exp(-x3*xx[i]);    hy2 := hy1-yy[i];
      zs := zs+sqr(hy2)
   end;
   sqsum := zs
end;    { sqsum }

procedure PartDeriv;
{-------------------}
var    diff,hx,hy : real;
       cnt1,cnt2,cntp : integer;

   procedure Allfcts;
   {-----------------}

   begin
      fnf1[1] := 1;    fnf2[1,1] := 0;    fnf2[1,2] := 0;
      fnf2[1,3] := 0;    fnf1[2] := exp(-k[3]*hx);    fnf2[2,2] := 0;
      fnf2[2,3] := -hx*fnf1[2];    fnf1[3] := k[2]*fnf2[2,3];
      fnf2[3,3] := -hx*fnf1[3]
   end;    { Allfcts }

begin
   for cnt1 := 1 to dim do
   begin
      fnfh[cnt1] := 0;    k[cnt1] := xh[cnt1];
      for cnt2 := 1 to dim do mata[cnt1,cnt2] := 0
   end;
   for cntp := 1 to mp do
   begin
      hx := xx[cntp];
      Allfcts;
      hy := k[1]+k[2]*fnf1[2];    diff := (yy[cntp]-hy);
      for cnt1 := 1 to dim do
      begin
         fnfh[cnt1] := fnfh[cnt1]-diff*fnf1[cnt1];
         for cnt2 := cnt1 to dim do
         mata[cnt1,cnt2] := mata[cnt1,cnt2]-diff*fnf2[cnt1,cnt2]
                            +fnf1[cnt1]*fnf1[cnt2]
      end
   end;
   for cnt1 := 1 to dim do
      for cnt2 := cnt1 to dim do mata[cnt2,cnt1] := mata[cnt1,cnt2]
end;    { PartDeriv }

procedure Optfac;
{----------------}
var    help : array[1..dim] of real;
       ff,df,fm : real;
       cnt1,cnt2 : integer;
```

```
begin
   for cnt1 := 1 to dim do
   begin
      help[cnt1] := 0;
      for cnt2 := 1 to dim do
        help[cnt1] := help[cnt1]+matw[cnt1,cnt2]*fnfh[cnt2];
      k[cnt1] := xh[cnt1]-ff*help[cnt1]
   end;
   ff := 0.6;   fm := 0;   df := 0.5;
   while (sqsum(k[1],k[2],k[3]) < sm) do
   begin
    sm := sqsum(k[1],k[2],k[3]);   fm := ff;   ff := ff+df;
    for cnt1 := 1 to dim do k[cnt1] := xh[cnt1]-ff*help[cnt1]
   end;
   ff := fm-df;
   for cnt1 := 1 to dim do k[cnt1] := xh[cnt1]-ff*help[cnt1];
   while (sqsum(k[1],k[2],k[3]) < sm) do
   begin
    sm := sqsum(k[1],k[2],k[3]);   fm := ff;   ff := ff-df;
    for cnt1 := 1 to dim do k[cnt1] := xh[cnt1]-ff*help[cnt1]
   end;
   for cnt1 := 1 to mp+1 do
   begin
      df := df/2;   ff := fm+df;
      for cnt2 := 1 to dim do k[cnt2] := xh[cnt2]-ff*help[cnt2];
      if (sqsum(k[1],k[2],k[3]) < sm) then
      begin
         sm := sqsum(k[1],k[2],k[3]);   fm := ff
      end
      else
      begin
         ff := fm-df;
         for cnt2 := 1 to dim do k[cnt2] := xh[cnt2]-ff*help[cnt2];
         if (sqsum(k[1],k[2],k[3]) < sm) then
         begin
            sm := sqsum(k[1],k[2],k[3]);   fm := ff
         end
      end
   end;
   ff := fm;
   for cnt1 := 1 to dim do xh[cnt1] := xh[cnt1]-ff*help[cnt1]
end;   { Optfac }

procedure Output;
{-----------------}
var    ss : real;
       cnt : integer;

begin
   ss := sqsum(k[1],k[2],k[3]);   writeln;
   for cnt := 1 to dim do
    writeln (' k(',cnt,') = ',xh[cnt]:12:8,' +/- ',
             sqr(abs(matw[cnt,cnt]*ss/(mp-dim))));
   writeln (' Sum of squared errors = ',ss)
end;   { Output }
```

```
begin
   Input;
   z := 1;
   repeat
    PartDeriv;
    InvertMatrix;
    Output;
    Optfac;
    z := z+1
   until (z > 10)
end.
```

```
                            NL_MULT
```

```
program nl_mult;
    { J. Kinkel, January 1989, based on Ebert/Ederer }

uses    crt;

const   dim = 3;   mp = 64;
        xx : array[1..64] of real
             = (0.000,0.005,0.010,0.015,0.020,0.025,0.030,
                0.035,0.040,0.045,0.050,0.055,0.060,0.065,
                0.070,0.075,0.080,0.085,0.090,0.095,0.100,
                0.105,0.110,0.115,0.120,0.125,0.130,0.135,
                0.140,0.145,0.150,0.155,0.160,0.165,0.170,
                0.175,0.180,0.185,0.190,0.195,0.200,0.205,
                0.210,0.215,0.220,0.225,0.230,0.235,0.240,
                0.245,0.250,0.255,0.260,0.265,0.270,0.275,
                0.600,0.605,0.610,0.615,0.620,0.625,0.630,
                0.635);
        yy : array[1..64] of real
             = (1.000000,0.854415,0.734266,0.627988,0.535742,
                0.459174,0.395416,0.337545,0.288678,0.249475,
                0.213886,0.185657,0.157447,0.135157,0.118451,
                0.099275,0.085432,0.074251,0.063035,0.054768,
                0.047787,0.039426,0.034518,0.030223,0.026509,
                0.023732,0.019806,0.014693,0.011894,0.009542,
                0.006593,0.005535,0.005540,0.007713,0.003393,
                0.004965,0.001981,0.003614,0.000934,0.0000215,
                0.0002637,0.0013670,0.0011571,-0.0006458,
                0.0019994,-0.0035763,-0.0026426,-0.0023520,
                -0.0026480,-0.0054493,-0.0026830,-0.0028068,
                -0.0018326,-0.0008719,-0.0006889,-0.0008531,
                -0.0052879,-0.0058611,-0.0047900,-0.0052341,
                -0.0022847,-0.0024500,-0.0030974,-0.0);

var    mata,matb,matu,matv,matw,fnf2 : array[1..dim,1..dim] of real;
       x,y,k,fnf1,fnfh,xh : array[1..dim] of real;
       sm : real;
       z : integer;
```

```
procedure Input;
{-----------------}
var   cnt : integer;

begin
   clrscr;   writeln;
   writeln (' Input of the estimated parameters :');   writeln;
   for cnt := 1 to dim do
   begin
      write (' k(',cnt,') = ? ');   readln (x[cnt])
   end;
   xh := x;   sm := 1e30
end;   { Input }

procedure InvertMatrix;
{-----------------------}

   procedure Fill;
   {---------------}
   var   row,col : integer;

   begin
      for row := 1 to dim do
      begin
         for col := 1 to dim do
         begin
            if (col = row) then
            begin
               matb[row,col] := 1;   matv[row,col] := 1
            end
            else
            begin
               matb[row,col] := 0;   matv[row,col] := 0
            end
         end
      end
   end;   { Fill }

   procedure Pivot;
   {----------------}
   var   pcnt,cnt,t : integer;
         s : real;

      procedure Exchange;
      {-------------------}
      var   tcnt : integer;

      begin
         for tcnt := 1 to dim do
         begin
            s:= mata[pcnt,tcnt];
            mata[pcnt,tcnt] := mata[t,tcnt];
            mata[t,tcnt] := s
         end;
         if (abs(mata[pcnt,pcnt]) <= 1e-30) then
         begin
            writeln (' No inversion is possible !');   Halt
         end
```

```
         else
         begin
            matv[pcnt,pcnt] := 0;    matv[t,t] := 0;
            matv[pcnt,t] := 1;    matv[t,pcnt] := 1
         end
   end;    { Exchange }

procedure GJ_Elimination;
{------------------------}
var    cnt1,cnt2 : integer;

begin
   for cnt1 := 1 to dim do
   begin
      for cnt2 := 1 to dim do
      begin
         if (cnt1 = pcnt) then
          if (cnt2 = cnt1) then
           matu[pcnt,pcnt] := 1/mata[pcnt,pcnt]
          else
           matu[cnt1,cnt2] := -mata[cnt1,cnt2]/mata[pcnt,pcnt]
         else
         begin
            if (cnt2 = pcnt) then
             matu[cnt1,pcnt] := mata[cnt1,pcnt]/mata[pcnt,pcnt]
            else
             matu[cnt1,cnt2] := mata[cnt1,cnt2]-mata[pcnt,cnt2]
                              *mata[cnt1,pcnt]/mata[pcnt,pcnt]
         end
      end
   end
end;    { GJ_Elimination }

procedure Multiplication;
{------------------------}
var    cnt1,cnt2,cnt3 : integer;

begin
   for cnt1 := 1 to dim do
   begin
      for cnt2 := 1 to dim do
      begin
         matw[cnt1,cnt2] := 0;
         for cnt3 := 1 to dim do
          matw[cnt1,cnt2] := matw[cnt1,cnt2]+matv[cnt1,cnt3]
          *matb[cnt3,cnt2]
      end
   end;
   matb := matw;
   for cnt1 := 1 to dim do
   begin
      for cnt2 :=  1 to dim do
      begin
         mata[cnt1,cnt2] := matu[cnt1,cnt2];
         if (cnt1 = cnt2) then matv[cnt1,cnt2] := 1
         else matv[cnt1,cnt2] := 0
```

```
                  end
               end
            end;  { Multiplication }

       begin
          for pcnt := 1 to dim do
          begin
             s := 0;
             for cnt := pcnt to dim do
              if (s <= abs(mata[cnt,pcnt])) then
              begin
                 s := abs(mata[cnt,pcnt]);    t := cnt
              end;
             Exchange;
             GJ_Elimination;
             Multiplication
          end
       end;   { Pivot }

       procedure Result;
       {------------------}
       var    cnt1,cnt2,cnt3 : integer;

       begin
          for cnt1 := 1 to dim do
          begin
             for cnt2 := 1 to dim do
             begin
                matw[cnt1,cnt2] := 0;
                for cnt3 := 1 to dim do
                matw[cnt1,cnt2] := matw[cnt1,cnt2]+mata[cnt1,cnt3]
                                      *matb[cnt3,cnt2]
             end
          end
       end;   { Result }

    begin
       Fill;
       Pivot;
       Result
    end;   { InvertMatrix }

    function sqsum(x1,x2,x3 : real) : real;
    {---------------------------------------}
    var    hy1,hy2,zs : real;
           i : integer;

    begin
       zs := 0;
       for i := 1 to mp do
       begin
          hy1 := x1+x2*exp(-x3*xx[i]);   hy2 := hy1-yy[i];
          zs := zs+sqr(hy2)
       end;
       sqsum := zs
    end;   { sqsum }
```

```
procedure PartDeriv;
{--------------------}
var   diff,hx,hy : real;
      cnt1,cnt2,cntp : integer;

   procedure Allfcts;
   {-----------------}
   begin
      fnf1[1] := 1;    fnf2[1,1] := 0;    fnf2[1,2] := 0;    fnf2[1,3] := 0;
      fnf1[2] := exp(-k[3]*hx);    fnf2[2,2] := 0;
      fnf2[2,3] := -hx*fnf1[2];    fnf1[3] := -k[2]*hx*fnf1[2];
      fnf2[3,3] := -hx*fnf1[3]
   end;    { Allfcts }

begin
   for cnt1 := 1 to dim do
   begin
      fnfh[cnt1] := 0;    k[cnt1] := xh[cnt1];
      for cnt2 := 1 to dim do mata[cnt1,cnt2] := 0
   end;
   for cntp := 1 to mp do
   begin
      hx := xx[cntp];
      Allfcts;
      hy := k[1]+k[2]*fnf1[2];    diff := (yy[cntp]-hy);
      for cnt1 := 1 to dim do
      begin
         fnfh[cnt1] := fnfh[cnt1]-diff*fnf1[cnt1];
         for cnt2 := cnt1 to dim do
           mata[cnt1,cnt2] := mata[cnt1,cnt2]+fnf1[cnt1]*fnf1[cnt2]
      end
   end;
   for cnt1 := 1 to dim do
    for cnt2 := cnt1 to dim do
     mata[cnt2,cnt1] := mata[cnt1,cnt2]
end;    { PartDeriv }

procedure Optfac;
{----------------}
var   help : array[1..dim] of real;
      ff,df,fm : real;
      cnt1,cnt2 : integer;

begin
   for cnt1 := 1 to dim do
   begin
      help[cnt1] := 0;
      for cnt2 := 1 to dim do
       help[cnt1] := help[cnt1]+matw[cnt1,cnt2]*fnfh[cnt2];
      k[cnt1] := xh[cnt1]-ff*help[cnt1]
   end;
   ff := 0.6;    fm := 0;    df := 0.5;
   while (sqsum(k[1],k[2],k[3]) < sm) do
   begin
    sm := sqsum(k[1],k[2],k[3]);    fm := ff;    ff := ff+df;
```

```
                 for cnt1 := 1 to dim do k[cnt1] := xh[cnt1]-ff*help[cnt1]
             end;
             ff := fm-df;
             for cnt1 := 1 to dim do k[cnt1] := xh[cnt1]-ff*help[cnt1];
             while (sqsum(k[1],k[2],k[3]) < sm) do
             begin
               sm := sqsum(k[1],k[2],k[3]);   fm := ff;    ff := ff-df;
               for cnt1 := 1 to dim do k[cnt1] := xh[cnt1]-ff*help[cnt1]
             end;
             for cnt1 := 1 to mp+1 do
             begin
                 df := df/2;    ff := fm+df;
                 for cnt2 := 1 to dim do k[cnt2] := xh[cnt2]-ff*help[cnt2];
                 if (sqsum(k[1],k[2],k[3]) < sm) then
                 begin
                     sm := sqsum(k[1],k[2],k[3]);    fm := ff
                 end
                 else
                 begin
                     ff := fm-df;
                     for cnt2 := 1 to dim do k[cnt2] := xh[cnt2]-ff*help[cnt2];
                     if (sqsum(k[1],k[2],k[3]) < sm) then
                     begin
                         sm := sqsum(k[1],k[2],k[3]);    fm := ff
                     end
                 end
             end;
             ff := fm;
             for cnt1 := 1 to dim do xh[cnt1] := xh[cnt1]-ff*help[cnt1]
         end;    { Optfac }

      procedure Output;
      {-----------------}
      var   ss : real;
            cnt : integer;

      begin
          ss := sqsum(k[1],k[2],k[3]);    writeln;
          for cnt := 1 to dim do
           writeln (' k(',cnt,') = ',xh[cnt]:12:8,' +/- ',
                     sqr(abs(matw[cnt,cnt]*ss/(mp-dim))));
          writeln (' Sum of squared errors = ',ss)
      end;    { Output }

   begin
      Input;
      z := 1;
      repeat
       PartDeriv;
       InvertMatrix;
       Output;
       Optfac;
       z := z+1
      until (z > 10)
   end.
```

```
                                    SIMPLEX
```

```
program simplex;
    { J. Kinkel, January 1989, based on Ebert/Ederer }

uses   crt;

const   dim = 3;   dim1 = 4;   mp = 64;
        xx : array[1..64] of real
            = (0.000,0.005,0.010,0.015,0.020,0.025,0.030,0.035,
               0.040,0.045,0.050,0.055,0.060,0.065,0.070,0.075,
               0.080,0.085,0.090,0.095,0.100,0.105,0.110,0.115,
               0.120,0.125,0.130,0.135,0.140,0.145,0.150,0.155,
               0.160,0.165,0.170,0.175,0.180,0.185,0.190,0.195,
               0.200,0.205,0.210,0.215,0.220,0.225,0.230,0.235,
               0.240,0.245,0.250,0.255,0.260,0.265,0.270,0.275,
               0.600,0.605,0.610,0.615,0.620,0.625,0.630,0.635);
        yy : array[1..64] of real
            = (1.000000,0.854415,0.734266,0.627988,0.535742,
               0.459174,0.395416,0.337545,0.288678,0.249475,
               0.213886,0.185657,0.157447,0.135157,0.118451,
               0.099275,0.085432,0.074251,0.063035,0.054768,
               0.047787,0.039426,0.034518,0.030223,0.026509,
               0.023732,0.019806,0.014693,0.011894,0.009542,
               0.006593,0.005535,0.005540,0.007713,0.003393,
               0.004965,0.001981,0.003614,0.000934,0.0000215,
               0.0002637,0.0013670,0.0011571,-0.0006458,0.0019994,
               -0.0035763,-0.0026426,-0.0023520,-0.0026480,-0.0054493,
               -0.0026830,-0.0028068,-0.0018326,-0.0008719,-0.0006889,
               -0.0008531,-0.0052879,-0.0058611,-0.0047900,-0.0052341,
               -0.0022847,-0.0024500,-0.0030974,-0.0);

type    region = array[1..dim] of real;
        region1 = array[1..dim1] of real;

var     simp : array[1..dim1,1..dim] of real;
        lsq : region1;
        koeff,op : region;
        allc : integer;

 procedure Input;
{----------------}
 var   cnt : integer;

 begin
    clrscr;   writeln;
    writeln (' Input of the estimated parameters');
    writeln (' These parameters must not be zero !');   writeln;
    for cnt := 1 to dim do
    begin
       write (' k(',cnt,') = ');   readln (koeff[cnt])
    end;
    op := koeff
 end;   { Input }
```

```
function sqsum(x1,x2,x3 : real) : real;
{--------------------------------------}
var   hy1,hy2,ts : real;
      i : integer;

begin
   ts := 0;
   for i := 1 to mp do
   begin
      hy1 := x1+x2*exp(-x3*xx[i]);   hy2 := hy1-yy[i];
      ts := ts+sqr(hy2)
   end;
   sqsum := ts
end;   { sqsum }

procedure InitSimp;
{------------------}
var   ssd : real;
      cnt1,cnt2 : integer;

begin
   for cnt1 := 1 to dim do simp[1,cnt1] := koeff[cnt1];
   lsq[1] := sqsum(koeff[1],koeff[2],koeff[3]);
   for cnt2 := 1 to dim do
   begin
      koeff[cnt2] := 1.2*koeff[cnt2];
      lsq[cnt2+1] := sqsum(koeff[1],koeff[2],koeff[3]);
      for cnt1 := 1 to dim do simp[cnt2+1,cnt1] := koeff[cnt1];
      koeff[cnt2] := op[cnt2]
   end
end;   { InitSimp }

procedure SortSimp;
{------------------}
var   lsqmin,tmp : real;
      cnt1,cnt2,cmin : integer;

begin
   for cnt1 := 1 to dim do
   begin
      lsqmin := lsq[cnt1];   cmin := cnt1;
      for cnt2 := cnt1 to dim1 do
      begin
         if (lsq[cnt2] <= lsqmin) then
         begin
            lsqmin := lsq[cnt2];   cmin := cnt2
         end
      end;
      for cnt2 := 1 to dim do
      begin
       . tmp := simp[cnt1,cnt2]; simp[cnt1,cnt2] := simp[cmin,cnt2];
         simp[cmin,cnt2] := tmp
      end;
      tmp := lsq[cnt1];   lsq[cnt1] := lsq[cmin];   lsq[cmin] := tmp
   end
end;   { SortSimp }
```

```
procedure Output;
{-----------------}
var   cnt : integer;

begin
   writeln;
   for cnt := 1 to dim do
     writeln (' k(',cnt,') = ',simp[1,cnt],' (',simp[dim1,cnt],')');
   writeln (' Sum of squared errors = ',lsq[1],' (',lsq[dim1],')')
end;   { Output }

procedure Calc (var k : region);
{------------------------------}
var   mean,refl,drefl,contr : region;
      lsqh,lrefl,ldrefl,lcontr : real;

   procedure MeanVector;
   {-------------------}
   var   cnt1,cnt2 : integer;

   begin
      for cnt1 := 1 to dim do mean[cnt1] := 0;
      for cnt1 := 1 to dim do
      begin
         for cnt2 := 1 to dim do
           mean[cnt1] := mean[cnt1]+simp[cnt2,cnt1];
         mean[cnt1] := mean[cnt1]/dim
      end
   end;   { MeanVector }

   procedure Reflection;
   {-------------------}
   var   cnt : integer;

   begin
      for cnt := 1 to dim do
      begin
         refl[cnt] := 2*mean[cnt]-simp[dim1,cnt];   k[cnt] := refl[cnt]
      end;
      lrefl := sqsum (k[1],k[2],k[3])
   end;   { Reflection }

   procedure DoubleRefl;
   {-------------------}
   var   cnt : integer;

   begin
      for cnt := 1 to dim do
      begin
         drefl[cnt] := 3*mean[cnt]-2*simp[dim1,cnt];
         k[cnt] := drefl[cnt]
      end;
      ldrefl := sqsum (k[1],k[2],k[3])
   end;   { DoubleRefl }
```

```pascal
   procedure Contraction;
   {----------------------}
   var    cnt : integer;

   begin
      for cnt := 1 to dim do
      begin
        contr[cnt] := (mean[cnt]+simp[dim1,cnt])/2;
        k[cnt] := contr[cnt]
      end;
      lcontr := sqsum (k[1],k[2],k[3])
   end;   { Contraction }

   procedure Decide;
   {-----------------}
   var    cnt : integer;

   begin
      if (lrefl <= lsq[dim1]) then
      begin
        for cnt := 1 to dim do simp[dim1,cnt] := refl[cnt];
        lsq[dim1] := lrefl
      end;
      if (ldrefl <= lsq[dim1]) then
      begin
        for cnt := 1 to dim do simp[dim1,cnt] := drefl[cnt];
        lsq[dim1] := ldrefl
      end;
      if (lcontr <= lsq[dim1]) then
      begin
        for cnt := 1 to dim do simp[dim1,cnt] := contr[cnt];
        lsq[dim1] := lcontr
      end
   end;   { Decide }

   procedure ContrBp;
   {------------------}
   var    cnt1, cnt2 : integer;

   begin
      for cnt1 := 2 to dim1 do
      begin
        for cnt2 := 1 to dim do
        begin
           simp[cnt1,cnt2] := (simp[1,cnt2]+simp[cnt1,cnt2])/2;
           k[cnt2] := simp[cnt1,cnt2]
        end;
        lsq[cnt1] := sqsum (k[1],k[2],k[3])
      end
   end;   { ContrBp }

begin
   MeanVector;
   Reflection;
   DoubleRefl;
```

```
    Contraction;
    lsqh := lsq[dim1];
    Decide;
    if (lsqh <= lsq[dim1]) then ContrBp
 end;    { Calc }

begin
   allc := 0;
   Input;
   InitSimp;
   repeat
    SortSimp;
    Output;
    Calc (koeff);
    allc := allc+1
   until (allc > 199)
end.
```

```
┌─────────────────────────────────────────────────────────────┐
│                          ER_KIN                             │
└─────────────────────────────────────────────────────────────┘
```

```
program er_kin;
    { J. Kinkel, January 1989, based on Ebert/Ederer }

uses    crt;

const    er : array[1..9] of string[80] =
               ('C2H6              ==>   CH3     + CH3',
                'C2H6  + CH3   ==>   C2H5    + CH4',
                'C2H5              ==>   C2H4    + H',
                'C2H6  + H     ==>   C2H5    + H2',
                'CH3   + C2H5  ==>   C3H8',
                'CH3   + H     ==>   CH4',
                'C2H5  + H     ==>   H2      + C2H4',
                'C2H5  + C2H5  ==>   C2H4    + C2H6',
                'H     + H     ==>   H2');
         dim = 9;

type    mat = array[1..40,1..4] of string[40];

var    lstr,rstr : mat;
       name : array[1..20] of string[40];
       left,right : array[1..40] of integer;
       cnt,ncnt : integer;

 procedure Decomposition;
{------------------------}
 var   hstr : string[80];
       i,z1,z2 : integer;

    procedure LeftSide;
   {-------------------}
    begin
       repeat
        if (hstr[i] = '+') then
```

```
       begin
          lstr[cnt,z1] := copy(hstr,z2,i-z2);
          i := i+1;   z1 := z1+1;   z2 := i
       end
       else i := i+1
     until (hstr[i] = '=');
     lstr[cnt,z1] := copy(hstr,z2,i-z2);
     left[cnt] := z1
   end;   { LeftSide }

   procedure RightSide;
   {-------------------}
   begin
     repeat
       if (hstr[i] = '+') then
       begin
          rstr[cnt,z1] := copy(hstr,z2,i-z2);
          i := i+1;   z1 := z1+1;   z2 := i
       end
       else i := i+1
     until (i = length(hstr));
     rstr[cnt,z1] := copy(hstr,z2,i-z2+1);
     right[cnt] := z1
   end;    { RightSide }

begin
   hstr := '';
   writeln (cnt:3,' ':10,er[cnt]);
   for i := 1 to length(er[cnt]) do
    if not(er[cnt][i] = ' ') then hstr := hstr+er[cnt][i];
   i := 1;   z1 := 1;   z2 := 1;
   LeftSide;
   i := i+3;   z1 := 1;   z2 := i;
   RightSide
end;   { Decomposition }

procedure Substances;
{--------------------}
var   i : integer;

   procedure Fillname (n:integer ; st:mat);
   {---------------------------------------}
   var   j,k : integer;
         flag : boolean;

   begin
      for j := 1 to n do
      begin
         flag := false;
         for k := 1 to ncnt do
            if (st[i,j] = name[k]) then flag := true;
         if not(flag) then
         begin
            ncnt := ncnt+1;
            name[ncnt] := st[i,j]
```

```
                end
             end
         end;    { Fillname }

   begin
      fillchar(name[1],800,chr($00));
      ncnt := 1;
      name[1] := lstr[1,1];
      for i := 1 to dim do
      begin
          Fillname (left[i],lstr);
          Fillname (right[i],rstr)
      end;
      for i := 1 to ncnt do writeln (' y(',i,')  =  ',name[i]);
      writeln
   end;   { Substances }

   procedure ReactionRates;
   {-----------------------}
   var   i,j,k : integer;

   begin
      for i := 1 to dim do
      begin
          write (' r(',i,')  =   k(',i,') * ');
          for j := 1 to left[i] do
           for k := 1 to ncnt do
            if (name[k] = lstr[i,j]) then
            begin
               write ('y(',k,')');
               if not(j = left[i]) then write (' * ')
            end;   writeln
      end;   writeln
   end;   { ReactionRates }

   procedure Diffeq;
   {----------------}
   var   i,j : integer;

      procedure ode (n:integer ; st:mat ; h:char);
      {-------------------------------------------}
      var   k : integer;

      begin
         for k := 1 to n do
          if (name[i] = st[j,k]) then write (h,'r(',j,')')
      end;   { ode }

   begin
      for i := 1 to ncnt do
      begin
          write (' d/dt y(',i,') = ');
          for j := 1 to dim do
          begin
             ode(left[j],lstr,'-');   ode(right[j],rstr,'+')
          end;   writeln
```

```
            end
      end;    { Diffeq }

begin
      clrscr;    fillchar(name[1],1600,chr($00));
      fillchar(lstr[1,1],6400,chr($00));    fillchar(rstr[1,1],6400,chr($00));
      fillchar(left[1],80,chr($00));    fillchar(right[1],80,chr($00));
      for cnt := 1 to dim do Decomposition;    writeln;
      Substances;
      ReactionRates;
      Diffeq
end.
```

```
                                    MM
```

```
program mm;
      { J. Kinkel, January 1989, based on Ebert/Ederer }

uses    crt;

const    z : array[0..9] of char =
                  ('a','b','c','d','e','f','g','h','i','j');

type    region = array[1..5] of char;

var    po : integer;
       vor : region;

 procedure Instruction;
{----------------------}
 begin
      clrscr;    writeln;
      writeln ('M A S T E R M I N D':48);    writeln;
      writeln ('    You have to guess a combination of');
      writeln ('    5 letters out of   a b c d e f g h i j .');
      writeln ('    The program tells you the number of');
      writeln ('    correct positions and the number of');
      writeln ('    additionally correct letters');    writeln;
      writeln ('    You have 15 guesses !');    writeln
 end;    { Instruction }

 procedure Selection;
{--------------------}
 var    cnt,x : integer;

 begin
      randomize;
      for cnt := 1 to 5 do
      begin
          x := random(9);    vor[cnt] := z[x]
      end
 end;    { Selection }
```

```
procedure Game;
{---------------}
var   h1,ein1,h2 : region;
      spz,cnt,le,zn : integer;

   function position : integer;
   {-------------------------}
     var   cp,cnt : integer;

   begin
      position := 0;   cp := 0;
      for cnt := 1 to 5 do
        if (h1[cnt] = h2[cnt]) then
        begin
           cp := cp+1;   h2[cnt] := '1';   h1[cnt] := '0'
        end;
      position := cp
   end;   { position }

   function letter : integer;
   {-------------------------}
     var   cb,cnt1,cnt2 : integer;

   begin
      letter := 0;   cb := 0;
      for cnt1 := 1 to 5 do
       for cnt2 := 1 to 5 do
        if (h1[cnt2] = h2[cnt1]) then
        begin
           cb := cb+1;   h2[cnt1] := '1';   h1[cnt2] := '0'
        end;
      letter := cb
   end;  { letter }

begin
   writeln;
   writeln ('   Now your guess of a combination :');   writeln;
   spz := 1;
   repeat
   h1 := vor;   zn := spz+14;
   if (zn > 25) then zn := 25;
   gotoxy (2,zn);   write (spz);
   for cnt := 1 to 5 do
   begin
      repeat
       gotoxy (cnt*5+4,zn);
       ein1[cnt] := readkey
      until (ein1[cnt] in ['a'..'j']);
      write (ein1[cnt]);   h2[cnt] := ein1[cnt]
   end;
   po := position;   le := letter;
   gotoxy (35,zn);   writeln (' Pos = ',po,'Letters = ':20,le);
   if (po = 5) then spz := 16 else spz := spz+1
   until (spz > 15)
end;   { Game }
```

```
procedure Result;
{------------------}
 var   cnt : integer;

 begin
    if not(po = 5) then
    begin
       writeln;   writeln (' Sorry - you didn''t guess');
       write (' The correct combination is :');
       for cnt := 1 to 5 do write (vor[cnt]:5);   writeln;   writeln
    end
    else writeln ('   My congratulations !')
 end;   { Result }

begin
   Instruction;
   Selection;
   Game;
   Result
end.
```

GOL

```
program gol;
   { J. Kinkel, January 1989, based on Ebert/Ederer }

uses    crt;

var   x,y : array[0..21,0..21] of byte;
      l : array[0..17] of byte;
      gen : integer;
      answer : char;

 procedure Input;
{----------------}
 begin
    clrscr;   writeln;
    writeln (' You have three different possibilities to');
    writeln (' choose the initial population :');   writeln;
    writeln ('  1  -  free programmed population');
    writeln ('  2  -  initial population with a mask');
    writeln ('  3  -  random initial population');   writeln;
    write (' Enter your selection : ');   readln (answer)
 end;   { Input }

 procedure Start;
{----------------}
 begin
    x[8,10] := 1;   x[8,11] := 1;   x[9,9] := 1;
    x[9,10] := 1;   x[10,10] := 1;   gen := 1
 end;   { Start }
```

```
procedure Mask;
{---------------}
var   mstr : array[1..20] of string[20];
      cnt1,cnt2 : integer;

begin
   mstr[1]   := '                    ';
   mstr[2]   := '                    ';
   mstr[3]   := '                    ';
   mstr[4]   := '                    ';
   mstr[5]   := '          **        ';
   mstr[6]   := '          **        ';
   mstr[7]   := '                    ';
   mstr[8]   := '         ****       ';
   mstr[9]   := '    ** *     *      ';
   mstr[10]  := '    ** **    *      ';
   mstr[11]  := '      *  * * **     ';
   mstr[12]  := '      * *  * **     ';
   mstr[13]  := '        ****        ';
   mstr[14]  := '                    ';
   mstr[15]  := '        **          ';
   mstr[16]  := '        **          ';
   mstr[17]  := '                    ';
   mstr[18]  := '                    ';
   mstr[19]  := '                    ';
   mstr[20]  := '                    ';
   for cnt1 := 1 to 20 do
    for cnt2 := 1 to 20 do
     if (mstr[cnt1][cnt2] = '*') then x[cnt1,cnt2] := 1
     else x[cnt1,cnt2] := 0;
   gen := 1
end;   { Mask }

procedure Rip;
{--------------}
var   p : real;
      cnt1,cnt2 : integer;

begin
   writeln;
   write (' Enter population density between 0 and 1 : ');   readln(p);
   randomize;
   for cnt1 := 1 to 20 do
    for cnt2 := 1 to 20 do
     if (random < p) then x[cnt1,cnt2] := 1 else x[cnt1,cnt2] := 0;
   gen := 1
end;   { Rip }

procedure livenletdie;
{---------------------}
begin
   l[3] := 1;   l[11] := 1;   l[12] := 1
end;   { livenletdie }
```

```
procedure Printgen;
{-------------------}
var   cnt1,cnt2 : integer;

begin
   clrscr;
   for cnt1 := 1 to 20 do
   begin
      writeln;
      for cnt2 := 1 to 20 do
        if (x[cnt1,cnt2] = 0) then write ('  ') else write (' *')
   end;
   writeln;   writeln (gen:15,'. generation');
   repeat
    gotoxy (5,25);   write ('Continue : <ENTER>... ');
   answer := readkey until (answer = chr($0D))
end;   { Printgen }

procedure Newgen;
{----------------}
var   i,j : integer;
      h : integer;

begin
   y := x;
   for i := 1 to 20 do
   begin
      y[i,0] := y[i,20];   y[i,21] := y[i,1]
   end;
   for j := 0 to 21 do
   begin
      y[0,j] := y[20,j];   y[21,j] := y[1,j]
   end;
   for i := 1 to 20 do
    for j := 1 to 20 do
     begin
       h := 9*y[i,j]+y[i-1,j-1]+y[i-1,j]+y[i-1,j+1]+y[i,j-1]+y[i,j+1];
       h := h+y[i+1,j-1]+y[i+1,j]+y[i+1,j+1];
       x[i,j] := l[h]
     end;
   gen := gen+1
end;   { Newgen}

begin
   Input;
   fillchar (x[0,0],484,chr($00));   fillchar (y[0,0],484,chr($00));
   fillchar (l[0],18,chr($00));
   case answer of
    '1' : Start;
    '2' : Mask;
    '3' : Rip
   end;
   repeat
    livenletdie;
    Printgen;
    Newgen
   until (gen > 100)
end.
```

```
                          PLAYING
```

```
program playing;
    { J. Kinkel, January 1989, based on Ebert/Ederer }

uses    crt,graph;

var    a1,a2,b1,b2,c1 : real;

 procedure SetPar;
{------------------}
 begin
    randomize;
    a1 := random;    a2 := random;   b1 := random;   b2 := random;
    c1 := random*2
 end;    { SetPar }

 procedure Calc;
{---------------}
 var    gd,gm : integer;
        cnt1,cnt2,x1,x2,y1,y2 : real;
        answer : char;

    procedure PlotData;
    {------------------}
    var    xp1,xp2,yp1,yp2 : integer;

    begin
       xp1 := round(((x1+3)*GetMaxX)/6);
       xp2 := round(((x2+3)*GetMaxX)/6);
       yp1 := round(((y1+3)*GetMaxY)/6);
       yp2 := round(((y2+3)*GetMaxY)/6);
       line (xp1,yp1,xp2,yp2)
    end;    { PlotData }
 begin
    cnt1 := 0;    gd := 0;    gm := 0;
    InitGraph (gd,gm,'');
    repeat
     cnt2 := cnt1+c1;
     x1 := 2*cos(a1*cnt1)-cos(a2*2*cnt1);
     x2 := 2*cos(a1*cnt2)-cos(a2*2*cnt2);
     y1 := 2*sin(b1*cnt1)-sin(b2*2*cnt1);
     y2 := 2*sin(b1*cnt2)-sin(b2*2*cnt2);
     PlotData;
     cnt1 := cnt1 + 0.05
    until (cnt1 > 12.6);
    repeat answer := readkey until (answer = chr($0D));
    CloseGraph;
    textmode(3)
 end;    { Calc }

begin
    SetPar;
    Calc
end.
```

```
                              REGLINP
```

```
program reglinp;
   { J. Kinkel, January 1989, based on Ebert/Ederer }

uses    crt,graph;

type    vector = array[1..100] of real;

var    x,y,w : vector;
       sl,ic,ss,dsl,dic,dev : real;
       ndp : byte;

 procedure Input;
{----------------}
 var    cnt : byte;
        answer : char;

 begin
    clrscr;   writeln;
    write (' How many data points ? ');    readln (ndp);
    for cnt := 1 to ndp do w[cnt] := 1;
    write (' Do you want to weight your data points (y/n) ? ');
    readln (answer);
    for cnt := 1 to ndp do
    begin
       write (' x(',cnt,') = ');   readln (x[cnt]);
       write (' y(',cnt,') = ');   readln (y[cnt]);
       if (answer in ['j','J','y','Y']) then
       begin
          write (' w(',cnt,') = ');   readln (w[cnt])
       end
       else w[cnt] := 1
    end
 end;   { Input }

 procedure Calc;
{---------------}
 var    sum1,sum2,sum3,sum4,sum5,dif1,dif2,den : real;
        cnt : byte;

    function sqsum : real;
   {----------------------}
    var   ts : real;
          cnt : byte;

    begin
       ts := 0;
       for cnt := 1 to ndp do
       ts := ts + w[cnt]*sqr(y[cnt] - ic - sl*x[cnt]);
       sqsum := ts
    end;   { sqsum }
```

```
begin
   sum1 := 0;    sum2 := 0;    sum3 := 0;    sum4 := 0;    sum5 := 0;
   for cnt := 1 to ndp do
   begin
      sum1 := sum1 + w[cnt];    sum2 := sum2 + x[cnt]*w[cnt];
      sum3 := sum3 + y[cnt]*w[cnt];
      sum4 := sum4 + sqr(x[cnt])*w[cnt];
      sum5 := sum5 + x[cnt]*y[cnt]*w[cnt]
   end;
   den := sum1*sum4 - sqr(sum2);
   dif1 := sum3*sum4 - sum5*sum2;    dif2 := sum1*sum5 - sum2*sum3;
   ic := dif1/den;    sl := dif2/den;    ss := sqsum;
   dev := sqrt(ss/(ndp-2));    dic := dev*sqrt(sum4/den);
   dsl := dev*sqrt(sum1/den)
end;    { Calc }

procedure Output;
{-----------------}
var    answer : char;

   procedure Compare;
   {------------------}
   var    cnt : byte;

   begin
      writeln;    writeln('x':7,'y(input)':21,'y(regression)':20);
      for cnt := 1 to ndp do
         writeln (x[cnt]:12:5,y[cnt]:14:5,ic+sl*x[cnt]:20:5)
   end;    { Compare }

   procedure Search;
   {-----------------}
   var    xvalue : real;

   begin
      repeat
         write (' x-value (Stop if x > 99) ? ');    readln (xvalue);
         writeln (' y(',xvalue:5:3,') = ',ic+sl*xvalue:12:8)
      until (xvalue > 99)
   end;    { Search }

begin
   writeln;    writeln (' The regression line for the data is');
   writeln ('      y = ',ic:12:8,' + ',sl:12:8,' * x');    writeln;
   writeln (' Sum of squared deviations = ',ss:12:8);    writeln;
   writeln (' Standard deviation = ',dev:12:8);    writeln;
   writeln (' Number of data points = ',ndp);    writeln;
   writeln ('      Slope = ',sl:12:8,' +/- ',dsl:12:8);
   writeln (' Intercept = ',ic:12:8,' +/- ',dic:12:8);    writeln;
   writeln (' Do you want to compare the input');
   write (' values with the regression values (y/n) ? ');
   readln (answer);
   if (answer in ['j','J','y','Y']) then Compare;    writeln;
   writeln (' Do you want to interpolate data');
   write (' using the regression line (y/n) ? ');    readln (answer);
   if (answer in ['j','J','y','Y']) then Search
end;    { Output }
```

```pascal
procedure PlotData;
{-------------------}
var   x1,y1 : vector;
      xmax,ymax,xmin,ymin : real;
      xp,yp,cnt,gd,gm : integer;
      answer : char;

   procedure Normalize;
   {-------------------}
   var   dx,dy : real;
         cnt : integer;

   begin
      x1 := x;   y1 := y;   xmax := 1e-30;   xmin := 1e30;
      ymax := xmax;   ymin := xmin;
      for cnt := 1 to ndp do
      begin
         if (x[cnt] >= xmax) then xmax := x[cnt];
         if (x[cnt] <= xmin) then xmin := x[cnt];
         if (y[cnt] >= ymax) then ymax := y[cnt];
         if (y[cnt] <= ymin) then ymin := y[cnt]
      end;
      dx := (xmax-xmin)*0.1;   dy := (ymax-ymin)*0.1;
      xmax := xmax+dx;   xmin := xmin-dx;
      ymax := ymax+dy;   ymin := ymin-dy;
      for cnt := 1 to ndp do
      begin
         x1[cnt] := (xmax-x1[cnt])/(xmax-xmin);
         y1[cnt] := (ymax-y1[cnt])/(ymax-ymin)
      end
   end;   { Normalize }

   procedure Cross (i,j : integer);
   {------------------------------}
   begin
      line (i-2,j,i+2,j);   line (i,j-1,i,j+1)
   end;   { Cross }

   procedure Plot;
   {--------------}
   var   xs1,xs2,ys1,ys2 : real;
         xa,xe,ya,ye : integer;

      procedure Setxy (xm,ym : real;var k : integer);
      {---------------------------------------------}
      begin
         x1[k] := (xmax-xm)/(xmax-xmin);
         y1[k] := (ymax-ym)/(ymax-ymin);   k := k+1
      end;   { Setxy }

   begin
      ys1 := ic+sl*xmin;   ys2 := ic+sl*xmax;
      if (sl = 0) then
      begin
         xs1 := 1e30;   xs2 := 1e30
      end
```

```
      else
      begin
         xs1 := (ymin-ic)/sl;   xs2 := (ymax-ic)/sl
      end;
      cnt := ndp+1;
      if (ys1 >= ymin) and (ys1 <= ymax) then Setxy (xmin,ys1,cnt);
      if (ys2 >= ymin) and (ys2 <= ymax) then Setxy (xmax,ys2,cnt);
      if (xs1 >= xmin) and (xs1 <= xmax) then Setxy (xs1,ymin,cnt);
      if (xs2 >= xmin) and (xs2 <= xmax) then Setxy (xs2,ymax,cnt);
      xa := GetMaxX-round(GetMaxX*x1[ndp+1]);
      xe := GetMaxX-round(GetMaxX*x1[ndp+2]);
      ya := round(GetMaxY*y1[ndp+1]);
      ye := round(GetMaxY*y1[ndp+2]);
      line (xa,ya,xe,ye)
   end;   { Plot }

begin
   Normalize;
   gd := 0;   gm := 0;
   InitGraph (gd,gm,'');
   rectangle (0,0,GetMaxX,GetMaxY);
   for cnt := 1 to ndp do
   begin
      xp := GetMaxX-round(GetMaxX*x1[cnt]);
      yp := round(GetMaxY*y1[cnt]);
      Cross(xp,yp)
   end;
   Plot;
   repeat answer := readkey until (answer = chr($0d));
   textmode (3)
end;   { PlotData }

begin
   Input;
   Calc;
   Output;
   PlotData
end.
```

```
┌────────────────────────────────────────────────────┐
│                      KINETIC                         │
└────────────────────────────────────────────────────┘
```

```
program kinetic;
   { J. Kinkel, January 1989, based on Ebert/Ederer }

uses   crt,graph;

const   dim = 4;   vk1 = 1;   vk2 = 1;

var   ya,y1 : array[1..20] of real;
      xa,xe,xd,cmax : real;
      answer : char;
```

```
procedure Input;
{----------------}
begin
   clrscr;
   xa := 0;   xe := 5;   xd := 0.05;   cmax := 1.5;
   y1[1] := 1.5;   y1[2] := 1;   y1[3] := 0;   y1[4] := 0;
end;   { Input }

procedure Calc;
{--------------}
var   y,yh,ode,k1,k2,k3,k4 : array[1..20] of real;
      be,en,x,h : real;
      subi,ccal : integer;
      abb : boolean;

   procedure Graphics;
   {------------------}
   var   gd,gm : integer;

   begin
      gd := 0;   gm := 0;
      InitGraph (gd,gm,'');   rectangle (0,0,GetMaxX,GetMaxY)
   end;   { Graphics }

   procedure SetOde;
   {----------------}
   var   cnt,z : integer;

      procedure diff;
      {--------------}
      begin
         ode[1] := -vk1*yh[1]*yh[2];          ode[2] := ode[1];
         ode[3] := vk1*yh[1]*yh[2]-vk2*yh[3]; ode[4] := vk2*yh[3]
      end; { diff }

   begin
      for z := 1 to dim do y[z] := y1[z];
      for cnt := 0 to subi-1 do
      begin
         x := x+cnt*h;
         for z := 1 to dim do yh[z] := y[z];
         diff;
         for z := 1 to dim do
         begin
            k1[z] := ode[z]*h;   yh[z] := y[z]+k1[z]/2
         end;
         diff;
         for z := 1 to dim do
         begin
            k2[z] := ode[z]*h;   yh[z] := y[z]+k2[z]/2
         end;
         diff;
         for z := 1 to dim do
         begin
            k3[z] := ode[z]*h;   yh[z] := y[z]+k3[z]
         end;
         diff;
```

```
        for z := 1 to dim do
        begin
            k4[z] := ode[z]*h;
            y[z] := y[z]+(k1[z]+2*k2[z]+2*k3[z]+k4[z])/6
        end
    end
end;    { SetOde }

procedure Output;
{----------------}
var   cnt : integer;

begin
    writeln;   writeln (be:5);
    for cnt := 1 to dim do writeln ('   y(',cnt,') = ',y[cnt]:12:8)
end;    { Output }

procedure PlotData;
{------------------}
var   cnt : integer;
      xp,yp : integer;

begin
    for cnt := 1 to dim do
    begin
        xp := round(be/xe*GetMaxX);
        yp := round(GetMaxY-y[cnt]*GetMaxY/cmax);
        MoveTo (xp,yp);   LineTo (xp,yp)
    end
end;

begin
    Graphics;
    for ccal := 1 to dim do
    begin
        y[ccal] := y1[ccal];   ya[ccal] := y1[ccal]
    end;
    subi := 1;   be := xa;
    repeat
      x := be;   en := be+xd;   h := xd/subi;
      SetOde;
      abb := false;
      repeat
        for ccal := 1 to dim do ya[ccal] := y[ccal];
        subi := subi*2;   h := xd/subi;
        SetOde;
        for ccal := 1 to dim do
        begin
            if (abs(y[ccal])+abs(ya[ccal]) >= 1e-6) then
            if (abs(ya[ccal]-y[ccal])/abs(ya[ccal]+y[ccal]) <= 0.01) then
              abb := true else abb := true
        end;
      until (abb);
    { Output;}
    PlotData;
```

```
      for ccal := 1 to dim do y1[ccal] := y[ccal];
      subi := round (subi/4);
      if (subi = 0) then subi := 1;
      be := be+xd
    until (be > xe)
  end;   { Calc }

begin
   Input;
   Calc;
   repeat answer := readkey until (answer = chr($0d));
   CloseGraph;
   textmode (3)
end.
```

```
                              IMPACT
```

```
program impact;
   { J. Kinkel, January 1989, based on Ebert/Ederer }

uses    crt,graph;

const   dim = 3;    nl = 6.022e23;    w = 1.245e12;    subi = 1000;

var   ya : array[1..20] of real;
      b,my,xa,xe : real;
      xv1,xv2,yv1,yv2 : integer;
      answer : char;

 procedure Input;
{----------------}
 var   cnt : integer;

 begin
    clrscr;   writeln;
    write (' Impact parameter in Angstroem ? ');   readln (b);
    my := 20/nl;   ya[2] := 16;
    ya[1] := arctan(b/sqrt(sqr(ya[2])-sqr(b)));
    ya[3] := -my*w*cos(ya[1]);   xa := 0;   xe := 1e-10
 end;   { Input }

 procedure Graphics;
{-------------------}
 var   gd,gm : integer;

 begin
    gd := 0;   gm := 0;
    InitGraph (gd,gm,'');
    Rectangle (0,0,GetMaxX,GetMaxY);
    xv1 := round(GetMaxX*0.05);   xv2 := round(GetMaxX*0.95);
    yv1 := round(GetMaxY*0.05);   yv2 := round(GetMaxY*0.95);
    SetViewPort (xv1,yv1,xv2,yv2,ClipOn);
    Circle (round((xv2-xv1)/2),round((yv2-yv1)/2),10)
 end;   { Graphics }
```

```
procedure Calc;
{---------------}
var   y,yh,ode,k1,k2,k3,k4 : array[1..20] of real;
      x,h : real;
      cnt,z : integer;

  procedure PlotData;
  {------------------}
  var   xp,yp : integer;

  begin
      xp := round((xv2-xv1)*(y[2]*sin(y[1])+20)/40);
      yp := (yv2-yv1)-round((yv2-yv1)*(y[2]*cos(y[1])+20)/40);
      MoveTo (xp,yp);   LineTo (xp,yp)
  end;   { PlotData }

  procedure diff;
  {--------------}

      function fnv (r : real) : real;
      {----------------------------}
      begin
          fnv := (-12*exp(13*ln(3.405/r))+6*exp(7*ln(3.405/r)))*194.3
      end;   { fnv }

  begin
      ode[1] := b*w/sqr(yh[2]);   ode[2] := yh[3]/my;
      ode[3] := my*sqr(b*w)/yh[2]/sqr(yh[2])-fnv(yh[2])
  end;  { ode }

begin
   h := (xe-xa)/subi;
   for z := 1 to dim do y[z] := ya[z];
   for cnt := 0 to subi-1 do
   begin
      x := xa+cnt*h;
      for z := 1 to dim do yh[z] := y[z];
      diff;
      for z := 1 to dim do
      begin
         k1[z] := ode[z]*h;   yh[z] := y[z]+k1[z]/2
      end;
      diff;
      for z := 1 to dim do
      begin
         k2[z] := ode[z]*h;   yh[z] := y[z]+k2[z]/2
      end;
      diff;
      for z := 1 to dim do
      begin
         k3[z] := ode[z]*h;   yh[z] := y[z]+k3[z]
      end;
      diff;
      for z := 1 to dim do
```

```
        begin
           k4[z] := ode[z]*h;
           y[z] := y[z]+(k1[z]+2*k2[z]+2*k3[z]+k4[z])/6
        end;
        PlotData
     end
  end;    { Calc }

begin
   Input;
   Graphics;
   Calc;
   repeat answer := readkey until (answer = chr($0d));
   CloseGraph;
   textmode (3)
end.
```

IMPACT2

```
program impact2;
   { J. Kinkel, January 1989, based on Ebert/Ederer }

uses    crt,graph;

const    dim = 8;    nl = 6.022e23;

var    ya,y1 : array[1..20] of real;
       xa,xe,xd,my1,my2 : real;
       xv1,xv2,yv1,yv2 : integer;
       answer : char;

procedure Input;
{----------------}
   var    cnt : integer;

   begin
      clrscr;    writeln;
      write (' M1 M2 ? ');    readln (my1,my2);
      my1 := my1/nl;    my2 := my2/nl;
      xa := 0;    xe := 1e-10;    xd := 1e-13;
      for cnt := 1 to dim do
      begin
         write (' y1(',cnt,') ? ');    readln (y1[cnt])
      end;
      y1[5] := y1[5]*my1;    y1[6] := y1[6]*my1;
      y1[7] := y1[7]*my2;    y1[8] := y1[8]*my2
   end;    { Input }

procedure Calc;
{---------------}
   var    y,yh,ode,k1,k2,k3,k4 : array[1..20] of real;
          be,en,x,h : real;
          subi,ccal : integer;
          abb : boolean;
```

```
procedure Graphics;
{-------------------}
 var    gd,gm : integer;

begin
    gd := 0;   gm := 0;
    InitGraph (gd,gm,'');
    Rectangle (0,0,GetMaxX,GetMaxY);
    xv1 := round(GetMaxX*0.05);   xv2 := round(GetMaxX*0.95);
    yv1 := round(GetMaxY*0.05);   yv2 := round(GetMaxY*0.95);
    SetViewPort (xv1,yv1,xv2,yv2,ClipOn)
 end;   { Graphics }

procedure SetOde;
{-----------------}
 var    cnt,z : integer;

    procedure diff;
    {---------------}
     var   rx,ry,radius,vrad : real;

     begin
        ode[1] := yh[5]/my1;   ode[2] := yh[6]/my1;
        ode[3] := yh[7]/my2;   ode[4] := yh[8]/my2;
        rx := yh[1]-yh[3];   ry := yh[2]-yh[4];
        radius := sqr(rx)+sqr(ry);
        vrad := -342.2*(exp(4*ln(11.6/radius))-2*exp(7*ln(11.6/radius)));
        ode[5] := vrad*rx;   ode[6] := vrad*ry;
        ode[7] := -ode[5];   ode[8] := -ode[6]
     end;  { diff }

 begin
    for z := 1 to dim do y[z] := y1[z];
    for cnt := 0 to subi-1 do
    begin
       x := xa+cnt*h;
       for z := 1 to dim do yh[z] := y[z];
       diff;
       for z := 1 to dim do
       begin
          k1[z] := ode[z]*h;   yh[z] := y[z]+k1[z]/2
       end;
       diff;
       for z := 1 to dim do
       begin
          k2[z] := ode[z]*h;   yh[z] := y[z]+k2[z]/2
       end;
       diff;
       for z := 1 to dim do
       begin
          k3[z] := ode[z]*h;   yh[z] := y[z]+k3[z]
       end;
       diff;
       for z := 1 to dim do
```

```
            begin
               k4[z] := ode[z]*h;
               y[z] := y[z]+(k1[z]+2*k2[z]+2*k3[z]+k4[z])/6
            end
         end
      end;    { SetOde }

   procedure PlotData;
   {-------------------}
   var   xp,yp : integer;

   begin
      xp := round((xv2-xv1)*(y[1]+16)/32);
      yp := (yv2-yv1)-round((yv2-yv1)*(y[2]+16)/32);
      MoveTo (xp,yp);   LineTo (xp,yp);
      xp := round((xv2-xv1)*(y[3]+16)/32);
      yp := (yv2-yv1)-round((yv2-yv1)*(y[4]+16)/32);
      MoveTo (xp,yp);   LineTo (xp,yp)
   end;    { PlotData }

begin
   Graphics;
   for ccal := 1 to dim do
   begin
      y[ccal] := y1[ccal];   ya[ccal] := y1[ccal]
   end;
   subi := 1;   be := xa;
   repeat
    x := be;   en := be+xd;   h := xd/subi;
    SetOde;
    abb := false;
    repeat
     for ccal := 1 to dim do ya[ccal] := y[ccal];
     subi := subi*2;   h := xd/subi;
     SetOde;
     for ccal := 1 to dim do
     begin
        if (abs(y[ccal])+abs(ya[ccal]) >= 1e-6) then
          if (abs(ya[ccal]-y[ccal])/abs(ya[ccal]+y[ccal]) <= 0.01) then
            abb := true else abb := true
     end;
    until (abb);
    PlotData;
    for ccal := 1 to dim do y1[ccal] := y[ccal];
    subi := round (subi/4);
    if (subi = 0) then subi := 1;
    be := be+xd
   until (be > xe)
  end;   { Calc }

begin
   Input;
   Calc;
   repeat answer := readkey until (answer = chr($0d));
   CloseGraph;
   textmode (3)
end.
```

```
                              PLOT3D
```

```
program plot3d;
    { J. Kinkel, January 1989, based on Ebert/Ederer }

uses    crt,graph;

var    dm1,dm2,gm : array[1..3,1..3] of real;
       mmin,mmax : array[1..251] of real;
       x,y : array[1..50] of real;
       z : array[1..50,1..50] of real;
       xmin,xmax,ymin,ymax,zmin,zmax,dx : real;
       cntx,cnty : integer;

 procedure Matrices;
{------------------}
 var    i,j,k : integer;
        pi,rh,ta : real;

 begin
    pi := 4*arctan(1);   rh := 30;   ta := 30;
    dm1[1,1] := cos(rh/180*pi);  dm1[2,2] := dm1[1,1];
    dm1[2,1] := sin(rh/180*pi);  dm1[1,2] := -dm1[2,1];    dm1[3,3] := 1;
    dm2[2,2] := cos(ta/180*pi);  dm2[3,3] := dm2[2,2];
    dm2[3,2] := -sin(ta/180*pi);  dm2[2,3] := -dm2[3,2];
    dm2[1,1] := 1;
    for i := 1 to 3 do
     for j := 1 to 3 do
     begin
        gm[i,j] := 0;
        for k := 1 to 3 do gm[i,j] := gm[i,j]+dm2[i,k]*dm1[k,j]
     end
 end;   { Matrices }

 procedure Calcxyz;
{------------------}
 var    xh,yh : real;

 begin
    zmin := 1e30;  zmax := -1e30;   cntx := 0;   xh := -10;
    repeat
     cntx := cntx+1;   cnty := 0;   yh := -10;
     repeat
     cnty := cnty+1;   x[cntx] := xh;   y[cnty] := yh;
     z[cntx,cnty] := 3*cos(sqrt(sqr(xh)+sqr(yh)));
     if (z[cntx,cnty] < zmin) then zmin := z[cntx,cnty];
     if (z[cntx,cnty] > zmax) then zmax := z[cntx,cnty];
     yh := yh+0.5
     until (yh > 10);
     xh := xh+0.5
    until (xh > 10)
 end;    { Calcxyz }
```

```
procedure MiniMax;
{------------------}
var   xn,yn : real;
      i : integer;

   procedure Ctrafo (k1,k2,k3 : real);
   {--------------------------------}
   var   tr : array[1..3] of real;
         j : integer;

   begin
      tr[1] := k1;   tr[2] := k2;   tr[3] := k3;   xn := 0;   yn := 0;
      for j := 1 to 3 do
      begin
         xn := xn+gm[1,j]*tr[j];   yn := yn+gm[2,j]*tr[j]
      end;
      if (xn < xmin) then xmin := xn;
      if (xn > xmax) then xmax := xn
   end;   { Ctrafo }

begin
   xmin := 1e30;   xmax := -1e30;
   Ctrafo (x[1],y[1],zmin);
   ymin := yn;
   Ctrafo (x[1],y[cnty],zmin);
   Ctrafo (x[cntx],y[cnty],zmax);
   ymax := yn;
   Ctrafo (x[cntx],y[1],zmax);
   dx := 250/(xmax-xmin);
   for i := 1 to 251 do
   begin
      mmin[i] := 1e30;   mmax[i] := -1e30
   end
end;   { MiniMax }

procedure Graphic;
{------------------}
var   grdr,grmo : integer;
      answer : char;
      xa,ya,xn,yn,sm : real;
      xp,yp,i,j,k : integer;

   procedure SetPen (px,py : real;ud : byte);
   {-----------------------------------------}
   begin
      xp := round(px/252*GetMaxX);
      yp := GetMaxY-round((py-ymin)/(ymax-ymin)*GetMaxY);
      if (ud = 0) then MoveTo (xp,yp) else LineTo (xp,yp)
   end;   { SetPen }

   procedure Ptrafo (k1,k2,k3 : real);
   {--------------------------------}
   var   tr : array[1..3] of real;
         cnt : integer;
```

```
begin
   tr[1] := k1;   tr[2] := k2;   tr[3] := k3;   xn := 0;   yn := 0;
   for cnt := 1 to 3 do
   begin
      xn := xn+gm[1,cnt]*tr[cnt];   yn := yn+gm[2,cnt]*tr[cnt]
   end;
   xn := int((xn-xmin)*dx+1.5)
end;   { Ptrafo }

procedure PlotData;
{-------------------}
var   iz,im : array[1..20] of integer;
      sl,yk : real;
      s1,il,cnt,xp,yp : integer;

   function sgn (diff : real) : integer;
   {-------------------------------------}
   begin
      if (diff > 0) then sgn := 1;
      if (diff < 0) then sgn := -1;
      if (diff = 0) then sgn := 0
   end;   { sgn }

   procedure Gko (px,py : real);
   {----------------------------}
   begin
      xp := round(px/252*GetMaxX);
      yp := GetMaxY-round((py-ymin)/(ymax-ymin)*GetMaxY)
   end;   { Gko }

   function fny(r : integer) : real;
   {--------------------------------}
   begin
      fny := (ya+sl*(r-xa))
   end;   { fny }

   procedure Iter;
   {--------------}
   var   yl,ko : real;

   begin
      yk := fny(cnt);   iz[il] := 0;
      if (cnt = xn) then yk := yn;
      if (yk <= mmin[cnt]) then
      begin
         iz[il] := -1;   yl := mmin[cnt];   mmin[cnt] := yk
      end;
      if (yk >= mmax[cnt]) then
      begin
         iz[il] := 1;   yl := mmax[cnt];   mmax[cnt] := yk
      end;
      if not(iz[il] = iz[il-1]) then
      begin
         im[il] := cnt;
         ko := s1*(fny(cnt-s1)-yl)/(fny(cnt-s1)-yk);
```

```
        if (abs(sl) > sm) and (abs(ko) < 1) then
         im[il] := round(cnt-sl+ko);
        il := il+1
     end
  end;    { Iter }

  procedure PlotorNot (chck : integer;kx,ky : real);
{------------------------------------------------}
  begin
     if (chck = -1) then SetLineStyle (1,0,1);
     Gko (kx,ky);
     if (abs(chck) = 1) then LineTo (xp,yp) else MoveTo (xp,yp);
     SetLineStyle (0,0,1)
  end;    { PlotorNot }

begin
  fillchar (im[1],40,chr($00));   fillchar (iz[1],40,chr($00));
  if not(xa = xn) then
  begin
     sl := (yn-ya)/(xn-xa);   im[1] := round(xa);
     s1 := sgn(xn-xa);   il := 2;   yk := ya;   iz[1] := 0;
     if (yk <= mmin[round(xa)]) then
     begin
        iz[1] := -1;   mmin[round(xa)] := yk    .
     end;
     if (yk >= mmax[round(xa)]) then
     begin
        iz[1] := 1;   mmax[round(xa)] := yk
     end;
     cnt := round(xa+s1);
     if (s1 > 0) then repeat
      Iter;
      cnt := cnt+s1
     until (cnt > xn)
     else
     repeat
      Iter;
      cnt := cnt+s1
     until (cnt < xn);
     im[il] := round(xn);
     if (il <= 2) then
     begin
        PlotorNot (iz[2],xn,yn);
        xa := xn;   ya := yn
     end
     else
     begin
        for cnt := 1 to il-2 do
         PlotorNot (iz[cnt],im[cnt+1],fny(im[cnt+1]));
        PlotorNot (iz[il-1],xn,yn);
        xa := xn;   ya := yn
     end
  end
  else
```

```
        begin
          iz[1] := 0;
          if (ya <= mmin[round(xa)]) then
          begin
             iz[1] := -1;   mmin[round(xa)] := ya
          end;
          if (ya >= mmax[round(xa)]) then
          begin
             iz[1] := 1;   mmax[round(xa)] := ya
          end;
          iz[2] := 0;
          if (yn <= mmin[round(xa)]) then
          begin
             iz[2] := -1;   yk := mmin[round(xa)];
             mmin[round(xa)] := yn
          end;
          if (yn >= mmax[round(xa)]) then
          begin
             iz[2] := 1;   yk := mmax[round(xa)];
             mmax[round(xa)] := yn
          end;
          if not(iz[1] = iz[2]) then
          begin
             PlotorNot (iz[1],xa,yk);
             PlotorNot (iz[2],xa,yn);
             ya := yn
          end
          else
          begin
             PlotorNot (iz[1],xn,yn);
             ya := yn
          end
        end
    end;   { PlotData }

begin
    grdr := 0;   grmo := 0;   sm := abs(ymax-ymin)/250*0.2;
    InitGraph (grdr,grmo,'');
    Ptrafo (x[1],y[1],z[1,1]);
    SetPen (xn,yn,0);
    xa := xn;   ya := yn;
    for i := 2 to cnty do
    begin
       Ptrafo (x[1],y[i],z[1,i]);
       PlotData
    end;
    Ptrafo (x[1],y[1],z[1,1]);
    SetPen (xn,yn,0);
    xa := xn;   ya := yn;
    for i := 2 to cntx do
    begin
       Ptrafo (x[i],y[1],z[i,1]);
       PlotData
    end;
    for i := 2 to cnty do
```

```
begin
   j := i;    k := 2;
   Ptrafo (x[1],y[i],z[1,i]);
   SetPen (xn,yn,0);
   xa := xn;    ya := yn;
   Ptrafo (x[k],y[j],z[k,j]);
   PlotData;
   j := j-1;
   repeat
    Ptrafo (x[k],y[j],z[k,j]);
    PlotData;
    k := k+1;
    if (k <= cntx) then
    begin
       Ptrafo (x[k],y[j],z[k,j]);
       PlotData;
       j := j-1
    end
    else j := 0
   until (j < 1) or (k > cntx)
end;
for i := 2 to cntx-1 do
begin
   j := cnty;    k := i+1;
   Ptrafo (x[i],y[j],z[i,cnty]);
   SetPen (xn,yn,0);
   xa := xn;    ya := yn;
   Ptrafo (x[k],y[j],z[k,j]);
   PlotData;
   j := j-1;
   repeat
    Ptrafo (x[k],y[j],z[k,j]);
    PlotData;
    k := k+1;
    if (k <= cntx) then
    begin
       Ptrafo (x[k],y[j],z[k,j]);
       PlotData;
       j := j-1
    end
   until (j < 1) or (k > cntx)
end;
Ptrafo (x[1],y[cnty],zmax);
Setpen (xn,yn,0);
Ptrafo (x[1],y[cnty],zmin);
Setpen (xn,yn,1);
Ptrafo (x[1],y[1],zmin);
SetPen (xn,yn,1);
Ptrafo (x[cntx],y[1],zmin);
SetPen (xn,yn,1);
xa := xn;    ya := yn;
Ptrafo (x[cntx],y[cnty],zmin);
PlotData;
Ptrafo (x[1],y[cnty],zmin);
PlotData;
```

```
      repeat answer := readkey until (answer = chr($0d));
      CloseGraph;
      textmode(3)
   end;   { Graphic }
begin
   fillchar (dm1[1,1],54,chr($00));   fillchar (dm2[1,1],54,chr($00));
   fillchar (gm[1,1],54,chr($00));
   Matrices;
   Calcxyz;
   MiniMax;
   Graphic
end.
```

```
┌──────────────────────────────────────────────────────────────┐
│                           CONTOUR                              │
└──────────────────────────────────────────────────────────────┘
```

```
program contour;
   { J. Kinkel, January 1989, based on Ebert/Ederer }

uses   crt,graph;

const   nh = 8;

var   x,y : array[1..50] of real;
      h : array[1..20] of real;
      z : array[1..50,1..50] of real;
      zmin,zmax : real;
      cntx,cnty : integer;

procedure calcxyz;
{------------------}
   var   xh,yh : real;
         cnt : integer;

   begin
      zmin := 1e30;   zmax := -1e30;   cntx := 0;   xh := -10;
      repeat
       cntx := cntx+1;   cnty := 0;   yh := -10;
       repeat
        cnty := cnty+1;   x[cntx] := xh;   y[cnty] := yh;
        z[cntx,cnty] := 3*cos(sqrt(sqr(xh)+sqr(yh)));
        if (z[cntx,cnty] < zmin) then zmin := z[cntx,cnty];
        if (z[cntx,cnty] > zmax) then zmax := z[cntx,cnty];
        yh := yh+0.5
       until (yh > 10);
       xh := xh+0.5
      until (xh > 10);
      for cnt := 1 to nh do
       h[cnt] := zmin+(cnt-0.75)*0.95*(zmax-zmin)/(nh-1)
   end;   { calcxyz }

procedure PlotData;
{----------------}
   var   xp,yp : array[1..4] of real;
         grdr,grmo,ls,ih,ix,iy,ip,px,py : integer;
         answer : char;
         z1,z2,z3,z4,hv : real;
```

```
        function sgn (num : real) : integer;
    {-----------------------------------}
    begin
        if (num > 0) then sgn := 1;
        if (num < 0) then sgn := -1;
        if (num = 0) then sgn := 0
    end;    { sgn }

begin
    grdr := 0;   grmo := 0;
    InitGraph (grdr,grmo,'');
    rectangle (0,0,GetMaxX,GetMaxY);
    for ih := 1 to nh do
    begin
        hv := h[ih];   ls := ih-round(ih/4)*4;
        SetLineStyle (ls,0,1);
        for ix := 2 to cntx do
        begin
            for iy := 2 to cnty do
            begin
                ip := 0;   z1 := hv-z[ix-1,iy-1];   z2 := hv-z[ix-1,iy];
                z3 := hv-z[ix,iy];   z4 := hv-z[ix,iy-1];
                if not(sgn(z1) = sgn(z2)) then
                begin
                    ip := ip+1;   xp[ip] := x[ix-1];
                    yp[ip] := y[iy-1]+(y[iy]-y[iy-1])*z1/(z1-z2)
                end;
                if not(sgn(z2) = sgn(z3)) then
                begin
                    ip := ip+1;   yp[ip] := y[iy];
                    xp[ip] := x[ix-1]+(x[ix]-x[ix-1])*z2/(z2-z3)
                end;
                if not(sgn(z3) = sgn(z4)) then
                begin
                    ip := ip+1;   xp[ip] := x[ix];
                    yp[ip] := y[iy-1]+(y[iy]-y[iy-1])*z4/(z4-z3)
                end;
                if not(sgn(z4) = sgn(z1)) then
                begin
                    ip := ip+1;   yp[ip] := y[iy-1];
                    xp[ip] := x[ix-1]+(x[ix]-x[ix-1])*z1/(z1-z4)
                end;
                if (ip = 2) then
                begin
                    px := round((xp[1]-x[1])/(x[cntx]-x[1])*GetMaxX);
                    py := GetMaxY-round((yp[1]-y[1])/(y[cnty]-y[1])*GetMaxY);
                    MoveTo (px,py);
                    px := round((xp[2]-x[1])/(x[cntx]-x[1])*GetMaxX);
                    py := GetMaxY-round((yp[2]-y[1])/(y[cnty]-y[1])*GetMaxY);
                    LineTo (px,py)
                end
            end
        end
    end;
```

```
    repeat answer := readkey until (answer = chr($0d));
    CloseGraph;
    textmode (3)
  end;    { PlotData }

begin
    fillchar (x[1],300,chr($00));   fillchar (y[1],300,chr($00));
    fillchar (h[1],120,chr($00));   fillchar (z[1,1],15000,chr($00));
    Calcxyz;
    PlotData
end.
```

```
┌─────────────────────────────────────────────────────────────────────────┐
│                                  MORSE                                     │
└─────────────────────────────────────────────────────────────────────────┘
```

```
program morse;
  { J. Kinkel, January 1989, based on Ebert/Ederer }

uses    crt,graph;

const    nh = 30;    lt : array[0..3] of word = ($FFFF,$F0F0,$CCCC,$AAAA);

var    x,y : array[1..50] of real;
       h : array[1..nh] of real;
       z : array[1..50,1..50] of real;
       zmin,zmax,zx,zy : real;
       cntx,cnty : integer;

 procedure calcxyz;
{------------------}
 var    xh,yh : real;
        cnt : integer;

begin
    zmin := 1e30;    zmax := -1e30;    cntx := 0;    xh := 1;
    repeat
     cntx := cntx+1;    cnty := 0;    yh := 1;
     repeat
     cnty := cnty+1;    x[cntx] := xh;    y[cnty] := yh;
     zx := exp(-2.07226*(xh-2.1163))-2*exp(-1.03613*(xh-2.1163));
     zy := exp(-2.07226*(yh-2.1163))-2*exp(-1.03613*(yh-2.1163));
     z[cntx,cnty] := (zx+zy)*0.1339;
     if (z[cntx,cnty] < zmin) then zmin := z[cntx,cnty];
     if (z[cntx,cnty] > zmax) then zmax := z[cntx,cnty];
     yh := yh+0.1
     until (yh > 5);
     xh := xh+0.1
    until (xh > 5);
    for cnt := 1 to nh do
     h[cnt] := zmin+(cnt-0.75)*0.95*(zmax-zmin)/(nh-1)
end;    { calcxyz }
```

```
procedure PlotData;
{-------------------}
var   xp,yp : array[1..4] of real;
      grdr,grmo,ls,ih,ix,iy,ip,
      xv1,xv2,yv1,yv2,px,py : integer;
      answer : char;
      z1,z2,z3,z4,hv : real;

   procedure Graphics;
   {-------------------}
   begin
      grdr := 0;   grmo := 0;
      InitGraph (grdr,grmo,'');
      rectangle (0,0,GetMaxX,GetMaxY);
      xv1 := round(GetMaxX*0.01);   xv2 := round(GetMaxX*0.99);
      yv1 := round(GetMaxY*0.01);   yv2 := round(GetMaxY*0.99);
      SetViewPort (xv1,yv1,xv2,yv2,ClipOn)
   end;   { Graphics }

   function sgn (num : real) : integer;
   {----------------------------------}
   begin
      if (num > 0) then sgn := 1;
      if (num < 0) then sgn := -1;
      if (num = 0) then sgn := 0
   end;   { sgn }

begin
   Graphics;
   for ih := 1 to nh do
   begin
      hv := h[ih];   ls := lt[ih mod 4];
      SetLineStyle (4,ls,1);
      for ix := 2 to cntx do
      begin
         for iy := 2 to cnty do
         begin
            ip := 0;
            z1 := hv-z[ix-1,iy-1];   z2 := hv-z[ix-1,iy];
            z3 := hv-z[ix,iy];       z4 := hv-z[ix,iy-1];
            if not(sgn(z1) = sgn(z2)) then
            begin
               ip := ip+1;   xp[ip] := x[ix-1];
               yp[ip] := y[iy-1]+(y[iy]-y[iy-1])*z1/(z1-z2)
            end;
            if not(sgn(z2) = sgn(z3)) then
            begin
               ip := ip+1;   yp[ip] := y[iy];
               xp[ip] := x[ix-1]+(x[ix]-x[ix-1])*z2/(z2-z3)
            end;
            if not(sgn(z3) = sgn(z4)) then
            begin
               ip := ip+1;   xp[ip] := x[ix];
               yp[ip] := y[iy-1]+(y[iy]-y[iy-1])*z4/(z4-z3)
            end;
            if not(sgn(z4) = sgn(z1)) then
```

```
               begin
                 ip := ip+1;    yp[ip] := y[iy-1];
                 xp[ip] := x[ix-1]+(x[ix]-x[ix-1])*z1/(z1-z4)
               end;
               if (ip = 2) then
               begin
                 px := round((xp[1]-x[1])/(x[cntx]-x[1])*GetMaxX);
                 py := GetMaxY-round((yp[1]-y[1])/(y[cnty]-y[1])*GetMaxY);
                 MoveTo (px,py);
                 px := round((xp[2]-x[1])/(x[cntx]-x[1])*GetMaxX);
                 py := GetMaxY-round((yp[2]-y[1])/(y[cnty]-y[1])*GetMaxY);
                 LineTo (px,py)
               end
             end
           end
       end;
     repeat answer := readkey until (answer = chr($0d));
     CloseGraph;
     textmode (3)
  end;   { PlotData }

begin
   fillchar (x[1],300,chr($00));   fillchar (y[1],300,chr($00));
   fillchar (z[1,1],15000,chr($00));
   Calcxyz;
   PlotData
end.
```

```
┌─────────────────────────────────────────────────────────────────┐
│                            TRAJECT                                │
└─────────────────────────────────────────────────────────────────┘
```

```
program traject;
   { J. Kinkel, January 1989, based on Ebert/Ederer }

uses    crt,graph;

const   dim = 6;

var    ya,y1 : array[1..20] of real;
       xa,xe,xd,my1,my2 : real;
       xv1,xv2,yv1,yv2,ma,mb,mc : integer;
       answer : char;

 procedure Input;
{----------------}
 var    cnt : integer;

 begin
    clrscr;   writeln;
    xa := 0;   xe := 1e5;   xd := 0.1;
    for cnt := 1 to dim do
    begin
       write (' y1(',cnt,') ? ');   readln (y1[cnt])
    end;
    ma := 1;   mc := 1;   mb := 12
 end;   { Input }
```

```pascal
procedure Calc;
{---------------}
var    y,yh,ode,k1,k2,k3,k4 : array[1..20] of real;
       be,en,x,h : real;
       subi,ccal : integer;
       abb : boolean;

  procedure Graphics;
  {------------------}
  var    gd,gm : integer;

  begin
      gd := 0;    gm := 0;
      InitGraph (gd,gm,'');
      Rectangle (0,0,GetMaxX,GetMaxY);
      xv1 := round(GetMaxX*0.01);    xv2 := round(GetMaxX*0.99);
      yv1 := round(GetMaxY*0.01);    yv2 := round(GetMaxY*0.99);
      SetViewPort (xv1,yv1,xv2,yv2,ClipOn)
  end;    { Graphics }

  procedure SetOde;
  {----------------}
  var    cnt,z : integer;

    procedure diff;
    {--------------}
    var    rx,ry,v1,v2 : real;

      function fnv (r : real) : real;
      {----------------------------}
      begin
          fnv := -4*(exp(-4*(r-2.1))-exp(-2*(r-2.1)))
      end;    { fnv }

    begin
        ode[1] := yh[4]/ma;    ode[2] := yh[5]/mb;
        ode[3] := yh[6]/mc;
        rx := abs(yh[1]-yh[2]);    ry := abs(yh[2]-yh[3]);
        v1 := fnv(rx);    v2 := fnv(ry);
        ode[4] := -(yh[1]-yh[2])*v1/rx;
        ode[6] := (yh[2]-yh[3])*v2/ry;    ode[5] := -ode[4]-ode[6];
    end;    { diff }

  begin
      for z := 1 to dim do y[z] := y1[z];
      for cnt := 0 to subi-1 do
      begin
          x := xa+cnt*h;
          for z := 1 to dim do yh[z] := y[z];
          diff;
          for z := 1 to dim do
          begin
              k1[z] := ode[z]*h;    yh[z] := y[z]+k1[z]/2
          end;
          diff;
```

```
              for z := 1 to dim do
              begin
                 k2[z] := ode[z]*h;   yh[z] := y[z]+k2[z]/2
              end;
              diff;
              for z := 1 to dim do
              begin
                 k3[z] := ode[z]*h;   yh[z] := y[z]+k3[z]
              end;
              diff;
              for z := 1 to dim do
              begin
                 k4[z] := ode[z]*h;
                 y[z] := y[z]+(k1[z]+2*k2[z]+2*k3[z]+k4[z])/6
              end
         end
   end;   { SetOde }

   procedure PlotData;
   {-------------------}
   var   xp,yp : integer;
         rx,ry : real;

   begin
      rx := abs(yh[1]-yh[2]);   ry := abs(yh[2]-yh[3]);
      xp := round((xv2-xv1)*(rx-1)/4);
      yp := (yv2-yv1)-round((yv2-yv1)*(ry-1)/4);
      MoveTo (xp,yp);   LineTo (xp,yp)
   end;

begin
   Graphics;
   for ccal := 1 to dim do
   begin
      y[ccal] := y1[ccal];   ya[ccal] := y1[ccal]
   end;
   subi := 1;   be := xa;
   repeat
    x := be;   en := be+xd;   h := xd/subi;
    SetOde;
    abb := false;
    repeat
     for ccal := 1 to dim do ya[ccal] := y[ccal];
     subi := subi*2;   h := xd/subi;
     SetOde;
     for ccal := 1 to dim do
     begin
        if (abs(y[ccal])+abs(ya[ccal]) >= 1e-6) then
        if (abs(ya[ccal]-y[ccal])/abs(ya[ccal]+y[ccal]) <= 0.01) then
          abb := true   else abb := true
     end;
    until (abb);
    PlotData;
    for ccal := 1 to dim do y1[ccal] := y[ccal];
    subi := round (subi/4);
```

```
         if (subi = 0) then subi := 1;
         be := be+xd
      until (be > xe)
   end;   { Calc }

begin
   Input;
   Calc;
   repeat answer := readkey until (answer = chr($0d));
   CloseGraph;
   textmode (3)
end.
```

LABDATA1

```
program labdata1;
   { J. Kinkel, January 1989, based on Ebert/Ederer }

uses   dos,crt,graph,auxinout;

var    wcil,rcil : text;
       dt,tmax : real;
       y,cnt : integer;
       answer : char;

 procedure rs232;
{----------------}

   procedure Initialize;
   {--------------------}
   var   f : text;
         res,baud,wl,par,stb : word;
         parin : char;
         flag : boolean;

      procedure Analyze;
      {------------------}
      begin
         case baud of
           9600 : baud := $e0;
           4800 : baud := $c0;
           2400 : baud := $a0;
           1200 : baud := $80;
            600 : baud := $60;
            300 : baud := $40;
            150 : baud := $20;
            110 : baud := $00
           else writeln (baud,' baudrate is not allowed here !');
           flag := false
         end;
         case parin of
          'n' : par := $00;
          'o' : par := $08;
          'e' : par := $18
```

```
            else writeln (parin,' as parity is not allowed here !');
            flag := false
        end;
        case wl of
            7 : wl := $02;
            8 : wl := $03
            else writeln (wl,' databits are not allowed here !');
            flag := false
        end;
        case stb of
            1 : stb := $00;
            2 : stb := $04
            else writeln (stb,' stopbits are not allowed here !');
            flag := false
        end;
        writeln;
        res := baud+par+wl+stb
    end;   { Analyze }

begin
    flag := true;
    clrscr;   writeln;
    write (' Baudrate ? ');                    readln (baud);
    write (' Parity (none/even/odd) ? ');      readln (parin);
    write (' Number of databits ? ');          readln (wl);
    write (' Number of stopbits ? ');          readln (stb);
    writeln;
    Analyze;
    if flag then
    begin
        assignaux (f,0,res);   rewrite (f);   close (f)
    end
    else
    begin
        writeln (' Your parameters were not correct !');
        repeat answer := readkey until (answer = chr($0d))
    end
end;    { Initialize }

begin
    clrscr;   writeln;
    write (' Do you want to configure your serial port COM1 (y/n) ? ');
    readln (answer);
    if (answer in ['j','J','y','Y']) then Initialize;
    assign (wcil,'com1');   rewrite (wcil);
    write (wcil,'A,IC1,F1'+chr($0d));   close (wcil)
end;   { rs232 }

procedure Input;
{----------------}
begin
    clrscr;   writeln;
    writeln (' What is the time interval between');
    write (' two data to be read into the computer (in sec.) ? ');
    readln (dt);
```

```
    write (' What is the overall measurement time (in min.) ? ');
    readln (tmax)
end;    { Input }

procedure Datacq;
{-----------------}
var   h,m,s,s100 : word;
      atime,endtime,mtime,htime : real;
      grdr,grmo,ax,ay : integer;

    procedure Cil;
    {--------------}
    const    send = 'G0,I1,T10';

    begin
       write (wcil,send+chr($0d));    read (rcil,y)
    end;    { Cil }

    function time : real;
    {---------------------}
    var   h,m,s,s100 : word;

    begin
       GetTime (h,m,s,s100);    time := h*3600+m*60+s
    end;    { time }

    procedure PlotData;
    {------------------}
    var   xp,yp : integer;

    begin
       xp := round((mtime-htime)/tmax/60*GetMaxX);
       yp := GetMaxY-round((y+32768)/65535*GetMaxY);
       LineTo (xp,yp)
    end;    { PlotData }

begin
    grdr := 0;   grmo := 0;
    InitGraph (grdr,grmo,'');
    assign (wcil,'com1');   assign (rcil,'com1');
    rewrite (wcil);          reset (rcil);
    htime := time;   atime := htime;
    Cil;
    ax := 0;   ay := GetMaxY-round((y+32768)/65535*GetMaxY);
    MoveTo(ax,ay);
    endtime := htime+tmax*60;
    repeat
     mtime := atime+dt;
     repeat until (time >= mtime);
     Cil;
     atime := time;
     PlotData
    until (time >= endtime);
    repeat answer := readkey until (answer = chr($0d));
    close (wcil);   close (rcil);
```

```
     CloseGraph;
     textmode (3)
  end;    { Datacq }

begin
   rs232;
   Input;
   Datacq
end.
```

```
                            GPCDATA
```

```
program gpcdata;
   { J. Kinkel, January 1989, based on Ebert/Ederer }

uses    dos,crt,graph;

var    wcil,rcil : text;
       dt,tmax,mmax : real;
       m,dm,pk : array[0..250] of real;
       value,cnt,grmo : integer;
       y : array[0..250] of integer;
       answer : char;

 procedure rs232;
{----------------}
 var    baud,local : word;

    procedure Initialize;
    {--------------------}
    var   f : text;
          wl,par,stb : word;
          parin : char;
          flag : boolean;

       procedure Analyze;
       {-----------------}
       begin
         local := 0;
         case baud of
         9600 : baud := $e0;
         4800 : baud := $c0;
         2400 : baud := $a0;
         1200 : baud := $80;
          600 : baud := $60;
          300 : baud := $40;
          150 : baud := $20;
          110 : baud := $00
         else writeln (baud,' baudrate is not allowed here !');
         flag := false
         end;
         baud := round(115200/baud);
         case wl of
         7 : local := 2;
         8 : local := 3
```

```
                else writeln (wl,' databits are not allowed here !');
                flag := false
              end;
              case stb of
              1 : local := local or $00;
              2 : local := local or $04
                else writeln (stb,' stopbits are not allowed here !');
                flag := false
              end;
              case parin of
              'n' : local := local or $00;
              'o' : local := local or $08;
              'e' : local := local or $18
                else writeln (parin,' as parity is not allowed here !');
                flag := false
              end
          end;    { Analyze }

    begin
        flag := true;
        clrscr;    writeln;
        write (' Baudrate ? ');                     readln (baud);
        write (' Parity (none/even/odd) ? ');    readln (parin);
        write (' Number of databits ? ');        readln (wl);
        write (' Number of stopbits ? ');        readln (stb);
        writeln;
        Analyze;
        if flag then
        begin
            port[$03F8] := lo(baud);   port[$03F9] := hi(baud);
            port[$03FB] := local
        end
        else
        begin
            writeln;
            writeln (' Your parameters were not correct !');
            repeat answer := readkey until (answer = chr($0d))
        end
    end;    { Initialize }

begin
    clrscr;    writeln;
    write (' Do you want to configure your serial port COM1 (y/n) ? ');
    readln (answer);
    if (answer in ['j','J','y','Y']) then Initialize;
    assign (wcil,'com1');   rewrite (wcil);
    write (wcil,'A,IC1,F1'+chr($0d));   close (wcil)
end;    { rs232 }

procedure Input;
{----------------}
begin
    clrscr;    writeln;
    writeln (' What is the time interval between');
    write (' two data to be read into the computer (in sec.) ? ');
```

```
       write (' What is the overall measurement time (in min.) ? ');
       dt := 5;
       tmax := 1250
end;   { Input }

procedure Datacq;
{-----------------}
var    h,m,s,s100 : word;
       atime,endtime,mtime,htime : real;
       grdr,ax,ay,cnt : integer;

   procedure Cil;
   {--------------}
   const    send = 'G0,I1,T10';

   begin
      write (wcil,send+chr($0d));    read (rcil,value)
   end;   { Cil }

   function time : real;
   {--------------------}
   var    h,m,s,s100 : word;

   begin
      GetTime (h,m,s,s100);    time := h*3600+m*60+s
   end;   { time }

   procedure PlotData;
   {----------------}
   var    xp,yp : integer;

   begin
      xp := round((mtime-htime)/tmax*GetMaxX);
      yp := GetMaxY-round((value+32768)/65535*GetMaxY);
      LineTo (xp,yp)
   end;   { PlotData }

begin
   grdr := 0;   grmo := 0;   cnt := 0;
   InitGraph (grdr,grmo,'');
   assign (wcil,'com1');    assign (rcil,'com1');
   rewrite (wcil);          reset (rcil);
   htime := time;   atime := htime;
   Cil;
   y[cnt] := value;
   ax := 0;   ay := GetMaxY-round((value+32768)/65535*GetMaxY);
   MoveTo(ax,ay);
   endtime := htime+tmax;
   repeat
    mtime := atime+dt;
    repeat until (time >= mtime);
    Cil;
    y[cnt] := value;   atime := time;
    PlotData
   until (time >= endtime);
```

```
      repeat answer := readkey until (answer = chr($0d));
      close (wcil);   close (rcil);
      grmo := GetGraphMode;
      RestoreCrtMode
   end;   { Datacq }

procedure BaseLine;
{--------------------}
var   cnt : integer;

begin
   for cnt := 0 to 250 do
   begin
      y[cnt] := y[cnt]-round(y[0]+(y[250]-y[0])/250*cnt);
      if (y[cnt] < 0) then y[cnt] := 0
   end
end;    { BaseLine }

procedure Mwd;
{--------------}
var   diff,sum1,sum2,sum3,sum4 : real;
      cnt : integer;

   function fnm (t : integer) : real;
   {---------------------------------}
   begin
      fnm := exp(13-0.005*t)
   end;   { fnm }

   procedure Output;
   {-----------------}
   begin
      clrscr;   writeln;
      writeln (' Number average Mn = ',sum2/sum1:10:0);
      writeln (' Weight average Mw = ',sum3/sum2:10:0);
      writeln (' Z   -   average Mz = ',sum4/sum3:10:0)
   end;   { Output }

begin
   mmax := 0;   sum1 := 0;   sum2 := 0;   sum3 := 0;   sum4 := 0;
   for cnt := 0 to 250 do m[cnt] := fnm(cnt*5);
   for cnt := 1 to 250 do
   begin
      diff := m[cnt-1]-m[cnt];
      dm[cnt] := (y[cnt]+y[cnt-1])/(2*diff);
      if (dm[cnt] > mmax) then mmax := dm[cnt]
   end;
   for cnt := 1 to 250 do
   begin
      pk[cnt] := (m[cnt]+m[cnt-1])/2;   sum1 := sum1+dm[cnt]/pk[cnt];
      sum2 := sum2+dm[cnt];   sum3 := sum3+dm[cnt]*pk[cnt];
      sum4 := sum4+dm[cnt]*sqr(pk[cnt])
   end;
   Output
end;   { Mwd }
```

```
  procedure PlotMwd;
{------------------}
  var   xp,yp,cnt,mincnt,maxcnt : integer;

    procedure Nlimits;
    {------------------}
    begin
       mincnt := 1;
       repeat mincnt := mincnt+1 until (dm[mincnt] > mmax/50);
       maxcnt := 250;
       repeat maxcnt := maxcnt-1 until (dm[maxcnt] > mmax/50)
    end;  { Nlimits }

  begin
    SetGraphMode(grmo);
    xp := 0;   yp := GetMaxY-round(dm[1]/mmax*1.01*GetMaxY);
    MoveTo (xp,yp);
    for cnt := 1 to 250 do
    begin
       xp := GetMaxX-round((m[250]-pk[cnt])/(m[250]-m[0])*GetMaxX);
       yp := GetMaxY-round(dm[cnt]/mmax*1.01*GetMaxY);
       LineTo (xp,yp)
    end;
    xp := 0;   yp := GetMaxY-round(dm[1]/mmax*1.01*GetMaxY);
    MoveTo (xp,yp);
    Nlimits;
    for cnt := mincnt to maxcnt do
    begin
       xp := GetMaxX-round((m[maxcnt]-pk[cnt])/(m[maxcnt]-m[mincnt])
                        *GetMaxX);
       yp := GetMaxY-round(dm[cnt]/mmax*1.01*GetMaxY);   LineTo (xp,yp)
    end;
    repeat answer := readkey until (answer = chr($0d))
  end;   { PlotMwd }

begin
   rs232;
   Input;
   Datacq;
   BaseLine;
   Mwd;
   PlotMwd
end.
```

```
┌────────────────────────────────────────────────────────────────┐
│                            QSORT                                 │
└────────────────────────────────────────────────────────────────┘
```

```
program qsort;
   { J. Kinkel, January 1989, based on Ebert/Ederer }

uses   crt;

var    x,y : array[1..256] of real;
       max : integer;
```

```pascal
procedure Input;
{----------------}
var    cnt : integer;

begin
   randomize;   clrscr;   writeln;
   write (' Number of data points ? ');   readln (max);
   for cnt := 1 to max do
   begin
      x[cnt] := random;   y[cnt] := x[cnt]
   end
end;   { Input }

procedure Sort;
{---------------}
var    abb,cnt1,cnt2 : integer;
       h : real;

begin
   abb := round(ln(max)/ln(2));   abb := round(exp(abb*ln(2))-1);
   repeat
    for cnt1 := 1 to max-abb do
    begin
       cnt2 := cnt1;
       repeat
        if (y[cnt2] > y[cnt2+abb]) then
        begin
           h := y[cnt2];   y[cnt2] := y[cnt2+abb];   y[cnt2+abb] := h;
           cnt2 := cnt2-abb
        end
        else cnt2 := 0
       until (cnt2 <= 0)
    end;
    if (abb/2 < 1) then abb := 0 else abb := round(abb/2)
   until (abb <= 0)
end;   { Sort }

procedure Output;
{-----------------}
var    cnt : integer;

begin
   writeln;                       writeln (' Unsorted data :');
   for cnt := 1 to max do
   begin
      write (x[cnt]:12:8);   if (cnt/6 = round(cnt/6)) then writeln
   end;   writeln;   writeln;   writeln (' Sorted data :');
   for cnt := 1 to max do
   begin
      write (y[cnt]:12:8);   if (cnt/6 = round(cnt/6)) then writeln
   end
end;   { Output }

begin
   Input;
   Sort;
   Output
end.
```

```
                              QQSORT2

program qqsort2;
   { J. Kinkel, January 1989, based on Ebert/Ederer }

uses   crt;

var   x,y : array[1..2000] of real;
      max : integer;

 procedure Input;
{----------------}
 var   cnt : integer;

 begin
    randomize;   clrscr;   writeln;
    write (' Number of data points ? ');   readln (max);
    for cnt := 1 to max do
    begin
       x[cnt] := random;   y[cnt] := x[cnt]
    end
end;   { Input }

 procedure Sort;
{---------------}
 var   left,right : array[1..50] of integer;
       h,d : real;
       cnt,le,ri,mi : integer;

   procedure Exchange (i,j : integer;var k : real);
   {-----------------------------------------------}
    begin
       k := y[i];   y[i] := y[j];   y[j] := k
    end;   { Exchange }

   procedure Arrange;
   {------------------}
    begin
       mi := round((ri+le)/2);
       if (d < y[le]) then
       begin
          if (d < y[mi]) then
          begin
             if (y[le] <= y[mi]) then Exchange (mi,ri,d)
             else Exchange (le,ri,d)
          end
       end
       else
       begin
          if (d > y[mi]) then
          begin
             if (y[le] <= y[mi]) then Exchange (mi,ri,d)
             else Exchange (le,ri,d)
          end
       end
    end;   { Arrange }
```

```
begin
   left[1] := 1;   right[1] := max;   cnt := 1;
   repeat
    le := left[cnt];   ri := right[cnt];   d := y[ri];
    if (ri-le >= 5) then Arrange;
    dec(le);
    while (ri > le) do
    begin
       inc(le);
       if (y[le] >= d) then
       begin
          repeat dec(ri) until (y[ri] < d) or (ri <= le);
          if (y[ri] < d) and (ri > le) then Exchange (ri,le,h)
       end
    end;
    ri := right[cnt];
    Exchange (le,ri,h);
    if ((le-left[cnt]) >= (right[cnt]-le)) then
    begin
       left[cnt+1] := left[cnt];   right[cnt+1] := le-1;
       left[cnt] := le+1
    end
    else
    begin
       left[cnt+1] := le+1;   right[cnt+1] := right[cnt];
       right[cnt] := le-1
    end;
    inc(cnt);
    if (right[cnt] <= left[cnt])
    then
    repeat dec(cnt) until ((cnt=0) or (left[cnt] < right[cnt]))
   until (cnt=0)
end;   { Sort }

procedure Output;
{-----------------}
var   cnt : integer;

begin
   writeln;                        writeln (' Unsorted data :');
   for cnt := 1 to max do
   begin
      write (x[cnt]:12:8);   if (cnt/6 = round(cnt/6)) then writeln
   end;   writeln;   writeln;   writeln (' Sorted data :');
   for cnt := 1 to max do
   begin
      write (y[cnt]:12:8);   if (cnt/6 = round(cnt/6)) then writeln
   end
end;   { Output }

begin
   Input;
   Sort;
   Output
end.
```

```
                        ADDRSORT

program addrsort;
   { J.Kinkel, January 1989, based on Ebert/Ederer }

uses   crt;

const   admax = 1000;

var   x,y,h : array[1..1001] of integer;
      max : integer;

 procedure Input;
{----------------}
 var   cnt : integer;

 begin
    randomize;   clrscr;   fillchar (x[1],2002,chr($00));
    fillchar (y[1],2002,chr($00));   fillchar (h[1],2002,chr($00));
    writeln;   write (' Number of data points ? ');   readln (max);
    for cnt := 1 to max do x[cnt] := random(1000)+1
 end;   { Input }

 procedure Sort;
{---------------}
 var   cnt1,cnt2,cnt3 : integer;

 begin
    cnt3 := 0;
    for cnt1 := 1 to max do h[x[cnt1]] := h[x[cnt1]]+1;
    for cnt1 := 1 to admax do
    if not(h[cnt1] = 0) then
     for cnt2 := 1 to h[cnt1] do
     begin
        cnt3 := cnt3+1;   y[cnt3] := cnt1
     end
 end;   { Sort }

 procedure Output;
{-----------------}
 var   cnt : integer;

 begin
    writeln;                   writeln (' Unsorted data :');
    for cnt := 1 to max do
    begin
       write (x[cnt]:6);   if (cnt/12 = round(cnt/12)) then writeln
    end;   writeln;   writeln;   writeln (' Sorted data :');
    for cnt := 1 to max do
    begin
       write (y[cnt]:6);   if (cnt/12 = round(cnt/12)) then writeln
    end
 end;   { Output }
```

```
begin
   Input;
   Sort;
```

```
                          REGSPL1
```

```
program regspl1;
   { J. Kinkel, January 1989, based on Ebert/Ederer }

uses    crt,graph;

const   dim = 20;
        x : array[1..20] of real
          = (0,1,2,3,4,5,6,7,8,9,10,11,12,13,14,15,16,17,18,19);
        y : array[1..20] of real
          = (0.5,1,3.5,4.2,5.5,8.2,8.5,10.2,10.9,11.5,11.6,11.5,
             10.6,9.5,7.2,5.5,3.2,2.1,0.9,0);

var   a,d : array[1..252,1..7] of real;
      b,c,z : array [1..250] of real;
      h,xmin,xmax,ymin,ymax : real;
      ngp : integer;

 procedure Input;
{----------------}
 begin
    clrscr;   writeln;
    write (' How many grid points ? ');   readln (ngp);
    if (ngp < 2) then ngp := 2;
    if (ngp >= dim-2) then ngp := dim-2
 end;   { Input }

 procedure Qsort;
{----------------}
 var   tmp : real;
       i,j,blo : integer;

 begin
    xmin := x[1];   xmax := xmin;   ymin := y[1];   ymax := ymin;
    for i := 2 to dim do
    begin
       if (x[i] < xmin) then xmin := x[i];
       if (x[i] > xmax) then xmax := x[i];
       if (y[i] < ymin) then ymin := y[i];
       if (y[i] > ymax) then ymax := y[i]
    end;
    blo := round(ln(dim)/ln(2));   blo := round(exp(blo*ln(2)))-1;
    repeat
     for i := 1 to dim-blo do
     begin
        j := i;
```

```
      repeat
        if (x[j] <= x[j+blo]) then j := 0
        else
        begin
          tmp := y[j];   y[j] := y[j+blo];   y[j+blo] := tmp;
          tmp := x[j];   x[j] := x[j+blo];   x[j+blo] := tmp;
          j := j-blo
        end
      until (j < 1)
    end;
    if (blo/2 < 1) then blo := 0 else blo := round(blo/2)
  until (blo <= 0)
end;   { Qsort }

function fnp (tmp : real) : real;
{-------------------------------}
begin
  fnp := 0.25*tmp*sqr(tmp)
end;   { fnp }

function fnq (tmp : real) : real;
{-------------------------------}
begin
  fnq := 1-0.75*(1+tmp)*sqr(1-tmp)
end;   { fnq }

procedure Fillb;
{---------------}
var   cnt : integer;

begin
  b[1] := x[1];   b[ngp] := x[dim];   h := (b[ngp]-b[1])/(ngp-1);
  for cnt := 2 to ngp-1 do b[cnt] := b[1]+h*(cnt-1)
end;   { Fillb }

procedure Filla;
{---------------}
var   ix,ib : integer;
      t : real;

begin
  ib := 1;   ix := 1;
  repeat
    if (x[ix] > b[ib+1]) then ib := ib+1
    else
    begin
      t := (x[ix]-b[ib])/h;
      a[ix,1] := fnp(1-t);   a[ix,2] := fnq(1-t);
      a[ix,3] := fnq(t);   a[ix,4] := fnp(t);
      a[ix,5] := ib;   ix := ix+1
    end
  until (ix > dim)
end;   { Filla }
```

```pascal
procedure Filldz;
{-----------------}
var   i,j,k,ih,jh : integer;

begin
   for i := 1 to ngp+2 do
   begin
      for j := 1 to 7 do d[i,j] := 0;
      z[i] := 0
   end;
   for k := 1 to dim do
   begin
      for i := 1 to 4 do
      begin
         for j := 1 to i do
         begin
            ih := round(i+a[k,5]-1);   jh := round(j+a[k,5]-1);
            d[ih,jh-ih+4] := d[ih,jh-ih+4]+a[k,i]*a[k,j]
         end;
         z[ih] := z[ih]+a[k,i]*y[k]
      end
   end;
   for i := 1 to ngp+2 do
   begin
      for j := 1 to 3 do
       if not(d[i,j] = 0) then d[j+i-4,8-j] := d[i,j]
   end
end;   { Filldz }

procedure SetEq;
{---------------}
var   i,j,k,l,l1,l2 : integer;
      dh,l3 : real;

begin
   for i := 1 to ngp+2 do
   begin
      dh := d[i,4];
      for j := 1 to 7 do d[i,j] := d[i,j]/dh;
      z[i] := z[i]/dh;   j := i+3;
      if (j > ngp+2) then j := ngp+2;
      if (i < ngp+2) then
      begin
         for k := i+1 to j do
         begin
            l3 := d[k,4-(k-i)];
            for l := 1 to 4 do
            begin
               l1 := l+3;   l2 := l1-(k-i);
               d[k,l2] := d[k,l2]-l3*d[i,l1]
            end;
            z[k] := z[k]-l3*z[i]
         end
      end
   end;
```

```
     for i := 1 to ngp+5 do c[i] := 0;
     i := ngp+2;
     repeat
      c[i] := z[i]-d[i,5]*c[i+1]-d[i,6]*c[i+2]-d[i,7]*c[i+3];
      i := i-1
     until (i < 1)
   end;   { SetEq }
procedure Interval(var jsp : integer;xsp : real);
{---------------------------------------------}
begin
     jsp := 0;
     if (xsp <= b[1]) then jsp:= 1
     else
     begin
        if (xsp >= b[ngp]) then jsp := ngp-1
        else
        begin
           jsp := jsp+1;
           if (xsp > b[jsp]) then repeat jsp:= jsp+1 until (xsp <= b[jsp]);
           jsp := jsp-1
        end
     end
end;   { Interval }

procedure Spline;
{-----------------}
var   xh,t,yh : real;
      i,j : integer;
      answer : char;

begin
     for i := 1 to dim do
     begin
        xh := x[i];
        Interval(j,xh);
        t := (xh-b[j])/h;
        yh := c[j]*fnp(1-t)+c[j+1]*fnq(1-t)+c[j+2]*fnq(t)+c[j+3]*fnp(t);
        writeln (i:5,x[i]:12:8,y[i]:12:8,yh:12:8)
     end;
     repeat answer := readkey until (answer =chr($0d))
end;   { Spline }

procedure Graphics;
{-------------------}
var   xspl,yspl,i,j,grdr,grmo : integer;
      diff,xh,yh,t : real;
      answer : char;

   procedure Cross (xp,yp : real);
   {----------------------------}
   var   px,py : integer;

   begin
      px := round((xp+diff)/xmax*GetMaxX);
      py := round((ymax-yp)/(ymax-ymin)*GetMaxY);
      Line (px-3,py,px+3,py);   Line (px,py-1,px,py+1)
   end;   { Cross }
```

```
procedure Supports;
{-------------------}
var   cnt,xp,yp : integer;

begin
   SetLineStyle (1,0,1);
   for cnt := 1 to ngp do
   begin
      xp := round((b[cnt]+diff)/xmax*GetMaxX);   yp := 0;
      MoveTo (xp,yp);   yp := GetMaxY;   LineTo (xp,yp)
   end;
   SetLineStyle (0,0,1)
end;   { Supports }

procedure Sspl;
{---------------}
begin
   Interval(j,xh);
   t := (xh-b[j])/h;
   yh := c[j]*fnp(1-t)+c[j+1]*fnq(1-t)+c[j+2]*fnq(t)+c[j+3]*fnp(t)
end;   { Sspl }

begin
   grdr := 0;   grmo := 0;
   diff := (xmax-xmin)/20;
   xmin := xmin-diff;   xmax := xmax+diff;
   diff := (ymax-ymin)/20;
   ymin := ymin-diff;   ymax := ymax+diff;
   InitGraph (grdr,grmo,'');
   rectangle (0,0,GetMaxX,GetMaxY);
   for i := 1 to dim do Cross(x[i],y[i]);
   Supports;
   xh := x[1];
   Sspl;
   xspl := round((xh+diff)/xmax*GetMaxX);
   yspl := round((ymax-yh)/(ymax-ymin)*GetMaxY);
   MoveTo (xspl,yspl);
   repeat
    xh := xh+(x[dim]-x[1])/100;
    Sspl;
    xspl := round((xh+diff)/xmax*GetMaxX);
    yspl := round((ymax-yh)/(ymax-ymin)*GetMaxY);
    LineTo (xspl,yspl)
   until (xh > x[dim]);
   repeat answer := Readkey until (answer = chr($0d));
   CloseGraph;
   textmode (3)
end;   { Graphics }

begin
   fillchar (a[1,1],10584,chr($00));   fillchar (d[1,1],10584,chr($00));
   fillchar (b[1],1500,chr($00));   fillchar (c[1],1500,chr($00));
   fillchar (z[1],1500,chr($00));
   Input;
   Qsort;
```

```
      Fillb;
      Filla;
      Filldz;
      SetEq;
      Spline;
      Graphics
end.
```

```
                              REGSPL2
```

```
program regspl2;
    { J.Kinkel, January 1989, based on Ebert/Ederer }

uses    crt,graph;

const   dim = 26;
        x : array[0..26] of real
          = (0,1,2,3,4,5,6,7,8,9,10,11,12,13,14,15,16,17,18,19,
              0.3,4.8,8.5,12.1,12.8,16.7,18.2);
        y : array[0..26] of real
          = (0.5,1,3.5,4.2,5.5,8.2,8.5,10.2,10.9,11.5,11.6,11.5,
              10.6,9.5,7.2,5.5,3.2,2.1,0.9,0,0.6,8.6,11.6,10.5,10,2.5,0.5);

var     d : array[0..300,0..10] of real;
        z,r : array[0..300] of real;
        a,b,c,h,ih : array [0..100] of real;
        xmin,xmax,ymin,ymax : real;
        max : integer;

 procedure Qsort;
{----------------}
 var    tmp : real;
        i,j,blo : integer;

 begin
      xmin := x[0];   xmax := xmin;   ymin := y[0];   ymax := ymin;
      for i := 1 to dim do
      begin
         if (x[i] < xmin) then xmin := x[i];
         if (x[i] > xmax) then xmax := x[i];
         if (y[i] < ymin) then ymin := y[i];
         if (y[i] > ymax) then ymax := y[i]
      end;
      blo := round(ln(dim)/ln(2));   blo := round(exp(blo*ln(2)))-1;
      repeat
       for i := 0 to dim-blo do
       begin
           j := i;
           repeat
           if (x[j] <= x[j+blo]) then j := -1
           else
           begin
               tmp := y[j];   y[j] := y[j+blo];   y[j+blo] := tmp;
               tmp := x[j];   x[j] := x[j+blo];   x[j+blo] := tmp;
```

```
               j := j-blo
            end
         until (j < 0)
      end;
      if (blo/2 < 1) then blo := 0 else blo := round(blo/2)
   until (blo <= 0)
end;   { Qsort }

procedure Fillh;
{----------------}
var   cnt : integer;

begin
   for cnt := 1 to dim do
      begin
         h[cnt] := x[cnt]-x[cnt-1];   ih[cnt] := 1/h[cnt]
      end
end;   { Fillh }

procedure Filld;
{----------------}
var   pp,qq : real;
      cnt1,cnt2 : integer;

   procedure derivm;
   {----------------}
   var   i,j : integer;

   begin
      d[0,5] := pp;   d[0,8] := -ih[1];   r[0] := pp*y[0];
      d[1,5] := pp;   d[1,7] := ih[1]+ih[2];   d[1,10] := -ih[2];
      r[1] := pp*y[1];   d[max-3,4] := -ih[dim-1];   d[max-3,5] := pp;
      d[max-3,7] := ih[dim]+ih[dim-1];   r[max-3] := pp*y[dim-1];
      d[max,4] := -ih[dim];   d[max,5] := pp;   r[max] := pp*y[dim];
      for i := 2 to dim-2 do
         begin
            j := 3*i-2;   d[j,5] := pp;   r[j] := pp*y[i];
            d[j,4] := -ih[i];   d[j,7] := ih[i]+ih[i+1];
            d[j,10] := -ih[i+1]
         end
   end;   { derivm }

   procedure deriva;
   {----------------}
   var   i,j : integer;

   begin
      d[2,5] := 2*qq*(h[1]+h[2]);   d[2,6] := (h[1]+h[2])/3;
      d[2,8] := qq*h[2];   d[2,9] := h[2]/6;
      d[max-2,2] := qq*h[dim-1];   d[max-2,3] := h[dim-1]/6;
      d[max-2,5] := 2*qq*(h[dim-1]+h[dim]);
      d[max-2,6] := (h[dim-1]+h[dim])/3;
      for i := 2 to dim-2 do
         begin
            j := i*3-1;   d[j,2] := qq*h[i];   d[j,3] := h[i]/6;
```

```
          d[j,5] := 2*qq*(h[i]+h[i+1]);   d[j,6] := (h[i]+h[i+1])/3;
          d[j,8] := qq*h[i+1];   d[j,9] := h[i+1]/6
      end
  end;   { deriva }

  procedure derivla;
  {------------------}
  var   i,j : integer;

  begin
      d[3,2] := -ih[1];   d[3,3] := ih[1]+ih[2];
      d[3,4] := (h[1]+h[2])/3;   d[3,6] := -ih[2];
      d[3,7] := h[2]/6;   d[max-1,0] := -ih[dim-1];
      d[max-1,1] := h[dim-1]/6;   d[max-1,6] := -ih[dim];
      d[max-1,3] := ih[dim]+ih[dim-1];
      d[max-1,4] := (h[dim-1]+h[dim])/3;
      for i := 2 to dim-2 do
      begin
          j := 3*i;   d[j,0] := -ih[i];   d[j,1] := h[i]/6;
          d[j,3] := ih[i]+ih[i+1];   d[j,4] := (h[i]+h[i+1])/3;
          d[j,6] := -ih[i+1];   d[j,7] := h[i+1]/6
      end
  end;   { derivla }

begin
    clrscr;   writeln;
    write (' pp (has to be lower than 1) ? ');   readln(pp);
    max := 3*dim-2;   qq := 1-pp;   pp := 2*pp;   qq := qq/3;
    for cnt1 := 0 to max do
    begin
        for cnt2 := 0 to 10 do d[cnt1,cnt2] := 0;
        r[cnt1] := 0
    end;
    derivm;
    deriva;
    derivla
end;   { Filld }

procedure SetEq;
{----------------}
var   i,j,k,l,l1,l2 : integer;
      dh,l3 : real;

begin
    max := 3*dim-2;
    for i := 0 to max do
    begin
        dh := d[i,5];
        for j := 5 to 10 do d[i,j] := d[i,j]/dh;
        r[i] := r[i]/dh;   j := i+5;
        if (j > max) then j := max;
        if (i < max) then
        begin
            for k := i+1 to j do
```

```
            begin
               13 := -d[k,5-(k-i)];
               if not(13 = 0) then
               begin
                  for 1 := 0 to 5 do
                  begin
                     11 := 1+5;    12 := 11-(k-i);
                     d[k,12] := d[k,12]+13*d[i,11]
                  end;
                  r[k] := r[k]+13*r[i]
               end
            end
         end
      end;
      for i := 0 to max do z[i] := 0;
      z[max] := r[max];    z[max-1] := r[max-1]-d[max-1,6]*z[max];
      z[max-2] := r[max-2]-d[max-2,6]*z[max-1]-d[max-2,7]*z[max];
      z[max-3] := r[max-3]-d[max-3,6]*z[max-2]-d[max-3,7]*z[max-1]
                  -d[max-3,8]*z[max];
      z[max-4] := r[max-4]-d[max-4,6]*z[max-3]-d[max-4,7]*z[max-2];
      z[max-4] := z[max-4]-d[max-4,8]*z[max-1]-d[max-4,9]*z[max];
      i := max-5;
      repeat
       z[i] := r[i]-d[i,6]*z[i+1]-d[i,7]*z[i+2]-d[i,8]*z[i+3];
       z[i] := z[i]-d[i,9]*z[i+4]-d[i,10]*z[i+5];
       i := i-1
      until (i < 0)
   end;    { SetEq }

   procedure CalcVec;
   {------------------}
   var   sum1,sum2 : real;
         cnt : integer;
         answer : char;

   begin
      a[0] := 0;    a[dim] := 0;
      for cnt := 1 to dim-1 do a[cnt] := z[3*cnt-1];
      ih[0] := z[0];    ih[dim] := z[max];
      for cnt := 1 to dim-1 do ih[cnt] := z[3*cnt-2];
      for cnt := 1 to dim-1 do d[cnt,0] := z[3*cnt];
      sum1 := 0;
      for cnt := 0 to dim do sum1 := sum1+sqr(y[cnt]-ih[cnt]);
      writeln;   writeln (' Sum of squared deviations = ',sum1);
      sum2 := 0;
      for cnt := 0 to dim-1 do
       sum2 := sum2+h[cnt+1]*(sqr(a[cnt])+a[cnt]*a[cnt+1]+sqr(a[cnt+1]));
      sum2 := sum2/3;
      writeln (' Integral on the square of the 2nd derivatives = ',sum2);
      for cnt := 0 to dim-1 do
      begin
         c[cnt] := ih[cnt]-a[cnt]*sqr(h[cnt+1])/6;
         b[cnt] := (ih[cnt+1]-ih[cnt])/h[cnt+1]-h[cnt+1]/6*(a[cnt+1]-a[cnt])
      end;
      repeat answer := readkey until (answer = chr($0d))
   end;    { CalcVec }
```

```
procedure Sspl (xh : real;var yh : real);
{----------------------------------------}
var   tmp1,tmp2 : real;
      j,j1,j2 : integer;

   procedure Interval;
   {------------------}
   begin
      j1 := 0;   j2 := dim;
      if (xh <= x[j1]) then j := 0
      else
      begin
         if (xh >= x[j2]) then j := dim-1
         else
         begin
            repeat
              j := round((j1+j2)/2);
              if (xh > x[j]) then j1 := j else j2 := j
            until ((j2-j1) <= 1);
            j := j1
         end
      end
   end;   { Interval }

begin
   Interval;
   tmp1 := x[j+1]-xh;   tmp2 := xh-x[j];
   yh := a[j]*tmp1*sqr(tmp1)/6/h[j+1]+a[j+1]*tmp2*sqr(tmp2)/6/h[j+1];
   yh := yh+b[j]*tmp2+c[j]
end;   { Sspl }

procedure Spline;
{----------------}
var   xspl,yspl : real;
      i : integer;
      answer : char;

begin
   for i := 0 to dim do
   begin
      xspl := x[i];
      Sspl(xspl,yspl);
      writeln (i:5,x[i]:12:8,y[i]:12:8,yspl:12:8)
   end;
   repeat answer := readkey until (answer = chr($0d))
end;   { Spline }

procedure Graphics;
{------------------}
var   xspl,yspl,i,j,grdr,grmo : integer;
      diff,xs,ys,tmp : real;
      answer : char;
```

```
procedure Cross (xp,yp : real);
{------------------------------}
var   px,py : integer;

begin
    px := round((xp+diff)/xmax*GetMaxX);
    py := round((ymax-yp)/(ymax-ymin)*GetMaxY);
    Line (px-3,py,px+3,py);   Line (px,py-1,px,py+1)
end;   { Cross }

begin
    grdr := 0;   grmo := 0;
    diff := (xmax-xmin)/20;
    xmin := xmin-diff;   xmax := xmax+diff;
    diff := (ymax-ymin)/20;
    ymin := ymin-diff;   ymax := ymax+diff;
    InitGraph (grdr,grmo,'');
    rectangle (0,0,GetMaxX,GetMaxY);
    for i := 0 to dim do Cross(x[i],y[i]);
    xs := x[0];
    Sspl(xs,ys);
    xspl := round((xs+diff)/xmax*GetMaxX);
    yspl := round((ymax-ys)/(ymax-ymin)*GetMaxY);
    MoveTo (xspl,yspl);
    repeat
     xs := xs+(x[dim]-x[0])/100;
     Sspl(xs,ys);
     xspl := round((xs+diff)/xmax*GetMaxX);
     yspl := round((ymax-ys)/(ymax-ymin)*GetMaxY);
     LineTo (xspl,yspl)
    until (xs > x[dim]);
    repeat answer := Readkey until (answer = chr($0d));
    CloseGraph;
    textmode (3)
end;   { Graphics }

begin
    clrscr;   fillchar (d[0,0],19866,chr($00));
    fillchar (z[0],1806,chr($00));   fillchar (r[0],1806,chr($00));
    fillchar (a[0],606,chr($00));   fillchar (b[0],606,chr($00));
    fillchar (c[0],606,chr($00));   fillchar (h[0],606,chr($00));
    fillchar (ih[0],606,chr($00));
    Qsort;
    Fillh;
    Filld;
    SetEq;
    CalcVec;
    Spline;
    Graphics
end.
```

```
                              ODESTIFF
```

```pascal
program odestiff;
    { J. Kinkel, January 1989, based on Ebert/Ederer }

uses    crt,graph;

const   k1 = 1e-2;   k2 = 1e3;   k3 = 1e2;   k4 = 1e8;
        dim = 6;    ddim = 12;

var    ya,y1,y : array[1..dim] of real;
       x,diff,xb,xe,deltax,deltav : real;
       answer : char;

procedure Input;
{----------------}
 var    cnt : integer;

 begin
    clrscr;   writeln;
    for cnt := 1 to dim do
    begin
       write (' Initial value of y(',cnt,') = ');   readln (y[cnt])
    end;
    ya := y;   y1 := y;
    writeln;
    write (' x-values (beg end) ? ');   readln (xb,xe);
    write (' Delta-x-value ? ');   readln (deltax);
    x := xb;   diff := xe-xb;   deltav := diff/50
 end;    { Input }

procedure Graphics;
{-------------------}
 var   grdr,grmo : integer;

 begin
    grdr := 0;   grmo := 0;
    InitGraph (grdr,grmo,'');
    rectangle (0,0,GetMaxX,GetMaxY);
    SetLineStyle (1,0,1)
 end;    { Graphics }

procedure Main;
{----------------}
 var   a : array[1..dim,1..ddim] of real;
       d : array[1..dim] of real;
       dx,dy : real;
       id,subi : integer;
       flag1,flag2 : boolean;

    procedure Output;
    {------------------}
     var   cnt : integer;
```

```
begin
  writeln;
  if flag1 then
    for cnt := 1 to dim do writeln (x:15:7,cnt:5,y[cnt]:30)
  else
  begin
    writeln (' No solution is possible !');    Halt
  end
end;    { Output }

procedure Deriv;
{-----------------}
begin
  d[1] := -k1*y[1]-k2*y[1]*y[2];
  d[2] := 2*k1*y[1]-k2*y[1]*y[2]+k3*y[4]-2*k4*sqr(y[2]);
  d[3] := k2*y[1]*y[2];    d[4] := k2*y[1]*y[2]-k3*y[4];
  d[5] := k3*y[4];    d[6] := k4*sqr(y[2])
end;    { Deriv }

procedure Matrix;
{------------------}
var   i,j : integer;

begin
  for i := 1 to ddim do
    for j := 1 to dim do a[j,i] := 0;
  for i := 1 to dim do
  begin
    a[i,i] := 1;    a[i,dim+i] := 1
  end;
  a[1,1] := a[1,1]-dx*(-k1-k2*y[2]);    a[1,2] := dx*k2*y[1];
  a[2,2] := a[2,2]-dx*(-k2*y[1]-4*k4*y[2]);
  a[2,1] := -dx*(2*k1-k2*y[2]);
  a[2,4] := -dx*k3;    a[3,1] := -dx*k2*y[2];
  a[3,2] := -a[1,2];    a[4,4] := a[4,4]-a[2,4];
  a[4,1] := a[3,1];    a[4,2] := a[3,2];
  a[5,4] := a[2,4];    a[6,2] := -dx*k4*2*y[2]
end;    { Matrix }

procedure Invert;
{-----------------}
var   s,t,abb,cnt,hlp : integer;
      fac : real;

  procedure Exchange;
  {--------------------}
  var   h : real;
        cnt : integer;

  begin
    for cnt := 1 to ddim do
    begin
      h := a[s,cnt];    a[s,cnt] := a[hlp,cnt];    a[hlp,cnt] := h
    end
  end;    { Exchange }
```

```
    procedure Normalize;
  {--------------------}
   var    fac : real;
          cnt : integer;

   begin
      fac := 1/a[s,s];
      for cnt := 1 to ddim do a[s,cnt] := fac*a[s,cnt]
   end;    { Normalize }

    procedure Subtract;
  {-------------------}
   var    fac : real;
          cnt : integer;

   begin
      fac :=a[t,s];
      for cnt := 1 to ddim do a[t,cnt] := a[t,cnt]-fac*a[s,cnt]
   end;    { Subtract }

 begin
    abb := 1;
    for s := 1 to dim do
    begin
       hlp := s;
       repeat
          if not(a[hlp,s] = 0) then
          begin
             Exchange;
             Normalize;
             for t := 1 to dim do
               if not(t = s) then Subtract;
               hlp := dim
          end
          else abb := hlp;
          hlp := hlp+1
       until (hlp >= dim);
       if (abb = dim) then
       begin
          flag1 := false;
          exit
       end
    end
 end;    { Invert }

 procedure SetY;
{----------------}
 var    cj1,cj2 : integer;

 begin
    x := x+dx;
    for cj1 := 1 to dim do
    begin
       dy := 0;
       for cj2 := 1 to dim do dy := dy+a[cj1,dim+cj2]*d[cj2];
```

```
          y[cj1] := y[cj1]+dy*dx
      end
  end;    { SetY }

  procedure Setflag;
  {------------------}
  var   ys : real;
        cnt,cf : integer;

  begin
      cf := 0;
      for cnt := 1 to dim do
      begin
          ys := abs(y[cnt])+abs(ya[cnt])+1e-6;
          if (abs(y[cnt]-ya[cnt])/ys <= 0.05) then cf := cf+1
      end;
      if (cf = dim) then flag2 := true
  end;    { Setflag }

  procedure PlotData;
  {-------------------}
  var   xp1,xp2,yp1,yp2,cnt : integer;

  begin
      for cnt := 1 to dim do
      begin
          xp1 := GetMaxX-round((xe-xb)/diff*GetMaxX);
          xp2 := GetMaxX-round((xe-x)/diff*GetMaxX);
          yp1 := GetMaxY-round(y1[cnt]*GetMaxY);
          yp2 := GetMaxY-round(y[cnt]*GetMaxY);
          MoveTo (xp1,yp2);   Line (xp1,yp1,xp2,yp2)
      end
  end;    { PlotData }

begin
    subi := 1;    flag1 := true;
  { Output; }
    repeat
      flag2 := false;   dx := deltax/subi;   x := xb;
      for id := 1 to subi do
      begin
          Deriv;
          Matrix;
          Invert;
          SetY
      end;
      repeat
        if (subi > 50) then
        begin
            deltax := deltax/1000;   subi := 1;
            dx := deltax/subi;   x := xb;
            for id := 1 to subi do
            begin
                Deriv;
                Matrix;
```

```
            Invert;
            SetY
         end
      end;
      ya := y;   y := y1;
      subi := subi*2;   dx := deltax/subi;   x := xb;
      for id := 1 to subi do
      begin
         Deriv;
         SetY
      end;
      Setflag
    until (flag2);
    PlotData;
    { Output; }
    xb := xb+deltax;   y1 := y;
    if (subi = 2) then deltax :=  deltax*2;
    subi := trunc(subi/5)+1;
    if (deltax > deltav) then deltax := deltav;
    if (deltax > (xe-xb)) then deltax := (xe-xb);
  until (xb >= xe)
end;   { Main }

begin
   Input;
   Graphics;
   Main;
   repeat answer := readkey until (answer = chr($0d));
   CloseGraph;
   textmode (3)
end.
```

General Index

BASIC Index

Program Index

The roman page numbers refer to the BASIC-programs, the slanted page numbers to the PASCAL-programs.